ROCK BLASTING
and
EXPLOSIVES ENGINEERING

Per-Anders Persson
Professor of Mining Engineering and
Director, Research Center for Energetic Materials
New Mexico Institute of Mining and Technology
Socorro, New Mexico, U.S.A.

Roger Holmberg
Vice President and Technical Director
Nitro Nobel AB
Gyttorp, Sweden

Jaimin Lee
Senior Research Scientist
Agency for Defense Development
Taejon, Korea

CRC Press
Boca Raton   London   New York   Washington, D.C.

Library of Congress Cataloging-in-Publication Data

Persson, Per-Anders.
  Rock blasting and explosives engineering / Per-Anders Persson, Roger Holmberg, Jaimin Lee.
    p. cm.
  Subtitle: A textbook for students and a handbook for scientists and engineers covering the science and engineering of the industrial use of explosives with major emphasis on rock blasting.
  Includes bibliographical references (p.) and index.
  ISBN 0-8493-8978-X
  1. Blasting. 2. Rock excavation. I. Holmberg, Roger. II. Lee, Jaimin. III. Title.
  1. Rock blasting. 2. Rock excavation. 3. Explosives. 4. Explosives engineering. I. Holmberg, Roger. II. Lee, Jaimin. III. Title.
TA748.P46 1993
624.1′52--dc20   93-28150
                                                                                   CIP

This book contains information obtained from authentic and highly regarded sources. Reprinted material is quoted with permission, and sources are indicated. A wide variety of references are listed. Reasonable efforts have been made to publish reliable data and information, but the author and the publisher cannot assume responsibility for the validity of all materials or for the consequences of their use.

Neither this book nor any part may be reproduced or transmitted in any form or by any means, electronic or mechanical, including photocopying, microfilming, and recording, or by any information storage or retrieval system, without prior permission in writing from the publisher.

The consent of CRC Press LLC does not extend to copying for general distribution, for promotion, for creating new works, or for resale. Specific permission must be obtained in writing from CRC Press LLC for such copying.

Direct all inquiries to CRC Press LLC, 2000 N.W. Corporate Blvd., Boca Raton, Florida 33431.

Trademark Notice: Product or corporate names may be trademarks or registered trademarks, and are used only for identification and explanation, without intent to infringe.

Visit the CRC Press Web site at www.crcpress.com

© 1994 by CRC Press LLC

No claim to original U.S. Government works
International Standard Book Number 0-8493-8978-X
Library of Congress Card Number 93-28150
Printed in the United States of America   7 8 9 0
Printed on acid-free paper

To May, Daga, and Yangmi

The Authors' Foreword

This text is intended to serve as a working handbook for mining, quarrying, and building construction engineers, as a textbook for self-study, and as a working text for university level classes in rock blasting and explosives engineering for students in mining, petrochemical, mechanical, and civil engineering.

Explosives are used extensively as a precision tool to fragment rock in the excavation for coal, metal ores, ballast materials, and minerals, to shape underground openings while leaving the remaining rock strong and undamaged, and to shape or accelerate metal in a variety of applications. The proper, controlled use of the explosive's power can save great amounts of money in support costs for blast-damaged underground openings, and in costs for repair of damaged buildings. Judicious use of explosive power applied correctly to produce an optimum degree of fragmentation can reduce wear on loading equipment and crushers, and increase considerably the throughput of a beneficiation plant or a ballast producer. The use and development of such techniques requires an understanding of the fundamentals of the functioning of explosives.

The text is intended to provide a firm foundation of understanding of the physical and chemical processes by which explosives release their energy and do work on the surrounding material. It also aims to set out the basic principles of the engineering methods for calculating the correct charge size and charge placement for a multitude of practical applications. It gives examples of how these methods can be used in production blasting, tunneling, and controlled blasting in a city. While the complicated structure of cracks and joints in a real rock mass constitutes a variable background against which to use such methods, recent developments have made it possible to select explosives that best fit different types of rock mass.

This text has grown over a long period of time. The beginning was a series of lectures given to senior undergraduates and graduate students at the Department of Materials Science and Mineral Engineering of University of California at Berkeley that one of the authors (PAP) gave during some rainy winter months of 1978. Further work was done jointly by two of the authors (PAP and RH), while occupying different positions in the Swedish Detonic Research Foundation and Nitro Nobel AB in Sweden, during the years from 1979 to 1984. Since then, one of the authors (RH) has had the fortune of being able to follow and contribute to the development of rock blasting techniques and products from the key position of the vice president for research, development and technology at Nitro Nobel AB, now part of the DYNO international group of explosives companies.

A further expansion and revision of the text was completed at the Center for Explosives Technology Research of the New Mexico Institute of Mining and Technology (New Mexico Tech), where the authors were given an opportunity to work together during the warm and sunny winter months of 1985. The final version of the text has grown into its present form while being used as a compendium for the several graduate courses in Explosives Engineering and Rock Blasting Engineering at New Mexico Tech taught by PAP. The typesetting of the original manuscript and many revisions to the theoretical material are due to JL.

The knowledge presented has been accumulated over many years by the authors and colleagues at the Swedish Detonic Research Foundation, SveDeFo, in the stimulating atmosphere of industrial cooperation between many different companies in many different countries, all involved in different stages of the work of providing the infrastructure for a modern industrial society.

Per-Anders Persson   Roger Holmberg   Jaimin Lee
December 28, 1992

The Authors' Foreword to the Third and Sixth Corrected Printings

For the third (1996) printing, the text, tables, and figures were carefully reviewed and some misprints, faulty equations, misspellings, and other minor errors were corrected.

For this sixth (2001) printing, eleven additional minor errors were corrected.

The authors are grateful to several readers, and in particular to professor Charles Fairhurst of University of Minnesota and William C. Davis of Los Alamos, who have provided helpful information leading to many of the additions and improvements in this third printing.

At the request of several readers, two historical notes have been added. Section 3.7 describes the invention and early developments of emulsion explosives and heavy ANFO. Section 5.13 gives a brief description of the invention of the NONEL shocktube and detonator system.

Chapter 4 "Shock Waves and Detonations, Explosive Performance" (especially Section 4.4 "Ideal and Non-Ideal Detonation") has been given a major face-lift which we hope will make this chapter more readable. Some material has been added to make the description of the subject matter in Section 8.7 "Rock Mass Damage and Fragmentation" more up-to-date.

The work of revising and collating the new and corrected text and the TeX typesetting for this printing was done by Richard R. Carlson, to whom we express our sincere thanks.

While regretting that these improvements and corrections were not included in the two first printings, we hope the book will now be even more useful as an engineering handbook, as a reference, and as a university textbook.

Per-Anders Persson   Roger Holmberg   Jaimin Lee
February 7, 2001

Acknowledgments

The authors are deeply grateful to the board of the Swedish Detonic Research Foundation, and to those Swedish industries and the Swedish Board for Technical Development that give economic support to the Foundation, for the opportunity given to one of the authors (PAP) to spend a period of sabbatical leave at the University of California, Berkeley; and also to the Department of Materials Science and Minerals Engineering for supplying the funds necessary to cover the costs incurred at Berkeley. They are also grateful to the Swedish Detonic Research Foundation that similarly allowed another of the authors (RH) to spend a period of sabbatical work at the Center for Explosives Technology Research of New Mexico Tech.

We are indebted to the State of New Mexico and New Mexico Tech for creating with the Rio Grande Research Corridor and the Center for Explosives Technology Research the scholarly climate and the economic means without which this textbook would still be an unfinished, unpublished manuscript, of use to no one.

The spirited use of the scientific and analytical approach to creating by experiment and theory, step by step, a deep understanding of the complicated engineering processes in rock blasting and explosives engineering was conscientiously pursued for many years at the Detonics Laboratory at Vinterviken, Stockholm, Sweden, under the inspired and inspiring leadership of two men who complemented and strengthened each other in a wonderful way. Carl Hugo Johansson and Ulf Langefors personified much that is outstanding in Swedish engineering science. Their work greatly influenced the rapid development of the Swedish technology of commercial explosives and rock blasting during the formative years from 1946 to 1965, and for a long time thereafter. PAP has a special debt of gratitude to them. Carl Hugo Johansson, the first research director of the laboratory at Vinterviken and founder of the Swedish Detonic Research Foundation, was co-author with PAP of the 1970 book *Detonics of High Explosives*, published in 1970. Ulf Langefors (co-author with Björn Kihlström of the book *The Modern Science of Rock Blasting*, published in 1963) who succeeded Carl Hugo Johansson as research director and later became president of Nitro Nobel AB, was the source of inspiration for PAP to invent the NONEL detonator system.

We are indebted to our former colleagues at the Swedish Detonic Research Foundation and to Nitro Nobel AB, Nils Lundborg, Algot Persson, Björn Kihlström, Sten Ljungberg, Gösta Larsson, Ingemar Persson, Gunnar Persson, Kenneth Mäki, Gert Bjarnholt, Bengt Niklasson, Bernt Larsson, Björn Engsbraaten, Anders Ladegaard-Pedersen, Finn Ouchterlony, Lars Granlund, Michael Cechanski, Gösta Lithner, Rune Gustafsson, Stig Olofsson, Conny Sjöberg, Göran Jidestig, Ingvar Bergqvist, Bernt Brunnberg, Sven Rosell, and to many others who have contributed to the corporate memory of explosives and rock blasting engineering which forms a major part of this book.

We are also indebted to many colleagues in the international explosives, rock fragmentation, mining, and civil engineering industries, Atlas Copco AB, Sandvik Rock Tools AB, Boliden AB, LKAB, Skanska, the Robbins Company, ICI Explosives and the former Atlas Powder Co., IRECO Incorporated, DYNO Industrier. Among such colleagues, we would particularly like to mention Sten Brännfors, Duri Prader, Gunnar Nord, Åke Kallin, Dick Robbins, John Grant, Bengt Lundberg, Gunnar Almgren, Ingemar Marklund, Berthold Johansson, and Norbert Krauland.

We have learned a lot from colleagues at the Los Alamos National Laboratory, who pioneered the use of explosives as a precision tool in dynamic working and forming of metals, and provided a wealth of solid experimental data and computational data and techniques which has served as a foundation for much of the technology reported here. We are particularly indebted to William C. Davis, a fine experimentalist and scientist, scientific mentor to all of us, who read and suggested numerous improvements to the manuscript, and to Charles L. Mader, creator of the improved BKW equation of state codes, the SIN hydrocode, and the HOM equation of state of reacting explosives. Bobby Craig, the superb experimentalist, helped in establishing the Eagle Laboratory of the Research Center for Explosives Technology at New Mexico Tech in 1984-85 and setting the high standards for the experimental work done there since then.

Present friends and colleagues at New Mexico Tech who have contributed material or the benefit of stimulating discussions to this text include Kay Brower, Pat Buckley, Andrew Grebe, Dennis Hunter, Vasant Joshi, Larry Libersky, Douglas Olson, Jimmie Oxley, Albert Petschek, Fred Sandström, Naresh Thadhani, and Pharis Williams.

Colleagues at other laboratories who have influenced our thinking about explosives and their effects are Milton Finger and Ed Lee of the Lawrence Livermore National Laboratory, and Jim Kennedy of the Sandia National Laboratories, now at Los Alamos.

Personal thanks are due to professors Tor Brekke, Neville Cook, Iain Finnie, Douglas Fuerstenau, Richard Goodman, and Michael Hood, who in many different ways aided and encouraged the initial work on this book in the Hearst Mining building at Berkeley in 1978.

A special, personal thanks is directed to the students taking the courses taught by PAP from which this book has grown, those in the TV studio on the Berkeley campus where the first course was held, as well as those at the Lawrence Livermore Laboratory and other institutes in the San Francisco area who followed the course on the TV link network. Similar thanks are due to all of the students at New Mexico Tech, Sandia National Laboratories, the Lyndon B. Johnson NASA Space Center White Sands Test Facility and Lockheed Inc., White Sands, NM, who have followed PAP's courses in Explosives Engineering, Shock Waves and Detonations (taught jointly by PAP and William C. Davis), Rock Blasting Engineering, and Commercial Explosives from 1984 onward.

The final revision and TeX typesetting of the camera-ready version of this book was done by Richard R. Carlson who gently improved the sometimes strange English written by foreigners and to whom great thanks are due from all three of us.

> We are very grateful for the unfailing support provided by our wives,
> May Persson, Daga Holmberg, and Yangmi Oh Lee, who endured the
> several years this text intruded on our family lives.

Table of Contents

Introduction 1
Chapter 1. Rock Strength and Fracture Properties 5
   1.1. Rock Strength Measurements 6
   1.2. Three-Dimensional Rock Strength and Deformation Energy . 11
   1.3. The Random Microcrack Model for Rock Strength 13
   1.4. Fracture Mechanics 15
   1.5. Rock Mass Strength and Structure 17
Chapter 2. Mechanical Drilling and Boring in Rock 25
   2.1. Bit Penetration, Chip Formation, and Force-Displacement Curves . 29
   2.2. Drillability and Boreability of Rock Materials 33
      2.2.1. Predicting Drillability 33
   2.3. The Stamp Test for Rock Drillability Classification 37
   2.4. Predicting Boreability in Full-Face Tunnel Boring 43
   2.5. Predicting Shift Capacity of a Disc Cutter Tunnel Boring Machine . 43
   2.6. Rock Strength and the Size Dependence
        of Drill/Bore Energy and Cost 46
   2.7. Comparison between Mechanical Tunnel Boring
        and Conventional Drill-Blast Tunneling 47
   2.8. Some Rules of Thumb 54
Chapter 3. Explosives 55
   3.1. Energetic Materials, Explosions, and Explosive Materials . . 56
      3.1.1. Energetic Materials 56
      3.1.2. Explosions and Explosive Materials 56
      3.1.3. Thermal Stability and Bulk Decomposition
          of Explosive Materials 58
      3.1.4. Thermal Decomposition, Hot Spots, Initiation, and Safety . 59
   3.2. Explosives, Propellants, and Pyrotechnics 64
      3.2.1. Explosives and Propellants 64
      3.2.2. Pyrotechnics 65
      3.2.3. Single Molecule and Composite Explosives 66
      3.2.4. Primary, Secondary, and Tertiary Explosives 66
   3.3. Liquid Mono-Propellants and Solid Composite Propellants . . 67
   3.4. Military Explosives 71
   3.5. Rock Blasting Explosives Systems 74
      3.5.1. Dynamite Explosives 75
      3.5.2. Watergels, Slurry Explosives, and Slurry Blasting Agents . 75
      3.5.3. Pumpable Blasting Agents 75
      3.5.4. Emulsion Explosives and Blasting Agents 76
      3.5.5. Permissible Explosives in Underground Coal Mining . . 77
      3.5.6. Aluminized Explosives 79
   3.6. Mechanized Explosives Charging 79
   3.7. Invention of Emulsion Explosives and Blasting Agents . . . 85

Chapter 4. Shock Waves and Detonations, Explosive Performance   87
  4.1. Historical Note 88
  4.2. The Rankine-Hugoniot Equations and
        the Chapman-Jouguet Theory 90
      The Rankine-Hugoniot Equations for Shock Waves . . . 91
      The Rankine-Hugoniot Equations for Detonation 92
      The Rayleigh Line 93
      The Hugoniot Equation 93
      The Chapman-Jouguet Detonation 95
  4.3. Stress Wave Transmission Through a Boundary 97
    4.3.1. Spalling 98
    4.3.2. Impedance Matching between Explosive and Rock . . . 99
  4.4. Ideal and Nonideal Detonation 100
    4.4.1. Equation of State of the Explosion Products 102
    4.4.2. Estimation of Detonation Pressure 106
    4.4.3. Estimation of Borehole Pressure 106
    4.4.4. Numerical Shock Wave and Detonation Data 107
    4.4.5. Computing Explosion Energy and Gas Volume 109
    4.4.6. Oxygen Balance 110
    4.4.7. Heat of Combustion 111
    4.4.8. Explosion Energy 111
    4.4.9. Detonation Energy 112
    4.4.10. Strength of Explosives 113
    4.4.11. Simple Calculation of Energy and Gas Volume
          of Oxygen-Balanced Compositions 114
  4.5. Explosive Performance in Rock Blasting 115
    4.5.1. Calculated Expansion Work 115
    4.5.2. Effective Expansion Work in Rock Blasting 116
    4.5.3. Early Performance Test Methods 119
      Ballistic Mortar 120
      Grade Strength 120
      Brisance 122
      Trauzl Lead Block Test 122
      Plate Dent Test 123
      Cylinder Expansion Test 126
      Underwater Detonation Test 126
      Crater Test 127
      Langefors' Weight Strength 127
      Breaking Index from Underwater Detonation Testing . . 129
    4.5.4. The New Explosive Weight Strength Concept:
        A Combination of Explosive and Rock Properties . . 130
    4.5.5. Computed Expansion Work 132
    4.5.6. Rock Blasting Performance Value s from the Bdzil-Lee
        Detonation Shock Dynamics Evaluation 134
    4.5.7. Explosive Properties of Some Commercial Explosives
        and Blasting Agents 140

Chapter 5. Initiation Systems 143
   5.1. Historical Note on Detonators 143
   5.2. Delay Detonators without a Primary Explosive 145
      5.2.1. Introduction 145
      5.2.2. Present-Day Detonators 146
      5.2.3. The New Non-Primary Explosive Detonator 148
          Conceptual Design 148
          Safety 149
   5.3. Electric Detonators 150
      5.3.1. Formulae for Electric Circuits 155
          Series Circuit 155
          Parallel Circuit 156
          Series-Parallel Circuit 158
   5.4. Semiconductor Bridge Ignition 158
   5.5. High-Precision Electric Detonating Caps 159
   5.6. Sequential Blasting Machine 160
   5.7. Detonating Cord 160
      5.7.1. NONEL 162
      5.7.2. Hercudet Initiation System 167
   5.8. Electromagnetic Firing Methods 169
      5.8.1. Nissan RCB System 169
   5.9. The Magnadet System 171
   5.10. Electronic Detonators 171
      5.10.1. Nitro Nobel's Electronic Detonator 175
          Detonator 175
          Blasting Machine 176
          System Characteristics 177
          Mode of Operation 177
   5.11. Safety Fuse 177
   5.12. Primer or Booster Charges 178
   5.13. Invention of the NONEL and Other Shock Tube Based
       Non-Electric Detonator Systems 181

Chapter 6. Principles of Charge Calculation For Surface Blasting   183
   6.1. Introduction 183
   6.2. Geometry Effects 184
      6.2.1. Degree of Fixation 184
      6.2.2. Size Relations 185
   6.3. Specific Charge 191
   6.4. Charge Calculations For Several Holes 194
      6.4.1. Influence of Explosive Performance 195
      6.4.2. Rock Constant 196
   6.5. Calculation of Burden 198
      6.5.1. Examples 198
   6.6. Drilling Deviations 201
      6.6.1. Swelling 207
      6.6.2. Practical Burden 208

Chapter 7. Charge Calculations For Tunneling **209**
  7.1. Introduction 209
  7.2. Development of Drilling Equipment 210
  7.3. Charge Calculations 213
    7.3.1. Dividing the Tunnel Face Area in Design Sections . . . 215
    7.3.2. Advance 217
    7.3.3. Burden in the First Quadrangle 218
    7.3.4. Charge Concentrations in the First Quadrangle 220
    7.3.5. The Second Quadrangle 221
    7.3.6. Lifters 224
    7.3.7. Stoping Holes 225
    7.3.8. Contour Holes 225
  7.4. Sample Charge Calculation 226
    7.4.1. Input Conditions 226
    7.4.2. Calculation 227
    7.4.3. Advance 227
    7.4.4. Cut 227
      First Quadrangle 227
      Second Quadrangle 227
      Third Quadrangle 228
      Fourth Quadrangle 228
    7.4.5. Lifters 228
    7.4.6. Contour Holes, Roof 228
    7.4.7. Contour Holes, Wall 229
    7.4.8. Stoping 229
    7.4.9. Initiation Sequence 230

**Chapter 8. Stress Waves in Rock, Rock Mass Damage,
  and Fragmentation** **233**
  8.1. Fracture Initiation 234
  8.2. Interaction Between Radial Cracks and
       the Returning Tensile Wave 237
  8.3. Stress Waves in Rock 237
  8.4. Rock Mass Damage and Fragmentation by Blasting 240
  8.5. Estimation of Near Region Vibration Particle Velocities . . . 244
  8.6. Influence of Nearby Free Surfaces 245
  8.7. Rock Mass Damage and Fragmentation 247
  8.8. The Rock Acceleration Process 248
  8.9. Fragment Size Distribution 248
    8.9.1. Short Delay Multiple-Row Rounds 250
    8.9.2. The Wide-Spacing Blasting Method 252
  8.10. Bench Blasting with Auxiliary Holes 254
  8.11. Coarse Fragmentation 255
  8.12. Influence of Geological Discontinuities upon Fragmentation . 256
  8.13. Influence of Void Volume upon Fragmentation 259

Chapter 9. Contour Blasting ... 265
9.1. Introduction ... 265
9.2. Smooth Blasting and Presplitting ... 265
9.3. Rock Damage in Blasting ... 267
9.4. Computed Damage Zones ... 268
9.5. Blast Planning ... 268
9.6. Experimental Observation of Rock Damage ... 270
9.7. Cautious Blasting in Open Pit Mines ... 272
9.8. Controlled Fracture Growth ... 274
9.9. Shaped Charges for Boulder and Contour Blasting ... 275
    9.9.1. Crack Initiation ... 277
    9.9.2. Design Parameters ... 279
    9.9.3. Liner Material ... 279
        Angle Between the Legs of the V-shaped Liner ... 279
        Stand Off Between the Linear Charge and
          the Target Material ... 279
        Explosive ... 279
        Borehole Pressure for a Decoupled Charge ... 281
        Penetration in Granite ... 281
    9.9.4. Comparison of Methods for Controlled Fracture Growth ... 281

Chapter 10. Computer Calculations for Rock Blasting ... 287
10.1. Introduction ... 287
10.2. Some Commercial Rock Blasting Computer Codes ... 287
10.3. The Fundamentals of Two Rock Blasting Codes ... 288
10.4. Computer Program for Tunneling ... 289
10.5. Computer Program for Bench Blasting ... 291

Chapter 11. Blast Performance Control ... 299
11.1. The Vertical Crater Retreat (VCR) Blasting Technique ... 302
11.2. The Cratering Theory ... 305
11.3. Choosing the Best Explosive for VCR Mining ... 307
11.4. Small-Scale Cratering Tests ... 311
11.5. Evaluation of Crater Results and Scaling ... 312
11.6. Application of the VCR Concept to Production Stopes ... 314
11.7. Detonation Performance Check Using Tracer Elements ... 314
11.8. Cast Blasting ... 317

Chapter 12. Flyrock ... 319
12.1. Introduction ... 319
12.2. Influence of the Specific Charge ... 319
12.3. Theoretical Relations Between Flyrock and Drillhole Diameter ... 324
12.4. Measurements of Flyrock Velocity and Maximum Throw ... 326
    12.4.1. Calculation of the Throw ... 328
    12.4.2. Precautions to Avoid Flyrock ... 329
    12.4.3. Effect of Rock Structure ... 330
    12.4.4. Drilling ... 330
    12.4.5. Charging and Initiation ... 330
    12.4.6. Covering ... 332

Chapter 13. Ground Vibrations ... 337
13.1. Introduction ... 337
13.2. Definitions and Basic Terms ... 338
    13.2.1. Types of Vibrations ... 338
    13.2.2. Parameters Describing Vibrations ... 338
    13.2.3. Parameters ... 339
    13.2.4. Different Types of Wave Motion ... 340
    13.2.5. Wave Propagation Velocities ... 342
13.3. Damage to Buildings and the Reasons for Annoyance ... 342
    13.3.1. Reactions to Vibrations ... 342
    13.3.2. The Origin of Damage to Buildings ... 345
    13.3.3. Primary Damage Criteria ... 345
13.4. Limiting Vibration Levels for Buildings and Installations ... 346
13.5. Guidance Levels for Buildings According to Swedish Standard ... 348
    13.5.1. Guidance Levels ... 350
        13.5.1.1 Uncorrected Vertical Peak Particle Velocity ... 351
    13.5.2. Construction Quality Factor ... 352
    13.5.3. Distance Factor ... 352
    13.5.4. Project Time Factor ... 353
    13.5.5. Three Sample Calculations of the Guidance Level ... 353
        Example 1. A Limestone Quarry Near Houses ... 353
        Example 2. Leveling Near a Brick Office Building ... 353
        Example 3. Construction Work Near a Brick Church ... 353
    13.5.6. Computers ... 354
13.6. Planning for Blasting Work ... 354
13.7. Vibration Level, Distance, and Charge Weight ... 356
13.8. Scatter of the Peak Particle Velocity ... 358
13.9. Cooperating Charges ... 361
13.10. Methods to Reduce the Vibration Level ... 365
    13.10.1. Costs When the Vibration Level Has to be Reduced ... 366
    13.10.2. Ground Excavation ... 366
13.11. General Methods to Reduce the Vibration Level Transmission of Vibration from Ground to Building ... 367
13.12. Model Studies with a Slot ... 367
13.13. Full-Scale Experiments with Slots in Rock ... 369
    13.13.1. Measures for Damping Vibrations which Reach Computers and Auxiliary Equipment ... 370
    13.13.2. Vibration Damping for Buildings ... 371
    13.13.3. Practical Views on Measurement Techniques ... 371
    13.13.4. Effect on Buildings ... 372
    13.13.5. Effects on Building Installations ... 372
    13.13.6. Degree of Disturbance to Man ... 373

Chapter 14. Air Blast Effects ... 375
14.1. Introduction ... 375
14.2. Characteristics of Pressure Waves in Air ... 375
14.3. Blast Scaling ... 377
14.4. Air Blast Induced Damage ... 380
14.5. Reduction of Air Blasts ... 382
14.6. Focusing Effects ... 384
14.7. Inform the Neighbors ... 384

Chapter 15. Toxic Fumes ... 387
15.1. Fume Classification of Explosives ... 388
15.2. Computer Calculations of Reaction Products ... 389
15.3. On-Site Measurements ... 389
15.4. Influence of Confinement ... 393
15.5. Recommendations ... 395

Chapter 16. Metal Acceleration, Fragment Throw, Metal Jets, and Penetration ... 397
16.1. The Gurney Equations for Metal Plate Acceleration ... 397
    16.1.1. Explosive Slab between Metal Plates of Equal Thickness ... 398
    16.1.2. Explosive Slab between Metal Plates of Unequal Thickness and Explosive Slab with a Single Metal Plate ... 399
    16.1.3. Explosive-Filled Metal Tube and Metal Sphere ... 399
    16.1.4. Summary of Gurney Equations and Resulting Velocities ... 400
    16.1.5. Specific Impulse Delivered to a Large Metal Object ... 401
    16.1.6. Direction of Motion of the Accelerated Metal Plate. ... 401
16.2. Shaped Charge Jet Formation ... 404
    16.2.1. Explosively Formed Fragments ... 406
16.3. Target Penetration ... 407
16.4. Safety Distance for Metal Fragments ... 410

Chapter 17. Explosive Art, Explosive Metal Forming, Welding, Powder Compaction, and Reaction Sintering ... 413
17.1. Explosive Art ... 413
    17.1.1. Verner Molin ... 415
    17.1.2. Evelyn Rosenberg ... 416
17.2. Explosive Forming ... 416
17.3. Explosive Welding ... 417
17.4. Explosive Powder Compaction and Reaction Sintering ... 418
    17.4.1. Explosive Powder Compaction ... 420
    17.4.2. Shock-Induced Chemical Reaction Synthesis and Sintering ... 420

Chapter 18. Safety Precautions, Rules, and Regulations ... 423
18.1. Personal Precautions ... 423
18.2. Inoffical Recommendations ... 424
18.3. Safe Operating Procedures (SOPs) ... 425
18.4. Government Regulations ... 425
    18.4.1. General References for Safe Storage, Handling, and Transportation of Explosives ... 425
    18.4.2. Military Standards, Publications, and Regulations for Safety in the Manufacture, Transportation, Storage, Handling, and Use of Explosives ... 427

Chapter 19. Safety in Production of Explosives 429
19.1. Introduction 429
19.2. Hazards Prediction 430
19.3. Waterbased Explosives 431
19.4. Mechanical Sensitivity Testing 432
    19.4.1. Shock and Impact Tests 432
        Air Gap Test 432
        Minimum Booster Test 433
        Cap Sensitivity Test 433
        BAM Fallhammer Test 434
        Projectile Impact Test 434
        BAM 50/60 Steel Tube Test 435
19.5. Heat and Friction Tests 435
    Koenen Test 436
    Princess Incendiary Spark Test 436
    External Fire Test for Hazard Division 1.5 436
    Deflagration to Detonation Transition (DDT) Test 436
    Woods Test 436
19.6. Detonation Stability Testing 437
    Velocity of Detonation (VOD) Test 437
    Critical Diameter Test 437
19.7. Testing of Unsensitized Emulsion Matrix for Transportation 438
    Type 1(a) Tests 438
    Type 1(b) Tests 439
19.8. Conclusion 439

Chapter 20. United Nations Recommendations on the Transport of Dangerous Goods 441
20.1. General Introduction 441
20.2. Classifying a Dangerous Substance or Article 442
20.3. Test Series 1, 2, and 3 444
    20.3.1. Test Series 1 445
    20.3.2. Tests Series 2 448
    20.3.3. Tests Series 3 448
20.4. Test Series 4 449
20.5. Test Series 5, 6, and 7 449
20.6. Decision-Making Agencies and Approved Laboratories 450

Exercises 451

Solutions to Selected Exercises 467

References 491

Units 505

Conversion Factors 506

List of Symbols 507

Glossary 511

Explosives Index 519

Name Index 525

Subject Index 533

Introduction

This text is about the civilized, creative, and productive use of explosives, primarily in mining and civil engineering, but also in other fields of engineering, and even in creating fine art. We will convey the message that explosives, just like all other materials, function according to the laws of physics and chemistry; that the effects of explosives can be precalculated and predicted quite accurately; and that rock blasting and explosive working of other materials are just branches of engineering, with equations and graphs, and good or bad design.

For many people, explosives are associated with destruction, fear, and sudden death. It is not surprising that there is such widespread fear of explosives. Military high explosives have been used, and are still being used as this is written, to spread death and devastation among people of different beliefs. In the less violent field of rock blasting, there is to most of us something awe-inspiring about the sight of a large expanse of hard, immobile, solid rock that has been in place for millions of years, suddenly coming to life and disintegrating into thousands of tons of rubble while the ground shivers, clouds of smoke emerge, and the rumble of thunder fills the air.

It is quite right, too, that detonators, booster charges, and explosives should be treated with due respect, in the same way that one has to be careful with a mousetrap set ready with a piece of cheese, or a loaded gun. One needs to know where the trigger is, safeguard it against inadvertent triggering, and keep out of the way when the trigger is squeezed. There are regulations, rules, and laws for the use of explosives which are meant to be such a safeguard against accidents. These are different in different countries, districts, and states, and we will deal only with the fundamental principles behind them. Remember that they exist, and go looking for them before you start handling explosives. But remember also that it is your personal responsibility to make sure you run a safe operation. The rules and regulations are a help, but nothing can replace your personal knowledge and understanding of the safety characteristics of explosives. This book is intended to supply some of that knowledge and understanding.

Rock blasting is used in applications ranging in size from small boulder blasting in the back garden, using perhaps 10 or 20 g of explosive, to large-scale open pit mining operations where perhaps 200 tons of explosive are consumed in a single round.

Rock blasting is big business. In 1990, the total consumption of commercial explosives in the USA was 2.1 billion kg (2.1×10^9 kg), equivalent to a volume of solid rock broken of about 12 m^3 per capita (total volume of rock broken divided by the total population). In 1993 prices in the USA, we may assume that the average cost of the explosives used in rock blasting is of the order of $1 per kg (44 cents per lb). The cost of accessories (detonators, booster charges, and specialty explosives) might add perhaps 25 % to that cost. The cost of drilling the holes in which to place the explosives could perhaps be $0.5 per kg explosive. The total average cost in a large-scale blasting operation would then be about $1.75 per kg explosive. The estimated total market value of commercial explosives, accessories, and blasthole drilling is then $3.5 to $4 billion. (The cost figures quoted are probably low by comparison on an international scale, reflecting the fact that the largest part — 90 % or more — of the US commercial explosives consumption is low cost ANFO and also the fact that a major part of the rock drilling for coal, which dominates large-scale blasting in the USA, is in soft, sedimentary rock.) Toward the end of the 1980s, the rock blasting explosives consumption in the former Soviet Union was about 2.7 billion kg, equivalent to perhaps 13 m^3 of solid rock broken per capita. Already in 1971, we estimated the total volume of rock and ore blasted in the world to be about 3.8 billion (3.8×10^9) m^3, equivalent to a volume of solid rock broken of not far from 1 m^3 per capita. In 1990, the commercial explosives consumption in Australia was more than 500 million kg, giving the world record for the largest volume of solid rock broken per inhabitant to that vast, sparsely populated continent, for an estimated 45 m^3 per capita. The Scandinavian countries have a long tradition of large-scale rock blasting operations; in the hard bedrock prevalent there, blasting is the normal first step in the construction of underground hydroelectric power plants, harbors, roads, airfields, and in metal and mineral mining. In 1990, the commercial explosives consumption in Sweden was 35 million kg, equivalent to a rock blasting volume of almost 6 m^3 per capita. We estimate corresponding values for the rock blasting volume in the Western part of Europe to be less than 2 m^3 per capita.

At the present time in the USA, coal mining consumes the greatest volumes of commercial explosives. Most of the coal mining in the USA is carried out as strip mining. In recent years, the explosives consumption, which closely reflects the general health of the economy, has stagnated or decreased. In a longer perspective, there is little doubt that the increasing population in the USA and worldwide, with its increasing demand for more roads, airfields, harbors, building materials, metals, mineral resources, and energy, will lead to further growth in the volume of rock blasting.

An understanding of the strength and deformation behavior of rock materials under dynamic loads and an appreciation of the forces generated by the detonating explosive charge are the basic elements of the engineering of rock blasting. It is a very exciting science – rock materials come in a wide variety of mechanical properties and they have a rich structure of bedding planes, cracks, fissures, or joints (Figure 1). The weak planes are oriented in all kinds of directions. This science demands more of fantasy and ability to think in a three-dimensional space than the simple science of strength of homogeneous materials such as metals or ceramics. The chemical reactions in detonating high explosives run at rates that are unusually high, and the detonation-wave travels through a stick of dynamite at a rate of several thousand meters per second. The pressures generated run into the range of a 100,000 atm.

In view of the complexity of the science of rock blasting, we have to simplify in order to solve its problems, and we have to simplify intelligently. Some problems we can solve by regarding rock as a homogeneous brittle solid. Often, when we are looking at what happens in the near regions of a borehole when the charge detonates, this is sufficient. With present-day personal computers, computational solutions to several problems of this kind can be obtained with ease.

Other problems of practical blasting can be attacked by the empirical method. Many years of industrial and experimental experience have been condensed into graphs or equations that tell us how much explosive is needed to obtain this or that size fragment distribution using this or that borehole diameter. Much of what we know about blasting pattern design falls into this category — experimental results summarized into an engineering technique of calculating charge sizes and borehole patterns to obtain the required fragmentation or the required strength of the remaining rock. Such blasting engineering calculations are increasingly being performed with the help of computers.

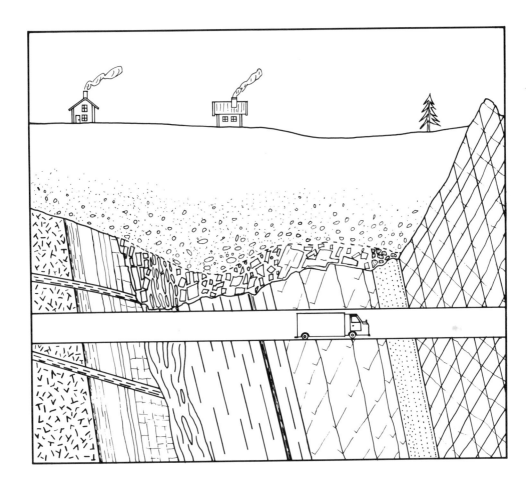

Figure 1. The rock mass structures and rock materials encountered along a typical road tunnel are often of very varied character. The tunneling methods must be able to deal with these varying conditions with a minimum of delay, and they therefore have to be generally applicable to a variety of rock mass and rock conditions.

Finally, just emerging at this time are the first efforts to combine a simplified description of the joint strength of the real rock structure with the dynamic stress or strain produced by the blast to find the rock damage or fragmentation.

All students of the science of rock blasting will meet the blast foreman or site engineer who declares that blasting is an art and not a science and that drilling patterns and charge weights cannot be predicted by engineers in an office, but need to be adjusted according to the feel of the material as the blasting job proceeds. It is good to know that there is no real conflict between the blasting engineering approach and the approach of the experienced blasting foreman, just as there is no real conflict between the mechanical engineering graduate and the mechanic who has spent many years turning pieces of brass in a lathe. The mechanic at the lathe can feel differences between different kinds of brass — all fitting into one class according to the ISO standard — that the engineering graduate has no way of distinguishing among.

A student of the engineering science of rock blasting will do well to listen to the blasting foreman who has seen more rock in his life, probably, than the engineer ever will. In his special area of experience, the foreman can add a great deal of insight and advice on the input data for the still rather crude blasting engineering calculations that we are now able to make. The engineering calculations, on the other hand, can be a great deal of help to the blasting foreman when some radical change occurs, such as a transfer to a new blasting site or the introduction of a new kind of explosive or a new borehole size. The blasting foreman is then out in deep water until he has performed a series of trial blasts. The drilling patterns that can be calculated in advance will be a good starting point for trial blasts that will lead quickly to an optimized blasting operation.

Particularly useful, of course, are the engineering calculations when we are faced with the task of predicting blasting costs, such as in planning or estimating or bidding for a new construction job or a future mining operation, in a site where no prior blasting experience is available.

Chapter 1

Rock Strength and Fracture Properties

Which properties of rock are important in the context of blasting, and how do we measure them? The two questions are not easily answered, because the physical and mechanical processes involved in the blasting process are not entirely clarified and understood. However, it is obvious that the structural characteristics of the rock mass and the strength and fracture properties of its ingredient rocks are important to know as a basis for understanding the blasting process. The following brief introduction to some selected areas of rock mechanics and geological engineering is intended to give a person not previously acquainted with these subjects a feeling for the way the rock materials and rock masses may react to the forces released by the detonation of explosive charges in drillholes.

Paradoxically, rock is both a weak and strong material. In tension, granite has only a small percentage of the strength of steel. The unconfined compressive strength is higher; for granite, about 5 or 10 times higher than the tensile strength. With increasing confinement, however, rock becomes very strong. For example, the strength of granite under conditions of uniaxial strain is very much above that of hard steel. Because there is an element of creep involved in the mechanism of failure of rock materials, there is also an influence of the time of loading, so that for a short time, rock can stand up to a higher stress than that equivalent to its normal static strength, measured with times of loading of the order of minutes. When we talk of shock wave loading, the typical times under stress are in the microsecond to millisecond range, and this results in about a factor of 2 increase in strength over the static value. Thus, for a plane shock wave, where the material cannot expand laterally, granite will stand a compressive stress of 3000–4000 MPa elastically before failing. This is 6 to 8 times the strength of steel.

The degree of brittleness of rock and the ability of the rock to release elastically stored deformation energy by crack propagation constitute another set of rock properties that we need to know and understand in order to understand the mechanism of rock blasting. The material property of fundamental importance in this context is the fracture toughness, and the tool for the mathematical treatment of crack propagation problems is called fracture mechanics.

An even more important property of rock in the context of blasting is the rich structure of fissures, bedding planes, cracks, flaws, or faults, generally called weak planes or joints, that are practically always present in a natural rock mass.

The result of a blast in a rock mass cannot be properly understood without reference to the character, strength, orientation, size, and frequency of joints. This is particularly true when we seek to explain the degree of fragmentation of the rock mass broken loose or the extent of damage caused to the remaining rock.

1.1 Rock Strength Measurements

Figure 1.1 shows schematically the different strength tests that are used to get a basis for understanding the strength and fracture properties of rock.

The weakness of rock is demonstrated by the tensile test. This test is made using a necked specimen to avoid fracturing at the jaws of the testing machine. Because the tensile strength is so low, for granite only about 1/10 that of steel, the deformation to fracture is quite small (on the order of 1 %) and, consequently, even small bending strains set up by the testing machine can cause large errors in the test result.

A simple and for some purposes quite useful test is the unconfined compressive test, in which a short cylinder ($L/d = 2$) is placed in a press that applies a uniform load to the cylinder's end surfaces. The test result is dependent on the uniformity of the load, and care must be taken to use a wafer material such as cardboard or leather between the contacting surfaces to accommodate the different moduli of the press platens and the rock.

The compressive strength of many rock materials is a factor of 5 or more greater than their tensile strength.

The scatter of the strength values from a series of test specimens of the same rock material is considerable. This is a result of the randomly distributed weak planes, microcracks, or flaws in the rock which greatly influence the rock strength. The flaws are often so small and the microcracks so fine that they are difficult to detect by the naked eye. There is also an element of creep in the strength and deformation characteristics of rock, as evidenced by the ability of rock to flow and deform plastically under tectonic stress over long periods of time (hundreds of thousands to millions of years).

Experiments have shown that the strength of rock is time dependent, so that the compressive strength when the load is applied in 1 msec is a factor of 2 or 2.5 greater than when the load is applied in 10 sec. (Figure 1.2.)

With confinement, that is when lateral expansion is restricted, rock in compression becomes stronger. This is because deformation to failure takes place as a shearing of weak planes. The action of confinement is to resist shearing, partly by creating lateral forces that resist the shear motion, and partly by increasing the friction on potential shearing surfaces by increasing the normal load thereby also increasing the rock strength. In the confined shear tests, the cylindrical test specimen, which is often necked to avoid the influence of end-effects, is sheathed by a thin rubber or copper cover to prevent penetration of hydraulic fluid into the pores of the rock. Axial load is applied by pistons inside a pressure vessel where the lateral confining load is by the hydrostatic pressure of oil surrounding the sheathed specimen. From tests of this kind, the compressive strength (which is equal to the major principal stress) can be mapped out as a function of lateral pressure. In these tests, the two minor principal stresses are equal and also equal to the lateral pressure.

Even more complicated tests have been devised to study rock strength behavior when all three principal stresses are different. The most interesting region from a practical engineering point of view, and also when we want to understand the fracture modes of rock in the complicated dynamic stress situation around a detonating charge in a drillhole in rock, is that in which the minor principal stress is small and the intermediate principal stress varies. In this region of the principal stress space, the compressive strength of rock varies with the magnitude of the intermediate stress.

One of the tests used in this region, commonly called the triaxial test, is the hollow cylinder test. A sheathed hollow cylinder of rock is exposed to different internal and external hydrostatic pressures at the same time as an axial piston load is applied. In a further complication of the test, torsion can be applied, often as an alternative to

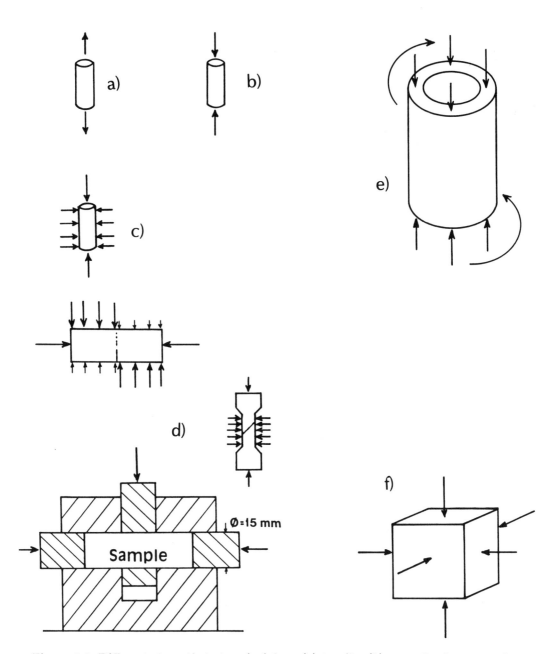

Figure 1.1. Different strength tests principles: (a) tensile; (b) unconfined compressive; (c) confined compressive; (d) confined shear; (e) triaxial using hollow cylinder under torsion; (f) triaxial using piston loads on faces of a cubic specimen.

having the internal pressure different from the external. Similar results have been obtained by piston-loading the square surfaces of a cubic rock specimen with opposing piston loads that are different in the three principal directions.

The information available on the influence of the intermediate principal stress on the strength of rock comes from experiments of this kind. These tests are, however, cumbersome and expensive to carry out.

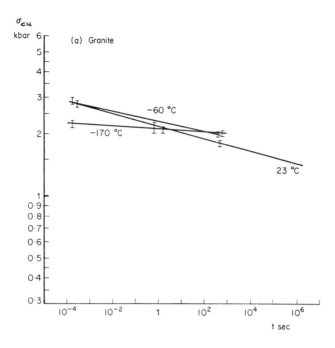

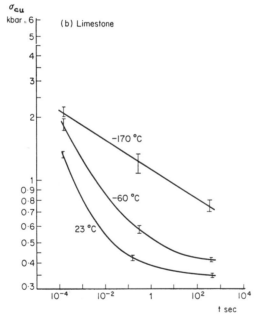

Figure 1.2. Unconfined compressive strength of granite (a) and limestone (b) at different temperatures vs. loading cycle time.

A comparatively simpler test which has proved very useful in the study of rock strength at very high confining pressures is the confined double shear test, developed by Lundborg (Figure 1.1d). A cylindrical test specimen is inserted in the yoke-like loading cell made of hardened steel. The normal load F_n is applied via pistons to the end surfaces of the cylinder. The cylinder is then sheared off by applying the shear load F_s to the hardened steel plate enclosing the middle part of the cylinder inside the yoke.

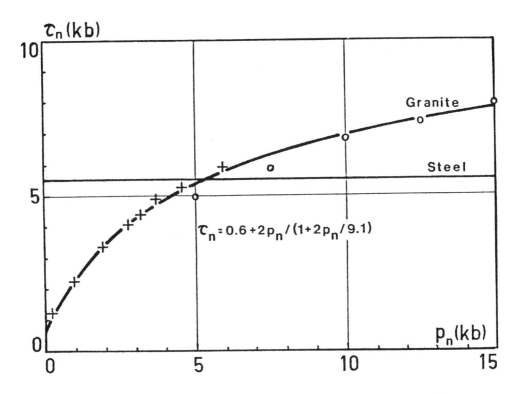

Figure 1.3. The shear strength of granite and steel as a function of normal pressures.

Although the stress concentrations at the edges of the shear surfaces are a complicating factor, the double shear test has given reproducible results that in general agree with other confined shear tests. It has a tendency to give strength values somewhat on the high side. This has been ascribed to the fact that the exact location of the two shear zones within the specimen volume is predetermined in this test. Therefore, shear is forced to occur in that location — which is not necessarily in the weakest shear plane. In other shear tests, the shear occurs at the weakest plane within the specimen volume. Figure 1.3 shows the shear strength of granite compared to that of steel.

Shear strength is thus composed of two parts: the friction between sliding crack surfaces characterized by a friction coefficient μ; and the fracturing or plastic deformation of the crystal grains, which approach a limiting shear strength τ_i when the deformation is entirely plastic. Based upon this simple assumption, Lundborg [1972] has succeeded in describing the confined double shear strength τ as a function of normal pressure σ_n by the three parameters, μ, τ_i, and the unconfined shear strength τ_0 by the expression

$$\tau = \tau_0 + \frac{\mu \sigma_n}{1 + \dfrac{\mu \sigma_n}{\tau_i - \tau_0}} \tag{1.1}$$

The confined double shear test has proved to be a useful source of strength data for use in rock blasting computations. Table 1.1 shows values of μ, τ_0, and τ_i for a number of rock materials.

Table 1.1. The constants μ, τ_0, and τ_i for different kinds of rocks and ores.

No.	Material	Locality	μ	τ_0 (kbar)	τ_i (kbar)
1	Granite I	Bohuslan	2.0	0.6	9.7
2	Pegmatite-gneiss	Valdemarsvik	2.5	0.5	11.7
3	Granite II	Bredseleforsen	2.0	0.4	10.2
4	Gneiss-granite	Valdemarsvik	2.5	0.6	6.8
5	Quartzite	Gautojaure	2.0	0.6	6.1
6	Grey Slate	Granbofosen	1.8	0.3	5.7
7	Skarnbreccia	Malmberget	1.5	0.4	20.4
8	Micagneiss	Vindelforsen	1.2	0.5	7.6
9	Limestone I	Granboforsen	1.2	0.3	8.7
10	Black Slate	Gautojaure	1.0	0.6	4.8
11	Lead Ore	Laisvall	2.5	0.6	8.1
12	Magnetite	Grangesberg	1.8	0.3	8.3
13	Leptite	Grangesberg	2.4	0.3	6.3
14	Iron Pyrites	Rutjebacken	1.7	0.2	5.5
15	Granite III	Rixo	1.8	0.3	11.9
16	Limestone II	Borghamn	1.0	0.2	10.2
17	Sandstone	Gotland	0.7	0.2	9.0
18	Flintstone	Skane	1.5	1.4	21.4
19	Glass		2.5	0.5	12.0

Table 1.2. Comparison of strength–granite and steel.

Strength		Typical Strength Values		
		Granite (MPa)	Granite (kbar)	Steel (kbar)
Tensile strength	σ_t	30	0.3	3
Compressive strength (unconfined)	σ_{cu}	200	2	3
Compressive strength (confined)	σ_{ci}	2000	20	3
Hugoniot Elastic Limit	HEL	3000–4000	30–40	8
Elastic Modulus	E	50000–80000	500–800	2100
Shear strength (unconfined)	τ_0	50	0.5	6
Shear strength (confined)	τ_i	1000	10	6
Coefficient of friction	μ		2	0.1–1

The ultimate degree of confinement (which occurs in the case of uniaxial strain in which no lateral expansion is allowed) can be produced in a triaxial compressive test. The two minor principal stresses are adjusted to keep the lateral strain at zero. The "infinite confinement" compressive strength σ_{ci} measured in this case is about an order of magnitude greater than the unconfined compressive strength σ_{cu}.

The case of compressive loading in uniaxial strain occurs in the shock compression in a plane shock wave. For small enough shock pressures, the uniaxial deformation is elastic. With increasing shock pressure, the material finally cannot accommodate the deformation elastically, and the strain is relieved by microscopic shear failure. The pressure when this happens is called the Hugoniot Elastic Limit (HEL). Because of the high loading rate in the shock wave that leads to a dynamic strength higher than the static strength, the HEL is another factor of 1.5 to 2 times greater than σ_{ci}.

Table 1.2 gives typical values for the different characteristic strengths of granite, compared to that of steel.

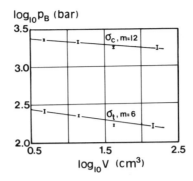

Figure 1.4. Strength-size relationship of granite.

The strength of rock is also size dependent, in that a large test specimen will fail at a lower stress than a small one. Weibull [1939] showed that several materials follow the formula

$$m \log_{10}\left(\frac{p_{b1}}{p_{b2}}\right) = \log_{10}\left(\frac{V_2}{V_1}\right) \qquad (1.1A)$$

where p_{b1} is the strength of a sample of volume V_1, p_{b2} that of a sample of volume V_2, and m is a constant. Figure 1.4 shows the size dependence of the tensile strength σ_t and compressive strength σ_c of granite on the test specimen volume V.

1.2 Three-Dimensional Rock Strength and Deformation Energy

The strength properties of rock when confined and stressed in different directions simultaneously have a profound importance on the understanding of the whole technology of rock fragmentation in drilling, boring, and blasting. Figure 1.5 shows a model of the strength of granite in the principal stress space. In the following, σ_x, σ_y, and σ_z are the three principal stresses; σ_1, σ_2, and σ_3 represent the principal stresses in order of decreasing size ($\sigma_1 > \sigma_2 > \sigma_3$). Along the space diagonal, all three principal stresses are equal. Thus, the diagonal is the axis of increasing hydrostatic pressure. Fracture occurs for all stress combinations that lie on the surface of the projectile; for those within it, the material holds. We can see that as long as at least one of the principal stresses are small, the strength of the material is small. But for large compressive stresses, the granite is strong. This is seen perhaps better in a plot of shear strength vs. normal pressure obtained from the Lundborg confined shear test (Figure 1.1d). It is interesting and thought provoking to compare, as in the figure, the strength of granite at a confining pressure of 10 kbar with that of hard steel. When we allow a rock cutting tool to work in such a geometry that it creates mainly compressive stresses, in effect it must cut a material that is considerably stronger than steel.

The apparent ease with which rock breaks in tension is only partly the result of the tensile strength of rock being small compared to the compressive strength. It is also the result of the small energy required for deformation to fracture in tension, which is much smaller than that in compression. In fact, the energies of deformation of granite to the point of fracture for the three stress systems — uniaxial tension, uniaxial compression,

12 Chapter 1. Rock Strength and Fracture Properties

Table 1.3. Strength and deformation energies of granite.

Strength	MPa	Bar	Psi
Tensile	30	300	4350
Compressive	200	2000	29000

Relative Energies of deformation to fracture	
Uniaxial tension	1
Uniaxial compression	100
Uniaxial strain	10,000

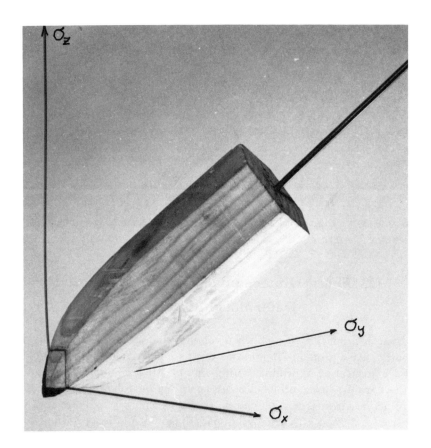

Figure 1.5. A model of the tri-axial strength of rock. σ_x, σ_y, and σ_z are the three principal stresses. The space diagonal represents hydrostatic pressure ($\sigma_z = \sigma_y = \sigma_x = p$).

and uniaxial strain — have the relative magnitudes of 1, 100, and 10000, respectively (Table 1.3). Thus, in order to fracture rock with minimum force and energy, the rock material should be loaded mainly in tension. The explosion gas pressure in a drillhole, working from the inside of the material, does create mainly tensile stresses, particularly when there is a free surface nearby. This explains why rock fragmentation by explosives is such an efficient, energy-conservative method.

1.3 The Random Microcrack Model for Rock Strength

The reasons for the strength behavior of rock and other brittle materials can be sought in the presence of microscopic cracks and flaws in the base material that is in itself strong. Most rock materials are aggregates in which separate crystal grains of different strength, different elastic and thermal moduli, and different size are cemented or grown together. Any deformation of sufficient magnitude will lead to local cracking or the development of microscopic flaws, pores, or weakened regions. Such flaws are also nearly always present in most natural rock materials because of the deformation the rock has undergone under the influence of tectonic forces and temperature changes.

In tension, microcracks grow, join, and ultimately lead to fracture at a low load. In compression, the friction on such microcracks that are stressed in shear leads to increased strength. With confinement, the crack growth is further restricted and friction is increased. This leads to a further increase in strength. In the limit, with increased hydrostatic confining pressure, we approach the real strength of the aggregate base material. It is conceivable, with a sufficiently high hydrostatic pressure, that the deformation of the weaker part of the aggregate grains will be plastic, while the hard grains still only deform elastically.

The random distribution in space, size, and direction of the microcracks or flaws is the reason for both the scatter of experimental strength measurement data and the dependence of strength on the size of the specimen. In a large specimen, it is more probable than in a small specimen that a sufficiently large flaw will have a direction favoring fracture in a given stress situation. Therefore, a large specimen or rock volume has a lower strength than a small specimen.

Lundborg [1972] has formulated a statistical theory of the strength of brittle materials based on the random microcrack model. This theory agrees very well with the available experimental data on tensile strength, compressive strength, triaxial strength, the strength/volume dependence, and the scatter of experimental strength data for several rock materials.

Lundborg assumes that the local shear strength of a brittle material is randomly distributed over both space and spatial angle of shear, in the same way as we may assume the microcracks to be randomly distributed within the piece of rock and to have a randomly varying size and direction in space. Therefore, fracture under a given load may not always occur in the direction where the shear stress is highest, but in a slightly different direction where the local shear strength happens to be low.

Lundborg used Weibull's statistical approach to model and describe mathematically the statistical nature of rock strength, putting the probability of fracture in the form

$$S(x) = 1 - e^{-kX} \tag{1.2}$$

where

$$X = \int \left(|\tau_n| - \mu \sigma_n \right)^M d\Omega \tag{1.3}$$

where Ω is the solid angle and k and M are constants for a given material, τ_n and σ_n are the shear and normal compressive stresses in and across a plane of a given direction, and the integration is carried out over all directions in space where $|\tau_n| > \mu \sigma_n$. μ can be regarded as an average of the coefficient of friction between the two surfaces of a microcrack. M represents the degree of statistical scatter of the shear strength as a function of solid angle. The subscript n refers to a stress in a direction normal to the shear direction.

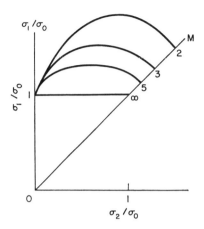

Figure 1.6. Strength as a function of the intermediate principal stress σ_2 at different M when $\mu = 1$ and $\sigma_3 = 0$. σ_0 is the strength σ_1 when $\sigma_2 = \sigma_3 = 0$.

By using experimental values of uniaxial compressive strength, say those for 50 % rupture, the value of kX can be found from Equation 1.2. Then, if μ and M are known, the strength at any combination of stresses can be calculated by letting X be constant. The calculation requires the use of a small computer.

Figure 1.6 shows the variation in strength according to Lundborg's theory. σ_1, σ_2 and σ_3 are the principal stresses in order of decreasing size, and the calculation is made for the case when $\sigma_3 = 0$, $\mu = 1$, and $M = \infty, 5, 3$, and 2. The classical strength criteria, such as those named after Mohr-Coulomb, von Mises, and Tresca, respectively, can be derived as special cases of the Lundborg theory (Figure 1.7).

There is a simple relationship between μ and M when the uniaxial and biaxial compressive strengths σ_{10} and σ_{20} are known:

$$\left(\frac{\sigma_{20}}{\sigma_{10}}\right)^M = \sqrt{1+\mu^2} + \mu \tag{1.4}$$

For high compressive stresses, μ is not constant, but decreases with the normal stress σ_n:

$$\mu = \frac{\mu_0}{1 + \frac{\mu_0 \sigma_n}{\tau_x}} \tag{1.5}$$

where τ_x approaches the limiting friction stress τ_i when σ_n tends to infinity.

Figures 1.8 through 1.11 show the comparison between experiment and the Lundborg theory for two different rock materials, (Lundborg [1974]). The agreement with experiment is excellent. The Lundborg theory is unique in that it contains as special cases the well-known fracture criteria widely used in engineering to explain the different fracture points under combined stresses for different types of materials. Each of the Mohr-Coulomb, von Mises, Tresca, and other strength criteria thus correspond to different sets of special values of the parameters μ and M in the Lundborg theory.

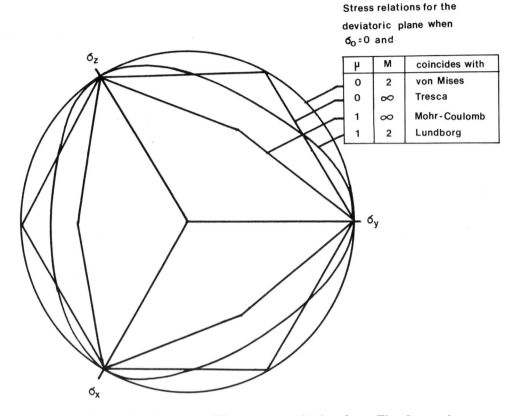

Figure 1.7. Comparison between different strength theories. The figure shows cuts through the strength projectile at right angles to the hydrostatic pressure axis. The four different strength criteria according to von Mises, Tresca, Mohr-Coulomb, and Lundborg represent different shapes of the projectile: von Mises represents a rotationally symmetric projectile, Tresca a symmetric hexagonal cross-section with sharp edges, while Mohr-Coulomb represents a deformed hexagonal cross-section which also has sharp edges. The Lundborg model for $\mu = 1$, $M = 2$ yields a projectile shape with rounded edges which has features of all these criteria.

1.4 Fracture Mechanics

The most important aspect of the strength of brittle materials is their ability to break by crack propagation. Because the tensile strength of these materials is so much lower than the compressive strength, and possibly also because they already contain microcracks, cracks form easily and, once formed, expand because of the concentration of tensile stresses at the crack tip (Figure 1.12). We will limit this discussion to cracking under biaxial stresses, that is, stress situations where two principal stresses are equal and the third is zero.

The stress concentration in front of the crack tip can be represented by the expression

$$\sigma(x) \approx \frac{K_I}{\sqrt{2\pi x}} \tag{1.6}$$

where the stress intensity factor K_I is a function of the crack length and the load σ_0. The critical value of K_I when the crack just starts moving is a material constant K_{IC}.

16   *Chapter 1. Rock Strength and Fracture Properties*

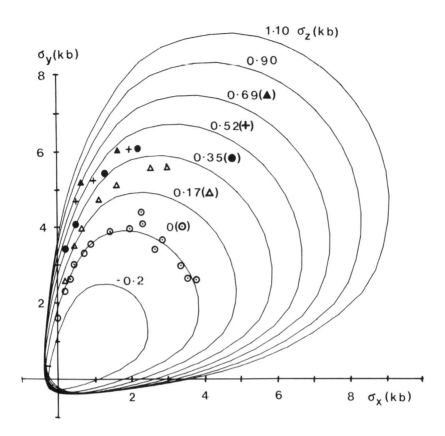

Figure 1.8. Comparison between calculated strength and Hoskins' [1969] experimental values on syenite. The symbols indicate different values of the smallest principal stress σ_z used by Hoskins corresponding to the calculated lines in the figure.

As the crack propagates, energy is absorbed by deformation work by the material at the crack tip or dissipated as elastic wave energy radiating out through the material from the crack tip. The work done per unit new crack surface is G_{IC} which is coupled to K_{IC} through the relation

$$G_{IC} = \frac{1-\nu^2}{E} K_{IC}^2 \tag{1.7}$$

where ν is Poisson's ratio and E is Young's modulus. G_{IC} is called fracture toughness and is the fundamental material constant. Some authors also refer to K_{IC} as the fracture toughness. Table 1.4 shows some typical fracture toughness values.

Fracture mechanics is at work in most processes of practical rock fragmentation (Figure 1.13).

1.5. Rock Mass Strength and Structure

Table 1.4. Typical fracture toughness values.

Material	E[GPa]	G_{IC}[J/m^2]	K_{IC} [MPa m$^{1/2}$]
Steel	210	10000	50
Aluminum	70	8000	25
Plexiglas	3	600	1.5
Granite I	50	150	3
Granite II	20	95	1.4
Marble I	50	10	0.7
Marble II	80	34	1.67
Sandstone	7	300	1.5
Limestone	15	65	1.0
Silicon Glass	50	1.2	0.25

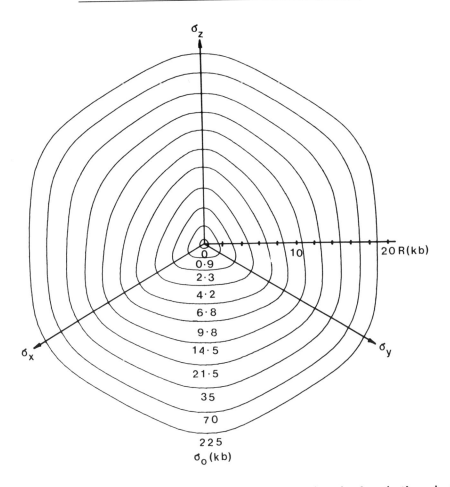

Figure 1.9 Calculated strength criteria of syenite for the deviatoric plane in the principal stress space at different mean stresses σ_0. The R-scale shows the radius of the yield surface, and the σ-axes the direction of the principal stresses.

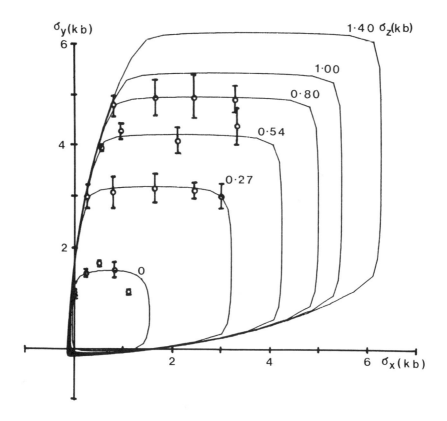

Figure 1.10. Comparison between calculated strength and experimental values from Akai and Mori [1967] on sandstone.

1.5 Rock Mass Strength and Structure

A rock mass, as distinguished from a rock strength test specimen, is a body of rock with its naturally occurring network of flaws and discontinuities, cracks, joints, and planes of weakness. These are important for a proper understanding of the real ability of a rock mass volume to withstand load, of how and why it fails, and of the resulting fragment size and shape (Figure 1.14).

Whether we want to know if an overhanging body of rock is likely to break and come down on a road to be constructed, or we want to know the stability of a tunnel or the stability of a pillar in a mine, we must of necessity know the major structure and orientation of weakness planes in that rock mass.

Mapping out the rock mass structure is not very difficult. Make a point of looking for features on rock faces along the road as you pass by; stop and draw a sketch of what you see. Quickly you begin to distinguish recurring discontinuities in the form of bedding planes, foliation partings, cracks, fissures, or joints. You begin to see intersecting groups of parallel planes, or random, irregular structures. Because we are dealing with a three-dimensional network of intersecting planes, the description of each plane has to include the compass bearing (called the strike) of its intersection with a reference plane (normally the horizontal), and the slope angle (dip) between the plane and the horizontal. For regular or recurring cracks we need two further descriptors, namely the average crack length and the average distance between parallel cracks. A good description of one

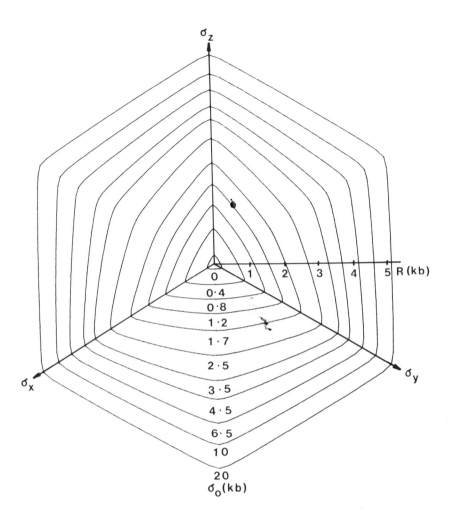

Figure 1.11. Calculated strength criteria of sandstone for the deviatoric plane in the principal stress space at different mean stresses σ_0. The R-scale shows the radius of the yield surface, and the σ-axes the direction of the principal stresses.

system of parallel cracks reads like this: strike N30E, dip 20°SE, crack length 2 m, crack spacing 0.3 m. For a statistical treatment, each of these four may be given a measure of the scatter around its mean value, usually the standard deviation.

The strength of joints is normally considerably less than that of the adjacent rock. It is described by two simple measures, the tensile or adhesive strength (often zero) at right angles to the plane, and the shear strength or friction angle along the plane. For a detailed understanding of the rock mass behavior under stress and vibration, we also need a measure of the elastic or plastic deformability of the joint (its "spring constant") and the way the shear strength or friction angle varies with shear deformation and crack separation.

Numerical modeling of rock mass strength and stability has to be based on simplified structures that retain the major and most important features of the real joint structure. Such models tend to become very cumbersome to deal with in three dimensions because of the complexity of the geometrical description. Therefore, wherever

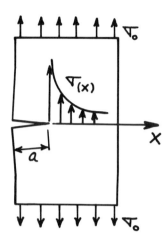

Figure 1.12. Stress concentration at the crack tip.

possible, two-dimensional modeling is preferred, using structure models such as in Figure 1.15.

Another very useful simplification in the modeling of rock mass behavior under low stress loads, such as in slope stability modeling, is to consider the rock material between the joints to be incompressible, so that all deformation is assumed to take place within the joint, as shear movement, elastic or plastic compression, tension, or separation. Cundall [1971] used this approach, which he called the block model, to calculate problems of slope stability, where the rock mass in a slope was under stress caused by gravity. The block model of rock mass deformation and fracture was later very successfully applied by Cooper [1981] to the dynamic stresses set up by the gas pressure in blasting. In Cooper's calculations (shown in Figure 1.16), the model yielded the throw trajectories and final positions of each block in the final muckpile.

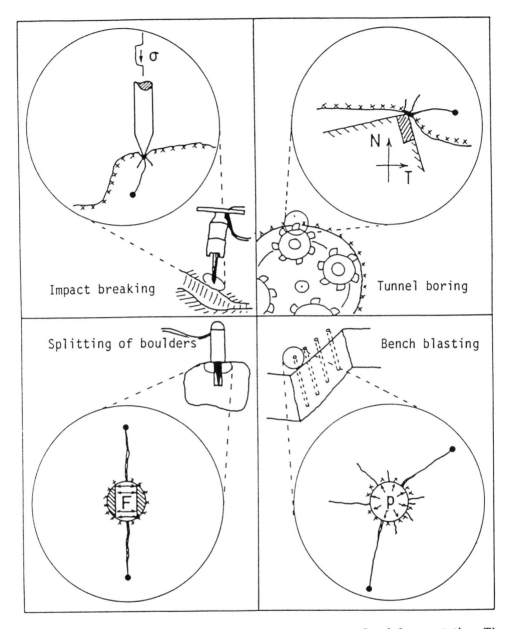

Figure 1.13. Fracture mechanics is at work in most processes of rock fragmentation. The active deformation areas at the crack tips are shown with black dots.

22   *Chapter 1. Rock Strength and Fracture Properties*

a, Material; Good, strong, competent granite.

b, Our boulder with planes of weakness drawn in.

Figure 1.14. A boulder of rock with its planes of weakness.

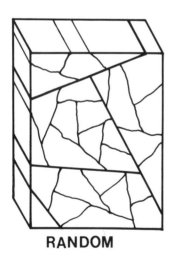
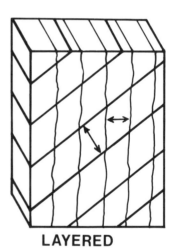

RANDOM   **LAYERED**

Figure 1.15. Two-dimensional rock structure models.

1.5. Rock Mass Strength and Structure   23

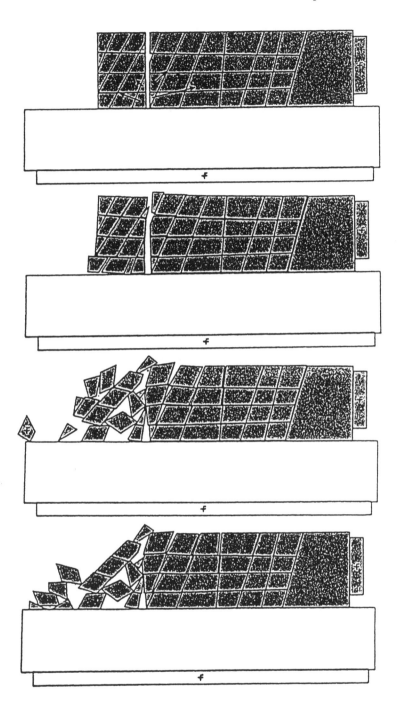

Figure 1.16. Computer modeling (in two-dimensions) [Cooper, 1981] of the dynamic fragmentation, throw, and back break due to the instantaneous explosion in a slit (using the Cundall [1971] rigid block model modified by Cooper to allow inertial effects in addition to gravity.)

Chapter 2

Mechanical Drilling and Boring in Rock

It is not possible to understand the economics of rock blasting in mining and construction engineering without reference to the economics of the production of the drillholes into which the explosive is loaded. The choice of explosive and the size of the most economic drillhole size and drilling depth is dictated by the cost of drilling. The cost of drilling depends on the penetration and wear characteristics of the rock material, and on the size and depth of the drillhole. In this chapter, we will seek to develop an overall understanding of the main factors that influence the cost of drilling, to lay a foundation upon which the optimization of the drilling and blasting process can be based.

In an even wider perspective, the entire operation of drilling, blasting, loading, transport, and crushing has to be considered as one system, to be optimized as a whole.

With the development in the 1950s of steel drill rods tipped with tungsten carbide cutting edges came the dominance of mechanical percussive drilling over all other methods of making small-diameter shotholes for rock blasting. Percussive drilling and blasting using conventional chemical explosives still dominates among today's rock excavation methods. The drilling machines have developed from the early simple hand-held pneumatic hammer machines (Figure 2.1), which drilled perhaps 0.2 m/min, to the present-day self-propelled multi-boom drilling jumbo rigs using heavy hydraulic hammer machines (Figure 2.2), which can drill hard rock at a rate exceeding 2 m/min each. The bit may be either a chisel bit having one, three, or four cutting wedges, or a stud (button) bit having several tungsten carbide buttons (Figure 2.3).

However, other means of mechanical rock cutting have been introduced as alternatives to percussive drilling. Mechanical raise (vertical shaft) boring has largely replaced the conventional drilling and blasting method, and mechanical full-size boring of long tunnels successfully competes with drilling and blasting in all but the hardest rock masses. In relatively soft rocks, rotary spiral drilling is used more and more for small to medium diameter shotholes in the diameter range 30 to 150 mm, with drill rods having fixed steel or tungsten carbide cutting edges like those on a spiral drill (Figure 2.4). Larger shotholes of diameters from 150 mm up are produced by rotary crushing drilling using tricone roller bits studded with tungsten carbide buttons (Figure 2.5). These cut by crushing the rock into chips as they are pressed successively into the rock surface at the hole bottom by the force of the rotating drill rod. Similar studded roller bits are used to produce raise boreholes even in hard crystalline rock. The rollers are held by bearings bolted on to a rotating shield which is pulled and rotated by a drill rod which passes through a smaller pilot hole up from an underground tunnel. Such raise boreheads have diameters from 0.5 to 3 m (Figure 2.6).

A major part of the world's long straight tunnels in sedimentary rock are excavated by full-diameter boring, using tunnel boring machines. For soft sedimentary rock materials, these may have an excavator shield type head such as that shown in Figure 2.7. For the intermediate and high hardness range, the disc cutter-type machines dominate. These have a rotating shield of a diameter approaching that of the tunnel. On the shield

Figure 2.1. One man operated light drilling machines (Atlas Copco BBC 24).

Figure 2.2. Electro-hydraulic drilling rig (Atlas Copco Boomer H 127).

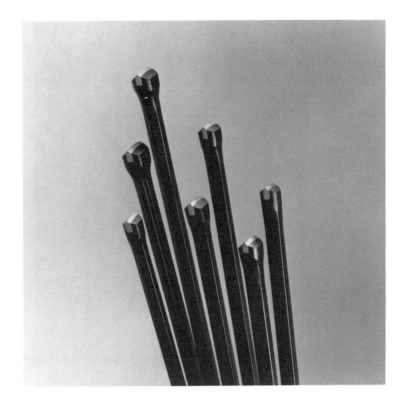

a

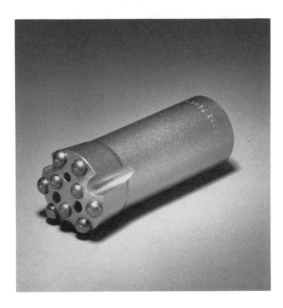

b

Figure 2.3. Percussion drill bits: a) single cutting edge chisel bits, b) button bit (R 32) (photo by Lars Lindgren).

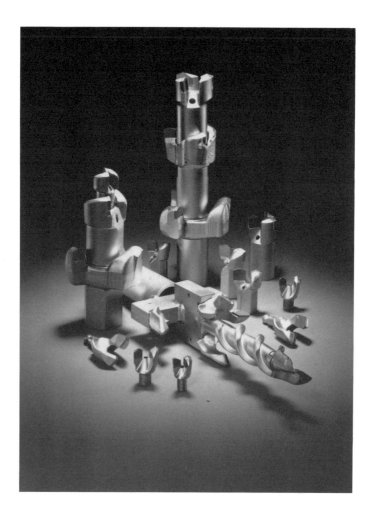

Figure 2.4. Spiral cutting drill bits (Sandvik Rock Tools).

are fastened heavy steel discs with bearings that allow the discs to roll heavily on the tunnel face as the rotating shield is hydraulically pressed against the tunnel face. With an average load per cutter disc of 10 to 30 tons, hand-sized flakes of rock are chipped off the rock face on either side of each cutter. The penetration per shield revolution may be of the order of 0.5 to 2 cm. The cutter discs are spaced over the shield so that the circular rolling grooves they leave on the rock face are evenly spaced 6 to 7 cm apart (Figure 2.8). Where the remaining rock is stable and dry, boring rates, including necessary stops for retooling, service and repairs, may be as high as 30 m per shift in limestone of compressive strength 200 MPa (2 kbar).

The fundamental mechanism of rock fragmentation under the action of a hard tool being pressed into a rock surface is similar, whether the tool is a blunt wedge, a hemispherical button, or the edge of a cutting disc. In the following section, we will go through the common features of this mechanism.

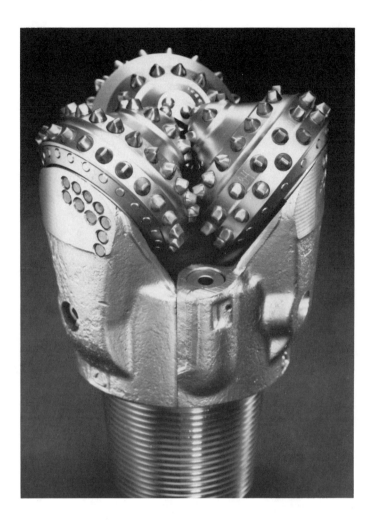

Figure 2.5. Tricone roller bit for rotary drilling (Sandvik Rock Tools).

2.1 Bit Penetration, Chip Formation, and Force-Displacement Curves

First, let us consider the simple case of a hard steel ball contacting the flat surface of a hard brittle solid such as glass (Figure 2.9).

The initial stage is the formation of a flat circular contact surface as the two solids deform elastically. Then a ring crack forms, the radius of which is some 10 % larger than that of the contact surface. Initially, the ring crack is perpendicular to the surface, but it quickly deviates to form a 90° conical central body which supports the load. In the third stage, the cone is crushed and the ball makes contact with the surface material outside of the ring crack. At this stage, the chip formation begins. Chips are formed by cracks originating in the crushed zone underneath the contact area and extending out and up toward the free surface, intersecting each other and the surface some distance from the contact area. In the fourth stage, the chips so formed come loose, and further penetration takes place by continued crushing and intermittent chip formation. As each

30   Chapter 2. Mechanical Drilling and Boring in Rock

Figure 2.6. A raise boring reaming head with tungsten carbide studded roller cutters, diameter 1.06 m (Atlas Copco MCT).

chip is released, there is a sudden drop in contact force corresponding to the decrease in the load-bearing area and in the degree of confinement of the material being crushed. A recording of the penetration force vs. displacement curve therefore has a typical sawtooth character, as shown in Figure 2.10.

Essentially the same process of gradual crushing and intermittent chip formation, accompanied by often sudden and violent release of cutter load, takes place within the

2.1. Bit Penetration, Chip Formation, and Force-Displacement Curves 31

Figure 2.7. Excavator shield tunnel borer for soft rock. Diameter 6.45 m. Project: Los Angeles Subway, CA (Robbins TBM 212 S-239).

Figure 2.8. Hard rock tunnel boring machine, diameter 6.5 m. Project: Klippen, Sweden (Atlas Copco Jarva MK 27).

Figure 2.9. Four stages in the penetration of a hard sphere into a hard, brittle solid: (a) elastic contact; (b) ring crack and start of conical crack; (c) crushing of cone formed by the ring crack and start of chip formation; and (d) continued crushing leads to departure of chips.

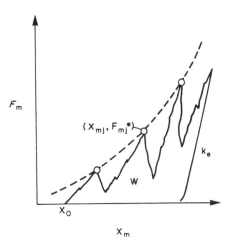

Figure 2.10. Force vs. displacement for a conical tungsten carbide indenter penetrating into a block of granite [B. Lundberg, 1974].

short time frame of each blow delivered to the rock by the cutting wedge of the bit of a percussive drill rod. The process repeats itself within a longer time frame during the penetration of a disc cutter as it rolls under load on a rock surface (Figure 2.11).

The sudden release of elastic energy, stored as tension in the mechanical parts of the boring machine, that comes with each chip formation causes severe machine vibrations. The vibrations are especially severe when boring in hard and brittle rock. These machine vibrations are a major factor in determining the life span of the bearings carrying the load of the tempered steel or tungsten carbide button bit equipped disc and other critical parts of the machine.

2.2 Drillability and Boreability of Rock Materials

The terms drillability and boreability are used, often somewhat loosely, to describe the degree of ease and economy with which a rock mass can be drilled or bored by a given machine. Thus the terms are mainly functions of the rock mass and its strength and structure, but they also depend on the drilling machine and the drill bit selected. To a certain degree, they also depend on the economy and characteristics of the machine chosen, as well as the economy and characteristics of other available machines and methods. Drillability refers to shothole drilling in the diameter range 25 to 500 mm; boreability refers to raise and tunnel boring in the diameter range 1 to 12 m. In the following sections, we will describe the techniques for predicting drillability and boreability developed at the Norway Institute of Technology (NTH), Trondheim, Norway, by Selmer-Olsen and Blindheim [1970] and Blindheim [1979]. The stamp test for rock drillability described by Wijk [1982] (a quick and useful test for estimating drillability) is also summarized. Predictions based on other tests would be equally useful; these are selected to illustrate the process and to show that a basis for judging the cost of drilling in a given rock mass quantitatively can be developed from test samples of the rock mass before actual drilling tests can be obtained.

2.2.1 Predicting Drillability

Drillability is determined by three factors:

1. Drilling rate (cm/min)
2. Bit wear (characterized by the length of borehole that can be produced in a given rock between two cutter grindings)
3. Bit life (the total length of hole drilled before the bit has to be scrapped)

The drilling rate can be estimated from a crushing test, such as the "Swedish Brittleness Test", in which an aggregate of the rock material to be tested is crushed in a cylinder by a falling weight (Figure 2.12); and a wear test, such as the Sievers test (Figure 2.13), in which the penetration into a rock specimen of a small rotary tungsten carbide chisel drill is measured under standardized conditions. There is a good linear relationship between the measured drilling rate using pneumatic percussion drills with 33 mm diameter tungsten carbide chisel bits and the drilling rate index (DRI), which is defined as a linear function of the brittleness value S from the Swedish brittleness test, with the wear value J from the Sievers wear test as a parameter (Figures 2.14 and 2.15).

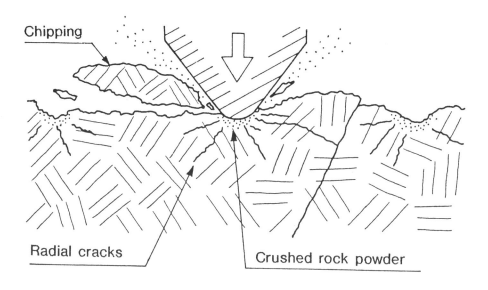

Figure 2.11. Disc cutting action in hard rock.

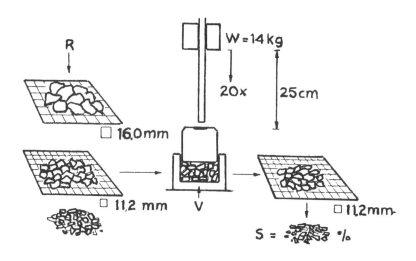

Figure 2.12. The Swedish brittleness test. S = Brittleness value. R = Crushed rock aggregate. W = Piston. V = Sample volume corresponding to 500 g at unit weight 2.65 g/cm^3.

For borehole diameters larger than 33 mm and heavier machines, the measured drilling rate (DRM) tends to be lower than that shown in Figure 2.15 at a given DRI. Reference must then be made to test results in a rock with known DRI. The new generation of hydraulic percussion drilling machines gives greatly increased drilling rates since they have a heavier bit load and a higher percussion frequency because they are tooled with stud bits with multiple tungsten carbide inserts instead of chisel bits.

2.2 Drillability and Boreability of Rock Materials   35

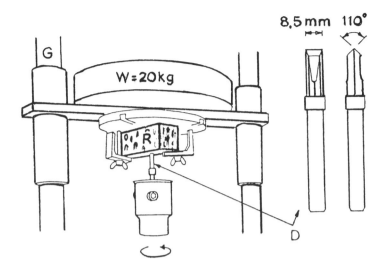

Figure 2.13. The Sievers J-value test. G = Guide. W = Weight. R = Rock specimen. D = Tungsten carbide drill. Sievers J-value (J) is measured as penetration in 1/10 mm after 200 rotations.

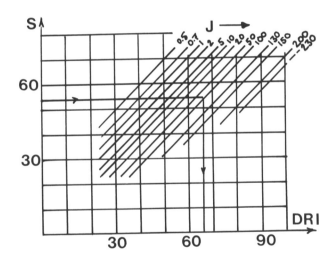

Figure 2.14. The drilling rate index (DRI) as compiled from the brittleness value (S) and the Sievers J-value (J).

The bit wear is the result of abrasion of the tungsten carbide insert by the hardest rock particles. The rate of bit wear increases with the content of quartzite or other equally hard minerals in the rock, and also with the drilling rate. The abrasiveness of different rocks on a given type of tungsten carbide bit can be determined by an abrasion test (Figure 2.16). It is found to depend not only on the quartz content, but also on the other constituents of the rock. The abrasion value (AV) is defined as the measured tungsten carbide slider weight loss in milligrams per 100 m sliding distance under a load of 10 kg. When testing cannot be performed, AV can be roughly estimated from Figure 2.17.

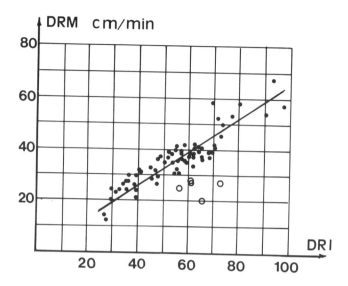

Figure 2.15. Correlation diagram for drilling rate index (*DRI*) and measured drilling rate in field tests (*DRM*) using light drilling equipment and chisel bits. Unfilled circles represent uncertain data.

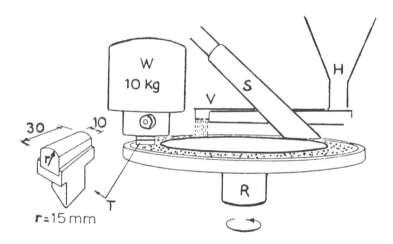

Figure 2.16. The tungsten carbide abrasion test. T = Tungsten carbide test specimen. W = Weight. V = Vibrating feeder. S = Vacuum drawoff. H = Rock powder funnel. R = Rotating steel disc. The abrasion value (*AV*) is measured as weight loss in milligrams of test specimen.

Rounded quartz grains are often found to be less abrasive than irregularly shaped ones. Grains with sharp edges are particularly abrasive. Generally, therefore, sedimentary rocks are less abrasive than igneous and metamorphic rocks at the same quartz content (Figure 2.18).

The bit wear can be expressed in terms of the bit wear index (*BWI*) which is a log/linear function of the abrasion value (*AV*) with the drilling rate index (*DRI*) as a parameter (Figure 2.19). The correlation between actual measured bit wear (*BWM*) and the drilling rate index (*DRI*) is shown in Figure 2.20. The measured bit wear is

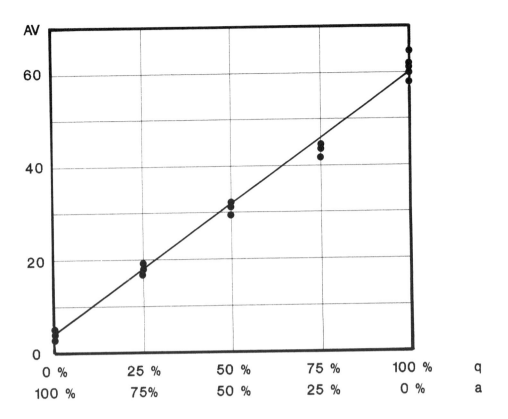

Figure 2.17. Reliability of the abrasion value test. AV = Abrasion value. q = Quartzite. a = Amphibolite.

expressed in the unit μm/m drilled length. For convenience, we often use the concept of bit life, which is expressed as the drilling length in meters that can be produced in a given rock by a given equipment between bit grindings, rather than the more scientific unit of bit wear in μm/m, which is the thickness of bit worn off in drilling 1 m of borehole.

The bit life is the total length of borehole than can be drilled by a given bit using a given machine configuration and a specific air or hydraulic pressure while drilling in a given rock mass. It is related to the bit wear in an obvious way mathematically, but the decisions made on site as to how often regrinding should be done, what is the practical number of regrindings, or when to discard a worn-out bit also influence the actual bit life value.

2.3 The Stamp Test for Rock Drillability Classification

A semiempirical test method for rock drillability classification has been used by the drilling manufacturer Atlas Copco AB. The method, based on a stamp test, in which a stamp diameter similar to the diameter of the worn button on a used drill bit, is described by Wijk [1982].

A computer code uses the information of rock parameters determined in laboratory tests and the available field experience data and analyzes these data with the help of stress wave theories and empirical data. Thereafter, a prediction is made of the drilling

38 Chapter 2. Mechanical Drilling and Boring in Rock

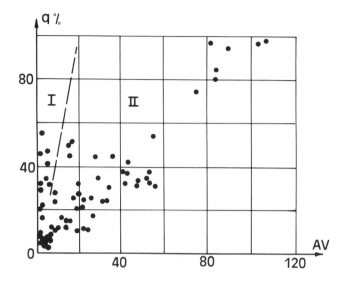

Figure 2.18. The relation between quartz content (q) and abrasion value (AV). I = sedimentary rocks; II = igneous and metamorphic rocks.

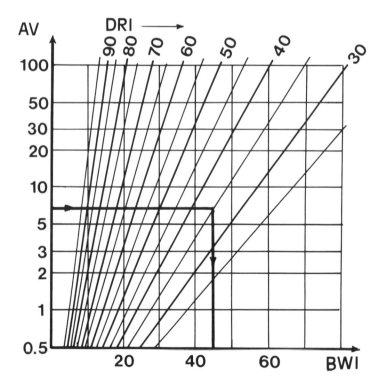

Figure 2.19. The bit wear index (BWI) as compiled from the abrasion value (AV) and the drilling rate index (DRI).

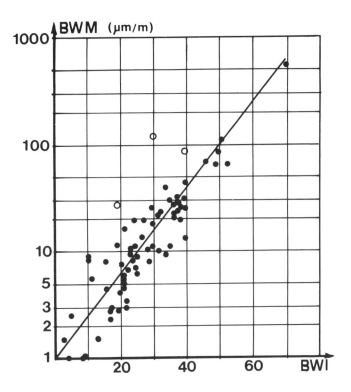

Figure 2.20. Correlation diagram for bit wear index (BWI) and bit wear measured in field tests as sum of front- and diameter- wear (BWM). Unfilled circles represent uncertain data.

rates that can be achieved if equipment parameters are changed. For percussive drilling, some important equipment parameters can be mentioned. They are:

  a. Piston mass and geometry
  b. Piston blow frequency and velocity, bit diameter and type
  c. Drill rod geometry
  d. Rotational speed
  e. Thrust force
  f. Kind of flushing

Figure 2.21 shows how the tested rock sample is grouted into a cylinder with a smooth top surface. The stamp has a diameter of 4 mm which corresponds to the contact areas between the used buttons on roller and percussive bits and the rock at the bottom of the drillhole.

The maximum force for crack initiation (Figure 2.21) is F_S, and the Stamp Test Strength Index σ_{ST} is defined as

$$\sigma_{ST} = \frac{F_S}{\pi a^2} \tag{2.1}$$

At the maximum force F_S, tensile cracks appear around the stamp, the rock underneath the stamp is crushed, and a crater is formed.

F_S equals the minimum static thrust force requirement for a roller bit button and gives the minimum piston blow velocity for a percussive drill. The displacement x_S at

Chapter 2. Mechanical Drilling and Boring in Rock

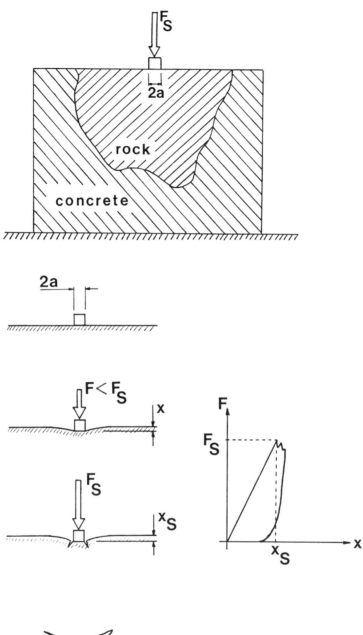

$$\sigma_{ST} = F_S / (\pi a^2)$$

Figure 2.21. The evolution of a Stamp Test on a brittle rock material [from Wijk, 1982].

2.3. The Stamp Test for Rock Drillability Classification

Table 2.1. Rock parameters for Bohus granite.

Stamp Test Strength	σ_{ST}	2390 MPa
Fracture Force	F_s	30 kN
Displacement	x_S	0.5 mm
Crater Volume	V	60 mm³

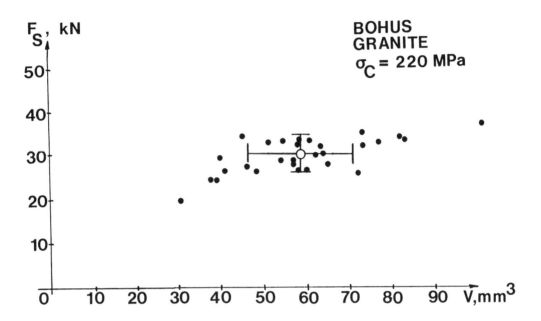

Figure 2.22. Stamp Test data (F_s as a function of V) for Bohus granite. Stamp radius a = 2 mm. σ_c is the rock material's unconfined compressive strength [from Wijk, 1982].

the fracture force F_S gives information about the piston length for percussive drilling. The crater volume V achieved is closely correlated with the drilling rate as it corresponds to the volume of crushed rock underneath the buttons on the drill bit.

For a brittle rock sample, the Stamp Test causes a deeper and wider crater than for a ductile rock, and the elastic prefracture deformation is normally smaller for the brittle rock. Ductile rocks are more effectively bored by rotary drag bit drilling, provided that the abrasiveness and strength of the rock do not result in too rapid wear of the drag bits.

Figure 2.22 shows Stamp Test data for Bohus granite, which is a good example of a brittle rock. Wijk [1982] gives data for other rocks as well.

The rock parameters σ_{ST}, x_S, and V are used for design of a drilling system or prediction of drilling rates if equipment parameters are changed. For Bohus granite, the approximate values in Table 2.1 can be used.

Consider a rock drill with a cylindrical piston, the cross-section area A of which is the same as for the drill rod on which it is striking with the blow velocity v and blow frequency f. The blows of the piston will generate compressive stress waves with the amplitude σ_o in the drill steel

$$\sigma_o = \frac{vE}{2c} \quad (2.2)$$

where $E = 210$ GPa and $c = 5200$ m/s are the Young's modulus and the stress wave velocity for steel, respectively. This stress wave also corresponds to a displacement Δ of the rod in the piston blow direction

$$\Delta = \epsilon L = \frac{vL}{c} \tag{2.3}$$

where L is the length of the piston. Thus, the mass m of the piston is

$$m = \rho A L \tag{2.4}$$

where $\rho = 7800$ kg/m^3 is the density of steel. E, ρ, and c are related by

$$c^2 = \frac{E}{\rho}. \tag{2.5}$$

Thus, for $v \approx 10$ m/s and $L \approx 0.5$ m, one finds $\sigma_o \approx 220$ MPa and $\Delta \approx 1$ mm. The dynamic force F_o in the drill steel is of course

$$F_o = \sigma_o A. \tag{2.6}$$

One finds $F_o \approx 150$ kN for $\sigma_o \approx 220$ MPa and $A \approx 700$ mm^2,

Assume that the drill bit has N buttons, each of which makes a circular contact area of radius $a = 2$ mm with the rock at the borehole bottom. Thus, the rock-bit configuration requires the dynamic force F_o to drill the rock

$$F_o = N F_S. \tag{2.7}$$

If the rock is only half as strong as Bohus granite ($F_S \approx 15$ kN), then the dynamic force $F_o \approx 150$ kN is sufficient for a bit with $N = 10$ buttons. However, one must also simultaneously require sufficient penetration; that is,

$$\Delta \geq x_S. \tag{2.8}$$

The value $x_S \approx 0.5$ mm for Bohus granite is also quite representative for other rock materials. Equations 2.3 and 2.8 yield

$$L \geq \frac{c x_S}{v}. \tag{2.9}$$

Thus, for a certain rock (i.e., a certain stamp strength σ_{ST}), a certain button wear (i.e., a certain radius a and consequently a certain fracture force F_S), and a certain bit and rod (i.e., a certain number N and a certain area A), Equations 2.1, 2.2, 2.6, and 2.7 yield the appropriate blow velocity v

$$v = \frac{2 c N \pi a^2 \sigma_{ST}}{E A}. \tag{2.10}$$

Equation 2.9 then yields the minimum piston length L and, accordingly, the required blow energy W is obtained

$$W = \frac{1}{2} m v^2. \tag{2.11}$$

A first rough approximation of the drilling rate D_r is given by

$$D_r = \frac{4 N V f}{\pi D^2} \tag{2.12}$$

where V is the average crater volume from the Stamp Test, D is the bit (and borehole) diameter, and f is the piston blow frequency.

The presentation above is, by necessity, extremely simplified in order to show how the rock parameters (σ_{ST}, x_S, and V), in principle, are used to design an appropriate drilling system. In practice, a multitude of other circumstances, some beneficial and others disadvantageous, must be taken into account. However, the overall effect is always such that Equation 2.12 is quite accurate provided that Equation 2.10 is satisfied, and provided that the piston length is "reasonably" larger than the minimum value given in Equation 2.9.

2.4 Predicting Boreability in Full-Face Tunnel Boring

Inasmuch as boring with disc cutter machines is the most frequently used technique, this section deals only with disc cutter boring.

Boreability is expressed in terms of the three factors:

1. Net penetration rate (m/hour of true boring time)
2. Cost of cutters (dollars/piece due to bit and bearing wear)
3. Total excavation costs (dollars/m tunnel length)

The net penetration rate is the length of tunnel that can be bored in a given rock mass by a given tunnel boring machine per hour of actual boring time using normal cutter load and normal cutter head rpm. (Note the distinct difference between the two terms penetration, P_e, which is expressed in mm/cutterhead revolution, and net penetration rate, P, which is expressed in m/hour. They are related through N, the number of cutterhead revolutions per minute.) The net penetration rate can be calculated by the equation

$$P = 60N \frac{P_e}{1000}. \qquad (2.13)$$

In tunnel boring much more than in shothole drilling, the presence of weakness planes or joints in the rock mass influences the net penetration rate, especially when the distance between weakness planes (joint spacing) is of the same order as that between the cutter grooves. The crushability of the rock material itself can be accounted for in boring by the same technique as in drilling, using the drilling rate index (DRI). Figure 2.23 shows the net penetration rate as a function of DRI with the distance between weakness planes as a parameter.

In the same way, the cutter cost per m^3 of solid rock bored can be estimated by using the bit wear index (BWI) derived as described above for drilling, again with the joint spacing as a parameter. Figure 2.24 shows the relationship.

With the great capital cost involved in acquiring a tunnel boring machine, the total excavation cost is greatly influenced by the ratio of productive to unproductive time during an average shift of working. The crucial figure is the number of new tunnel-meters produced in a shift. In the following section, we will present the very clear calculation of that figure given by an experienced Swiss contractor, Duri Prader [1977], reproduced here with his permission with only minor editorial changes.

2.5 Predicting Shift Capacity of a Disc Cutter Tunnel Boring Machine

This section describes a systematic way of dealing with the engineering problem of estimating boreability of a given rock mass with respect to a given boring machine with disc cutters.

We may make the following remark as price estimators concerning boreability whatever may be understood by this word.

Let us suppose that we as contractors are placed in a situation in which we know:

- The rock to be bored
- The diameter of the tunnel
- That we would use a disc cutter machine

From these facts, we should be able to predict the daily advance in fullface boring. This is a major factor influencing our bid.

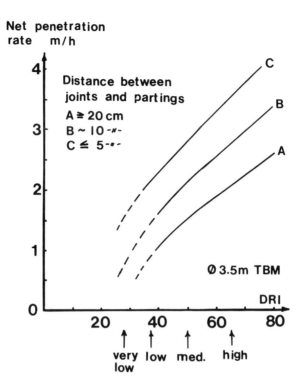

Figure 2.23. Net penetration as a function of the drilling rate index and the distance between joints and partings.

Before we call in some of our field engineers to ask them for facts and figures, we need to determine how much time will be needed for the heading to advance 1 meter. We will distinguish four different time-elements, one representing productive time in hours per meter (Equation 2.14) and three being unproductive times (Equations 2.15, 2.16, and 2.17), expressed in *clock hours per tunnelmeter*.

Productive:

$$\frac{100}{P_e} \times \frac{1}{N} \times \frac{1}{60} \qquad (2.14)$$

where P_e is the penetration (cm) per revolution, and N is in revolutions per minute (rpm).

Unproductive: (change of cutters)

$$+L \times \frac{100}{P_e} \times \frac{t_c}{\lambda} \times \frac{1}{1000} \qquad (2.15)$$

where L is the path length all cutters per revolution, t_c is the number of hours to change one cutter, and λ is the lifetime of one cutter (in km of pathlength).

2.5. Predicting Shift Capacity of a Disc Cutter Tunnel Boring Machine

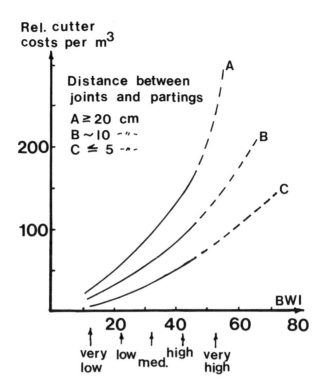

Figure 2.24. Relative cutter costs per solid m³ as a function of the bit wear index and the distance between joints and partings.

Unproductive: (meter-dependent)

$$+ t_1 \tag{2.16}$$

where t_1 is hours per tunnelmeter.

Unproductive: (fixed time loss per shift)

$$+ \frac{t_{sh}}{\text{tunnelmeter per shift}} \tag{2.17}$$

where t_{sh} is hours per shift.

Penetration means advance per revolution of the cutterhead. The distance each disc cutter rolls over the rock during one revolution of the cutterhead is called the path length of the respective cutter. Loss of time that is proportional to the meters of advance (meter-dependent unproductive time) is caused by such things as changing trains; retracting and advancing the grippers; regripping; laying tracks, pipes, and cables; or surveying.

Before our next step, our equipment department says that a fairly good estimate on the values of N and L would be:

$$N \approx \frac{24}{D}$$

46   Chapter 2. Mechanical Drilling and Boring in Rock

and
$$L \approx 19.5 \times D^{1.75}$$

where D is the tunnel diameter (m). Thus, the anticipated advance per shift can be described as

$$\text{Tunnelmeters per shift} = \frac{T_{sh} - t_{sh}}{0.07D + t_1 P_e + 2\frac{t_c}{\lambda}D^{1.75}} P_e \qquad (2.18)$$

where T_{sh} = hours nominal shift time.

From this equation, we can observe three interesting facts:

1. When P_e increases, the increase in the value for tunnelmeters per shift is less than proportional.
2. When a bigger diameter is chosen, the value for the tunnelmeters decreases less than in proportion to $D^{1.75}$, perhaps even slower than D increases.
3. Only one term is influenced by the average lifetime of the cutters. The weight of this term becomes greater as the diameter increases.

Now we know which questions to ask our field engineers. From experience, they should be able to provide values for t_{sh}, t_1, t_c, P_e, and λ. We learn the following:

t_{sh} may be 10% to 15% of T_{sh}.
t_1 may be 0.3 to 0.7 hours per meter.
t_c may be 1 to 1.5 hours per cutter.

Reasonably accurate values for P_e and λ can be obtained by the machine manufacturer from laboratory tests which they run on samples we have provided.

Finally, it seems to us the word *boreability* means the higher or lower degree to which a rock mass lends itself to be bored by a given boring machine in an economic way, economic compared to tunneling with other available methods.

2.6 Rock Strength and the Size Dependence of Drill/Bore Energy and Cost

In Chapter 1, we discussed the reasons for the way rock strength is influenced by specimen size and confinement. We showed how the strength of rock is normally governed by the presence of randomly oriented and spaced microcracks. In a large rock volume, there is a higher probability than in a smaller one of finding a sufficient number of microcracks, suitably oriented to cause fracture. The average strength of small specimens of a given rock is therefore higher than that of large specimens of the same rock.

We also showed how greatly the strength of rock increases with increasing confinement, and particularly how the energy of deformation to fracture increases as we go from uniaxial tension through uniaxial compression to uniaxial strain.

These fundamental strength properties of rock materials lead to an interesting and useful relationship between the force; the energy requirement; and the consequent cost for drilling, boring, or blasting on the one hand; and the size of the cavity produced on the other. Counted per unit volume of excavated rock, it is less costly in terms of dollars, tool stresses, and energy expenditure to make a 250 mm drillhole than a 25 mm drillhole. It is even less costly, still per unit volume of excavated rock, to bore a 2.5 m diameter tunnel, less costly still to blast a 10 m diameter tunnel, and least costly of all, to make a 1000 m diameter open pit in the same rock. In the following, we will derive an approximate mathematical model for predicting the volume cost of excavation which is based on the fundamental strength properties of the brittle rock material.

A geometry that produces a stress system approaching that of uniaxial compression or even uniaxial strain is difficult to avoid when excavating rock at the circumference of the bottom surface of an extended cavity such as a drillhole or a tunnel. The energy required for that part of the excavation is dominant over the energy required for excavation over the rest of the surface. We may then, in the first approximation, assume the energy required to excavate a unit length of hole or tunnel to be roughly proportional to the length of the circumference of the bottom surface. Thus, we have

$$E \approx d$$

where d is the characteristic diameter of the excavation. It follows that the average energy per unit excavated volume e is inversely proportional to the diameter

$$e \approx \frac{1}{d}.$$

Surprisingly, it turns out that the excavation cost per unit volume of rock excavated is also approximately inversely proportional to the diameter. In Figure 2.25, the dashed line that approximates the excavation cost per unit volume excavated rock K, for widely different methods at widely different diameters has the equation

$$K = \frac{k_1}{d} \tag{2.19}$$

where the cost per unit hole cross sectional area k_1 is a constant. The relation in Equation 2.19 is valid approximately for drilling, full face boring, and drill and blast tunneling. Thus, for practical purposes, although the cost of energy is usually small compared to the excavation cost, the energy consumption may still give an important indication of the probable excavation cost, because it reflects the drilling stresses, the consequent size and cost of the tool and machine, and the rate of wear.

The constant k_1 has to have a value of $80/m^2$ to agree with the dashed line in Figure 2.25. The cost to excavate 1 m^3 of hard rock in a 1 m diameter hole is then $80; a 100 mm diameter drill hole is $800; and a 100 m diameter quarry is $0.80. Although this is a rough average and a very simple approximation, it gives a useful rule of thumb for estimating the order of magnitude of the cost that is likely to occur. (The cost figures in this paragraph and in Figure 2.25 are in 1972 dollars. Present cost data indicate that the cost has increased so that the constant k in 1993 dollars should have a value of $118/m^2$, that is, a cost increase of nearly 50 % — much less than inflation.)

It would appear that the technical developments almost compensate for a major part of the inflation to give only a slowly changing (almost time-independent) unit cost in actual dollars. For example, the cost of 1 kg of dynamite has essentially stayed constant, between $1 and $2, since the late 1800s, and a similar development has occurred for the cost of construction steel, and for tungsten carbide. Naturally, the rule of thumb represented by Equation 2.19 and Figure 2.25 is mainly for orientation and comparison purposes. Wherever an actual estimate for commercial purposes needs to be made, up-to-date costs references have to be acquired.

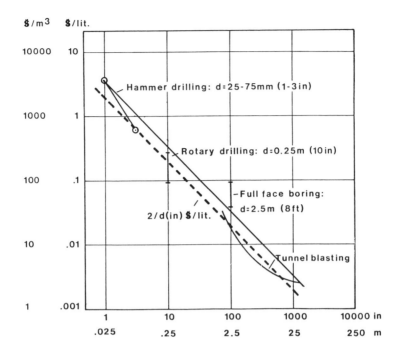

Figure 2.25. Approximate excavation costs for different size openings in rock. The original cost figures (experimental points and the dashed line) are in 1972 dollars. The solid line represents approximate 1993 costs. (The original idea of this curve was given by the late Ake Kallin of Atlas Copco AB, a fine and knowledgeable man and an unsurpassed source of knowledge in blasting.)

2.7 Comparison between Mechanical Tunnel Boring and Conventional Drill-Blast Tunneling

Faced by the necessity to make a choice between mechanical boring and conventional drill-blast techniques to make a tunnel, one may find the following rules of thumb helpful as a starting point:

Long (> 2 km) straight tunnel in

Sedimentary rock — definitely mechanical boring
Hard igneous rock — possibly mechanical boring

Short (< 2 km) crooked (curved) tunnel in

Sedimentary rock — probably drill-blast
Hard igneous rock — definitely drill-blast

Where rock instability and/or water influx problems may be expected, the greater flexibility of the drill-blast method may decide in its favor. On the other hand, where there is a bonus to be gained for exact tunnel profile or straight smooth walls to reduce the cost of lining, then the balance may swing in favor of mechanical boring. For example, where hydraulic flow is involved, a 6.2 m^2 circular tunnel with smooth walls is equivalent to a 10 m^2 unlined, conventionally excavated tunnel.

2.7. Mechanical Tunnel Boring vs. Conventional Drill-Blast Tunneling

The ready availability of a particular set of the heavy and costly drilling or boring equipment within the company, or for sale or rental nearby, is also often a major deciding factor in the choice of equipment for a particular project, even if different equipment would have been able to drill or bore nominally at a lower cost per unit volume.

To make a more detailed comparison between the two methods, the following four factors must be considered:

1. Advance rates
2. Labor costs
3. Equipment costs and supplies
4. Indirect costs

The results of a detailed sample calculation of these four factors are given in Tables 2.2 to 2.6 for each of four simple tunnels, covering the four combinations of mechanical boring (6.2 m^2) and drill-blast tunneling (7.5 m^2), in limestone and in granite. The tunnel length in each case was 3 km, and the tunnels were assumed to be driven with a single heading with a spoil dump near the portal. The rock would be sound and would require no support in either case. The labor costs shown on the 6th line of the summarizing Table 2.6 are based on the average total labor costs of $25 per hour.

The equipment rental rate for mechanical boring was calculated by an experienced contractor using a formula that has been tested by experience. It is based on a machine life calculated in meters rather than in hours, and it assumes that there will be sufficient work for the machine to permit complete depreciation within a 5-year period. Other input data in the table were based on experience dated around 1975.

Clearly, the outstanding figures are the high advance rate of the mechanical boring machine in limestone (8.3 m/shift), nearly double that of drill-blast tunneling in granite rock, and the very high cost of mechanical boring in granite, $589 per tunnelmeter, almost exactly 50% higher than that of drill-blast tunneling in granite.

While cost calculations such as these quickly lose their dollar relevance because of changing prices and developing technology, the methods and logics used above are less bound to a particular era, and they can be applied with different numbers characteristic of the actual present-day situation. We have included them here for that reason, as a guidance in the technique of cost estimating.

Table 2.2. Calculation of advance rates.

	Item	Units	Machine Bored (MB)		Drill and Blast (DB)	
			Limestone	Granite	Limestone	Granite
1.	Compressive Strength	kg/cm^3	1700	2500	1700	2500
2.	Tunnel Dimensions					
	a. Diameter	m	2.8[a]	2.8[a]		
	b. Area	m^2	6.2	6.2	7.5	7.5
	c. Length	km	3	3	3	3
3.	Penetration Rate	m/hr	2.55	1.5		
	(using 14 tons per disc cutter)					
4.	Total Boring Time	hr	1200	2000		
5.	Time to Re-set Mole	hr	50	50		
	(1 minute per 1 m stroke)					
6.	Time Required for Cutter Changes[b]					
	a. (L) Total rolling path for all cutters as a function of cutterhead revolution	m/rev	120			
	b. (P_e) Depth of cutter penetration per cutterhead revolution	cm	0.49	0.29		
	c. Total rolling path for all cutters (as a function of tunnel advance)	km/m	24.5	41.5		
	d. (λ) Length of rolling path per (life of) cutter ring	km	400	150		
	e. Cutter changes required per length of tunnel excavated	changes/m	0.06	0.27		
	f. (t_c) time required to change 1 cutter	hr	1	1		
	Total time required for cutter changes (excavated length dependent)	hr	200	890		
7.	Empty transfer cars (excavated length dependent) (3000 times @ 2 minutes each)	hr	100	100		
8.	Current cumulative time required for tunnel excavation	hr	1550	3040		
9.	Repair and maintenance as a function of total hours (total time dependent)	%	31	25		
10.	Total required excavation	hr	2250	4040		
11.	Total number of excavation shifts (assume 6.2 hrs is actual working time, exclusive of travel and meal time for MB only)		360	650		
12.	Advance rate	m/shift	8.3[a]	4.6[a]	5.3[a]	4.8[a]
13.	TBM utilization	%	41.5[a]	38[a]		
14.	Total number of excavation shifts				590	645

[a] Figures used to summarize estimate are in Table 2.6.
[b] Calculations based on formulae in Section 2.5.

Table 2.3. Calculation of labor costs.

Item	Machine Bored (MB)		Drill and Blast (DB)	
	Limestone	Granite	Limestone	Granite
1. Construction of switch, shifts	10	10	10	10
2. Set-up and dismantle, shifts	20	20	10	10
3. Remove TBM from tunnel	5	5	—	—
4. Sum of 1, 2, and 3 above	35	35	20	20
5. from Table 2.2, sections 11 and 14	360	650	570	625
6. Total number of shifts	395	685	590	645
7. Number of men/shift	8[1]	8[1]	6[1]	6[1]
8. Labor cost per man shift, $	130	130	130	130
9. Labor cost per shift, $/shift ($8 \times 130$, 6×130)	1040	1040	780	780
10. Labor cost per man shift for entire project, $	$410,000	$715,000	$460,000	$505,000
11. Labor cost per meter of tunnel, $/m	137	238	151	168
12. Labor cost per meter of tunnel, $/m^3	227	39	20	22

[1] Figures used to summarize estimate are in Table 2.6.

Table 2.4. Calculation of equipment costs and supplies.

Item		Machine Bored (MB)		(DB)
		Limestone	Granite	—
1. TBM depreciation based on:				N/A
a. TBM (in tunnel length) life = 15 km				
b. TBM (in time) life = 5 years				
c. Interest rate = 10%				
d. Initial TBM cost (including 20% extra for parts) as a function of TBM diameter	$/m		370000	
e. Total costs as a function of tunnel length per unit of dia.				
$\left[\dfrac{\text{initial cost, \$/m}}{\text{tunnel length, km}}\right] \times \left[\dfrac{\text{int. rate, \%/yr} \times \text{time, yrs}}{2} + 1\right]$	$/km		30833	
Interest cost: $/km/initial cost, $	%/km		0.083	
f. Cost per unit of diameter: $\left[\dfrac{370,000 \times D \times 0.08}{(\pi/4) \times D^2 \times 1000}\right]$	$/m		38.3	
g. For $D = 2.8$ – Cost per unit volume excavated	$/m^3		13.7	
Cost per unit length excavated	$/m		86.0[1]	
2. Cutter costs for TBM based on:				N/A
a. Cost of a replacement disc	$		165	
b. Cutter changes per tunnel length	changes/m	0.06	0.27	
c. Total costs for discs	$/m	10	46	
d. Cost of hub assembly	$		1060	
e. Life expectancy of hub as function of disc life		5.0 ×	7.5 ×	
f. Hub changes required as a function of tunnel length	changes/m	$\dfrac{0.06}{5.0} = 0.012$	$\dfrac{0.28}{7.5} = 0.037$	
g. Cost of hubs	$/m	13.00	38.00	
h. Total tool cost	$/m	23[1]	84[1]	
3. TBM spare parts:				N/A
a. Main bearing ($6000/7.5 km)	$/m	0.8	0.8	
b. Other parts	$/m	6.2	9.3	
c. Total	$/m	7.0	10.1	
4. Final equipment (TBM) repair:	$	12000	18000	N/A
	$/m	4[1]	6[1]	
	$/m^3	0.66	1	

(continued)

[1] Figures used to summarize estimate are in Table 2.6.

52 Chapter 2. Mechanical Drilling and Boring in Rock

Table 2.4. (Continued)

Item	Machine Bored (MB)		Drill and Blast (DB)	
	Limestone	Granite	Limestone	Granite
5. Equipment depreciation (DB):				
a. Drill Jumbo (9 SEK/m^3)				
b. Compressor (5 SEK/m^3)				
c. Loader and miscellaneous (7 SEK/m^3)				
d. Total: $\dfrac{(21 \text{ SEK/m}^3 \times 7.5 \text{ m}^3)}{(4.2 \text{ SEK/\$})} = \$/m$			33[1]	37[1]
6. Equipment depreciation (MB and DB):				
a. Shuttle cars	3.4	6.7	4.1	8.3
b. Locomotives	3.4	5.0	5.4	6.0
c. Pumps and fans	2.1	3.8	3.4	3.8
d. Total	9.0[1]	15.0[1]	13.0[1]	18.0[1]
7. Equipment operation and maintenance:				
a. Pipes				
b. Tracks				
c. Ventilation				
d. Electrical cables				
e. Transformers				
f. Total	37[1]		37[1]	
8. Drill steel and explosives:			41[1]	50[1]
9. Electricity:				
a. Rate of \$0.03/HP hr				
b. TBM hours	1200	2000		
c. HP/hr	350	250		
d. Cost $	12600	15000		
e. Total cost (1.33 × cost) $	16800	20000		
$/m	5[1]	7[1]		
$/m^3	0.9	1.1		
f. 50 KW hr/m^3 for pneumatic drilling $/m			11.0	10.0

[1] Figures used to summarize estimate are in Table 2.6.

Table 2.5. Calculation of indirect costs.

Item	Machine Bored (MB)		Drill and Blast (DB)	
	Limestone	Granite	Limestone	Granite
1. Staff/personnel (10% of total direct costs)				
2. Plant set-up (3–4% of total direct costs)				
3. Overhead, risks (8% of total direct costs)				
4. Total sum of direct costs (from Table 2.6), $/m	309	483	287	323
5. Total indirect costs (22% of direct costs), $/m	68[1]	106[1]	65[1]	72[1]
6. Indirect costs, $/m				
a. Cost dependent	41	62	32	36[2]
b. Time dependent	27	44	33	36[2]

[1] Figures used to summarize estimate are in Table 2.6.
[2] Note: Base for time dependent calculation.

2.7. Mechanical Tunnel Boring vs. Conventional Drill-Blast Tunneling

Table 2.6. Estimate summary.

Table	Item	Machine Bored (MB)		Drill and Blast (DB)	
		Limestone	Granite	Limestone	Granite
2.2.2a	Area, m^2	6.2 (ϕ 2.8)	6.2 (ϕ 2.8)	7.5	7.5
2.2.12	Advance, m/shift	8.3	4.6	5.3	4.8
2.3.6	Total shifts (including set-up and dismantle)	395	685	590	645
2.3.7	Total men per shift	8	8	6	6
2.2.13	Net operating time as percentage of total time	41	38		
		U.S. Dollars/meter of tunnel			
2.3.11	Labor costs	137	238	151	160
2.4	Equipment costs and supplies:				
2.4.1d	TBM depreciation	86	86		
2.4.2h	Cutters for TBM	23	84		
2.4.3c	Spare parts for TBM	7.0	10.1		
2.4.4	Repair (final) for TBM	4	6	2	2
2.4.5d	Jumbo, compressor, loader, and misc.			33	37
2.4.6d	Shuttle car, locomotive, pumps, and fans	9	15	13	18
2.4.7f	equipment, operation, and maintenance	37	37	37	37
2.4.8	Drill steel, explosives			41	50
2.4.9	Electricity	5	7	10	11
	TOTAL DIRECT COSTS	309	483	287	323
	Indirect Costs (staff, plant, restoration)	68	106	63	71
	TOTAL COSTS	**377**	**589**	**350**	**394**

Table 2.7. Drill hole volumes and costs.

Hole diameter		Hole volume	Hole cost (granite)	
inch	mm	liter/m	$/liter	$/m
1	25.4	0.51	3	1.5
2	50.8	2.03	1.5	3
3	76.2	4.56	1	4.6
4	101.6	8.11	0.75	6.1
6	152.4	18.2	0.5	9.1
10	254	50.7	0.30	15.2
15	381	114	0.20	22.8
20	508	203	0.15	30.5

2.8 Some Rules of Thumb

As a person involved in the explosives end of drilling and blasting, you may find it useful to express drilling costs in dollars per liter drillhole volume rather than the conventional dollars per meter drillhole length. This is because, in the first approximation, the consumption of explosive for blasting is a given weight of explosive for each unit of volume of rock to be blasted. In other words, if we disregard rock density changes, the "powder factor" is essentially a constant as a first approximation. To make room for this given weight of explosive inside the rock, we require, again in the first approximation, a given volume of drillhole, whether the diameter is large or small. This is the background to the following simple rules of thumb:

Specific charge   0.5–1 kg/m³ solid rock

Specific drilling   0.5–1 liter/m³ solid rock

Cost of drilling $K = \dfrac{3}{d}$ $/liter drillhole volume

where K is expressed in US dollars/liter drillhole volume and d is the hole diameter expressed in inches (25.4 mm).

There are of course extreme cases with very soft or very hard rock where the specific charge is as small as 0.1 kg/m³ or as high as 3 kg/m³; but comparing blasting operations similar in size, the specific charge and the specific drilling keep surprisingly steady even with large changes in hole diameter and rock materials.

Table 2.7 is a summary of drillhole volumes and costs, in 1993 US dollar values.

Practical life sometimes requires a person to roughly estimate the volume of a drillhole while he is standing underground in a badly lit tunnel with water up to the top of his boots and also dripping from above. It is then well to know that the liter volume of 1 m length of a borehole is found with surprising accuracy (about 1.3% high) by dividing the square of the borehole diameter expressed in inches (25.4 mm) by two:

$$V = \frac{d^2}{2} \qquad \text{(in liters/m)}. \qquad (2.20)$$

Chapter 3

Explosives

From a practical point of view, explosives are simply materials that are intended to produce an explosion, i.e., to have the ability to rapidly decompose chemically, thereby producing hot gas which can do mechanical work on the surrounding material. To be useful in practical applications, explosives also have to have enough chemical stability to not decompose spontaneously under any of the stimuli, such as friction, impact, or limited heating that may occur in often quite rough normal handling and storage. From a fundamental point of view, the thermal stability of energetic materials and their reaction to fire are important deciding factors in classifying them into groups of different "sensitivity". We will introduce the fundamental groupings of primary, secondary and tertiary explosives, based on these factors, and also the practical grouping of materials into explosives, propellants, and pyrotechnics, and their subgroups, which reflect the use for which they were produced.

The overall focus in this chapter is on commercial or rock blasting explosives. Rock blasting demands different types of explosives than bombs and shells. We will describe different types of rock blasting explosives and the methods used for conveying them to the blasting site and putting them into the drillhole. We will also develop an understanding of how and why these explosives are different from explosives intended for military use and from propellants, pyrotechnics, and other energetic materials.

In Chapter 4, we will discuss in detail the related topics of detonation, which is the rapid process by which the explosive's chemical energy is released, and shock waves, by which the chemical reaction is ignited and spread along the charge in the drillhole and by which the pressure developed in the drillhole is transmitted into the rock. A major part of that chapter is devoted to the prediction and determination of the performance of explosives, again with the main emphasis on rock blasting applications.

In all explosives, the rapidity of the chemical reaction depends on the strength of the shock wave propagating the detonation reaction. In high energy molecular explosives, where the chemical reaction is the decomposition of a single molecular species, the detonation reaction is very rapid and its rate decreases only slowly with decreasing shock amplitude. In contrast, in the lower energy commercial explosives, which are often composites in which the fuel and oxidizer component are contained in different grains, the chemical reaction occurring in the explosion or detonation is quite strongly dependent on the charge size and on the amount of confinement or resistance offered to the expansion of the reaction products. Conversely, the wave motion at close range in the surrounding rock is closely related to the kind of explosive and its rate of reaction. For example, the explosive used in the greatest quantity in rock blasting is ANFO, which is a mixture of crystalline or prilled ammonium nitrate (AN) with about 5.5% fuel oil (FO). ANFO in this form, having a density in the range 0.8 to 1 g/cm^3, is generally unable to support a steady detonation in an unconfined cartridge of diameter smaller than about 50 mm. However, the same explosive, loaded into a reasonably long 38 mm diameter borehole in rock, will readily be initiated to detonation by a standard number 8 detonator. In the drillhole, as a result of the confinement by the surrounding rock, the high pressure and temperature generated in the explosive by the detonator is maintained for a longer time than it is in an unconfined charge, which allows more time for the chemical reaction to occur. A heavy-gauge steel tube provides a confinement that is nearly as strong as that of rock, but even a 52/60 mm inner/outer diameter steel tube is less effective than the rock surrounding a 35 mm borehole in rock. This is evidenced by the somewhat lower detonation velocity and higher amounts of toxic reaction products obtained in the steel tube than in the borehole from the detonation of the same quality ANFO explosive (Figure 3.1). Most other rock blasting explosives show a similar interdependence between detonability, charge size, and confinement.

3.1 Energetic Materials, Explosions, and Explosive Materials

3.1.1 Energetic Materials

The term *energetic material* is mostly used to comprise all materials that can undergo exothermal chemical reaction releasing a considerable amount of thermal energy. A wider definition is sometimes used to include inert materials at high pressure and/or high temperatures. It includes in the concept of energetic material any material that has a high internal energy because it is compressed to a high pressure and/or heated to a high temperature. A high pressure gas in a pressure vessel by this definition is an energetic material, which seems logical. The white-hot tungsten filament in a lamp bulb could also, with the wider definition, be termed an energetic material. It would be an example of an energetic material that is not explosive because, even at that high temperature, tungsten is still a solid and does not expand, except very slightly due to thermal expansion. By contrast, the nickel bridge wire in an exploding bridge wire (EBW) detonator, heated into its gaseous supercritical state by a short-duration, high-energy, high-voltage electric discharge, would be an example of an explosive energetic material, although no chemical reaction is involved in releasing the energy. The vaporized bridge wire material expands explosively, shock initiating the secondary explosive of the detonator, pressed into contact with the bridge wire. Yet another example of an energetic material that is not explosive in itself is a pyrotechnic material mix. An example of a pyrotechnic material is thermite, an intimate finely powdered mix of iron oxide, Fe_2O_3, with aluminum. Thermite reacts chemically with the release of thermal energy sufficient to heat the resulting mix of Fe and Al_2O_3 to an intensively luminous white hot, but even the Fe is in the condensed state, and the hot reaction products therefore do not expand explosively. In the following, we will mostly use the term *energetic material* to mean explosives, propellants, and pyrotechnics.

3.1.2 Explosions and Explosive Materials

An explosion is basically any rapid expansion of matter into a volume much larger than the original. The rapid expansion of the gas when a gas-filled latex balloon bursts, or when a pressurized gas tank ruptures, are explosive events. An explosion also results from the rapid burning of black powder enclosed in a container that bursts, or from the detonation of a charge of TNT.

In general, we use the term *explosive* to mean a material that can undergo an exothermal chemical reaction resulting in a rapid expansion of the reaction products into a volume larger than the original. The energy released by the chemical reaction (the explosion energy) appears as thermal and potential (compression) energy of the

3.1. Energetic Materials, Explosions, and Explosive Materials

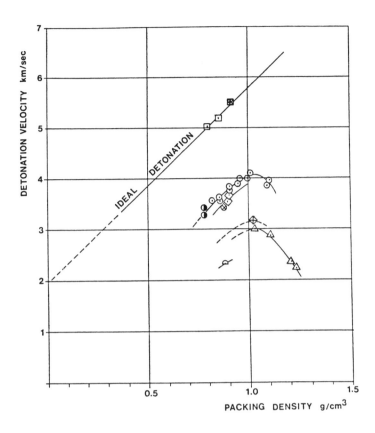

⊡	crushed prills	268	mm in rock
⊙	crushed prills	52	mm in steel tube
⬦	crushed prills	36	mm in steel tube
◐	crushed prills	34	mm in rock
⊕	crushed prills	21.5	mm in steel tube
⌒	crushed prills	12	mm in steel tube
⊞	whole prills	268	mm in rock
⊗	whole prills	52	mm in steel tube
△	crystalline ANFO	52	mm in steel tube

Figure 3.1. ANFO detonation velocity as a function of packing density and charge diameter.

reaction products. The explosion energy of most explosives is typically of the order of 4 MJ/kg. The energy of 4 MJ released by the explosion of 1 kg of explosive, if completely transformed into thermal energy, could heat 10 kg of water from 4 to 100°C. The resulting temperature before expansion (the explosion temperature) is typically in the range 2000–5000 K and the resulting pressure is in the range 1–20 GPa (10–200 kbar). Because of the high initial pressure, a considerable proportion of the explosion energy will be transformed into mechanical work by the expansion of the reaction products to a larger volume.

3.1.3 Thermal Stability and Bulk Decomposition of Explosive Materials

It is characteristic for all chemical reactions that the rate at which they proceed increases very rapidly with increasing temperature. A simple rule of thumb is that a temperature increase of 10°C will approximately double the reaction rate. It applies approximately to reactions in energetic materials at room temperature as well as at higher temperatures (see, for example, Figure 3.5 in the next section). With explosive materials, the chemical decomposition reaction can proceed in two radically different ways. One is the homogeneous decomposition which proceeds in an energetic material at any given nearly constant temperature. It proceeds at the same rate throughout the material, but is faster at a higher temperature. The other is the more rapid decomposition that takes place locally, mostly at a free surface in contact with a hot gas, such as the flame, which heats the surface, causing the material there to decompose while the rest of the material remains cool and stable. The former type of process is responsible for the deterioration of an explosive or a propellant in long term storage, or may cause catastrophic self-heating, for example if a kettle of molten explosive is kept at too high a temperature for too long. The latter type of process is what we normally refer to as burning, or surface burning. It is the process by which propellants are intended to react upon ignition, producing a steady flow of hot gas, and it may also be the reaction process in shock initiation and detonation in porous explosives.

In both of these processes, we mean by the reaction rate that rate at which a fraction of the original mass is consumed by the chemical reaction. If λ is the fraction of the original mass that is consumed by the reaction, then the bulk or mass reaction rate $\mathcal{R}$ is given by

$$\mathcal{R} = \frac{d\lambda}{dt} \tag{3.1}$$

where t is the time.

If an explosive material is over-heated locally, the chemical reaction at that point rapidly leads to local self-heating, producing more heat than can be conducted away to the surrounding, cooler explosive. Ultimately, a gas pocket forms, in which the main energetic reactions take place. These most often involve the gaseous intermediate reaction products from the initial reaction. The further reaction may lead to flame burning at the interface between the gas and the reacting explosive, or, if the cause of the local heating is of very short duration, the reaction may die out. We will discuss flame burning in more detail later in this chapter; here, we will review some of the main features of the slow thermal decomposition or thermal stability of a mass of energetic material at a homogeneous temperature.

The crucial question in deciding whether the slow thermal decomposition will lead to run-away self-heating or not is: *Can the heat produced by the thermal decomposition reaction be carried away by conduction or convection to the surrounding?* Obviously, the bigger the mass of reacting material, the more difficult it will be for the heat to be removed because the surface area (which determines how much heat can be removed per unit time,) is proportional to the size squared, while the mass (and the total thermal energy produced in it per unit time) is proportional to the size cubed. The temperature at which the heat generated by chemical reaction just equals the heat conducted away is called the critical temperature T_c. This is often close to the explosion temperature T_{expl} — at which a sudden run-away self-heating occurs, followed by an explosion, if the material is enclosed. By measuring T_c or T_{expl} for samples of an energetic material having different sizes, fundamental information on the energetic materials response to heating can be gained. This information is often for convenience expressed in terms of the reaction rate coefficient, defined under the assumption that the reaction follows a first-order Arrhenius equation, i.e., that the mass rate of reaction can be expressed by the equation

$$\frac{d\lambda}{dt} = (1 - \lambda) Z \, e^{-E_a/RT} \tag{3.2}$$

where Z is the preexponential factor, E_a the activation energy, T the absolute temperature, and R the gas constant. The term

$$K = Z \, e^{-E_a/RT} \tag{3.3}$$

is often referred to as the decomposition rate coefficient K. The logarithm of the rate coefficient determined in this way often is found to be an approximately linear function of the inverse absolute temperature.

Figure 3.2 shows the rate coefficients for ammonium nitrate and ammonium perchlorate derived in this way from experiments covering a large range of temperatures; Figure 3.3 shows the rate coefficients for a number of single molecule explosives. The critical temperatures T_c for a given size of explosive can be derived from such data, provided other thermal characteristic data for the material (such as density, specific heat, and thermal conductivity) are known. The critical temperature is greater, the greater the thermal conductivity. The critical temperature for a given energetic material is higher for a melt that is stirred, and thus is provided with better thermal contact with the surrounding, than for an unstirred or solid sample of the same size. Figure 3.4 shows the critical temperature as a function of sample size for TNT, stirred and unstirred.

The decomposition rate of a given energetic material is often quite strongly dependent on the presence of other materials, energetic or not, in the form of additives or contaminants. Therefore, whenever an accurate determination of the critical temperature needs to be made to in order to avoid an accidental overheating situation in processing or heated storage, the rate coefficient should be determined experimentally with the actual material to be used, and with any possible contaminants or additives present.

The thermal decomposition information contained in the above diagrams gives one part of the information needed for classifying energetic materials as more or less sensitive to heating. However, another (at least equally important and crucial part) is the relative magnitude of the input heat needed for breaking the first chemical bonds in the chain of chemical decomposition reactions. For primary explosives, this is usually small; for secondary explosives, it is larger; and it is largest for tertiary explosives (see below for definitions of these terms). The first-bond breakage energy is much larger for TNT than for PETN, and still larger for TATB than for TNT. An emulsion explosive containing AN, fuel oil, and water has a relatively low first-bond breakage energy, characteristic of the breaking of the hydrocarbon chain; however, because of the higher first-bond breakage energy of AN (splitting the AN molecule into ammonia and nitric acid), and because of the additional heat consumed in evaporating the water, such an explosive is one of the most thermally rugged explosives in practical use today.

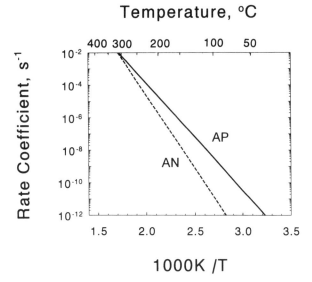

Figure 3.2. The thermal decomposition rate coefficient $K = Z e^{-E_a/RT}$ as a function of the inverse absolute temperature $1/T$ for ammonium nitrate and ammonium perchlorate [D. Olson, 1992, private communication].

3.1.4 Thermal Decomposition, Hot Spots, Initiation, and Safety

So far, we have dealt with the slow decomposition of explosives without much mention of the mechanisms at very high rates of reaction. We will now discuss the extension of data from slow decomposition measurements to events occurring within the very short time intervals during which explosive decomposition reaction takes place. Let us first look at the decomposition of a single explosive substance, nitroglycerin. As mentioned already, the rate of decomposition is highly temperature-dependent.

Figure 3.5 shows very approximately how the calculated half-life (the time needed for half of the original mass of explosive to undergo thermal decomposition) of nitroglycerin depends on the temperature. The available experimental data has been extrapolated over many orders of magnitude, and the decomposition rates at the extremes of the curve may be orders of magnitude off. However, the curve does give a good idea of how quickly the decomposition rate increases and the half-life decreases with increasing temperature.

At room temperature, neat nitroglycerin is very stable, with a calculated half-life of 1 million years. At 800°C, the half-life is only about 1 millionth of a second. We see also that increasing the temperature from, say, 800 to 900°C increases the rate of reaction by about a factor of ten. A shock wave in nitroglycerin with velocity 7.58 km/sec (equal to the detonation velocity) would have a front pressure of 36 GPa and a particle velocity of about 3 km/sec. The internal energy of the compressed material, assuming no reaction took place during compression, would be on the order of 4.5 MJ/kg, and consequently the temperature would be perhaps 3000 K, assuming the specific heat to be about 1500 J/(kg K). We can see that the rate of reaction in such an explosive is probably high enough for an almost immediate complete reaction to take place at the detonation front. The reaction zone length would be very short.

ARRHENIUS PLOT OF ROGER'S KINETICS AND PYX

* Rate constant expression extrapolates to k = 10^3 s^{-1} at a temperature of 770 K (500°C). Possibly, this exceptionally high activation energy is correct. A larger PYX cook-off test is needed.

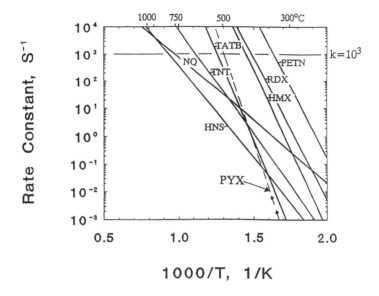

Figure 3.3. The thermal decomposition rate coefficient $K = Z e^{-E_a/RT}$ as a function of the inverse absolute temperature $1/T$ for several secondary explosives [Zinn and Rogers, 1962].

In most commercial explosives, however, the detonation velocity is lower, and the shock compression energy is also correspondingly lower, so that the mechanism of homogeneous shock heating cannot provide a high enough temperature for immediate homogeneous reaction to start at the detonation front. These explosives rely on the local heating of deliberately created reaction centers, hot spots, for functioning. The most frequently used reaction centers are air or gas bubbles introduced into the explosive. By the mechanism of bubble collapse, energy is focused within a small volume of the material close to and downstream of the bubble. At such a "hot spot", the material can reach a temperature considerably in excess of that of the rest of the explosive, and the hot spot can then act as a center for the continued chemical reaction. From the hot spot, the reaction can then spread. Conceivably, The reaction may spread partly by burning at the interface between the hot reaction products and the unreacted surrounding material, and partly by interaction of shock waves in the unreacted material originating from the rapidly burning hot spots. In addition, the turbulence of the flow caused by the closure of voids may also contribute to the spreading of the chemical reaction. Inherent in the process is the fact that the rate of burning at a gas/solid interface at a given temperature increases with pressure. An important part of the success of present-day cap sensitive water gel explosives are the refined methods of introducing

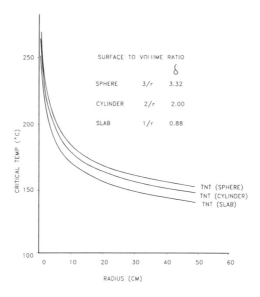

Figure 3.4. The critical temperature T_c for run-away self-heating due to thermal decomposition of TNT as a function of sample size [Zinn and Rogers, 1962].

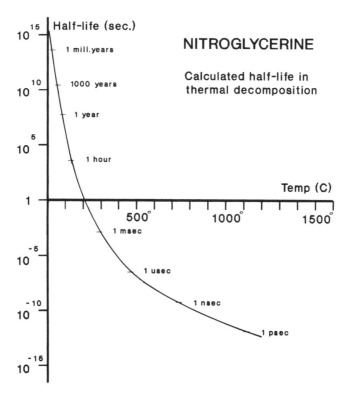

Figure 3.5. Calculated half-life in thermal decomposition of NG as a function of temperature.

and stabilizing a sufficient number of effective air or gas inclusions into these explosives, which are otherwise extremely difficult to detonate in compact form. A similar mechanism of inhomogeneous heating upon sudden compression of a powder is responsible for the reliable functioning of ANFO explosives, provided they are sufficiently porous. At high initial density, such as 1.4 kg/liter or more, ANFO is very difficult to initiate.

Another mechanism for creating a hot spot by bubble collapse can occur in slower processes, such as in low-velocity impact of a bubble containing explosive. As the bubble is compressed, the air or air/vapor mix in the bubble volume is adiabatically compressed and heated, and may be hot enough to ignite burning at the inside surface of the bubble.

Hot spots can also originate from the shock interaction adjacent to high-density or high-strength particles mixed into the explosive, and from a hazard point of view, one should always be on guard against the risk of getting grit of one sort or the other into the explosive.

Another form of hot spot is that developed by frictional heating between hard surfaces rubbing against each other. The temperature at the points of contact between two metals can quite easily reach the melting point of the lower-melting metal. A low thermal conductivity makes it easier to reach high temperatures. Caught between two extended, rubbing metal surfaces, localized high-temperature regions can easily occur in the explosive, sufficiently high for a high-rate reaction to get started. The confining effect of the contacting surfaces prevents the reaction products from escaping and causes the pressure to build up, further increasing the rate of reaction.

The conditions that can generate this latter type of hot spot are part of the classical risk situation that has probably caused 90 % of all explosive accidents in the explosives and rock blasting industry over the years. The blow of a hammer on a surface not cleaned of explosive, a metal tool dropped accidentally into an explosives kneading machine, explosives caught between the cement floor and the foot of a kneading machine, unreacted explosive left in a borehole and caught between the rock and the drill-steel when drilling in an old borehole, unreacted explosive caught between two rock boulders in loading—these are all situations that bring a risk of accidental initiation with them. With the now widespread use of very friction-insensitive waterbased explosives, this type of hot spot does not pose as serious an accident threat as it did with the more sensitive dynamite-type explosives. In a very friction-insensitive explosive, a small hot spot cannot cause detonation of a larger quantity of the explosive because there is not sufficient heat to sustain the reaction until it has become self-supporting. The less sensitive modern explosives (emulsion explosives, ANFO, and slurry explosives) have greatly reduced these risks in the field.

In spite of this, and mainly because their great safety margin against initiation by heating has been more than offset by production units of ever-increasing size, waterbased slurry and emulsion explosives in the last decade have been involved in more explosion accidents involving over-heating and accidental fire than any other explosive in large-scale use today. Table 3.1 lists explosion accidents involving waterbased explosives and blasting agents in the time period from 1979 to 1990.

Table 3.1. Accidents with waterbased explosives.

Manufacture or in-operations related to manufacture and at-burning			
1966	USA	Premix	2 killed
1966	USA	Slurry, pumping	
1973	Norway	Premix	5 killed
1974	USA	MMAN, transport	2 killed
1975	Canada	Powermex, cartridging	8 killed
1976	USA	MMAN, pumping	
1976	Sweden	Slurry, burning	
1977	Japan	Slurry, burning	
1981	Switzerland	MMAN	
1984	France	Slurry, fire	
1987	South Africa	Tovex, manufacturing	5 killed
1988	Sweden	Emulsion, burning	
1988	Canada	Emulsion, pumping	4 killed
1989	France	MMAN-slurry, mixing	
1990	Canada	Emulsion, pumping	
1990	South Africa	Emulsion, pumping	
1990	Russia	Emulsion	16 killed
1990	South Africa	Emulsion, pumping	
1991	China	Emulsion, manufacturing	7 killed
1993	South Africa	Emulsion, burning	
1994	Papau New Guinea	Emulsion, pumping (?)	11 killed
1995	Sweden	Emulsion, burning	
Total 1966-1995			60 killed

3.2 Explosives, Propellants, and Pyrotechnics

3.2.1 Explosives and Propellants

The explosives used in rock blasting or in military warheads and solid propellants used in guns and rockets are all explosive energetic materials. If heated by a flame or by contact with a hot object, they can all burn without the addition of external oxygen from the air. The burning, if initiated on the surface of the explosive grain, may proceed as surface burning, at a rate which is roughly directly proportional to the surrounding gas pressure. The burning may also be initiated by shock heating and will then proceed through the grain at a much higher velocity.

The difference between an explosive and a propellant is functional rather than fundamental. Explosives are intended to function by detonation following shock initiation. Propellants are intended to burn steadily at a rate determined by the design pressure of the rocket or gun breech, and they are ignited to burning by a flame that provides a spray of hot burning particles. Most explosives, however, can also burn steadily if ignited as a propellant. (At a given moderately large charge size, some explosives burn at atmospheric pressure, others only at a somewhat elevated pressure.) Many high-energy solid propellants may be made to detonate if exposed to a sufficiently strong shock-wave, if the pressure of the gas surrounding the burning propellant is radically increased, or if the burning area increases suddenly by fracturing of the grain.

Except at the very highest pressures, burning proceeds by evaporation or endothermic partial decomposition of the solid explosive or propellant at the burning surface. The main release of heat occurs in a flame off the surface, and the rate of burning is determined by how rapidly heat can be transmitted back into the surface material to continue the evaporation or initial decomposition. Because the heat transfer increases with increasing pressure, and the thickness of the flame zone also decreases, the linear rate of burning perpendicular to the burning surface also increases, as it turns out almost linearly with pressure. Equation 3.4 describes the linear burning rate r as a function of pressure p. The exponent ν has values for different propellants and explosives in the range 0.9–1.1. The heat transfer also increases with gas flow along the burning surface. This effect, often referred to as *erosive burning*, has to be taken into account when designing long hollow solid propellant grains for rockets. (The term *propellant grain* was originally used for the individual particles or grains of a gun propellant charge, but is now generally applied also to solid rocket propellant charges, even the very large solid propellant charges used in the space shuttle booster rockets).

$$r = r_0 \left(\frac{p}{p_0}\right)^\nu \quad (3.4)$$

where r_0 is the linear burning rate (measured at right angles to the burning surface) at a reference pressure p_0. Many insensitive propellants and explosives cannot sustain burning at atmospheric pressure except, possibly, when a very large quantity is exposed to a very large fire.

A condition for stable burning in a rocket with constant burning surface area and constant nozzle area is that the pressure exponent ν in the linear burning rate equation is less than 1. Very insensitive propellants are almost impossible to detonate, since their critical diameters for steady detonation is larger than the dimension of the grain. Typical values of linear burning rates are a few centimeters per second at typical rocket combustion chamber pressures (0.1–0.3 GPa or 100–300 bars); a few tens of centimeters per second at typical gun barrel pressures (0.2–0.5 GPa or 2000–5000 bars). For detonating explosives at pressures in the range 1–20 GPa, the linear burning rate may range from 0.5 m/sec to close to the detonation velocity of 2–8 km/sec. In the lower detonation pressure regime typical of many commercial composite explosives, we believe that the burning is initiated at a large number of points (hot spots) where the explosive is heated locally by the passage of the detonation shock front. Burning then appears to proceed by a surface burning or grain burning mechanism. Because the burning surface is much larger than the cross-sectional area of the charge, the burning can keep up with the detonation front even though the linear rate of burning is considerably lower than the detonation velocity. Only at very high detonation pressures in high-energy explosives such as pure nitroglycerin (NG) may the linear rate of burning in a homogeneous explosive approach or equal the detonation velocity.

The rate at which reaction product gases are released from a given mass of propellant in a slow-burning mode of reaction is of course proportional to the size of the total burning surface. Finely powdered propellants, such as black powder or a pistol powder, have such a large specific burning surface area (area per unit weight) that they may have a violently explosive action when allowed to burn in a confined volume such as a drillhole in rock, or in the gun. Therefore, the propellant black powder was widely used in rock blasting for several hundred years before the emergence of detonating explosives.

3.2.2 Pyrotechnics

Pyrotechnic materials are most often used in the form of a compacted powder mix of two or more materials that can react exothermally with each other producing mainly solid or liquid reaction products. The reaction proceeds through the material at a steady rate dependent on the density of the compact, the heat released in the reaction, and the thermal diffusivity of the powder compact. Any small amount of gas released by the reaction also strongly influences the rate at which the reaction proceeds through the compact. The delay element in commercial time interval detonators is a pyrotechnic composition, such as lead oxide Pb_3O_4 mixed with a reactive metal such as Ti or Zr, pressed into a lead or aluminum tube. Ignited by the flame and hot particles emitted from the matchhead of an electric detonator, or by the flame and hot particles emitted from the end of the NONEL tube, the pyrotechnic reacts at a steady rate, determined by the composition, density, and intimacy of mixing between the ingredients, of 0.01–0.2 m/sec, giving a range of delays between 10 msec and several seconds for lengths of the pyrotechnic filled tube between 5 and 50 mm.

Other pyrotechnics are used for welding, for generating light in tracer ammunition, for parachute candles, and for fireworks.

An interesting new development in pyrotechnics is the self-propagating high-temperature synthesis (SHS) in which a hard material such as titanium boride TiB_2, a possible material for ceramic armor applications, is formed by an exothermic pyrotechnic reaction (propagating in a mode similar to that in the delay element) between the powder ingredients Ti and B. A more recent development is the shock-induced reaction synthesis (SRS) in which a pyrotechnic reaction is initiated in a Ti/B powder mix by a shock wave. In this case the reaction forming the TiB_2 proceeds at a much higher velocity than normal pyrotechnic reactions, attached to the shock front which moves with velocities in the range 1–5 km/sec. As a result of the considerable temperature increase, of the order of 1000°C caused by the shock compression of the powder, and the accompanying intense mixing of the ingredient materials, the reaction appears to go to completion within a fraction of a microsecond. The very high temperature reached in the reacted material can be moderated by adding TiB_2 powder to the reactants mix of Ti and B powder.

A wide range of intermetallic and ceramic materials with unique properties have been made using these methods, still being developed into industrial scale.

3.2.3 Single Molecule and Composite Explosives

There are two fundamentally different kinds of explosive materials, namely single explosive substances and composite explosive mixtures. Single explosives are chemical substances that contain in one well-defined molecule all that is needed for an explosion. The molecule decomposes into mainly gaseous reaction products, such as CO_2, N_2, and H_2O. The solid explosive trinitrotoluene, TNT, ($C_7H_5(NO_2)_3$), and the liquid explosives nitromethane, NM, (CH_3NO_2), and nitroglycerin, NG, ($C_3H_5(NO_3)_3$), are examples of single explosive substances.

A composite explosive can be a mixture of two single explosive substances, a mixture of a fuel and an oxidizer, or an intermediate mixture containing one or more single explosive substances together with fuel and/or oxidizer ingredients.

The explosive mixture of carbon and oxygen obtained by impregnating a special kind of porous charcoal with liquid oxygen (an explosive used formerly in coal mining) is the extreme example of a composite explosive, a mixture of a fuel and an oxidizer only.

Most rock blasting explosives are composites containing both single explosive substances and several fuels and oxidizers. In addition, they often contain ingredients such as ballast materials or water, that do not add energy to the chemical reaction, but are used to modify the explosive's consistency or flow properties. Rock blasting explosives systems are described in detail under a separate heading below.

3.2.4 Primary, Secondary, and Tertiary Explosives

Single molecule explosives (Table 3.2) range with respect to the strength of the stimulus required for initiation or a self-supporting chemical decomposition reaction: from primary explosives, such as lead azide (PbN_6), which are used as igniting charges in detonators; through the secondary explosives, of which NG, NM, and TNT are examples; to tertiary explosives, of which ammonium nitrate, AN (NH_4NO_3), is an example. The primary explosives are able to transit from surface burning to detonation within very small distances. A 0.2 mm thick grain of lead azide when ignited will transit from burning to detonation within a distance less than the grain thickness. This is because the lead azide molecule is very simple, decomposing in a very simple two-step reaction, and also because the reaction products have a high molecular weight. Reaction products are generated at the surface faster than they can expand away from the surface, which results in a quick build-up of pressure at the burning surface. The secondary explosives, too, can burn to detonation, but only in relatively large quantities. For example, a stick of dynamite can burn as a candle, slowly, if ignited with a flame (although the authors strongly warn against performing a demonstration of this without due precautions against the chance event of a detonation), whereas a truckload of dynamite may burn to detonation. Under normal conditions, tertiary explosives are extremely difficult to explode and are in fact officially classed as nonexplosives provided that certain conditions are observed (such as that an oxidizer not be mixed with fuels or sensitizers, and that the grainsize exceeds a certain minimum size.) They are nonetheless explosive, as demonstrated by some of the largest accidental explosions in history, such as the (April 16, 1947) Texas City explosion of ammonium nitrate [see Biasutti, 1985] , and the recent (May 4, 1988) explosion of a large quantity of ammonium perchlorate in Henderson, Nevada, USA.

3.3 Liquid Mono-Propellants and Solid Composite Propellants

Propellants, like explosives, can consist either of a single molecular species, in which case they are often called *mono-propellants*, or they can be composed of a mix of several ingredients which react with each other. The latter are called *composite propellants*. An example of a mono-propellant is hydrazine H_2N_2, which decomposes spontaneously into H_2 and N_2 when brought in contact with a catalyst such as a platinum wire mesh. By spraying hydrazine into a combustion chamber where it can contact the catalyst, the hydrazine will burn at a rate corresponding to the pumping rate and the burning will stop when the flow of propellant is stopped. Mono-propellants are all liquid—somewhat incongruously, a liquid solution mix of two or more propellant components is also referred to as a mono-propellant.

Solid propellants are traditionally divided into three groups, all of which contain more than one molecular species. In single-base propellants, of which nitrocellulose gunpowder (Gun-cotton) is an example, the main ingredient, the base, is nitrocellulose, compacted and formed into grains by adding a solvent, ethyl ether, which is then evaporated, leaving a hard grain of the required shape and size. Small amounts of stabilizers, such as dimethyl phthalate, and burn rate modifiers are added. Two examples of double-base propellants are ballistite and cordite. Ballistite, the first smokeless gun propellant, was invented by Alfred Nobel in 1887. Cordite, the similar smokeless gun propellant, was "invented" in Great Britain two years later by Nobel's friend and later adversary, Professor Frederick Abel, following Nobel's disclosure to him of the details of the ballistite invention. (See Bergengren [1960], pages 113-119, for details of this bitter

Table 3.2. Some single chemical explosive substances.

Common name, Composition	Symbol	Mol. weight	ρ_{max} g/cm^3	D_i (km/sec) 1 g/cm^3	D_i (km/sec) ρ_{max}	p_i kbar	Q_d kJ/g	V_{gas} cm^3/g
Primary Explosives:								
Cupric azide, $Cu(N_3)_2$		147.6						
Hydrazoic acid, HN_3		43.0						
Mercury fulminate, $HgC_2N_2O_2$		284.7	4.42		5.05	5.4	1.79	315
Lead styphnate, $PbC_6H_3N_3O_9$		468.3	3.10			5.2	1.54	407
Lead azide, PbN_6		291.3	4.71		5.1	5.6	1.53	231
Silver azide, AgN_3		149.9	5.1		5.90^a			224
Tetrazene, $C_2H_8N_{10}O$ (guanylnitrosoaminoguanyltetrazene)		188.2	1.7				2.75	1190
Diazodinitrophenol, $C_6H_2N_2O(NO_2)_2$	DDNP	210.1	1.63		7.10		3.43	856
Secondary Explosives, Liquids:								
Methylnitrate, CH_3NO_3		77.0	1.22		6.30		6.17	909
Nitroglycerin (glyceroltrinitrate), $C_3H_5(NO_3)_3$	NG	227.1	1.60		7.58	220	6.30	715
Ethyleneglycoldinitrate, $C_3H_4(NO_3)_2$	EGDN	152.1	1.48		7.30		6.83	737
Tetranitromethane, $C(NO_2)_4$	TNM	196.0	1.65		6.36	159	2.29	686
Nitroform (trinitromethane), $CH(NO_2)_3$		151.0	1.60					
Dinitromethane, $CH_2(NO_2)_2$		106.1						
Nitromethane, CH_3NO_2	NM	61.0	1.13		6.29	141	6.4	723
Ethylenedinitramine, $C_2H_6N_2(NO_2)_2$	EDNA	150.1	1.5		7.61		4.42	908
Isopropyle nitrate, $C_3H_7NO_3$	IPN	105.1	1.04		5.4	85	2.36	

^{a} Detonation velocity at initial density of 4.113 g/cm^3.

(table is continued on next page)

patent fight.) These propellants consist of a rubbery solid solution of nitrocellulose in nitroglycerin, which constitute the two bases. Both in single-base and in double-base propellants, the bases are single molecular species which are energetic materials in themselves. In composite propellants, the ingredients are a mix of particulate oxidizers, of which the most frequently used is ammonium perchlorate, particulate high-density single molecular energetic ingredients such as HMX (which is also used as an explosive), held together by a rubbery plastic binder (often hydroxy-terminated polybutadiene, HTPB) which is polymerized following the mixing and pouring of the propellant into its grain form. In high-performance composite propellants, nitroglycerin is often added to the composition, and in recent developments, the plastic binder, which is essentially a low-grade hydrocarbon fuel component, adding little to the energy and specific impulse of the propellant, is replaced by an energetic binder, which in itself is an energetic material.

Figure 3.6 shows a breakdown of some common energetic materials into the groups described above.

Table 3.2. Some single chemical explosive substances (continued).

Common name, Composition	Symbol	Mol. weight	ρ_{max} g/cm^3	D_i (km/sec) 1 g/cm^3	D_i (km/sec) ρ_{max}	p_i kbar	Q_d kJ/g	V_{gas} cm^3/g
Secondary Explosives, Solids:								
Mannitolhexanitrate (Nitromannit), $C_6H_8(NO_3)_6$	MHN	452.2	1.73		8.26		6.02	755
Trinitroazidobenzene, $C_6H_2N_3(NO_2)_3$		254.1	1.75					
Nitrocellulose, $C_{24}H_{40-x}O_{20-x}(NO_3)_x$	NC	1188.6^a	1.66		7.30		10.60	875
Pentaerythritoltetranitrate, $C_5H_8(NO_3)_4$	PETN	316.2	1.77	5.55	7.98	300	6.12	780
Ditrinitroethylurea, $C_5H_6N_2O(NO_2)_6$	DiTeU	386.2	1.79					
Trinitrophenylmethylnitramine (Tetryl), $C_7H_5N(NO_2)_4$		287.2	1.70	5.60	7.56		3.22	
Nitrostarch (penta), $C_{12}H_{15}O(NO_3)$		549.2	1.60		6.19			
Hydrazine nitrate, $N_2H_4 \cdot HNO_3$		95.1	1.63		8.69		3.83	1001
Hexanitrohexaazaisowurzitan $C_6H_2N_6(NO_2)_6$	HNIW	434	2.1		10.2	450	(calculated)	
Cyclotrimethylenetrinitramine (Hexogen), $C_3H_6N_3(NO_2)_3$	RDX	222.1	1.80	6.08	8.75	347	5.46	908
Cyclotetramethylenetetranitramine (Octogen), $C_4H_8N_4(NO_2)_4$	HMX	296.2	1.90		9.10	393	5.46	908
Diaminotrinitrobenzene, $C_6H_5N_2(NO_2)_3$	DATB	243.2	1.79		7.52	259	4.67	625
Hexanitrobenzene (hexyl), $C_6N_6O_{12}$	HNB	252.1	1.97		9.30	355		
Hexanitrostilbene, $C_{14}H_6(NO_2)_6$	HNS	450.2	1.79		7.14	221	4.12	700
Triazidotrinitrobenzene, $C_6N_9(NO_2)_3$	TNTAB	336.2	1.74		8.58			
Picric acid, (trinitrophenol) $C_6H_3O(NO_2)_3$		229	1.80		7.35		4.40	675
Picrylaminodinitropyridine, $C_{17}H_7N_3(NO_2)_8$	PYX	621.3	1.75		7.38	237		
Ammonium picrate, $C_6H_6NO(NO_2)_3$		246	3.44		7.15	214	2.88	800
Nitrotriazolone, $C_2H_2ON_3(NO_2)$	NTO	130	1.93		8.42	316		
Trinitroazetidine, $C_3H_4N_4O_6$	TNAZ	192.1	1.84		8.8^b	380^b		
Nitroguanidine, $CH_3N_3NO_2$	NQ	105	1.70	5.46	8.20		2.88	1075
Trinitrobenzene, $C_6H_3(NO_2)_3$	TNB	213	1.76		7.30		4.81	678
Ethylenedinitramine, $C_2H_6N_2(NO_2)_2$	EDNA	150.1	1.5		7.57		4.42	908
Monomethylamine nitrate	MMAN	94.1	1.42				3.00	1189
Ethylenediaminedinitrate, $C_2H_{10}N_4O_6$	EDD	186.1	1.58		6.80		3.25	1083
Trinitrotoluene (Trotyl), $C_7H_5(NO_2)_3$	TNT	227	1.64	5.01	6.95	190	4.10	690
Triaminotrinitrobenzene, $C_6H_6N_3(NO_2)_3$	TATB	258.2	1.90		7.98	315	3.12	
Dinitrotoluene, $C_7H_6(NO_2)_2$	DNT	182.1	1.54				3.27	800
Tertiary Explosives:								
Mononitrotoluene, $C_7H_7NO_2$	MNT	137.1	1.16					
Ammonium nitrate, NH_4NO_3	AN	80.1	1.73	5.76	(8.51)^c		1.59	980
Ammonium perchlorate, NH_4ClO_4	AP	117.5	1.95	4.04	(6.49)^c	176	2.05	803

^{a} $x = 12$. ^{b}Calculated using $\gamma = 2.76$. ^{c}Because of their low detonation temperature, AN and AP are unlikely to sustain detonation at crystal density even at very large charge diameters.

70 Chapter 3. Explosives

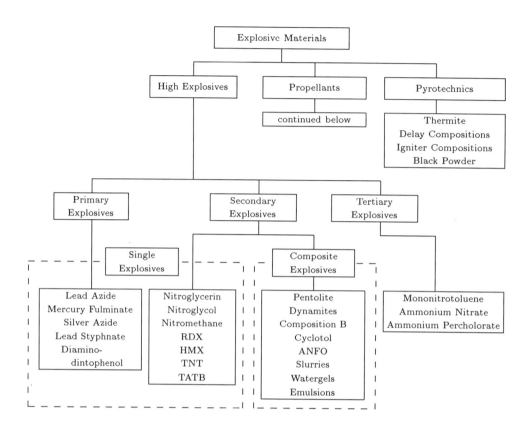

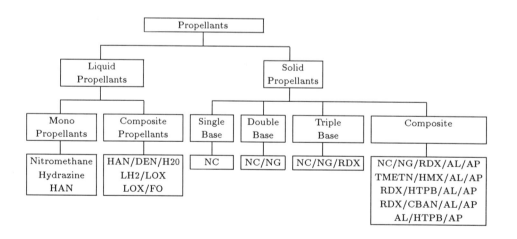

Figure 3.6. Classification of energetic materials with respect to their ingredients and intended use.

3.4 Military Explosives

Explosives for military purposes have developed in a direction totally different from those intended for rock blasting. As will be seen later in this chapter, commercial rock blasting explosives, in peacetime, can be transported to the place of their use in large tank-trucks. There, they are simply pumped, or augered, at a rate of 250 kg/min or more, into drillholes costing anywhere between 10¢ and $1 per liter drilled-out hole-volume. The least expensive rock blasting explosives can be sold profitably for less than $1 per kg (1993 US prices). Raw materials, processing, transportation, mixing and loading into the hole, and a small profit margin are all included in this price. The performance required in rock blasting tends to favor explosives such as waterbased ammonium nitrate emulsions or ANFO, which have a low flame temperature and consequently expend their limited energy early on in their expansion. In hard-rock blasting, the effective expansion work is done by the time the explosive reaction products have expanded to no more than 5 or 10 times the original volume occupied by the explosive. There is little premium to be gained from striving for a high fill-density of the explosive in the drillhole, when doubling the size of the drillhole may cost less than a dollar per liter. The density of ANFO in the drillhole is typically just over 0.8 g/cm^3, and its energy slightly less than 4 MJ/kg. Most commercial explosives used for large-scale blasting are produced near the point of use, or even on site. The requirements for storage life therefore is not very long. The classical dynamites could be safely stored for many years, provided they were kept dry. Cartridged emulsion explosives may have a shelf life of 1 or 2 years. Bulk blasting agents should be capable of "sleeping" loaded in the drillhole for a week or even a month; but in most cases, the blast is fired within a day of being loaded.

In contrast, military explosives may have to be stored for 10 to 20 years. In fact, some gun ammunition from the First World War is still in storage in many explosives magazines, and the supervision of storage stability of propellants by "surveillance testing" is an important part of military explosives science. The economic background of military explosives usage is also totally different from that of commercial explosives. The bulk of all military explosives are carried to the place of their use by the most expensive of all transportation vehicles, a modern military aircraft costing perhaps $50 or even $200 million each, having a load carrying capacity often less than half that of a small ANFO truck. The plane may be based on an aircraft carrier, a technically complicated city at sea, or may have to be tanked in the air with fuel flown halfway around the world, at night, several times en route. The home-run part of the journey is most likely over hostile enemy territory, with sophisticated ground-to-air missiles and anti-aircraft fire waiting to fire. The pilot is literally putting his life on the line each time he carries a load of explosive to its target. It is obvious that any additional explosive effect that can be squeezed into the limited space of a bomb or projectile, the diameter and length of which is often predetermined, is worth almost any price. Similar conditions can be applied to other weapon systems, whether they be artillery munitions, missiles, or underwater warfare devices.

Therefore, military explosives developers strive for the maximum density combined with the maximum explosion energy. In order to concentrate a maximum of damage effect at the target, specialized explosives with high detonation velocities and high detonation pressures are used to accelerate a metal shell to form a high-velocity metal jet or metal fragment. The optimization of this process requires very high precision in density, grainsize, and composition.

A further important requirement increasingly demanded by ammunitions buyers is that the munitions have to be insensitive to initiation of mass detonation of pallets of munitions in storage or at the front line. The modern "insensitive munitions" should be demonstrably not mass detonable by projectile impact, fire, or a host of other stimuli as mandated by the hostile environment of war.

All these requirements lead to expensive, specialized explosives, loaded into sophisticated, expensive warheads. However, the actual cost of the explosive — even if very expensive compared to that of ANFO — is a quite small, although not negligible, part of the cost of the warhead, and an even smaller part of the entire delivery system.

As a consequence, military explosives may cost between $5 and $100 per kg, have densities in the upper part of the range from 1.5 to 1.9 g/cm^3, and explosion energies in the range 5 to 6.5 MJ/kg.

The most common military explosive still in very large-scale use is TNT, which replaced picric acid, the first military high explosive, during and after the First World War. TNT, an arene with density 1.64 g/cm^3 and a melting point of about 83°C if pure, is castable at temperatures between 80 and 100°C. TNT also serves as a carrier and energetic binder for higher density, higher energy nitramine explosives, first and foremost RDX, with a density of 1.8 g/cm^3, increasingly also HMX, of density 1.9 g/cm^3.

The common names adopted for the variety of compositions of TNT with added RDX or HMX are confusing. The background is the following: The compound cyclo-trimethylenetrinitramine, first identified as a chemical compound in Germany in 1920, was first produced in Italy before 1939 and then on a larger scale in Great Britain from 1942 during the Second World War. For security reasons, it became known as the Research Department Explosive, RDX. Another name for the same compound, Hexogen, is used in Europe, Russia, and many other parts of the world. Information about the new, powerful explosive was shared with the US, and production of RDX was soon begun in the USA at Holston Defense Corporation. They used a different process than that originally employed by the British, namely the Bachmann process [Bachmann and Sheenan, 1949], in which acetic anhydride is used in the nitration process. The chemical compound cyclotetraethylenetetranitramine, which is a natural byproduct of the Bachmann process for RDX, was discovered at Holston in 1949 and identified by Wright [1949]. After it was discovered that the new compound had a higher melting point than RDX, it was given the code-name HMX, High Melting Explosive. Another name for this compound, Octogen, is widely used in China, Europe, and Russia and many other parts of the world.

In Great Britain, the first practical military explosive based on RDX (a mix of 91/9 wt% RDX/Beeswax) was given the name Composition A, commonly known as Comp. A. Later developments during the Second World War included Comp. B (60/39/1 wt% RDX/TNT/wax). All of these compounds contained small amounts of wax, to reduce the friction and impact sensitivity of the hard crystalline RDX. Also in Great Britain, Comp. C was developed, a hand-moldable "plastic" explosive, consisting of 90/9/1 RDX/polyisobutylene/wax. Comp. C-1, C-2, C-3, and C-4 (produced in the USA) are later modifications of the original Comp. C, using different binders and different RDX grainsize distributions to provide the best moldability for the purpose, demolition, for which this explosive is most often used.

Composition B, a castable mix of RDX with TNT and a small addition of desensitizing beeswax, soon became the most frequently used high-energy military explosive on both sides of the Atlantic. Today, Comp. B. (although gradually being replaced in new ammunition by castable plastic-bonded RDX) is the most common projectile fill in ammunition magazines. Very large quantities of old TNT-filled ammunition are still in store in most countries. Precision-cast Composition B has been widely used for shaped charges.

Cyclotol became the generic name for RDX/TNT mixes in the nuclear weapons community in the USA. It was found that by using a bimodal grainsize distribution of RDX, it was possible to cast mixes with a larger percentage of RDX, and also, that by cooling from the bottom of the charge and using a large riser, in which excess TNT could rise, explosives with as large an RDX content as 75 wt % in TNT could be produced, starting from a nominal composition of 70/30 RDX/TNT.

(The early work in the USA on such castable RDX/TNT compositions was made with the classical Comp. B, having a nominal composition of 60/39/1 RDX/TNT/wax, but giving a cast material of composition 64/36 RDX/TNT, the subject of a great many studies on detonation physics.)

In the USA, Octol became the generic name for HMX/TNT mixed explosives. The most common composition of this kind was Octol 75/25, consisting of 75/25 wt % HMX/TNT, used for very demanding purposes. These compositions have now been replaced by hot-pressed compositions containing 94 to 96 wt % HMX (see below).

Although already having the most negative oxygen balance of all common explosives, TNT mixed with added high-energy fuel in the form of aluminum powder, an explosive commonly called Tritonal, and TNT/RDX compositions with large percentages of aluminum, commonly called Torpex, are used for increased blast effect and heave in underwater mines, torpedoes, naval artillery shells, and deep penetrating bombs. To better utilize the excess fuel in these types of explosives, repeated efforts have been made to include ammonium nitrate and ammonium perchlorate, and several AN/TNT/Al compositions have been fielded. However, the long-term storage properties of these have not been encouraging.

Using a curable plastic as a binder, RDX or HMX cast cure explosives such as PBXN-107 are now in large-scale production as a pourable fill for artillery shells and missile warheads. The binder in such compositions is essentially a ballast fuel, which reduces the density and energy of the composite explosive below that of the pure explosive. Experimental compositions using energetic binders and plasticizers such as FEFO (bis (2-fluro-2,2-dinitroethyl) formal) and GAP (glycidyl azide polymer) have partially overcome this problem.

For the most demanding applications, the pressed nitramines RDX or HMX, with no more than 4 to 6 wt % energetic binder, have been developed. Of these, PBX-9404 is the most powerful, although it is now considered too sensitive for most military applications. It is largely replaced by PBX-9501 which has equal performance but with lower shock and impact sensitivity than PBX-9404.

The most sensitive of the secondary explosives which have found military use in large quantity is PETN, for long the preferred secondary explosive in detonating cord, plastic explosive (sheet explosive), detonators, and booster charges, the latter often in the form of a castable mix (50/50 or 60/40 wt % PETN/TNT) for somewhat reduced sensitivity.

Probably the most insensitive, but still relatively high performance military explosives are based on TATB, which is replacing HMX in very demanding applications where an extreme insensitivity to shock, friction, and fire is required. PBX-9502 is an example of such a very insensitive military explosive, with TATB as its main ingredient. TATB has a crystal density of 1.90 g/cm^3.

The research and development of new military explosives now increasingly focuses on the search for single molecule explosives that have even higher density than HMX and an equally high energy, and on explosives with an extreme insensitivity to heat. In the former group are DINGU, NTO, and HNIW. The latter is an example of a new, emerging class of explosive molecules, called *cage molecules*, which appear to combine high energy (similar to that of HMX) with even higher density than the nitramines. In

Table 3.3. Properties of a selected group of common military explosives, arranged roughly in order of used tonnage.

Common name	Approximate composition	Initial density (g/cm^3)	Detonation velocity (m/sec)	Detonation pressure (GPa)	Explosive energy (MJ/kg)
Picric acid	100% Picric Acid	1.77	7350	23.9	4.396
TNT	100% TNT	1.65	6900	22.2	4.396
Comp. B	60/40 RDX/TNT	1.65	7800	24.4	4.995
Comp. C-4	90/10 RDX/polyisobutylene	1.72	8040	25.7	5.852
Cyclotol	75/25 RDX/TNT	1.74	8252	24.6	5.120
Octol	76/24 HMX/TNT	1.81	8476	34.3	4.497
PBXN-107	86/14 RDX/plastic binder[a]	1.64	8120	27.0	
PBX-9404	94/3/3 HMX/Nitrocellulose/ chloroethylphosphate	1.84	8800	36.5	
PBX-9501	95/2.5/2.5 HMX/estane/BDNP-F[b]	1.84	8830	36.5	
PBX-9502	95/5 TATB/Kel-F[c]	1.89	7710	28.5	
HNS	100% HNS	1.72	7000	26.2	5.695

[a] 5/3.9/3.7/1.5 2-ethylhexylacrylate/dioctyl maleate/n-vinyl-2-pyrolidone/additives
[b] 50/50 bis-dinitro-propyl-acetal/bis-dinitro-propyl-formal
[c] 75/25 chlorotrifluoroethylene/vinylidine fluoride copolymer

the latter group are PYX and HNS, which have found use as replacements for PETN in high-temperature applications, such as in primacord and for boosters and detonators intended for use at elevated temperatures.

The ultimate hoped-for military explosive is exemplified by the compact molecule octanitrocubane, a cubic array of carbon atoms with an NO_2 group attached at each corner. If the molecule could be synthesized, which has proven to be a very difficult if not impossible task, octanitrocubane would have a predicted density of 2.2 g/cm^3 and an energy exceeding that of HMX.

Table 3.3 shows the properties of a selected group of common military explosive compounds, arranged roughly in order of used tonnage.

3.5 Rock Blasting Explosives Systems

The extremely sensitive primary explosives are never handled in practical rock blasting except safely contained inside the detonator. Even so, a small glow from a lighted cigarette that is allowed to fall into the open end of an ordinary fuse cap detonator containing lead azide is enough to set off a detonation in the cap — a detonation sufficiently strong to mutilate beyond repair the hand holding the cap.

Although less sensitive to accidental initiation than lead azide, most secondary single explosive substances are nowadays considered too sensitive to be handled in pure, loose powder form under the normal rough conditions of loading drillholes for rock blasting. Both primary and some pure secondary explosives, such as PETN, RDX, and HMX, are transported only as a paste, obtained by mixing the explosive powder with 10 wt % of a water/ethylene glycol solution which serves as a desensitizer.

3.5.1 Dynamite Explosives

The classical dynamite explosives consist of a mixture of NG and the similar liquid explosive EGDN (ethylene glycol dinitrate, nitroglycol, $C_3H_4(NO_3)_2$) as sensitizer and plasticizer, mixed with a blend of fuel and oxidizer powder consisting of fuels such as wood-meal and oxidizers such as ammonium nitrate (AN) or sodium nitrate (SN). The NG/EGDN liquid is made into a gel by means of 1 or 2 % added nitrocellulose, which dissolves in the NG/EGDN liquid in the same manner as gelatin does in water to make a jelly. Adding EGDN to NG lowers the freezing point from that of NG from 3°C to 13°C to -25°C to -60°C dependent on the mixture ratio. The low freezing point dynamite made with such an NG/EGDN mix is safer to use at sub-freezing temperature.

Modern dynamites often contain additives intended to desensitize the NG/EGDN. Almost all dynamite explosives are cap sensitive — they can be initiated unconfined by a blasting cap containing less than 1 g of high explosive.

3.5.2 Watergels, Slurry Explosives, and Slurry Blasting Agents

Before the introduction of emulsion explosives, which now dominate the rock blasting explosives market, dynamite was for many purposes replaced by cap sensitive watergel explosives. In these, a gelled water solution, which contained oxidizer salts such as AN and SN together with water soluble fuels and/or explosive ingredients such as monomethylamine nitrate (MMAN), replace NG as plasticizer and sensitizer. The rest of the mixture, suspended in the gel to form a thick grainy fluid, consists of granular TNT or finely powdered aluminum as a sensitizer; undissolved AN and SN oxidizers; and added solid fuel components, of which granular aluminum is the one most frequently used to compensate for the low energy due to the cooling action of the water ballast.

The addition of finely powdered aluminum also increases the reaction rate of other explosive ingredients (by increasing the detonation temperature and increasing the number of hot spots from which the chemical reaction can start). Very fine-grained aluminum (paint grade) actually can be used as a replacement for self-explosive sensitizing ingredients.

3.5.3 Pumpable Blasting Agents

The introduction of water as an ingredient in rock blasting explosives greatly increased the safety in processing, storage, and loading, and opened the way for the introduction of mechanical pumping as a means of conveying the waterbased blasting agents into large-diameter drillholes. This brought about a revolutionary decrease in the cost of the explosives loading operation, by allowing delivery to the blasting site in bulk trucks from which the blasting agents were pumped into the holes at rates up to 200 kg/min. In comparison to dynamites, the waterbased explosives showed a greatly reduced friction and impact sensitivity and reduced risk for burning to detonation in the open. Most dynamites burn when exposed to a flame in the open; waterbased explosives generally do not. However, under confinement, such as in a pump, even waterbased explosives can burn to detonation, as exemplified by at least one major accident, attributed to the ignition of a watergel explosive accidentally pumped while the exit opening of the pump was blocked.

3.5.4 Emulsion Explosives and Blasting Agents

Emulsion explosives and blasting agents are the latest development away from ingredients that are in themselves explosive substances. By using suitable emulsifiers, a concentrated, hot solution of AN in water can be dispersed to form micron-size droplets in a fuel oil. In the resulting margarine- or vaseline-like, smooth mixture, the AN solution stays as a supercooled liquid without crystallizing even upon cooling to sub-zero temperatures. In contrast to ANFO which cannot be used in water-filled drillholes because of the high water solubility of AN, emulsion explosives have excellent water resistance since each AN/water droplet is surrounded by a thin film of oil which repels water. The extremely small droplet size, and the submicron thickness of the oil film gives a very large contact area between the fuel and the oxidizer solution; the intimacy of mixing of the fuel and oxidizer approaches that of a solution. Nonetheless, the emulsion matrix is essentially nondetonable as is, at densities above 1.35 to 1.40 g/cm^3, because of the low temperature increase obtained when it is compressed by a shock wave. By distributing in it finely dispersed voids in the form of glass microballoons or gas bubbles, that can act as hot spots to initiate the chemical reaction upon shock compression, a variety of emulsion explosives or blasting agents of different sensitivity levels can be produced. As they contain no ingredient that is an explosive in itself, and also because of the desensitizing effect of the water content, all such emulsions have a high degree of inherent safety.

For use in small-diameter boreholes, cap sensitive compositions are needed. As the name indicates, these are usually sensitive enough to be initiated in the drillhole by a standard Number 8 blasting cap. In the US, they are termed "Class A Explosives". For increased safety of handling on site, with the exception of some dynamite-type explosives for special purposes, most of the Class A explosives used in practice today are all emulsions, used by themselves or mixed with AN or ANFO into various mixtures or composites, each carefully balanced to be just sensitive enough to be initiated by a detonator in the temperature range and the drillhole diameters recommended.

The development of equipment to drill large-diameter boreholes of diameters 200—400 mm (8—16 in) has made it possible to use extensively a class of explosives which are called blasting agents in the USA. These are compositions that cannot be initiated unconfined by a single blasting cap (standard number 8 cap). They normally need a booster charge for reliable initiation in a large-diameter drillhole in rock. Blasting agents can be of different kinds, such as dry or moist powder, porous prills mixtures such as ANFO, or water-containing mixtures, which are generally emulsions, slurries, or watergels, or mixtures of ANFO and emulsions.

ANFO and emulsion-based blasting agents now dominate the market for commercial rock blasting materials because of their high energy efficiency and moderate price. ANFO by itself has a low packing density in drillholes (the pour density is about 800 kg/m^3, the density when jet-loaded by compressed air may reach 900—1,000 kg/m^3). By mixing ANFO with an emulsion explosive, a higher density product, commonly called heavy ANFO, having a higher energy content per unit mass than the emulsion explosive can be produced. The interstices between the ANFO grains are at least partly filled with emulsion explosive.

Figure 3.7 shows the density, the explosion energy per unit mass, and the bulk strength defined as the product of density and explosion energy as functions of the the mixing ratio ANFO/emulsion. Blends containing more than about 60 wt % emulsion are pumpable, blends with lower emulsion contents are usually fed into the drillhole by an auger, which can also serve as a blender for the two ingredients.

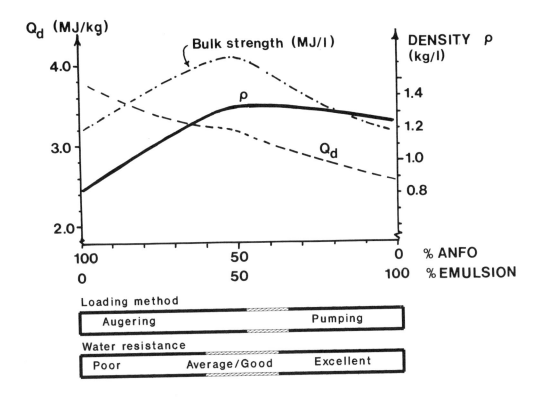

Figure 3.7. Density, explosion energy, and bulk strength of ANFO/emulsion mixes (heavy ANFO) as a function of the emulsion content.

ANFO and the emulsion are often carried to the blasting site by combination trucks, having separate tanks for the two ingredients. These are mixed as they are conveyed into the drillhole by pump or auger. Some pumpable emulsion blasting agents contain a small amount of glass microballoons that have a diameter of 50 to 100 μm and a wall thickness less than 1 μm. These can be used either mixed with ANFO or, in water-filled drillholes, by themselves. The nonsensitized emulsion from the plant is nowadays most often sensitized by chemical gassing. The gassing agent is fed into the emulsion stream just before it enters the pump conveying the material into the auger or into the drillhole.

The blasting agents have increased the safety against accidental initiation by friction. The development of mechanized equipment for safe bulk handling and loading of large quantities of blasting agents has opened the way for lowering the cost of the entire blasting operation. Without the waterbased emulsions, or earlier the watergel or slurry blasting agents, the modern large-scale open pit and strip mining operations would not be economically feasible.

3.5.5 Permissible Explosives in Underground Coal Mining

In underground coal mines, and in some tunneling, the risk of methane/air or coal dust/air explosions mandates a wide spectrum of precautions on all underground activities. These cover the use of specially designed no-spark lights, lamps, electric equipment, and special detonators and explosives, and also special procedures and rules of operation and working. All of these are designed to minimize the risk of mine fires and methane/air (firedamp) or coal dust/air explosions, which almost invariably cause death and devastation on a tragically large scale. To amend, monitor, and enforce these rules and regulations, each coal mining nation has established a board and an inspectorate which usually have very considerable powers to regulate and control underground operations. The permission to use a particular explosive in a particular mine in a particular country is granted by the national coal board, bureau of mines, or inspectorate of coal mining after the explosive has been extensively tested in a test gallery that is normally part of a special national laboratory for safety in coal mining. Each nation has its own test procedure and the final decision is made by the inspector of coal mining after he has taken into consideration the particular conditions and procedures prevailing in the mines in his country.

Therefore, the "permissibility" of an explosive is not only a property of the explosive; it is considerably influenced by the local conditions, regulations, and methods of charging and shotfiring, mine ventilation, and, naturally, also by the past experience of the inspector himself and the accident record of his country's coal mining industry.

All this makes it impossible to define and classify permissible explosives generally. Each country uses a different set of classes to distinguish between their different permissible explosives with respect to the degree of "permissibility". Hence, each mine or even mine area is similarly graded with respect to the severity of the risk situation.

To be permitted in a particular class, an explosive has to be designed to give a sufficiently low reaction product temperature and expansion rate that its detonation in a drillhole does not ignite methane or coal dust in the test prescribed for that class of permitted explosives. It is also important that the reaction products do not contain burning solid fragments that could cause ignition.

For easy conditions, such as in a well-ventilated modern mine with large coal seams, large tunnels, and a low methane content in the coal, a weak normal explosive may be permitted. For more severe conditions, compositions containing heat-absorbing ballast materials, such as sodium chloride, are used. For deep, narrow seams, gassy and difficult-to-ventilate mines, special ion exchange permissible explosives are necessary. These contain a mixed pair of salts, such as ammonium chloride (NH_4Cl) and sodium nitrate ($NaNO_3$). These react endothermally with each other in the afterburning phase of the detonation, resulting in a very low flame temperature.

In the methane (CH_4) oxidation reaction,

$$CH_4 + 2O_2 \longrightarrow CO_2 + 2H_2O$$

two requirements must be met to cause ignition. First, the methane/oxygen mixture must have an ignitable methane concentration. Second, a minimum induction period is needed. Since the induction period is dependent upon the temperature and the time, coolants such as $NaCl$ are added to the explosive composition. The addition of coolants lead to lower explosion temperatures and shorter free flame periods.

The explosive lower limit for methane is 5.5% (volume percent) in air; the upper limit is 15.5%. If the methane-air mixture is richer, it will not explode but will burn if heated. Methane concentration is controlled by proper ventilation with a high-enough air velocity. The ventilation regulations, which normally are designed to remove toxic fumes as well as methane, and therefore depend on the number of persons working in a particular tunnel, vary considerably from country to country. The ventilation air input can vary from 1 to 5 m^3 per minute and man. If diesel equipment is used, several cubic meters per minute per brake horsepower is added. The methane concentration should be kept less than 20% of the lower explosive limit to maintain a high safety factor.

3.5.6 Aluminized Explosives

Aluminum powder is often added to watergel, slurry, and emulsion explosives in order to improve the energy output. As can be seen from Table 4.4, the heat of formation is very high per mole of the formed solid detonation product aluminum oxide, Al_2O_3. Thus, with the addition of a small amount of aluminum to the slurry or emulsion, the theoretical potential energy release can be increased considerably. However, experiments indicate that the aluminum powder does not react completely within the detonation reaction zone; therefore, not all of the aluminum combustion energy is released during the important early expansion phase where the rock fragmentation occurs. Generally, all of the aluminum reacts; part of it, however, in the low-pressure afterburning phase, where it only contributes to heating the escaping reaction products. The use of high percentages of aluminum in the explosive composition is costly and, considering that the only a fraction of the added energy is utilized for the rock fragmentation, aluminum contents in commercial blasting agents and class A explosives are generally kept in the range between 0 and 5 wt %, where the increased temperature in the vicinity of the burning aluminum grains appears to increase the reaction rate of the composition as a whole.

The reaction kinetics is strongly influenced by the shape, the type of surface coating used, and the grain size of the aluminum powder. It is not unusual for post-detonation explosions to occur in the muck pile when high percentages of aluminum are used as a fuel in explosives. This can be due to the following: When aluminum is used, the reaction product temperature is high, which shifts the equilibria to favor formation of hydrogen, H_2, which later may mix with atmospheric air and cause a secondary explosion. An additional or alternative explanation may be incomplete reaction of the AN or other nitrates in the explosive, leading to large H_2 (and CO) contents in the reaction products.

High hydrogen contents in the reaction products of explosives having a high aluminum content have been suspected to be the cause of secondary explosions occurring after the initial blast in underground mining where large hole diameters and inverted benching or vertical crater retreat type shots were used. Aluminized slurries had been used in a blast, and when miners dropped a stone into the hole to see if the hole was open, a spark allegedly ignited the H_2/O_2 mixture in the hole, causing an explosion that sent the stone back with high velocity.

3.6 Mechanized Explosives Charging

The process of charging explosives into boreholes is increasingly done mechanically by special equipment. This is done both for the sake of safety and to relieve the loading crew of the hard work of carrying explosives to the drillhole and of packing the explosive into the hole. Charging equipment for horizontal or vertical-up holes is usually pneumatic, blowing the powdered or cartridged explosive through a flexible plastic hose (Figure 3.8). For vertical-down dry holes, dry powdered explosives such as ANFO can be allowed to fall into the hole from an auger feed (Figure 3.9), whereas pumping liquid or paste-like explosives through a hose is the preferred method for wet vertical-down or horizontal holes (Figure 3.10).

Further developments toward increased safety, at least with respect to safeguarding against the accidental explosion of large quantities of explosives in transit from the explosives factory to the blasting site, are the on-site-mix or in-the-drillhole-mix explosive systems. The carbon/liquid oxygen system was an early example of such an explosive

80 Chapter 3. Explosives

Figure 3.8. Pneumatic loading of ANFO. The ANOL equipment is a stainless steel compressed air driven unit for charging ANFO into blastholes with diameters between 25 and 150 mm (Nitro Nobel AB).

Figure 3.9. Auger feed loading of ANFO in a $d = 330$ mm diameter drillhole (IRECO Incorporated).

system, the nonexplosive ingredients of which could be transported separately and safely to the site. They were brought together to form a very sensitive explosive just prior to loading into the drillhole.

A present-day on-site-mix system for charging large-diameter boreholes with typically 500 to 1000 kg slurry explosive per drillhole consists of a mix-truck with two or more separate tanks for powder and liquid ingredients. These ingredients are mixed in a continuous on-line mixer as the mixture is poured or pumped into the drillhole. The simplest system of this kind is the ANFO mix truck, which has a large container of prilled AN and a separate tank of fuel oil. The fuel oil is sprayed from a metering pump into the stream of AN conveyed by an auger which dumps the mixture into a chute or funnel placed over the opening of the drillhole. A more sophisticated system for on-site mixing of slurry or emulsion blasting agents is in effect a small, mobile explosives factory placed on a truck, with separate tanks for powder and water-bearing liquid explosive or nonexplosive ingredients, sensitizers, density modifiers, and gelling and cross-linking agents. The metered components of the explosive, some of which may in themselves be premixed at a base plant, are fed by auger or pumped continuously through the mixer. The mixed explosive is fed into a charging pump which feeds through the loading hose down into the drillhole, where the quick-acting cross-linking agents of the watergel slurry, or the viscoelastic properties of the emulsion, transfer the easy-flowing liquid into a viscous fluid or, in the case of the watergels, into a rubber-like, water-resistant substance. Figure 3.11 shows an example of the on-site-mix truck.

Bringing the explosives factory out onto the blasting site in this way has disadvantages as well as advantages. It is normally easier to keep a high product quality when working in a stationary, well-organized explosives plant. Wherever restrictions against

Figure 3.10. Early mechanical pumping of a TNT-slurry at the Boliden Aitik mine (1968).

road transportation do not hinder, it is often preferable to bring the quality-tested, ready-made bulk explosive to the site in a pump truck rather than risk failures due to missing or wrongly-metered components. The majority of the world's large open-pit mines and strip mining operations each have a sufficiently large consumption of bulk explosives to have their own stationary bulk explosives plant situated conveniently close to the mine to minimize transportation costs. Figure 3.12 shows a bulk explosives plant. Figure 3.13 shows a pump truck in the process of loading boreholes.

In underground mining, packaged explosives like ANFO, emulsions, or slurries can be loaded into large-diameter drillholes pointing upwards (upholes) using mechanical loading techniques such as the HALF-PUSHER technique developed by Nitro Nobel AB. This charging method relies on special, relatively long, stiff explosives cartridges equipped with fins which make the charge jam in the drillhole so that the charge can only move in one direction, further up into the hole. The HALF-PUSHER is an air-powered unit, initially fastened to the charge, which pushes the charge a short distance uphole, uses the charge to pull itself the same distance into the hole, again pushes the charge uphole, and repeats this operating cycle until the charge is in position. At that point in time, the HALF-PUSHER separates itself from the charge and can be removed from the hole (Figure 3.14). The HALF-PUSHER method is quick and reliable if the rock is not too jointed.

Prilled or crystalline ANFO can be loaded pneumatically into small-diameter drillholes by being blown through a stiff plastic hose using a high-speed stream of compressed air. The AN prills or grains impact on arrival and partly crush and jam each other into the hole. The friction against the drillhole wall prevents the charge from falling out of the hole even if it is a vertical hole loaded from below. In vertical upholes larger

3.6. Mechanized Explosives Charging 83

Figure 3.11. A site-mixed-slurry (SMS) explosives truck at work (IRECO Incorporated).

Figure 3.12. A bulk emulsion explosive plant ready for delivery to the Sishen Mine (Nitro Nobel AB).

Figure 3.13. Pumping emulsion explosive (Emulite) or mixed emulsion/ANFO (Emulan) into 380 mm diameter water-filled drillholes (Nitro Nobel AB).

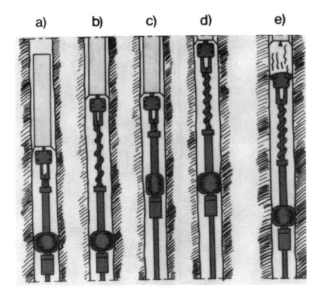

"Half-pusher"

Figure 3.14. The HALF-PUSHER technique.

than 4 in diameter, however, the charging of bulk ANFO, bulk slurries, or bulk emulsions can be troublesome because the friction against the hole wall is not always enough to keep the heavy charge in place. By adjusting the nozzle geometry, air pressures, and water content, it is possible to load the holes so that the explosive stays in, but the method must be performed with accuracy and skill in order to work.

Figure 3.15 shows another approach, called "snow-charging", for bulk loading ANFO into large-diameter up-holes, which was first used at the Mufilira Mine in Zaire. An air-bleeder hose with a larger diameter than the loading hose is inserted into the hole. At the collar, the bleeder hose is hooked into its position as the air gap between the bleeder hose and the borehole wall is sealed off. Afterwards, the loading hose is inserted and pushed to the top of the borehole; and the ANFO is loaded in the normal way. The ANFO falls like snow to fill the space between the bleeder hose and the drill-hole wall, while the air escapes through the bleeder hose. In this way, upholes as large as 8 in diameter or larger could be loaded with ANFO.

3.7 Invention of Emulsion Explosives and Blasting Agents

The invention history of emulsion explosives and blasting agents began with the invention by Egly and Neckar [1964] in the United States that a relatively water resistant blasting agent, albeit with relatively limited lifetime could be produced by mixing crystalline AN with a water-in-oil emulsion made by high shear mixing of a water solution of AN into a mix of oil and emulsifier. Gehrig [1965] produced an all-solution emulsion explosive by high-shear mixing a solution of nitric acid, sodium nitrate, and water into a fuel phase consisting of mineral oil plus an emulsifier such as sorbitan mono-oleate

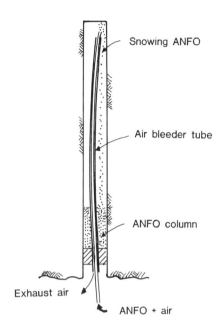

Figure 3.15. Charging 4 in holes with the "snow-charging" method.

(tradename Span 80). The resulting explosive called "Aquanite" had limited stability and was chemically reactive with some ores because of its acid content.

The first stable, practically useful all emulsion blasting agents were invented by Bluhm [1969]. He emulsified a hot water solution of AN and SN in a fuel phase consisting of mineral oil and emulsifier, and made the emulsion detonable in charge diameters down to 75 mm (3 in) by introducing gas bubbles by aerating in a high shear rate mixer, by chemical gassing, or by adding hollow glass microballoons. The products, which were not cap sensitive, were called "Aquaram" or, with added aluminum, "Aquanal".

Wade [1973a, 1973b, 1978] finally made cap-sensitive AN emulsion explosives by initially using a combination of carefully optimized air inclusions such as glass microballoons plus the catalytic action of heavy metal ions such as copper or chromium, and ultimately showed that all-AN cap-sensitive emulsion without catalysts could be produced by carefully balancing the water and microballoon content.

Clay [1978] was the first to prepare a blasting agent by mixing ANFO and/or crystalline AN into an emulsion matrix containing no air bubbles or other sensitizing agent. This invention has led to the development of a whole range of very cost effective blasting agents consisting of blends of prilled ANFO with various weight percentages of emulsion.

Chapter 4

Shock Waves and Detonations, Explosive Performance

A person can get in the car and drive to work every morning without understanding the inner functioning of an internal combustion engine. He will have to trust and rely on the judgement of the car manufacturer that the engine is of a sound design initially. He has to take the advise of his car mechanic whenever something goes wrong. If he does not like the advise, he may go to another mechanic and very likely will hear a different story. In the end he has to pay to get the car fixed, one way or the other. Many people survive and prosper that way. Others learn a little about cars and engines and get satisfaction out of knowing how to keep the engine running. Certainly there are people working for the car manufacturer who have to be expert at internal combustion, and most car mechanics nowadays must have a good understanding of the subject.

In the same way, a drill and blast engineer can design blasts and set them off every day of the year without understanding the difference between combustion and detonation. He or she will have to rely upon the advice and designs of an explosives supplier for selecting the best explosive for different rock conditions, and he or she will need a consultant when something goes wrong. If he or she does not like what the consultant says, another consultant may be brought in, and will undoubtedly have a different story.

Others need to know more: students of explosives engineering; engineers and explosives scientists working for an explosives company or for a mining company that has its own explosives plants; blasting consultants; and blasting engineers who want to understand more of what goes on in and around the drillhole.

This chapter is for them. The sudden release of energy and reaction products at high pressure by a rapid chemical reaction in explosive contained in a drillhole in rock gives rise to compression waves in the explosive and in the surrounding rock material. These waves are called detonation waves or shock waves depending on whether they involve chemical reaction or not. They play a central role in the functioning of the explosive and in the fragmentation of the rock. The chapter describes the simple theory of shock waves and detonation, and attempts to describe the effect the explosion has on the rock, i.e., the performance of the explosive. It describes in some detail the older theories, concepts, and performance tests still used in many places to describe and give numbers for the rock blasting performance of explosives. It also attempts to introduce the reader to the present theories and concepts of explosive performance as they are currently being developed. These parts of the chapter are by necessity somewhat incomplete, because the last word in these developments is not yet said.

The concepts of shock waves and detonation waves are treated together because a detonation wave is really a shock wave, supported by the explosive reaction that the shock wave ignites and propagates as it travels through the explosive material. In the context of rock blasting, furthermore, there is justification for treating shock waves and detonation waves together because the mechanical effect of the detonation in an explosive in a drillhole in rock is propagated and its effects are spread throughout the surrounding rock by means of a shock wave. The overall performance of an explosive in rock blasting or any other application is of course dependent on how much thermal energy is released in the chemical reaction that occurs when the material detonates, and on how much useful mechanical work the reaction products can do as they expand. A major portion of this chapter deals with the prediction and measurement of explosive performance, a subject which is now still in a process of rapid development.

4.1 Historical Note

In Alfred Nobel's early efforts in 1863 and 1864 to reliably make nitroglycerin and nitroglycerin mixed with black powder explode in drillholes he clearly understood the importance of the shock (concussion) generated by his "detonator" in producing the "especially violent explosions" that allowed nitroglycerin to be used for practical rock blasting. Later, F.A. Abel [1869] demonstrated that unconfined charges of guncotton, nitroglycerin, dynamite, and mercury fulminate burned if ignited by a flame but "detonated" if exposed to the blow of a Nobel Detonator. The real physical nature of the phenomenon of detonation was, however, first clearly recognized as a propagating explosive wave by Berthelot and Vieille in 1881 and Mallard and Le Chatelier in 1881 the same year during their parallel investigations of gaseous explosions. (See Oppenheim [1960] for more details of the history of these classical discoveries.) In 1893, Shuster suggested an analogy between detonation waves and the nonreactive shock waves which had been discussed by Riemann and others in 1860. Shuster pointed out that the detonation wave is headed by a shock front which advances with constant velocity into the unreacted explosive and is followed by a zone of chemical reaction which supports the detonation wave. This suggestion that the detonation wave was a reactive shock laid the groundwork for Chapman [1899] and Jouguet [1905] to develop the classical Chapman-Jouguet (CJ) hydrodynamic theory of steady-state detonation. (See also a "Chronology of Early Research on Detonation Wave" by Bauer, Dabora, and Manson [1990] for additional historical details.)

The Chapman-Jouguet theory is described by a one-dimensional model in which the chemical energy is instantaneously released in the discontinuous shock front across which the conservation laws apply — the detonation velocity was assumed to be the minimum velocity given by the conservation laws.

During World War II, the CJ theory was refined independently by the Russian Zeldovich [1940], the American Von Neuman [1942], and the German Döring [1943]. They assumed that the chemical energy release did not take place in the shock front, but at a finite zone behind the front of the detonation wave.

Research in military explosives technology has provided the commercial explosive research with accurate thermochemical data for explosive ingredients and detonation products. New and more accurate equations of state for gaseous and solid reaction products have been developed. Thermodynamic-hydrodynamic computer programs contribute to the understanding of the explosive performance. Studies of detonation phenomena help develop a deeper understanding of how the energy is released.

In commercial explosives, the ingredients' particle size is an important factor which determines the rate at which the ingredients react with each other. Figure 4.1 shows a comparison of the characteristic size of an ANFO grain and that of a cell in an emulsion explosive. Even though ANFO and an aluminized emulsion may contain the same chemical energy, their field performances may differ. This is due to differences

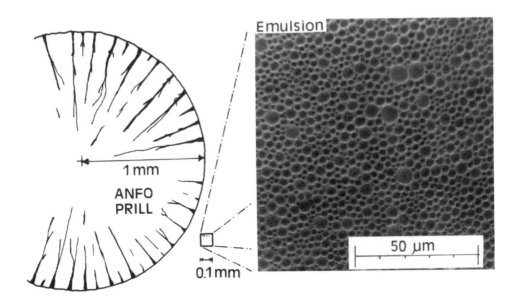

Figure 4.1. The cell (droplet) size in an emulsion explosive is more than two orders of magnitude smaller than the size of an AN prills grain. The emulsion particle size, less than 10 μm, is intermediate between the ANFO prills grain, of approximate radius 1 mm, and the size of the molecules of typical molecular explosives (photo by Sten Ljungberg, Nitro Nobel AB).

in detonation velocities, to differences in the ratio between shock and heave energy, and to differences in the expansion work done before the rock breaks and the gaseous products are ventilated through the formed cracks. The solid oxidizer AN grains, called prills, in ANFO have a diameter around 2 mm and are impregnated with the liquid fuel oil. The emulsion oxidizer is a liquid salt solution made up of 0.005 mm droplets, with each droplet surrounded by a thin oil film. It is easy to understand that the specific contact surface between oxidizer and fuel is many orders of magnitude larger for the emulsion than for the ANFO. These large differences in reaction surface area will obviously influence the reaction rates for the two explosives, and at least partially compensate for the presence of a considerable percentage of water in the emulsion.

The size of a typical single explosive molecule is six orders of magnitude smaller than an AN prills grain and four orders of magnitude smaller than an emulsion droplet. However, most single explosives, including PETN and NG, are dependent on the presence of hotspots for their critical detonation and other detonation characteristics.

In the next section, we will concentrate on describing the classic CJ theory that is the necessary first step toward understanding the detonation phenomenon.

For students or engineers who would like to penetrate deeper into the science of detonation and detonation theory, we recommend books written by Taylor [1952], Cook [1958, 1974], Johansson and Persson [1970], Brinkley and Gordon [1972], Mader [1979], Fickett and Davis [1979], and later articles by Bdzil [1981] and Bdzil, Fickett, and Stewart [1989].

90   Chapter 4. Shock Waves and Detonations, Explosive Performance

Figure 4.2. Schematic picture of retracting shock wave [after Courant and Friedrichs, 1950].

4.2 The Rankine-Hugoniot Equations and the Chapman-Jouguet Theory

The mechanism of shock compression of a material is illustrated in an excellent way by the image of a row of skiers piling up in front of an obstacle (Figure 4.2). A less exciting schematic picture of the process is shown in Figure 4.3, where a rigid piston with velocity u drives into an undeformable tube containing a material at rest of density ρ_0, pressure p_0, temperature T_0, and specific internal energy E_0. The "piled-up" shock compressed material moving with the same velocity as the piston has density ρ_1 ($> \rho_0$), pressure p_1 ($> p_0$), temperature T_1 ($> T_0$), and specific internal energy E_1 ($> E_0$). The sudden shock transition from state 0 to state 1 takes place at the shock front which moves into the material at rest with velocity D and gives the shocked material the velocity u.

The equations governing the one-dimensional (linear and laminar) flow of a material through a plane shock front oriented at right angles to the flow direction can be written if we consider that mass, momentum, and energy must be conserved as material passes from one side of the shock front to the other.

Conservation of mass requires that the amount of mass per unit time and unit area which passes from the low pressure region into the shock front must be equal to the amount of mass per unit time and unit area which passes from the shock front into the

4.2. The Rankine-Hugoniot Equations and the Chapman-Jouguet Theory

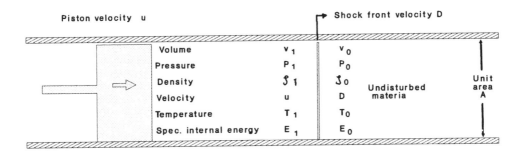

Figure 4.3. Constant velocity piston-generated shock transition from state 0 to state 1.

high pressure region. This requires that

$$\rho_0 D = \rho_1 (D - u). \qquad (4.1)$$

Conservation of momentum requires that the change of momentum of mass per unit area must be equal to the resultant force acting on the unit area. This implies that

$$\rho_0 D^2 - \rho_1 (D - u)^2 = p_1 - p_0. \qquad (4.2)$$

Finally, the internal energy changes by $\rho_0 D (E_1 - E_0)$. The kinetic energy will change by $\rho_0 D \left[\frac{(D-u)^2}{2} - \frac{D^2}{2} \right]$, and the work done on the material is $p_0 D - p_1 (D - u)$. The conservation of energy requires that the change in internal energy plus the change in kinetic energy is equal to the work done. Using Equation 4.1, we have

$$\rho_0 D (E_1 - E_0) + \rho_0 D \left[\frac{(D-u)^2}{2} - \frac{D^2}{2} \right] = p_0 D - p_1 (D - u)$$

then dividing by $\rho_0 D$ — which equals $\rho_1 (D - u)$ — gives

$$E - E_0 + \frac{(D-u)^2}{2} - \frac{D^2}{2} = \frac{p_0}{\rho_0} - \frac{p_1}{\rho_1}. \qquad (4.3)$$

A. The Rankine-Hugoniot Equations for Shock Waves

The conservation Equations 4.1, 4.2, and 4.3, which are named the Rankine-Hugoniot shock relations, can be simplified and presented as follows:

Mass: $\qquad \rho_0 D = \rho_1 (D - u). \qquad (4.4)$

Momentum: $\quad p_1 - p_0 = \rho_0 D u. \qquad (4.5)$

Energy: $\qquad p_1 u = \rho_0 D \left(E_1 - E_0 + \frac{u^2}{2} \right). \qquad (4.6)$

When $p_1 \gg p_0$, Equations 4.5 and 4.6 combine to give

$$u^2 \approx E_1 - E_0 + \frac{u^2}{2}$$

or

$$E_1 - E_0 \approx \frac{1}{2} u^2. \tag{4.6a}$$

That is, the increase in internal energy equals the increase in kinetic energy in a high amplitude shock wave. In other words, the piston work put into the system is equipartitioned into internal energy and kinetic energy. Equation 4.6a provides an exceptionally simple and elegant way of finding the increase in internal energy when the shock particle velocity is known. If the average specific heat were known, we would then also know the temperature increase $T_1 - T_0 = (E_1 - E_0)/c_v$. Unfortunately, very little is usually known about the specific heat of matter at the high pressures and high temperatures prevailing in high amplitude shock waves.

B. The Rankine-Hugoniot Equations for Detonation

In the case of a pure shock wave, the only source of energy is the piston moving against pressure p_1 with velocity u. We can see from Equation 4.6 that the energy delivered by the piston is equal to the sum of the changes in internal and kinetic energy.

In the case of a detonation wave, the chemical decomposition is assumed to take place close to the shock front, which is now the detonation front moving with the velocity of detonation D. To the right of the front in Figure 4.3, we have the unreacted explosive and, to the left, we have the completely reacted detonation products.

The conservation equations have the same form for detonation waves as for shock waves. However, we must remember that the chemical reaction changes the material from an explosive to reaction products. This is a profound change often involving, for example, a change in the total number of moles. Therefore, the initial internal energy of the explosive and the internal energy of the reaction products are not exactly comparable, except in the case when the explosive and the reaction products are both polytropic gases. For a shock wave or a detonation wave in a polytropic gas (ideal gas with constant specific heat c_v), in the case of a detonation having a heat of reaction Q, the energy conservation equation (Equation 4.6) can be written:

$$\text{Detonation:} \quad p_1 u + \rho_0 D Q = \rho_0 D \left[c_v (T_1 - T_0) + \frac{u^2}{2} \right]. \tag{4.7}$$

$$\text{Shock:} \quad p_1 u = \rho_0 D \left[c_v (T_1 - T_0) + \frac{u^2}{2} \right]. \tag{4.8}$$

For a condensed explosive, the specific internal energy $E_1 - E_0$ will not only be a function of the intensive variable temperature, but will also be a function of pressure and specific volume. We need the additional information provided by an equation of state (see Section 4.4.1) for the explosive as well as for the reaction products to resolve the exact relationship between the internal energy and the shock particle velocity. Therefore, it is not possible to express the exact internal energy for the detonation in a condensed material in the simple way used in Equations 4.7 and 4.8. The difference is not large, however, and as a rough approximation, for $p_1 \gg p_0$, we can write in analogy with Equation 4.6a

Detonation: $$E_1 - E_0 \approx Q + \frac{u^2}{2}. \qquad (4.7a)$$

Shock: $$E_1 - E_0 \approx \frac{u^2}{2}. \qquad (4.8a)$$

The heat of reaction Q in Equation 4.7a is equal to the heat evolved when the detonation products are formed from the explosive and the products are then brought to STP. Q is defined as

$$Q = -\left[\Sigma n_i (\Delta H_f^o)_i - \Sigma n_j (\Delta H_f^o)_j\right] \qquad (4.9)$$

where n_i equals the number of moles of the ith species of detonation products whose standard enthalpy of formation at 298.15 K is $(\Delta H_f^o)_i$. The number of moles in the unreacted explosive of the jth ingredient is n_j, and its standard enthalpy of formation at 298.15 K is $(\Delta H_f^o)_j$. Water is in its gaseous phase.

C. The Rayleigh Line

Solving for u in the mass conservation equation (Equation 4.4) and combining it with the momentum conservation (Equation 4.5) yields the relation

$$-\rho_0^2 D^2 = \frac{p_1 - p_0}{v_1 - v_0} \qquad (4.10)$$

which describes the so-called Rayleigh line in the p–v diagram (Figure 4.4). The slope of the line is obviously negative and equal to $-\rho_0^2 D^2$. If the pressure (p_1) is higher than the initial pressure (p_0), the volume must be less than the original volume (v_0). There will be two extreme limits for the slope of the line. The horizontal line which corresponds to zero-velocity describes the constant pressure $(p_1 = p_0)$ explosion state.

The vertical line (which would correspond to an infinite detonation velocity, $D = \infty$, when $v_1 = v_0$) describes the constant volume explosion state, which is at the intersection of the Hugoniot curve with the vertical line.

D. The Hugoniot Equation

By eliminating D and u in the Rankine-Hugoniot shock relations (Equations 4.4, 4.5, and 4.6), we can derive the Hugoniot equation for an inert shocked material

$$E_1 - E_0 = \frac{1}{2}(p_1 + p_0)(v_0 - v_1) \qquad (4.11)$$

where E_1 is the specific internal energy of the shock compressed material at (p_1, v_1) and E_0 is the specific internal energy for the uncompressed material at (p_0, v_0).

For a detonation wave the Hugoniot equation, derived from Equations 4.4, 4.5, and 4.7a, becomes approximately

$$E_1 - E_0 = \frac{1}{2}(p_1 + p_0)(v_0 - v_1) + Q \qquad (4.11a)$$

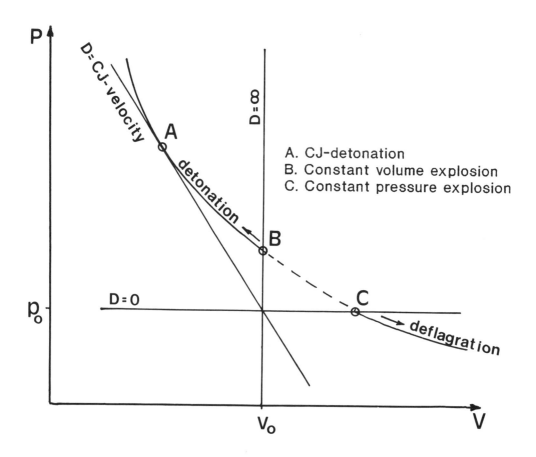

Figure 4.4. Diagram of Rayleigh lines and the Hugoniot curve for the reaction products.

where E_1 is the specific internal energy for the reaction products at (p_1, v_1) and E_0 is the specific internal energy for the unreacted explosive at (p_0, v_0). With thermal data and an equation of state for the reaction products, the quantities E_1 and E_0 can be determined.

Equations 4.11 and 4.11a can be rewritten into equations expressed in enthalpy. We have the familiar relation

$$H = E + pv \qquad (4.12)$$

which, for an inert shock, together with Equation 4.11 gives

$$H_1 - H_0 = \frac{1}{2}(p_1 - p_0)(v_0 + v_1) \qquad (4.13)$$

and, for a detonation, together with Equation 4.11a gives

$$H_1 - H_0 = \frac{1}{2}(p_1 - p_0)(v_0 + v_1) + Q \qquad (4.13a)$$

E. The Chapman-Jouguet Detonation

Expressing E_1 as a function of p and v in the Hugoniot equation gives the Hugoniot curve in the p–v diagram for the reaction products (Figure 4.4). This Hugoniot curve is the locus for all states that can be reached by single shocks followed by completed exothermic chemical reactions. The conservation equations require that the point associated with complete reaction in the p–v diagram lies on the Hugoniot curve and on the Rayleigh line. This condition is not enough to determine the detonation wave velocity D because there are several possible intersections between the Rayleigh line and the Hugoniot curve for the reaction products. Figure 4.4 shows that there is a lowest possible detonation velocity, represented by the point of tangency A between the Hugoniot curve and the Rayleigh line of slope $-(\rho_o)^2 (D_{CJ})^2$. This point is called the Chapman-Jouguet point, and its significance will become clear later. There are no solutions to the one-dimensional, constant velocity detonation Rankine-Hugoniot equation (Equation 4.11) for detonation velocities lower than D_{CJ}. The Chapman-Jouguet detonation thus appears to be the lowest velocity stable detonation permitted by the conservation equations. (Later in this chapter, we will see that lower detonation velocities are possible in detonations with a curved front, but at this stage, we are only discussing the one-dimensional detonation with a plane front.) A very high detonation velocity will result in a steep Rayleigh line intersecting the Hugoniot curve at two points — one at a higher pressure where it can be shown that the flow is subsonic relative to the front, and one at a lower pressure where the flow is supersonic. The former is an *overdriven detonation* that can be thought of as a detonation driven from behind by a piston with velocity greater than the particle velocity at the CJ point. The latter, a weak (lower pressure) high velocity detonation could result if by some means the explosive reaction could be made to propagate at a higher velocity than D_{CJ}. Such would be the case, for example, if a laser-beam that was sufficiently intense to ignite the explosive to immediate reaction were sweeping along a charge encased in a stiff glass tube with velocity greater than D_{CJ}, igniting the explosive as it becomes illuminated. There do not appear to be any physically meaningful solutions to the Hugoniot equation in the region between points B and C. Points beyond point C represent burning or deflagration, in which the reacted material expands to greater volumes (depending on the ambient pressure).

With reference to Figure 4.5, let curve 1 represent the Hugoniot curve for the unreacted explosive. The straight line OA is the Rayleigh line for the particular, piston-driven shock transition to state A on the shock adiabat 1. Were the unreacted material allowed to expand without chemical change from state A, it would follow the isentrope 2 which is less steep than the shock adiabat.

If the material were to react chemically, releasing a reaction energy, the end states at different piston velocities would be on the detonation products' shock adiabat or Hugoniot curve 3, above the shock adiabat. When the material is allowed to expand from state 3 on this curve, it follows the isentrope 4 which also is less steep than the adiabat 3. With decreasing piston velocity, the overdriven detonation state, i.e., the upper intersection of the Rayleigh line with the Hugoniot curve, will be at lower and lower pressures until point C is reached, the CJ point. At this point, the Rayleigh line is tangent to the isentrope 5 through C, and also to the Hugoniot curve 3. This has an interesting consequence. The sonic velocity c at any state point is given by the slope of the isentrope passing through that point as defined by the thermodynamic equality

$$\rho^2 c^2 = -\left(\frac{\partial p}{\partial v}\right)_s. \qquad (4.14)$$

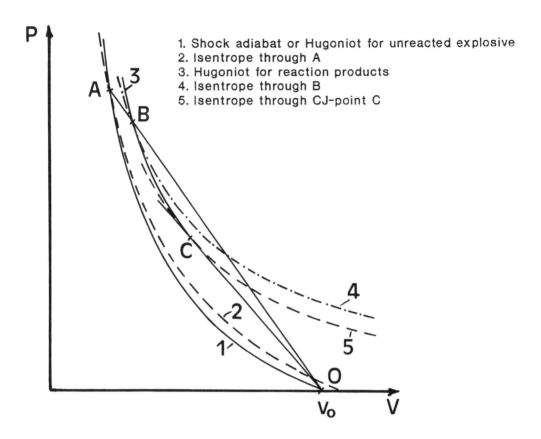

Figure 4.5 Shock and detonation transition in the p,v plane.

Specifically, at the Chapman-Jouguet point C, we have

$$\rho_1^2 c_1^2 = -\left(\frac{\partial p}{\partial v}\right)_{s_1}. \tag{4.15}$$

Also, the Rayleigh line is a tangent

$$\rho_0^2 D^2 = -\left(\frac{\partial p}{\partial v}\right)_{s_1}. \tag{4.16}$$

It follows that

$$\rho_1 c_1 = \rho_0 D_{CJ} \tag{4.17}$$

which, combined with the mass conservation equation (Equation 4.4), gives the Chapman-Jouguet condition

$$c_1 = D_{CJ} - u. \tag{4.18}$$

Equation 4.18 states that the sonic velocity c_1 is equal to the difference between the detonation front velocity D_{CJ} and the particle velocity u. This explains the unique stability of the detonation velocity observed already by Berthelot and others. The Chapman-Jouguet detonation propagates with a velocity that leaves the reaction states (at pressures higher than the Chapman-Jouguet pressure) essentially undisturbed by rarefaction waves coming from behind. These can reach, but not pass through, the sonic point which, in the ideal Chapman-Jouguet detonation, coincides with the Chapman-Jouguet point. If the supporting piston is stopped, the wave will still propagate with the stable detonation velocity D_{CJ}, which depends on the properties of the explosive and particularly its initial density. By contrast, the inert shock wave (A in Figure 4.5) or the overdriven detonation wave (B in Figure 4.5) will both slow down if the piston is stopped. The overdriven detonation will slow down gradually to the steady Chapman-Jouguet detonation where the detonation velocity has its minimum and the flow is sonic.

4.3 Stress Wave Transmission Through a Boundary

As a shock wave is transmitted into the medium surrounding the charge in a drillhole the shock wave amplitude and energy decrease with distance from the hole. For very high shock pressures, the deformation of the material accompanying the one-dimensional shock compression is plastic. But as the shock propagates radially out from the drillhole, the amplitude decreases very quickly and soon reaches the limit for plastic deformation in one-dimensional compression, called the Hugoniot elastic limit (HEL). From then on, the deformation is purely elastic. Such elastic compression waves are called *stress waves*, and they propagate with the velocity of sound. When the stress wave travels into a new medium with a different impedance, a fraction of the energy will be reflected and another fraction will be transmitted. This is, for example, the case when a charge detonates in a drillhole in a high-density iron ore, and the shock wave then travels into the lower density host rock.

The impedance of a medium is given by the product of its density ρ and its sound wave velocity c. Consider an elastic infinite medium through which a plane stress wave passes. The stress induced σ is the product of the density ρ, the sound velocity c, and the particle velocity u

$$\sigma = \rho c u \tag{4.19}$$

In general, if a plane compressive stress wave reaches a boundary which is not parallel to the wave front, four waves are generated. Two of these are reflected waves, moving back into the medium from which the original wave came, a shear wave and a compression or expansion wave; the other two waves, also a shear wave and an expansion or compression wave, are transmitted into the new medium. This general case of wave interaction with a boundary has been treated by Kolsky [1953] and Rinehart [1975].

Let us consider the simpler, special case when the shock wave has a normal incidence to the boundary between medium a and medium b. Then, a wave with stress level σ_R and particle velocity u_R is reflected back into medium a. Another wave is transmitted into medium b which we will assume has density ρ_b. It has stress level σ_T, particle velocity u_T, and shock wave velocity c_b.

According to Equation 4.19, we have

$$\begin{aligned} u_a &= \frac{\sigma_a}{\rho_a c_a} \\ u_R &= \frac{-\sigma_R}{\rho_a c_a} \\ u_T &= \frac{\sigma_T}{\rho_b c_b}. \end{aligned} \tag{4.20}$$

The following conditions must be fulfilled as we assume that the two materials are in contact with each other during the shock wave passage:

$$\sigma_a + \sigma_R = \sigma_T$$
$$u_a + u_R = u_T. \tag{4.21}$$

Combining Equation 4.20 with Equation 4.21 gives the following expressions for the stress levels of the reflected and the transmitted wave:

$$\frac{\sigma_R}{\sigma_a} = \frac{1-\mu}{1+\mu} \tag{4.22}$$

$$\frac{\sigma_T}{\sigma_a} = \frac{2}{1+\mu} \tag{4.23}$$

where μ is equal to the ratio between the impedances

$$\mu = \frac{\rho_a c_a}{\rho_b c_b}. \tag{4.24}$$

From Equations 4.22 and 4.23, we can see what happens if the ratio μ between the impedances varies. If the stress wave travels toward a medium with the same impedance ($\mu = 1$), no reflection occurs ($\sigma_R/\sigma_a = 0$). When a stress wave passes from rock to air ($\rho_a c_a \gg \rho_b c_b$), i.e., when μ is very large, almost no energy is transmitted. If $\rho_a c_a > \rho_b c_b$, i.e., $\mu > 1$, then the reflected compression wave will appear as a tensile wave. Finally, if $\mu < 1$, then the reflected wave is a compression wave.

Figure 4.6 shows the relation between the reflected or transmitted wave and the incoming wave for various impedance matching conditions.

With the help of dynamic photoelasticity, experimental studies can be carried out, in which stress wave interaction with interfaces, wave induced fracturing, and spalling can be investigated. Such experimental studies have been described by Rossmanith and Knasmillner [1983].

4.3.1 Spalling

If the stress wave in rock strikes a free boundary with normal incidence, we will have a totally reflected wave. The P-wave generated from the detonating charge will hit the free surface as a compression wave and will be reflected as a tensile wave, capable of breaking the rock in thin slices if the tensile stress in the reflected wave exceeds the rock material's tensile fracture strength. For regular rock blasting in competent rock, spalling does not generally occur, because the charge is too far away from the free surface, and the P-wave has attenuated too far by the time it reaches the free surface. However, in very weak rock materials, and in stronger rock materials when the charge is close to the free surface (such as is the case in crater blasting) spalling does occur. Multiple spalling, when several parallel slices of rock are broken, occurs when the incoming wave has a stress more than twice the tensile fracture strength of the rock material.

Figure 4.7 shows a simplified case when a triangular-shaped stress wave with a stress level σ_s equal to twice the fracture strength σ_t strikes a free boundary and reflects. When the first slice is formed, $(\sigma > \sigma_t)$, then a new free surface is formed where the remaining compressive wave can reflect. As the tensile stress level once again exceeds the fracture strength, a new slice is formed. In the figure, λ denotes the wave length, t the time ($t = 0$ when the wave hits the boundary), and c is the P-wave velocity.

4.3. Stress Wave Transmission Through a Boundary   99

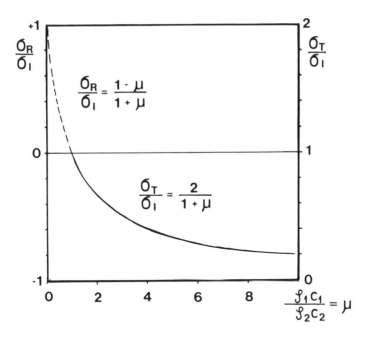

Figure 4.6. Relation between reflected or transmitted wave and the incoming wave as a function of the impedance match μ.

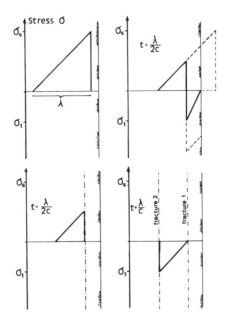

Figure 4.7. Spalling due to the interaction of a triangular stress wave with a free surface.

Table 4.1. Detonation wave and p-wave velocities, densities, and wave impedances for several materials and rock materials (after Hatheway and Kirsch, 1982).

	Detonation velocity D (m/s)	p-wave velocity c (m/s)	Density ρ (kg/m^3)	Impedance 10^6 kg m^{-2}s^{-1}
Explosive:				
ANFO ($d = 3$ in)	3200	—	900	2.9
Emulsion	5200	—	1200	6.2
Dynamite (low velocity)	2500	—	1450	3.6
Dynamite (high velocity)	5500	—	1450	8.0
Rock Material:				
Basalt, dense	—	5560	2761	15.4
Granite, Lithonia, GA	—	2710	2640	7.2
Granite, Valencia County, NM	—	5230	2800	14.6
Hematite, ore, Soudan, MN	—	6280	5070	31.8
Sandstone, Berea, Amherst, OH	—	2640	2182	5.8
Marlstone, Rifle, CO	—	3280	2310	7.6
Limestone, Marly, Rifle, CO	—	2380	2250	5.4
Limestone, Chalky, Pickstown, SD	—	1340	1410	1.9
Concrete	—	4580	2220	10.2

4.3.2 Impedance Matching between Explosive and Rock

It is well known in the rock blasting industry that explosives with a high brisance (or a high shock energy), such as a high density, high detonation velocity dynamite, are well suited for blasting in hard competent rock. A low brisance explosive such an ANFO (which produces a lower shock wave energy, but a relatively high bubble expansion or heave energy) is better suited for soft, porous, or heavily fractured rock. The best matching for optimum shock wave transmission to the rock occurs when the detonation impedance $\rho_a D$ equals the impedance of the rock material $\rho_b c_b$, i.e., when μ for the explosive/rock combination equals one.

Table 4.1 gives some impedances for several explosives and rock materials. Due to seismic anisotropy, it is possible that the same rock material might have considerably different P-wave velocities if the measurements are made across or along the bedding or foliation planes. Oil shale is anisotropic in this way. Many authors have claimed that the impedances of explosive and rock should match for a good fragmentation result. As can be seen from Table 4.1, dynamite may have a low or high detonation velocity. Experience from Swedish limestone and granite quarries indicates that one cannot judge by eye any difference in fragmentation whether the detonation velocity is low or high. However, this may be due to the fact that, in both cases, $\rho_a D < \rho_b c_b$.

There are undoubtedly additional factors other than the impedance match that influence the blasting result. Some of these will be discussed in Chapter 7.

4.4 Ideal and Nonideal Detonation

In the foregoing treatment of detonation, it was assumed that the chemical reaction was completed immediately behind the detonation shock front. Such *ideal* detonations propagate with the well-defined Chapman-Jouguet detonation velocity D_{CJ}. At the detonation front, the shock compression heats the explosive to a high temperature, where the rates of the chemical decomposition reactions are much higher than those encountered in lower temperature decomposition, such as in a fire. However, the detonation reaction rates in real explosives

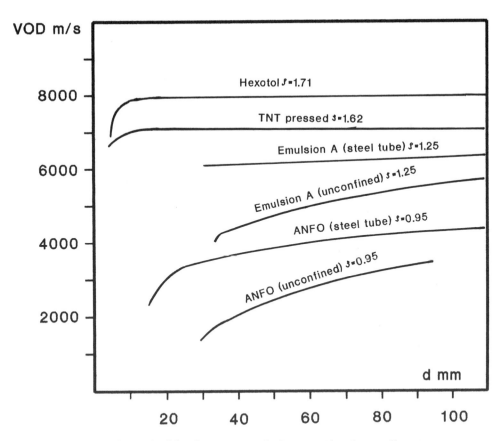

Figure 4.8. Detonation velocities for some explosives vs. the charge diameter. Densities are in kg/l.

are not infinitely high. Therefore, in detonations with limited charge dimensions, the rate of energy release by chemical reaction is counteracted by energy loss due to expansion of the reacting explosive. In such *nonideal* detonations, the detonation velocity may approach, but never reaches, the Chapman-Jouguet detonation velocity.

Many commercial explosives have a reaction zone which extends 10 mm or more behind the front. Military high explosives like HMX and TNT, in contrast, have a very short reaction zone, less than a millimeter. For high density pressed HMX and TNT, the detonation velocity is close to D_{CJ} for charge diameters larger than about 10 mm (Figure 4.8). A watergel/TNT slurry might need a charge diameter of 10 cm and rock confinement in order to detonate at all, and the detonation velocity may be considerably below D_{CJ} even in drillholes of 300 mm diameter in rock.

In a steady-state detonation, the rate of energy added to the reaction products by the chemical reaction balances the rate of energy lost by expansion. Information about increased or decreased energy at any point within the reaction zone is propagated into the surrounding material by pressure waves. In the ideal detonation, the information of all the energy released within the reaction zone will be able to reach and influence the pressure at the detonation shock front and help drive it forward. In a nonideal detonation, some of the energy is released so far behind the shock wave that the corresponding pressure waves never reach the shock front. Therefore, steady-state nonideal detonations have velocities lower than that of the steady-state ideal detonation.

The effect of confinement is to slow down the decrease of pressure and temperature behind the detonation front, thereby increasing the reaction rates. Increased confinement thus has a similar effect to increased charge diameter. Without confinement, which is nearly the case for a small diameter charge in a larger diameter drillhole, the charge diameter may be too small for the detonation to propagate at all, whereas the same explosive may detonate properly when completely filling the borehole. This "coupling" of the detonation to the drillhole wall is a phenomenon well known to the rock blasting practitioner, who refers to the charge in a drillhole completely filled with explosive as being "fully coupled", and calls a charge not filling the hole a "decoupled" charge.

4.4.1 Equation of State of the Explosion Products

In a detonation, the front pressure can reach over 10 GPa and the reaction product temperature may exceed 5000 K, while the density of the reaction products is up to 25% greater than the initial density of the explosive. At these extremely high energy densities, the ideal gas law

$$pv = nRT \tag{4.25}$$

is not a good way to describe the intensive variables p, v, and T for the reaction products.

A simple improvement is to assume that a certain part of the total volume (called the *co-volume*) is occupied by the molecules. This co-volume could be expected to be much more resistant to compression than the rest of the volume. Abel wrote his simple equation of state in the form

$$p(v - \kappa) = nRT \tag{4.25a}$$

where the co-volume κ is a constant.

Cook assumed the co-volume to be a function of temperature and specific volume v

$$p[v - \kappa(T, v)] = nRT. \tag{4.25b}$$

The virial equation of state [see Taylor, 1952], was obtained by adjusting the coefficients in a truncated series expansion of the Abel equation of state. Assuming ideal mixing of the gases, the co-volume can be written as the sum of the individual co-volumes

$$\frac{pv}{nRT} = 1 + \left(\frac{\kappa}{v}\right) + 0.625 \left(\frac{\kappa^2}{v^2}\right) + 0.287 \left(\frac{\kappa^3}{v^3}\right) + 0.193 \left(\frac{\kappa^4}{v^4}\right) \tag{4.26}$$

$$\kappa = \sum_i n_i k_i$$

where n_i is the number of moles of the ith gaseous product, and k_i is its co-volume.

The Becker-Kistiakowski-Wilson (BKW) equation of state [Mader 1979] has found widespread use for describing detonation states. It has the form

$$pv_g = nRT \left(1 + xe^{\beta x}\right)$$

$$x = \frac{\kappa}{v_g (T + \theta)^\alpha} \tag{4.27}$$

$$\kappa = \sum_i n_i k_i$$

where v_g is the volume of gas, n_i is the number of moles in the ith gaseous product, α, β, and θ are empirical constants, and k_i represents their co-volumes. The co-volumes for each gas can be estimated from shock compression data for each individual gas species (CO_2, H_2O, N_2) starting from its condensed state. However, shock compressed states attained in this way have lower temperatures than the detonation states, therefore covolume values must be fitted to give the best agreement with measured detonation data. The BKW equation of state is most reliable at high density states, since the co-volume values were fitted to detonation data for high density military explosives.

The Jacobs-Cowperthwaite-Zwisler (JCZ3) equation of state has the form

$$p = \frac{nRT}{M_0} \widehat{\rho} \Phi \qquad (4.28)$$

where $\widehat{\rho}$ is the initial density, M_0 is the mass of the system including condensed phases, $v = M_0/\widehat{\rho}$, and Φ is the complex imperfection term which is equal to 1 for the ideal gas.

All of the above equations of state were originally calibrated against detonation states, and were carefully fitted to give the best agreement with measured detonation velocities. The two original sets of BKW equation of state parameters, the BKW and BKW-RDX parameter sets, were calibrated to fit detonation data for TNT and RDX, respectively. Later, efforts to fit broader ranges of explosives of different kinds have resulted in the BKW-R and BKW-S parameters, still most accurate in reproducing detonation states.

The Jones-Wilkins-Lee (JWL) equation of state was developed to allow fitting to lower pressure data obtained using the cylinder expansion test (See later in this chapter for a detailed description of this test). The JWL equation has been successfully used to model metal acceleration in a cylindrical geometry, such as for accelerating a projectile shell. It has also been used extensively to model other application of explosives, in the design of explosive devices, and in the modeling of non-ideal detonations. It has to be remembered, however, that the more different the application is from the experiment upon which the calibration was based, the more uncertain is the prediction. Specifically, when using the cylinder expansion test to calibrate the behavior of non-ideal detonations, the equation of state parameters may be inadvertently adjusted to fit data which really depend on the rate of reaction.

Efforts have been made to employ complex theoretical equations of state in which molecular models are involved. So far, it has always been found necessary to calibrate these equations of state with values achieved from detonation experiments. The velocity of detonation can easily be measured for various charge diameters, and the CJ velocity can be estimated through extrapolation to an infinite diameter velocity, as shown in Figure 4.9. CJ pressures and particle velocities can be estimated by measuring the free-surface velocity of an inert material of various thickness in contact with a plane detonation wave. The detonation velocity and the density of the explosive determines the Rayleigh line, and together with the information of CJ pressure and particle velocity, we can determine the CJ volume and the CJ point.

The temperature is very difficult to measure experimentally. CJ temperatures calculated using different calibrated equations of state differ widely from each other.

An equation of state coupled to the hydrodynamic equations is needed for describing temperature, pressure, volume, and thermodynamic equilibrium conditions within the reaction products. Because some of the equations are nonlinear, generally no analytical solutions can be found, and the solutions have to be sought using numerical techniques and computer programs which have been developed to a high level of sophistication. These programs can calculate the detonation product composition during the isentropic expansion from the detonation or explosion state.

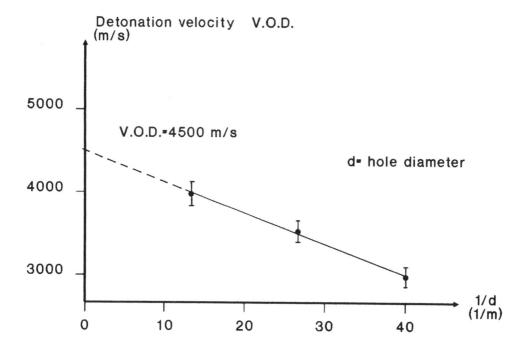

Figure 4.9. Example of how to use experimental detonation velocity measurements to estimate the ideal CJ velocity.

A major obstacle to the correct calculation of detonation reaction product states is the uncertainty about the real reaction product gas composition. Particularly difficult to predict is the balance between free carbon (which can occur in two crystal structures with different energy, graphitic and diamond-like structures), carbon monoxide, and carbon dioxide in oxygen-deficient explosives. TNT, the most frequently used military explosive even today, is particularly difficult to deal with because of its large oxygen deficiency.

Throughout this book, the oxygen balance is expressed (as a mass percentage of the original composition) as the excess of oxygen in a composition above that needed to achieve complete oxidation. It is calculated (see Equation 4.43) with carbon dioxide as the product which is assumed to contain all oxidized carbon – this is correct for explosives near oxygen balance. In treatises on rocket propellants, where larger percentages of aluminum and other high-energy fuels are preferred, and consequently the oxygen balance is shifted toward the negative side, oxygen balance is often calculated with carbon monoxide as the product assumed to contain all oxidized carbon. Either system makes sense in its setting, but the reader should be aware that different systems are used in different treatises and should check which system an author is using.

Commercial explosives are usually close to oxygen-balanced, so that the main detonation products are water, carbon dioxide, and nitrogen. For these materials, the reaction product gas composition, and the explosion energy, calculated with the various equations of state will not differ significantly. In fact, the explosive engineer who is interested in the available explosion energy of a watergel slurry or aluminized ANFO-product can calculate this with pen and paper if the explosive formulation is known. This subject will be discussed in more detail later in this chapter.

Table 4.2. Comparisons between calculated results obtained with three equations of state.

Explosive	TNT BKW	TNT JCZ3	TNT NITRODYNE	NG BKW	NG JCZ3	NG NITRODYNE	ANFO[1] BKW	ANFO[2] JCZ3	ANFO[1] NITRODYNE
ρ_0 (g/cc)	1.64	1.64	1.64	1.59	1.59	1.59	0.9	0.9	0.9
P (Kbar)	206	185	91	247	214	109	74	64	23
D_{CJ} (m/s)	6950	6833		7699	7500		5531	5156	
V (cc/g)	0.451	0.463	0.61	0.464	0.478	0.63	0.811	0.816	1.11
T (K)	2937	3525	3719	3217	4476	4932	2286	2856	2844
Products (moles/kg)									
CO_2	7.30	9.26	5.96	13.20	13.06	13.21	3.82	3.81	3.82
CO	0.83	1.98	3.64	0.02	0.15			0.07	0.02
C_{solid}	22.69	19.56	21.22						
CH_4		0.01							
NO				0.22	0.79			0.02	
N_2	6.60	4.95	6.60	6.50	6.19	6.61	11.81	11.78	11.81
NH_3	0.13	3.30			0.03			0.02	
O_2				1.00	0.81	1.10			
H_2O	10.99	5.91	10.86	11.01	10.96	11.01	27.77	27.72	27.77
H_2		0.12	0.15		0.01			0.06	0.01
Gas volume STP (liter/kg)	579	572	610	716	717	716	973	975	973
Q (MJ/kg)	5.36	5.18	5.11	6.25	6.18	6.27	3.91	3.89	3.91
Weight strength s rel. LFB	1.01	0.98	0.97	1.18	1.17	1.19	0.84	0.84	0.84

[1] ANFO 94.55/5.45   [2] ANFO 94.5/5.5

The available computer programs, such as the BKW code [Mader, 1962] and the TIGER code [Cowperthwaite and Zwisler, 1973], are ideal tools for calculating the available work during the expansion of the detonation products. The programs are useful for understanding ideal and nonideal explosive performance. By comparing the calculated ideal detonation velocity of a given explosive with the measured detonation velocity, we may make predictions as to the degree of non-ideality. A measured detonation velocity that is markedly lower than the calculated ideal value at the same density indicates that part of the reaction occurs too late to influence the detonation velocity.

Table 4.2 gives computer-calculated data for three explosives: TNT, nitroglycerin, and ANFO. These have oxygen balances -73.98%, +3.52%, and -0.04%, respectively. The computer codes used are the BKW code, the TIGER code, and the NITRODYNE code [Holmberg, 1977]. The CJ values are given by the BKW code and the TIGER code. The constant volume explosion values are reported from the NITRODYNE code. Note that the energies differ more for the strongly oxygen-deficient explosive TNT than for the others.

The BKW code [Mader, 1979] is named after the Becker-Kistiakowski-Wilson equation of state which it uses to describe the relationship between pressure, specific volume, and temperature of the reaction products.

The TIGER code is similar to the BKW code, but has several alternative equations of state. In the calculations reported in Table 4.2, the calculations were made using the JCZ3 equation of state. The TIGER code is now also available in a user-friendly version for personal computers running under Windows 3.1, 3.11, or Windows 95. This code is called TIGERWIN or *Tiger for Windows*.

The NITRODYNE code uses a virial type of equation of state [see Taylor, 1952].

4.4.2 Estimation of Detonation Pressure

The symbol γ is used to define the negative of the slope of an isentrope in the $\log p$ vs $\log v$ diagram:

$$\gamma = -\left(\frac{\partial \ln p}{\partial \ln v}\right)_s \\ = -\frac{v}{p}\left(\frac{\partial p}{\partial v}\right)_s. \tag{4.29}$$

At the CJ point, we know that the Rayleigh-line is a tangent to the Hugoniot and the isentrope. The slope of the Rayleigh-line gives

$$\left(\frac{\partial p}{\partial v}\right)_s = \frac{p_1 - p_0}{v_1 - v_0}. \tag{4.30}$$

Using the two equations above,

$$\gamma = \frac{v_1}{p_1}\frac{p_1 - p_0}{v_0 - v_1}. \tag{4.31}$$

For a condensed explosive, $p_0 \ll p_1$. So, if p_0 is neglected, we have

$$\gamma = \frac{v_1}{v_0 - v_1}. \tag{4.32}$$

Combining this with the equation for the Rayleigh-line gives

$$\rho_0^2 D^2 = \frac{p_1}{v_0 - v_1}$$

and we will obtain a most useful relation for estimating detonation pressures:

$$p_{CJ} = \frac{\rho_0 D^2}{\gamma_{CJ} + 1}. \tag{4.33}$$

For engineering calculations with explosives having $\gamma > 1$ g/cm^3, γ_{CJ} is close to 3 (though γ decreases with decreasing density and approaches 1.25–1.4 when the density is considerably below 1). The assumption $\gamma_{CJ} = 3$ gives the well-known expression

$$p_{CJ} = \frac{\rho_0 D^2}{4} \tag{4.33a}$$

which is very useful for estimating the ideal CJ pressure when detonation velocity and density are known from measurements.

Commercial explosives often behave in a nonideal way due to a long reaction zone which sharply influences the detonation velocity. By measuring the detonation velocity for several different charge diameters and plotting the velocities against the reciprocal of the diameter, the ideal velocity for an infinite diameter can be estimated through extrapolation (Figure 4.9). For this example, the D-value to use in Equation 4.33 would be $D = 4500$ m/s. With a density $\rho_0 = 1000$ kg/m^3, the estimated CJ pressure becomes 5.06 GPa.

4.4.3 Estimation of Borehole Pressure

The borehole pressure p_B for a fully-coupled hole can be estimated through

$$p_B = \frac{p_{CJ}}{2}. \tag{4.34}$$

When decoupled charges are used with an air gap (in a smooth blasting operation, for example), the detonation products could expand to a volume several times larger than the initial volume of the charge. This, of course, results in a very low borehole pressure initiating a low stress field, which we can assume is not capable of enlarging the original borehole diameter. If the constant volume explosion pressure is given by the explosive manufacturer, a simple estimation of the borehole pressure can be made with the help of the gamma law

$$p_E v_E^{\gamma_1} = p_B v_B^{\gamma_1} \tag{4.35}$$

where E is the suffix for the constant volume explosion conditions, and B is related to the borehole condition. γ_1 is the ratio between c_p and c_v, and it can be assumed that

$$\gamma_1 = \frac{c_p}{c_v} \approx 1.5. \tag{4.36}$$

Equation 4.36 gives

$$p_B = p_E \left(\frac{d_1}{d}\right)^{2\gamma_1} \tag{4.37}$$

where d is the hole diameter, and d_1 is the charge diameter.
Finally,

$$p_B \approx p_E \left(\frac{d_1}{d}\right)^3. \tag{4.38}$$

The borehole pressure can be calculated more exactly using numerical codes in which proper equations of state for the detonation products and the rock are used.

4.4.4 Numerical Shock Wave and Detonation Data

It has been found from experiments that many materials have a linear relationship between the shock wave front velocity D and the particle velocity u

$$D = C + Su. \tag{4.39}$$

This relationship, coupled with the shock relations, makes it possible to evaluate the state variables for any given shock strength. The parameters C and S for some materials are given in Table 4.3. Shock adiabats for the same materials are given in Figure 4.10.

The shock pressure and internal energy can be considered as consisting of three parts: a cold compression (0 K isotherm) part; a thermal part; and at very high pressures in metals, an electronic excitation part. With $\mu = \frac{\rho}{\rho_0} - 1 = \frac{v_0}{v} - 1$, Al'tshuler et al. [1962] gave the relation

$$p = p_k(\mu) + \rho_0 \Gamma(\mu)(\mu+1)c_v T + \frac{\rho_0 \beta_0}{4}(\mu+1)^{1/2} T^2 \tag{4.40}$$

$$E = E_k(\mu) + c_v T + \frac{\beta_0}{2}(\mu+1)^{-1/2} T^2 \tag{4.41}$$

Table 4.3. Parameters in the Relation $u_s = C + Su_p$.

Material	ρ_0 (g/cm^3)	C (km/sec)	S
Uranium	18.90	2.60	1.45
Palladium	12.00	4.05	1.50
Brass	8.41	3.75	1.45
Iron[a]	7.84	3.80	1.62
95 AL/4.5 Cu	2.79	5.25	1.40
Marble[b]	2.70	4.00	1.32
Granite, Westerley[c]	2.63	2.10	1.63
Limestone, Solenhofen[c]	2.60	3.50	1.43
Plexiglas	1.18	2.75	1.30
Water	1.00	1.70	1.70

For pressures [a] < 400 kbar, [b] < 130 kbar, and [c] < 160 kbar.

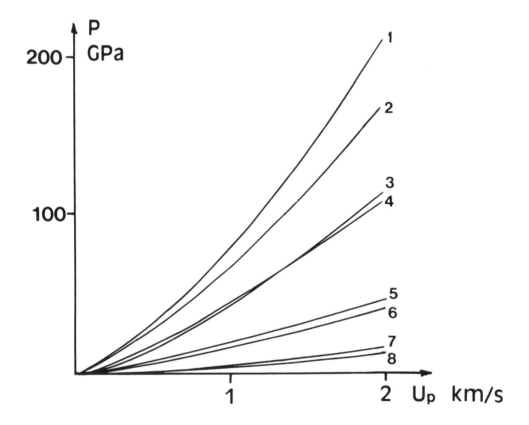

1. Uranium, $\rho_0 = 18.9$ g/cm^3
2. Palladium, $\rho_0 = 12.0$ g/cm^3
3. Brass, $\rho_0 = 8.41$ g/cm^3
4. Fe (0.1 % C), $\rho_0 = 7.84$ g/cm^3
5. 95.5 Al/4.5 Cu, $\rho_0 = 2.79$ g/cm^3
6. Marble, $\rho_0 = 2.70$ g/cm^3
7. Plexiglas, $\rho_0 = 1.18$ g/cm^3
8. H_2O, $\rho_0 = 1$ g/cm^3

Figure 4.10. Experimental shock adiabats. Data for Fe are after Al'tshuler [1965] and for Al, H_2O are after Rice et al. [1958]. Other materials after Persson and Persson [1964 and 1965].

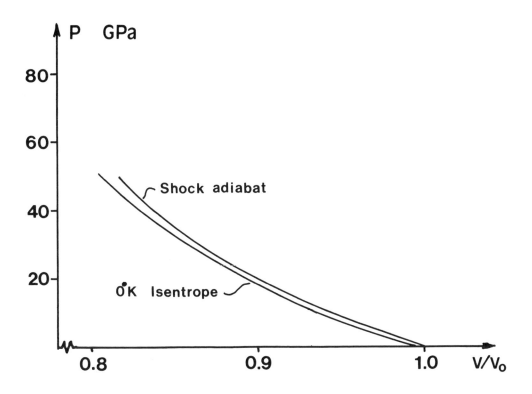

Figure 4.11. The shock adiabat and the 0 K isentrope for copper, showing the small proportion of thermal pressure in the shock adiabat total pressure for metals [after Rice et al., 1958].

where
- $p_k(\mu)$ is the pressure along the 0 K isotherm
- $E_k(\mu)$ ($= \int_0^{p_k} p_k\, dv$) is the internal energy along the 0 K isotherm
- $\Gamma(\mu)$ is the Grüneisen coefficient for the lattice
- ρ_0 the initial density of the metal (density at $T = 0$ K and $p = 0$)
- β_0 is a coefficient in the electronic contribution to the specific heat
- T is the absolute temperature

(Note that β_0 here is different from the coefficient of thermal expansion.)

For metals, the thermal part of the internal energy is insignificant at shock pressures between 100 and 500 kbar (Figure 4.11). This is also the case for hard rock materials. The increase in internal energy is then mainly caused by the compression of the lattice.

For more compressible, organic materials (such as most explosives), the cold compression term is often negligible in the 100 kbar shock pressure region. The expression

$$E_1 - E_0 \approx c_v(T_1 - T_0) \qquad (4.42)$$

is then a good approximation, useful for estimating the shock temperature. For inhomogeneous materials, of course, the expression only gives an average temperature, and the local hot spot temperature can be a great deal higher.

4.4.5 Computing Explosion Energy and Gas Volume

When the composition of an explosive is known, and provided it does not contain ingredients with unknown chemical or thermodynamic properties, the composition, pressure, temperature, and other thermodynamic properties of the reaction products can be computed with some degree of certainty. In such calculations, the basic assumption is that the chemical reactions have reached equilibrium, and that the reaction products all have the same temperature (thermal equilibrium). These conditions are not always fulfilled, particularly not close to the detonation front where the time for completion of the reactions is short. Another case is when solid detonation products, such as Al_2O_3, or inert materials, like SiO_2, occur and do not cool quickly enough. However, in the much longer time-scale of the expansion of the borehole and fracturing of the rock around it, the assumption that thermal equilibrium is achieved is often a good approximation to reality.

When the explosive composition has many different kinds of atoms, the computational problem becomes complex. It then involves the solution of a matrix of a large number of coupled conditional equations that govern the fight between many different reaction product species for the available atoms of different kinds. Such calculations have to be left to the computer, and there are several such programs available. They generally give good agreement with experimentally determined results, with some notable exceptions. One exceptional situation is when there is an excess of carbon in the composition; parallel to the fight for the meager supply of oxygen between hydrogen (to make H_2O) and carbon (to make CO_2 and CO), there is then the added uncertainty of the distribution of the surplus carbon atoms between CO_2, CO, and C (molecular carbon).

Present-day theories are not sufficiently advanced to solve these problems, and we are not able to compute very well the explosion energy and product gas composition of explosives that have a great excess of carbon.

Fortunately, and understandably, however, the greatest energy output is obtained when there is no excess fuel in the explosive composition. Also, for explosives intended for use underground, an oxygen-balanced composition produces the least total amounts of toxic gases such as CO and NO (because the allowable concentration of NO is lower than for CO, in practice, an explosive for underground use should balanced to produce less NO than CO). Most compositions in practical use in rock blasting, therefore, are nearly oxygen-balanced. These give relatively uncomplicated reaction products and are amenable to thermochemical calculations. To illustrate the calculation, we shall consider some simple examples. But first, let us give some useful definitions.

4.4.6 Oxygen Balance

If a single molecular explosive or a composite explosive composition contains just enough oxygen to oxidize all the hydrogen atoms in the composition to water, all the carbon to carbon dioxide, and all other fuel components to their highest degree of oxidation (e.g., aluminum to Al_2O_3), while all nitrogen becomes dinitrogen N_2, then that composition is said to be oxygen balanced, and its oxygen balance is said to be zero. Explosives with more or less oxygen than this are said to be oxygen rich or oxygen deficient, respectively, or to have a positive or negative oxygen balance.

The concept of oxygen balance is particularly useful as a first guideline when formulating explosives to produce a minimum of toxic fumes. An explosive with excess oxygen produces toxic NO and NO_2, an explosive with an oxygen deficiency produces toxic CO. Explosives for use underground with poor ventilation should be formulated to produce a minimum total toxic effect, and the part of the packaging material, paper or plastic wrapper which takes part in the chemical reaction should be included in the composition. It is difficult if not impossible to calculate accurately the production of toxic fumes, because most of the toxic fumes content from real explosives in rock come from regions of the charge that react incompletely, such as at the surface of the charge. Therefore, explosives are often formulated on the basis of field experience from fumes measurements in actual blasting situations.

Mathematically, oxygen balance is given as the ratio between the mass of oxygen which must be added to or removed from the composition to achieve oxygen balance and the mass of the composition. It is positive for oxygen-rich explosives and is expressed as a fraction or percentage of the explosive formula mass, for example grams O_2/100 grams explosive. For an explosive which contains only some or all of the atoms: aluminum, boron, carbon, calcium, chlorine, fluorine, hydrogen, potassium, nitrogen, sodium, and oxygen (with the formula $Al_{al}B_bC_cCa_{ca}Cl_{cl}F_fH_hK_kN_nNa_{na}O_o$), the oxygen balance will be

$$\text{oxygen balance} = \qquad\qquad\qquad\qquad\qquad\qquad\qquad\qquad\qquad (4.43)$$
$$-\frac{32\left(\tfrac{3}{4}al + \tfrac{3}{4}b + 1c + \tfrac{1}{2}ca - \tfrac{1}{4}cl - \tfrac{1}{4}f + \tfrac{1}{4}h + \tfrac{1}{4}k + 0n + \tfrac{1}{4}na - \tfrac{1}{2}o\right)}{\text{explosive molecular weight}} \times 100\,\%.$$

where the indices al, b, c, ca, cl, f, h, k, n, na, and o denote the number of atoms of each element in a mole of the explosive composition. The contribution of nitrogen to the oxygen balance is zero since it does not bind to the other elements. It is assumed that there is enough hydrogen in the formulation to bind chlorine and fluorine to hydrochloric and hydrofluoric acids.

4.4.7 Heat of Combustion

Heat of combustion ($-\Delta H^o_{C,298.15}$) is the heat evolved when we have a complete combustion of a substance with enough oxygen available to form completely reacted products at 298.15 K and constant pressure with all reactants and products in their appropriate standard states. Water is then in its liquid phase.

4.4.8 Explosion Energy

The explosion energy Q_v for an explosive is the energy released in a constant volume explosion. Reference is made to the chemical equilibrium of the decomposition products at the constant volume explosion state. No $p\Delta v$ work is done on the surroundings as the volume is constant before and after the explosion. Assuming we use the ideal gas law, the $p\Delta v$-term can be replaced by ΔnRT, where Δn is the difference of gaseous moles for products and reactants, R is the gas constant (8.3143 J/mole), and T is the temperature (298.15 K),

$$Q_v = -\left[\sum_{i=1}^{k} n_i(\Delta H^o_f)_i - \sum_{j=1}^{\ell} n_j(\Delta H^o_f)_j - \Delta nRT\right] \qquad (4.44)$$

where n_i = number of moles of the ith species of detonation products, n_j = number of moles of the jth explosive ingredient, and ΔH^o_f = standard enthalpy of formation at 298.15 K. Water is in its gaseous phase.

Table 4.4. Heats of formation for some explosives and reaction products at temperature 298.15 K and 1 atm pressure.

Compound	Formula	Formula weight (g/mole)	Heat of Formation at constant pressure (ΔH_f^o) (kJ/mole)	Heat of Formation at constant volume (ΔE) (kJ/mole)
HMX	$C_4H_8N_8O_8$	296.16	+74.9	+104.6
RDX	$C_3H_6N_6O_6$	222.12	+61.5	+83.8
TNT	$C_7H_5N_3O_6$	227.14	−74.5	−57.1
EGDN, Nitroglycol	$C_2H_4N_2O_6$	152.07	−243.5	−228.6
NM, Nitromethane	CH_3NO_2	61.04	−113.0	−105.5
PETN	$C_5H_8N_4O_{12}$	316.15	−530.1	−500.4
TNM, Tetranitromethane	CN_4O_8	196.04	+54.4	+69.3
NG, Nitroglycerin	$C_3H_5N_3O_9$	227.09	−379.9	−358.8
DATB, Diaminotrinitrobenzene	$C_6H_5N_5O_6$	243.14	−154.4	−134.6
AN, Ammonium nitrate	NH_4NO_3	80.04	−365.7	−354.5
Dodecan (fuel oil)	$C_{12}H_{26}$	170.34	−351.0	−318.8
Carbon dioxide	CO_2	44.01	−393.7	−393.7
Carbon monoxide	CO	28.01	−110.5	−111.7
Carbon (s)	C	12.01	0.0	0.0
Oxygen	O_2	32.00	0.0	0.0
Nitrogen	N_2	28.01	0.0	0.0
Water (l)	H_2O	18.02	−285.8	−282.1
Water (g)	H_2O	18.02	−241.8	−240.6
Ammonia	NH_3	17.04	−45.9	−43.4
Aluminum oxide(s)	Al_2O_3	101.96	−1670.0	−1666.3
Methane	CH_4	16.05	−74.9	−72.4

Sources of ΔH_f^o are Brinkley and Gordon [1972]; *Handbook of Chemistry and Physics*, 64th Edition, CRC Press; and *JANAF Thermochemical Tables*.

If the heats of formation at constant volume (ΔE) are known, Q_v can be determined by simply adding the heats of formation for the products and subtracting the heats of formation for the explosive reactants (Table 4.4).

$$Q_v = -\left[\sum_{i=1}^{k} n_i(\Delta E)_i - \sum_{j=1}^{\ell} n_j(\Delta E)_j\right]. \qquad (4.45)$$

4.4.9 Detonation Energy

The detonation energy Q_d is the heat of reaction where reference is made to the chemical equilibrium of the decomposition products at the Chapman-Jouguet point. Water is in its gaseous phase.

$$Q_d = -\left[\sum_{i=1}^{k} n_i(\Delta H_f^o)_i - \sum_{j=1}^{\ell} n_j(\Delta H_f^o)_j\right]. \qquad (4.46)$$

In the literature, $-\Delta H_d$ is used for the heat of detonation and is determined in a bomb calorimeter. ΔH_d is determined calorimetrically when confined explosives are detonated and the detonation products are cooled by a shielded, surrounded water volume. By measuring the temperature increase of this water, the released energy can be measured. However, this $-\Delta H_d$ will most likely not be the same as Q_d since the pressures are not the same. There is a pressure drop in the bomb calorimeter when the gases are cooled and liquid water is formed. This shifts the equilibria and thereby changes the amount of released energy (especially for oxygen-deficient explosives). However, the heat of combustion can be determined calorimetrically in an accurate way as the products formed are stable.

4.4.10 Strength of Explosives

The strength of an explosive is of course related, although not exclusively, to the theoretical available chemical energy in the explosive composition. When we speak of the rock blasting strength of an explosive, we in fact speak of a very complicated relationship, in which not only the chemical energy of the explosive enters, but also the detonation properties, the rates of the chemical reactions, and the shock wave propagation and strength characteristics of the rock material to be blasted. In the absence of a well-defined simple relationship, different explosives manufacturers use different expressions to indicate their explosives' strength. These expressions variously include energy, total energy, theoretical energy, absolute weight strength, absolute strength value, bulk strength, absolute bulk strength, and are not always very well defined in the brochures provided by all manufacturers.

Some manufacturers assume that all of the heat or only part of the heat in the solid products is released during the expansion of the detonation products to atmospheric pressure. The explosive user must be aware of exactly how the theoretical energy has been calculated so he does not compare "pears with apples" when he is trying to rate various explosives.

A *CHNO* explosive, i.e., an explosive composition containing only the atoms carbon, hydrogen, nitrogen, and oxygen, gives as its main reaction products CO_2, H_2O, and N_2. We can allow a small deviation from oxygen balance by allowing small amounts of CO, H_2 or O_2. Approximately then, for excess oxygen,

$$C_x H_y N_z O_w \longrightarrow x\,CO_2 + \frac{y}{2} H_2O + \frac{z}{2} N_2 + \left(\frac{w}{2} - x - \frac{y}{4}\right) O_2. \tag{4.47}$$

Nitroglycerin, for example, gives

$$C_3 H_5 N_3 O_9 \longrightarrow 3\,CO_2 + 2.5\,H_2O + 1.5\,N_2 + 0.25\,O_2. \tag{4.48}$$

We obtain the energy released by the constant volume explosion by adding the heats of formation at constant volume of the reaction products and subtracting the heat of formation at constant volume of the nitroglycerin. These are given in Table 4.4.

$$Q_v = -\left[3(-393.7) + 2.5(-240.6) - (-358.8)\right] = 1423.8 \text{ kJ/mole}$$

or, per kg explosive,

$$Q_v = \left(\frac{1423.8}{227.09}\right) 1000 = 6270 \text{ kJ/kg}.$$

If the enthalpies ΔH_f^o were used for calculation of Q_v, a correction term ΔnRT must be used (see Equation 4.44).

The heat of formation is the amount of thermal energy developed when a gram-molecule of a chemical compound is created from its elements. The heat of formation is negative if the system has given off heat during the reaction of formation (exothermic reaction). The compound and the elements should be given at their respective standard states, that is, the stable state at STP. Oxygen then is a molecular gas, water a liquid, and aluminum a solid.

In the literature, one can find values of the heat of formation (at constant volume or pressure at, for example, 25°C) in tables under the heading "Enthalpy of Formation", $\Delta H^{o}_{f,298.15}$. The heat of formation at constant pressure of water in gas phase at 25°C is $\Delta H_f = -241.826$ kJ; that is,

$$H_2 + \frac{1}{2}O_2 \longrightarrow H_2O + 241.826 \text{ kJ}. \qquad (4.49)$$

To transfer from ΔH^o_f (constant p) to ΔE (constant v), we can make use of either of the following expressions:

$$-\Delta E = -\Delta H^o_{f(p)} - p\Delta v \qquad (4.50)$$

$$-\Delta E = -\Delta H^o_{f(p)} - \Delta n RT \qquad (4.51)$$

where Δv is the change in volume due to the difference Δn in the number of gas molecules between the compound and its elements, R is the gas constant (8.3143 J/(mol K)), and T is the temperature (K). Thus, for the formation of water in gas phase at 25°C, $\Delta n = -0.5$, and

$$-\Delta E = 241,826 - 0.5(298.15)(8.3143) = 240,587 \text{ J}.$$

Table 4.4 gives heats of formation of some common compounds and elements at $T = 298.15$ K.

4.4.11 Simple Calculation of Energy and Gas Volume of Oxygen-Balanced Compositions

For routine calculations of the explosion energy of oxygen-balanced explosives, it is convenient to use a table where the contribution to the explosion energy and the contribution to the gas volume from each component is precalculated.

Table 4.5 gives such values for a few common ingredients. For example, take the oxygen balanced mixture of ammonium nitrate and fuel oil (we may use the hydrocarbon dodecan, $C_{12}H_{26}$, as an approximation of diesel oil).

From Table 4.5, we find that the oxygen-balanced composition (by weight) is

$$\frac{NH_4NO_3}{\text{Dodecan}} = \frac{3.48}{0.2000} = \frac{17.4}{1} = \frac{94.6}{5.4}$$

that is, 54 g oil and 946 g AN for 1 kg explosive.

	AN	+	FO	=	Sum
The contributions of:					
to energy:	946(1.583)	+	54(44.226)	=	3885.72 kJ/kg
to gas volume:	946(0.8400)	+	54(3.288)	=	972.19 l/kg.

Table 4.6 gives some computed and experimental values of explosion energy, gas volume, and weight strength (as defined below in Equation 4.54) for a number of common explosives.

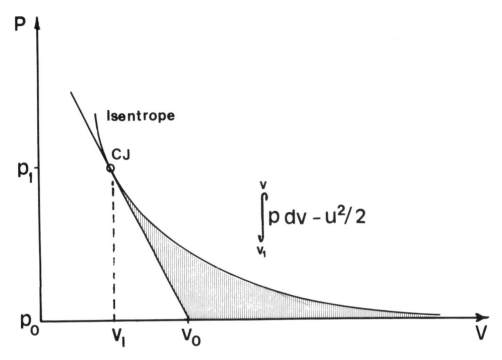

Figure 4.12. Definition of the expansion work.

4.5 Explosive Performance in Rock Blasting

In a well-performing explosive, the chemical energy released should be effectively converted to mechanical work. We call this the expansion work, and it is the work that the released gaseous detonation products exert on the borehole wall. The expansion work W can be defined as

$$W = \int_{v_1}^{v} p\,dv - \frac{u^2}{2} \tag{4.52}$$

where the index 1 indicates the state at the CJ point. (See Figure 4.12.)

4.5.1 Calculated Expansion Work

If the reaction products can be expanded all the way down to atmospheric pressure, the expansion work will almost be equal to the detonation energy Q_d. A slight discrepancy will occur because the temperature of the expanded detonation products at atmospheric pressure will probably not be exactly 298.15 K, and because of the changes in chemical equilibrium and partial retention of nonequilibrium reaction product compositions that are likely to take place as the reaction products expand along the isentrope. The theoretical Q_d is calculated using the equilibrium reaction product composition at the Chapman-Jouguet state.

116 Chapter 4. Shock Waves and Detonations, Explosive Performance

Table 4.5. Data for computing explosion energy and gas volume of oxygen balanced explosives. H_2O is in gas phase.

Substance	Formula	Molecular weight (g/mole)	Oxygen balance g O/g substance	Contribution to energy at oxygen balanced composition (kJ/g)	Contribution to gas volume (STP) at oxygen balanced composition (liter/g)
Oxidizers:					
Ammonium nitrate	NH_4NO_3	80.04	+0.2000	1.583	0.8400
Sodium nitrate	$NaNO_3$	84.99	+0.4706	−1.155	0
Potassium nitrate	KNO_3	101.11	+0.3956	−1.146	0
Calcium nitrate	$Ca(NO_3)_2$	164.09	+0.4876	−1.793	0.1364
Barium nitrate	$Ba(NO_3)_2$	261.35	+0.3061	−1.625	0.0858
Potassium chlorate	$KClO_3$	122.55	+0.3917	+0.4025	0
Sodium chlorate	$NaClO_3$	106.44	+0.4509	+0.6519	0
Potassium perchlorate	$KClO_4$	138.55	+0.4619	+0.0603	0
Ammonium perchlorate	NH_4ClO_4	117.49	+0.3404	+1.456	0.5725
Sodium nitrite	$NaNO_2$	69.00	+0.3478	+1.657	0
Balance substance and fuels:					
Aluminum	Al	26.98	−0.8889	30.880	0
Silicon	Si	28.09	−1.140	30.093	0
Carbon	C	12.01	−2.667	32.791	1.867
Sulphur	S	32.06	−1.000	9.272	0.700
Calcium silicide	$CaSi_2$	96.25	−0.8315	22.582	0
Glycerine	$C_3H_8O_2$	76.10	−1.216	16.151	1.703
Ethylene glycol	$C_2H_6O_2$	62.07	−1.291	17.219	1.806
Starch (cellulose)	$C_6H_{10}O_5$	162.15	−1.185	16.226	1.521
Woodmeal (cork)	$C_7H_{12}O_5$	176.17	−1.364	18.226	1.654
Paraffin	$C_{25}H_{52}$	352.69	−3.45	43.849	3.245
Water (liq-gas)	H_2O	18.02	0	−2.322	1.244
Urea	CH_4ON_2	60.06	−0.7987	9.184	1.490
Diesel oil (Dodecan, FO)	$C_{12}H_{26}$	170.34	−3.48	44.226	3.288
Guargum (Jaguar 100)	$C_6H_{10}O_5$	162.15	−1.185	16.226	1.521
Hydrazine	N_2H_4	32.05	−1.000	16.820	2.10
Hydrazine hydrate	$N_2H_4 \cdot 4H_2O$	104.11	−0.640	12.762	1.79
Sugar (sucrose)	$C_{12}H_{22}O_{11}$	342.31	−1.122	15.146	1.508
Teepol and BP 474			−1.95	21.548	0.777

(continued)

4.5.2 Effective Expansion Work in Rock Blasting

In a bench blasting operation, all this expansion energy cannot be used. As the rock starts to break, cracks open up and the detonation products propagate and escape through them. These events, which depend on the rock properties, will happen long before the lower pressure region has been reached. Typically, this happens for a pressure on the order of 1000 bar and when the products have expanded to 10–20 times the initial volume. At this stage, only approximately 50–70% of the detonation energy Q_d corresponds to the expansion work used for breaking the rock (Figure 4.13). The higher the flame temperature of the explosive, the greater a fraction of the explosion energy remains as thermal energy at this stage and is wasted.

Obviously, all of the maximum available energy in the explosive is not used in rock breaking. A considerable part of the available energy is still there in the reaction

4.5. Explosive Performance in Rock Blasting

Table 4.5. (continued)

Substance	Formula	Molecular weight (g/mole)	Oxygen balance g O/g substance	Contribution to energy at oxygen balanced composition (kJ/g)	Contribution to gas volume (STP) at oxygen balanced composition (liter/g)
Explosives:					
Nitroglycerin (NG)	$C_3H_5O_9N_3$	227.09	+0.0352	6.269	0.6907
Nitroglycol (EGDN)	$C_2H_4O_6N_2$	152.07	0	6.841	0.7368
NG/EGDN 50/50			+0.0176	6.586	0.7137
Diethylene glycol dinitrate	$C_4H_8O_7N_2$	196.12	−0.4081	10.862	1.0288
PETN	$C_5H_8O_{12}N_4$	316.15	−0.1012	7.687	0.7795
Mannitolhexanitrate	$C_6H_8O_{18}N_6$	452.17	+0.0708	8.795	0.6432
Nitrocellulose 12% N			−0.3850	9.753	0.9327
Nitrocellulose 12.5% N			−0.3517	9.485	0.9090
Nitrocellulose 13.0% N			−0.3184	9.213	0.8850
DiTeU	$C_5H_6O_{13}N_8$	386.16	0	6.766	0.6962
Hexogen (RDX)	$C_3H_6O_6N_6$	222.12	−0.2161	8.983	0.9076
TNT	$C_7H_5O_6N_3$	227.14	−0.7398	14.530	1.0850
BNT (78.4% 2.4 and 21.6% 2.6)	$C_7H_6O_4N_2$	182.14	−1.1420	18.887	1.3532
MNT	$C_7H_7O_2N$	137.14	−1.8102	26.163	1.7974
BNT mixture (TNT/BNT/MNT/NB 32/8/55/5)			−1.4056	21.778	1.526
Nitrobenzene (NB)	$C_6H_5O_2N$	123.11	−1.625	24.134	1.638
Tetryl	$C_7H_5O_8N_5$	287.15	−0.4736	11.879	0.936
Monomethylamine nitrate (MMAN)	$CH_6O_3N_2$	94.07	−0.340	8.243	1.270
Nitromethane (NM)	CH_3NO_2	61.04	−0.393	10.586	1.104
Isopropylnitrate	$C_3H_7NO_3$	105.10	−0.990	15.816	1.490
Hydrazine nitrate (HN)	$H_5N_3O_3$	95.06	+0.0840	3.828	0.940
Octogen (HMX)	$C_4H_8N_8O_8$	296.16	−0.2161	8.920	0.9076

Table 4.6. Explosion energy, gas volume, and weight strength s relative to LFB-dynamite and ANFO for some explosives.

Explosive	Q (MJ/kg)	V_g (m³/kg)	Weight Strength s rel. LFB	rel. ANFO
Dynamite I	5.00	0.850	1.00	1.19
Dynamite II	4.42	0.904	0.91	1.08
ANFO	3.89	0.973	0.84	1.00
TNT-Al slurry	4.50	0.700	0.89	1.06
Light slurry	3.44	0.900	0.75	0.89
ANFO 10% Al	5.56	0.800	1.09	1.30
TNT	4.1	0.690	0.82	0.98
RDX	5.54	0.908	1.09	1.30
PETN	6.12	0.780	1.17	1.39
Nitroglycerin	6.27	0.715	1.19	1.42
Nitromethane	6.4	0.723	1.21	1.44

See Equation 4.54 for a definition of weight strength.

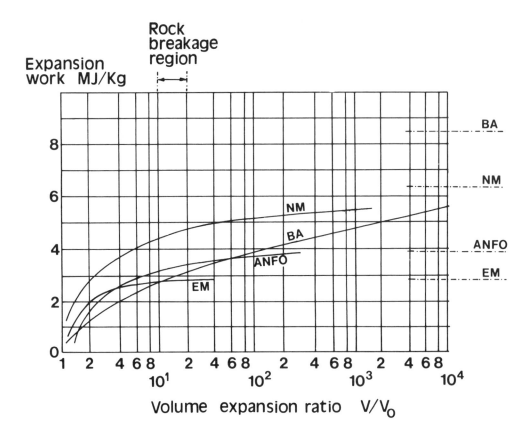

Figure 4.13. Comparison of TIGER BKW-R calculated expansion work as a function of volume expansion ratio for four different explosives: ANFO, an emulsion explosive (EM), nitromethane (NM), and a highly aluminized slurry explosive (BA). The horizontal lines to the right indicate the level of the chemical explosion energy for each of the four explosives. The effective volume expansion ratio for rock blasting is in the region 10 to 20, the lower value for hard rock, the higher for weaker rock materials. In that V/V_0 range, the rock breaks loose and the remaining expansion work is lost through expansion in the air. The end-point of each curve is the point at which the calculated temperature of the reaction products has reached 100°C.

products as heat at the time the gas begins to escape to the atmosphere through opening cracks; and even after expansion to atmospheric pressure, the reaction products may be quite hot, particularly for high-energy explosives. This fact must be considered when explosives are compared on the basis of their detonation energy content. A heavily aluminized ANFO, for example, does not have a breakage performance much higher than ANFO, in spite of having a much higher detonation energy than ANFO. The values of the useful rock blasting expansion work, up to the point when the rock fractures to vent the gas pressure in the drillhole, for ANFO and aluminized ANFO differ less than do the values of the expansion work delivered by the two explosives in the underwater test where expansion proceeds to a very low pressure region. The difference between the useful rock blasting work of ANFO and aluminized ANFO may even be smaller than shown in Figure 4.14 since the solid aluminum oxide Al_2O_3 formed does not necessarily cool quickly enough for the extra thermal energy released in the aluminum reaction to

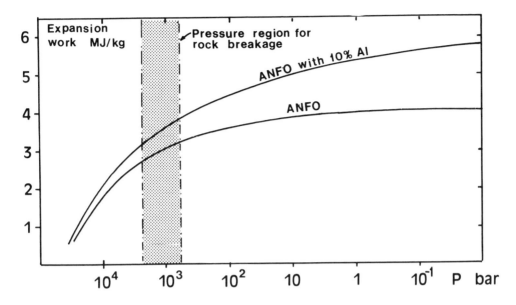

Figure 4.14. Expansion work performed by ANFO with and without Al.

appear in the expansion work. In large diameter charges with large burdens of rock, however, the expansion is certainly slow enough for thermal equilibrium between the fine particulate alumina and the reaction product gas.

It is suggested that in their product specifications, explosive manufacturers should give the following values:

1. Detonation energy, Q_d
2. Shock wave and bubble pulse energy from underwater detonation tests
3. Expansion work performed at $v/v_0 = 10, 15,$ and 20

The expansion work should be given for $v/v_0 = 10, 15,$ and 20 as this can be used for comparing the blasting performance in hard, medium, and weak rock. A standardized equation of state, like the BKW or the JCZ3 equation of state, should be used for calculating the expansion work. It would definitely be easier for the consumer to compare the various explosive products if the above specifications were given. However, in the fierce competition in the marketplace, it may be difficult to get an agreement on standardized specifications. There may also be differences between similar products from different manufacturers, important for the customer's choice, that would not necessarily be reflected in a simple standardized specification.

4.5.3 Early Performance Test Methods

A number of various test methods have been developed over the years since Alfred Nobel's days in repeated efforts to describe and compare the breakage performance of different explosives. In the following, short descriptions of a number of such test methods are given. With the widespread use of blasting agents, which do not detonate in small laboratory sample sizes, the usefulness of several of these small scale laboratory test methods for explosive strength has become very limited. Some of the methods therefore

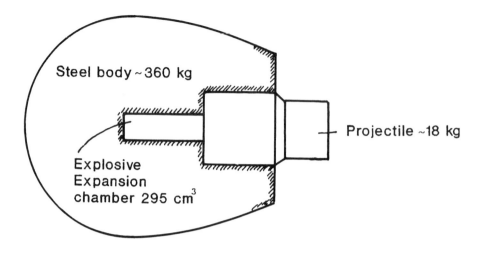

Figure 4.15. Ballistic mortar. The mortar is suspended at the end of a light pendulum rod assembly.

now have mainly historical interest, but have been included here for reference. The following ten tests are described:

1. Ballistic mortar   2. Grade strength   3. Brisance
4. Trauzl lead block test   5. Plate dent test   6. Cylinder test
7. Underwater detonation test   8. Crater test   9. Langefors' weight strength
10. Breaking Index from Underwater Detonation Testing

Ballistic Mortar

A ballistic mortar is made of steel and is provided with a hole where an explosive can be detonated (Figure 4.15). Approximately 10–20 g of explosive can be detonated in a volume of about 20–30 times the explosive volume. A steel projectile with a mass of about 18 kg is accelerated when the detonation products are expanded. The mortar is suspended at the end of a pendulum rod which allows measurement of the deflection due to the reaction force when the projectile is accelerated. This deflection, a measure of the energy, is expressed as the percent of the deflection from a reference explosive.

However, it is not easy to compare the explosive performance for rock blasting by this experiment as the explosive products in the ballistic mortar experiment are allowed to expand inside the explosion chamber to as much as 20–30 times the initial volume before any work is done to accelerate the piston. This is not representative of any real-life rock blasting situation. In normal rock blasting, the explosive usually fills the drillhole volume completely. Even in smooth blasting with charge diameters much smaller than the drillhole situation, for example, the products expand to a maximum of 10 times the initial volume before they exert pressure on the borehole wall. Because of the small test sample size, the ballistic mortar test may not be meaningful for blasting agents. In normal rock blasting, the explosive generally fills the drillhole completely.

4.5. Explosive Performance in Rock Blasting   121

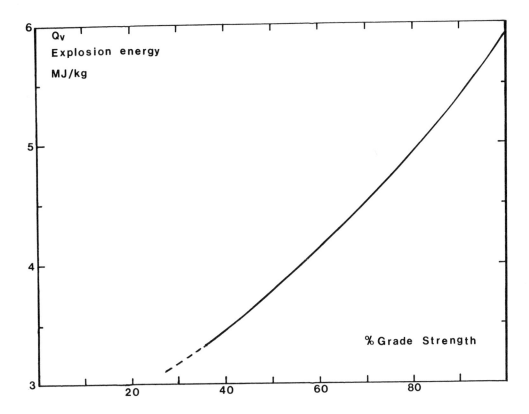

Figure 4.16. Relation between % grade strength and explosion energy.

Grade Strength

The grade strength relation makes it possible to relate the explosive performance for various explosives to ballistic mortar values corresponding to straight dynamites with a certain percentage of nitroglycerin. An explosive giving the same pendulum deflection as a 60% dynamite has a 60% grade strength.

A straight dynamite contains nitroglycerin and an oxygen-balanced oxidizer-fuel mix of sodium nitrate and wood meal. It should be pointed out that the grade strength relation is not a linear function of the calculated explosion energy. Cook [1958] gave a relation between the grade strength and the explosion energy (Figure 4.16). It is likely that, instead of using the ballistic mortar, many manufacturers simply made the translation to grade strength from the calculated explosion energy.

Table 4.7. Brisance values for some explosives [From Feodoroff and Sheffield, 1962].

	Brisance (by Kast) 10^6(Kpm/cm^2)/s	Relative %TNT	Density ρ_0 (g/cm^3)	Deton. velocity D (m/s)
Ammonium nitrate, AN	17	19.7	1.3	2800
Lead azide	74.9	87	1.3	5300
TNT	86.1	100	1.59	6970
Nitroglycerin	145.9	170	1.6	7700
PETN	172.8	200	1.69	8300

Brisance

In the 1920s, Kast introduced the *brisance value* as a measure of the ability of an explosive to fragment or demolish a solid object when fired in direct contact with it. The brisance value B_v was expressed as

$$B_v = f \rho_0 D \tag{4.53}$$

where D is the detonation velocity, ρ_0 is the loading density, and f is the specific pressure calculated from the ideal gas law at the explosion temperature defined by

$$f = p_0 v_0 \frac{T}{273}$$

where p_0 is the atmospheric pressure, and v_0 is the specific volume of gaseous reaction products (cm^3/g) at temperature 273 K and pressure p_0. The explosion temperature T (K) is given by

$$T = \frac{Q_v}{c_v} + 273$$

where c_v is the mean specific heat of the explosion products. In the past, the brisance value was used by explosive manufacturers to estimate the blasting performance.

Brisance is not a reliable measure of bench blasting performance. B_v would be more reliable for predicting the explosive ability in boulder blasting by mudcapping, meaning that the brisance value merely is a description of the detonation pressure.

Table 4.7 shows brisance values for some explosives. The values were obtained from Feodoroff and Sheffield [1962].

Trauzl Lead Block Test

In this test, a lead cylinder (20 cm × 20 cm) is used (Figure 4.17). 10 g of the explosive to be tested is inserted into the 2.5 cm hole, sand stemmed, and detonated. The volume of the cavity formed is determined. This volume is a measure of the expansion work performed by the well-coupled explosive.

Meyer [1977] reports lead block excavation values for several explosives (Table 4.8).

4.5. Explosive Performance in Rock Blasting

Table 4.8. Lead block Excavation values after Meyer [1977]. Charge size is 10 g.

Explosive	Cavity size (cm^3)
Nitroglycol	610
Nitroglycerin	530
PETN	520
TNT	300
Guhr Dynamite	412
ANFO	316

Table 4.9. Plate dent test results [From Smith, 1967]).

Explosive Type	Density (g/cm^3)	Explosive diameter (mm)	Explosive length (mm)	Dent Depth (mm)
HMX	1.730	41.3	203	10.07
PETN	1.665	41.3	203	9.75
RDX	1.537	41.3	203	8.20
TNT	1.629	12.7	25.4	1.93
TNT	1.629	12.7	50.8	1.93
TNT	1.640	41.3	203	6.86

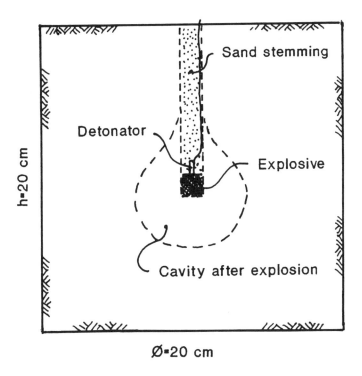

Figure 4.17. Trauzl lead block test.

124 Chapter 4. Shock Waves and Detonations, Explosive Performance

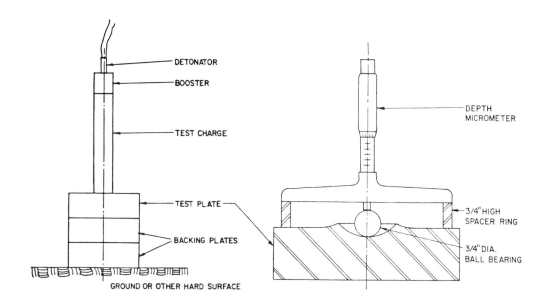

Figure 4.18. Plate dent test configuration [after Smith, 1967].

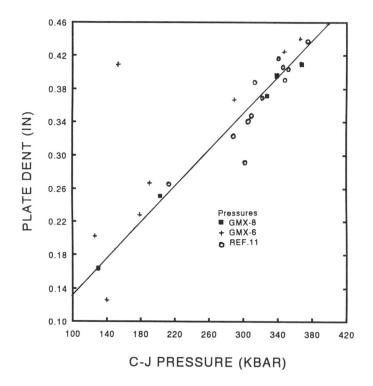

Figure 4.19. Plate dent test data compared to detonation pressure data.

4.5. Explosive Performance in Rock Blasting

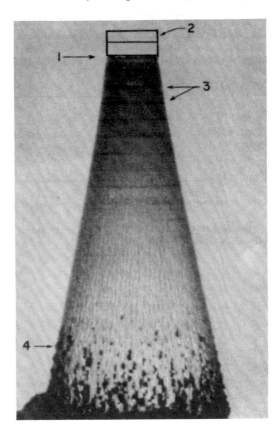

1. Detonation front
2. Undisturbed copper cylinder
3. Markers every 10 mm
4. Reaction products begin to leak out through fractured copper cylinder wall

Figure 4.20. Kerr cell camera picture of a copper cylinder wall being accelerated by the explosive Composition B (RDX/TNT 60/40) at density 1.68 g/cm^3. Exposure time 6×10^{-8}s. Copper cylinder diameter 29.9/25.0 outside/inside mm (photo by G. Bjarnholt).

Plate Dent Test

In the plate dent test, an unconfined charge in contact with a steel witness plate (Figure 4.18) is detonated. When the detonation wave reaches the plate, a dent is produced in the plate; its depth is then measured. For many explosives, there is a linear relationship between dent depth and detonation pressure. Thus, the test can be relevant for grading explosives to be used in mudcapping. Figure 4.19 shows data from the plate dent test plotted against measured detonation pressure data.

Plate dent test results reported by Smith, L.C. [1967] are given in Table 4.9.

Table 4.10. Underwater detonation test data and calculated energies.

Explosive	Charge weight (kg)	Shock energy (MJ/kg)	Bubble energy (MJ/kg)	Detonation[a] pressure (GPa)	Loss[b] factor μ	Exper. expansion work (MJ/kg)	Q_d BKW code (MJ/kg)	Q_v NITRODYNE code (MJ/kg)
EGDN	0.392	1.50	2.77	19.7	2.07	5.99	6.70	6.82
ANFO	0.360	0.92	2.01	2.0	1.25	3.22	3.78	3.89
ANFO	2.30	0.92	2.08	2.8	1.34	3.31	3.78	3.89
ANFO	10.05	1.12	2.43	2.3	1.28	3.86	3.78	3.89
TNT	4.99	0.97	2.11	18.8	2.04	4.09	5.33	4.95
ANFO/Al 90/10	0.356	1.13	2.80	2.0	1.25	4.19	5.45	5.56

[a] Detonation pressure is estimated from $p = \dfrac{\rho D^2}{4}$.

[b] The shock energy loss factor describes the heat loss to surrounding water, $\mu = (Q_d - \text{bubble energy})/(\text{shock energy})$.

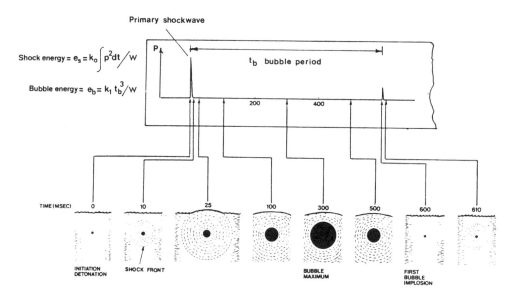

Figure 4.21. Underwater detonation test. Pressure time history from the detonation of a 100 kg charge at a depth of 20 m. The pressure gauge (not shown) is 20 m away from the charge at the same depth as the charge.

Cylinder Expansion Test

This test method was developed at the Lawrence Livermore National Laboratory in order to compare dynamic performance of explosives and to derive empirical equations of state for their detonation products. A copper tube is filled with the tested explosive. When it is detonated, the cylinder wall will accelerate outwards due to the expansion of the detonation products. The cylinder wall acceleration is registered with a high speed camera until the cylinder has expanded to about three times the original diameter (Figure 4.20). The pressure at this point has dropped to some hundred bars. Although the test gives a good measure of the expansion work, when compared to the rock blasting situation, the confinement is low and not at all comparable to the rigid borehole in a bench blast.

4.5. Explosive Performance in Rock Blasting

Underwater Detonation Test

By detonating an explosive underwater at a known depth while measuring the shock wave energy and the bubble pulse energy, it is possible to evaluate the expansion work performed by the explosive. Figure 4.21 shows what happens when pressure vs. time-history is measured by a piezo-electric pressure gauge. After initiation of the 100 kg charge at a depth of 20 m, a shock wave is sent out, and the reaction products start to expand against the surrounding water. After 300 ms, the bubble has expanded to its maximum and starts to implode because the hydrostatic pressure is larger than the pressure inside the bubble. At a time of 600 ms, the first implosion occurs and can be recorded as a shock wave. Now the bubble starts to expand again. The oscillation continues until the bubble reaches the surface and the gaseous products are vented into the atmosphere.

This test has the advantage that the confinement of the charge has a large mass which forces the expansion time to be long. The explosive performance can also be tested at various hydrostatic pressures which helps grade the explosive for its performance in deep water-filled boreholes or in underwater blasting operations. Several explosive manufacturers are using this test continuously for production control of their products.

Still, this method does not give the ultimate solution for estimating the rock breaking capacity of the explosive. Water has a much lower acoustic impedance than rock, and water does not have the mechanical strength of rock.

Methods of calculating the expansion work have been reported by Cole [1948], Bjarnholt and Holmberg [1976], Vestre [1987], and Harries and Beattie [1988].

Table 4.10 gives underwater detonation test data.

Crater Test

The cratering method described in Chapter 11 can also be used for rating the blasting performance of various explosives in a given formation. In the crater test, a given size charge (height equal to six times the hole diameter) is buried at various depths and fired. The achieved crater volumes and critical depth where no breakage occurs are measured. From this data, a Strain Energy Factor can be calculated.

There are some drawbacks with the crater test. They include:

  a. Quite a large number of tests must be performed to achieve the wanted Strain Energy Factor. This is both costly and cumbersome.
  b. Although the explosive detonation products are performing an expansion work toward rock, the explosive geometry is not the same as the one we have in bench blasting or in tunneling (the cut withheld) where two free surfaces are available instead of just one.

When VCR blasting (vertical crater retreat) is performed, the crater test is of course the best available today for judging the explosive performance.

Table 4.11 gives Strain Energy Factors for some explosive-rock combinations.

Langefors' Weight Strength

Different explosives have different explosion energies and produce different amounts of gaseous reaction products. Assuming the explosion energy and the volume of the gaseous reaction products to be the major factors influencing the rock blasting strength of explosives, Langefors made a field experiment study of a number of different explosives

Table 4.11. Strain Energy Factors for Some Explosive-Rock Combinations [after Bauer et al., 1965].

Rock	Explosive	Strain Energy Factor		Optimum depth ratio (Δ_0)	Density (tons per cu. yd.)
		(ft·lbs$^{-1/3}$)	(m·kg$^{-1/3}$)		
Frozen Treat Rock	Slurry	1.8	0.71	0.85	2.3
Frozen Overburden	ANFO	1.95	0.77	0.93	1.5
Frozen Overburden	50% Forcite	2.35	0.93	0.85	1.5
Frozen Yellow Ore	ANFO	2.65	1.05	0.85	2.0
Frozen Blue Ore	Slurry	2.75	1.09	0.83	2.4
Slate	ANFO	3.2	1.27	0.88	1.6
Blue Ore (soft)	ANFO	3.2	1.27	0.90	2.4
Paintrock	ANFO	3.3	1.31	0.90	1.6
Quartzite	ANFO	3.4	1.35	0.80	1.7
Blue Ore (medium)	ANFO	3.45	1.37	0.80	2.4
Iron Formation (decomposed)	ANFO	3.4	1.35	0.70	2.3
Quartzite	50% Forcite	3.5	1.39	0.75	1.7
Blue Ore (hard)	ANFO	3.7	1.47	0.65	2.4
Quartzite (hard)	Slurry	3.7	1.47	0.62	1.7
Greywacke (low phos.)	60% Giant Gelatin	4.0	1.59	0.49	
Granite	60% Giant Gelatin	4.15	1.65	0.55	2.2
Iron Formation	Slurry	4.26	1.69	0.53	2.3
Specularite (bedded)	Slurry	4.3	1.71	0.55	2.7
Magnetite	Slurry	4.3	1.71	0.55	2.7
Magnetite (low phos.)	60% Giant Gelatin	4.35	1.73	0.48	2.7
Magnetite	60% Giant Gelatin	4.60	1.82	0.45	2.53

used in 7 m deep, 32 mm boreholes in bench blasting rounds. The rock was Swedish fissured granite. The breaking capacity was determined relative to the performance of a standard explosive with regard to the capacity of the explosive to make a clean cut-off at the bottom of the hole. Langefors and his co-workers concluded that the influence on the strength of the explosion energy is four to five times greater than that of the gas volume (at STP). They gave the expression for calculating the strength of an explosive s_{calc}

$$s_{calc} = \frac{5e}{6} + \frac{v_g}{6} \qquad (4.54)$$

where $e = \dfrac{Q}{Q_0}$ is the ratio of the explosion energy of the explosive to that of the standard explosive, and $v_g = \dfrac{V_g}{V_{g0}}$ is the ratio of the volume (at STP) of the gaseous reaction products of the explosive to that of the standard explosive, LFB Dynamite. The weight strength determined for an arbitrary explosive with known explosion energy and gas volume using Equation 4.54 has to be used with care. The experiments upon which the equation is founded were made with dynamite-like explosives, most of which had a high density and relatively high explosion energies. It has been used extensively to compare different dynamite type explosives with each other. The equation is not well suited to describe the behavior of low density, low explosion energy explosives such as ANFO or ammonium nitrate water-in-oil emulsion explosives. And it would indicate that Thermite, with $V_g = 0$ would be a powerful rock blasting explosive, which it is not. For completeness, Table 4.12 gives the data of the explosives used and a comparison of the observed with the calculated strength.

The explosive used as a standard both for the observed breaking capacity and for the explosion energy and the gas volume factors used for calculating s was the normal

Table 4.12. Rock-breaking Capacity of Explosives.

Explosive	Observed breaking capacity s_{obs}	Explosion energy ratio e	Gas volume ratio v_g	s_{calc} $\left(\frac{5}{6}e + \frac{1}{6}g\right)$	Number of shots
LFB	1	1	1	1	
LFIV	1.04±0.05	1.06	0.98	1.04	
G-Borenit	0.96±0.07	0.94	0.71	0.90	
Nitrolit	0.80±0.06	0.80	0.98	0.83	
304[1]	1.30±0.07	1.52	0.36	1.33	
LFB	1	1	1	1	15
G.D.	1.23	1.35	0.84	1.27	2
Securit	0.93±0.03	0.85	1.01	0.88	4
Nabit	0.88±0.04	0.90	0.77	0.88	3
Nabit 2	0.91	0.89	1.02	0.91	1
Amatol	ca 0.90	0.86	1.05	0.88	2
Na 01	ca 0.88	0.88	0.70	0.85	1
Na 12	1.00±0.05	1.02	0.81	0.99	2
Imatrex	(0.62±0.02)	0.98	0.37	0.89	8

[1] 304 is a high energy aluminized explosive with a very small gas volume.

plastic dynamite used in Sweden about 1952, the LFB Dynamite. It had an explosion energy and a gas volume (STP) of

$$Q_0 = 5.00 \text{ MJ/kg}.$$

$$V_{g0} = 0.850 \text{ m}^3/\text{kg}$$

Nowadays, weight strengths are better referred to a more universally-known explosive such as ANFO, the s_{calc} of which is 0.84 relative to LFB. To find the strength of an explosive relative to ANFO, we make use of the expression

$$s_{calc}(\text{ANFO}) = \frac{1}{0.84}\left(\frac{5e}{6} + \frac{v_g}{6}\right) \qquad (4.55)$$

where e and v_g must still be related to the Q_0 and V_{g0} of LFB Dynamite.

Although originally based on limited experimental material, the method of calculating weight strength from explosion energy and gas volume has been used extensively on a wide range of explosives under different blasting conditions. We have found that it grades different nitroglycerin-based explosives relative to each other in good agreement with the practical experience of their relative capacity of breaking rock. Because it does not take the shock wave loss effects into account, it tends to overrate high-density, high-energy explosives when they are loaded to completely fill the borehole. It also underrates the explosives used in the largest quantities in modern rock blasting, ANFO, heavy ANFO, and emulsion explosives. These explosives have a low flame temperature and therefore expend their mechanical work early on in the volume expansion, before the reaction products vent to the atmosphere. They therefore have a higher efficiency in rock blasting than the higher-temperature nitroglycerin explosives (this appears to compensate for the lower explosion energy which is much lower for ANFO and emulsions than for most nitroglycerin explosives).

Table 4.13. Comparison of calculated weight strength values s_{calc} with "breaking index" values B_i derived from experimental shock wave and bubble energies from the underwater explosion test.

Explosive	Density (kg/m³)	Detonation Velocity (m/sec)	s_{calc} (rel. to ANFO)	B_i (rel. to ANFO)
Slurry 1	1560	4600	1.06	1.01
Slurry 2	1330	4600	0.95	0.89
Slurry 3	1250	4200	0.93	0.84
Slurry 4	1250	4200	0.89	0.81
ANFO	920	4500	1.00	1.00
ANFO 10% Al	920	4400	1.30	1.32

Breaking Index from Underwater Detonation Testing

Bjarnholt [1977] has suggested using the data obtained from the underwater detonation test to describe the relative strength of explosives. The major part of the useful rock fragmentation work is done in the high-pressure region of the expansion process, which has a greater influence on the shock wave energy than on the gas bubble oscillation energy. Bjarnholt therefore weighted the bubble energy e_b with a factor 0.2 and the shock energy e_s with a factor 1.0.

He proposed a "breaking index" B given by the expression

$$B = \frac{e_s + 0.2\, e_b}{(e_s + 0.2\, e_b)_{\text{ANFO}}} \tag{4.56}$$

where e_s is the shock wave energy per unit weight of explosive, and e_b is the bubble energy per unit weight of explosive.

The "breaking index" gives results in reasonable agreement with the Langefors weight strength as seen from Table 4.13, but we have as yet too little experience to know whether it is an improvement over the Langefors weight strength or not.

4.5.4 The New Explosive Weight Strength Concept: A Combination of Explosive and Rock Properties

The first step in a charge calculation for engineering purposes is to describe the rock blasting strength of the explosive to be used. This is best done by comparing the rock blasting performance of the explosive to be used with that of a standard explosive. The rock blasting performance of the standard explosive in the specific rock mass to be blasted must be known. The performance may be defined as the mass of explosive required per unit volume of solid rock to obtain a specified end result of the blast, such as a given fragment size distribution when blasting using a specific drillhole diameter in a specific drill pattern and with a specific initiation sequence.

Because of the many variables that influence the end result of the blast, determining the performance of any explosive is no easy task. We will approach this task here by determining the fraction of the explosive's available explosion energy which is transformed into useful fragmentation and acceleration of the blasted rock. This requires that we be able to determine the energy losses in blasting. Let us first discuss what these energy losses are.

When an explosive charge is detonated in a drillhole in rock, the expanding reaction products deform and crush the rock material close to the drillhole — the more energy is expended in this way, the less energy there is left for the subsequent fracturing and acceleration of the major part of the blasted rock. The magnitude of this loss will depend mainly on the packing density of the explosive, the detonation pressure, and the high-pressure expansion characteristics of the explosive reaction products. However, the energy loss is also a function of the shock transmission and deformation properties of the surrounding rock material and the fracture toughness of the rock mass further away from the hole. The rock blasting strength of a given explosive thus depends not only on the properties of that particular explosive, but also on the properties of the specific rock material or rock mass that surrounds the charge. For many commercial explosives in the range of drillhole diameters used in present-day rock blasting, the chemical reactions that take place in the detonation are not instantaneous, but take place during the time the explosive does work on the rock material surrounding the drillhole. The stronger the surrounding rock is, and the higher its density, the higher is the detonation velocity. Also, for such relatively slowly reacting explosives, the detonation velocity and pressure increase with increasing charge diameter. Therefore, as an added complication, we have to take into account the fact that the gas pressure inside the drillhole in a given rock mass is greater for a larger drillhole diameter. For a given explosive and drillhole diameter, the pressures are greater, the more resistance the rock mass offers to hole expansion.

Given the use of high-power computers, it is now possible to do a complete computer calculation of the entire process of detonation of the explosive (the expansion of the reaction product gases, and the shock propagation, fracturing, acceleration, and throw of the rock mass) while taking all these factors into account. Such an approach is useful for research purposes but would be impractical for the practicing mining engineer. Several major explosives companies, and some major mining and building construction companies, have the ability to carry out such calculations, and are in the process of calibrating their main explosives products against experiment and field production blasts in different rock materials. The computer codes used are closely guarded and proprietary to each company. They are used as instruments of competition to advise the customer and user of explosives as to the most profitable explosive or the most profitable blasting technique to use in different applications. In the following, we will describe a simpler approach that provides a realistic measure of the strength of the explosive without the use of extensive experimentation and massive computer calculations.

In the past, the rock blasting weight strength of an explosive was considered to be only a function of the explosive itself. Langefors' weight strength s concept is an example of this. For charge calculations where we wish to examine the effect of limited changes in drillhole patterns, changes in the charge weight or drillhole length (while keeping to similar types of rock, within a limited range of similar explosives and within a limited range of drillhole diameters) using the simple weight strength s will give a completely satisfactory answer. Most of the routine charge calculations required in an ongoing mining or quarrying operation are of this kind.

However, we may be faced with moving to a completely and radically different set of conditions, such as when considering a replacement of small-diameter holes with large-diameter holes, for example replacing dynamite in 50 mm diameter drillholes with ANFO or heavy ANFO in 300 mm diameter holes, or when bidding for a construction job in hard rock when our previous experience has been gained entirely from blasting in soft sedimentary materials. Then, there is a need for more elaborate charge calculations, or for acquiring the experience of others to get a starting point for new, simplified charge calculations within the range of these new conditions.

To provide a background for the development of more elaborate charge calculations, in the following we will outline the requirements and the technique for developing a new weight strength number. We will take into account the computed expansion work at (a) the ideal, maximum detonation velocity, (b) the actual, lower detonation velocity at which the explosive detonates in the drillhole and the corresponding reactive expansion work, and (c) the rock material's properties. We will use the same symbol s for the strength of the explosive defined in this new way as was used in the past by Langefors because the new weight strength number will replace Langefors' weight strength number in the set of equations originally developed by Langefors and Kihlström.

4.5.5 Computed Expansion Work

For calculating the explosive's available expansion work as a function of volume expansion ratio at ideal, or CJ detonation conditions, we select the TIGER equilibrium thermochemical computer code, developed by SRI International for Lawrence Livermore National Laboratory and the US Army Armament Research Development and Engineering Center (ARDEC). The compiled (executable) Fortran TIGER code is available in an IBM PC-compatible version. Although the distribution of this code is somewhat restricted by the US traffic in arms regulations, it is available in most parts of the world, and several consulting companies and laboratories will provide calculations on a fee basis. We standardize the equation of state used in the TIGER code to the BKW equation of state, with parameters fitting RDX, or the slightly different set of parameters called readjusted parameters (BKW-R). An alternative to the TIGER code is the BKW code, a public domain predecessor to the TIGER code, which, if the same parameters in the BKW equation of state are used, will give identical results to those of the TIGER code. The TIGER code calculates, based on a weight percent listing of the ingredients of the explosive and a proposed listing of possible reaction products, the Chapman-Jouguet detonation data, i.e., detonation velocity, pressure, density, and internal energy of the explosive's reaction products at the CJ point. It also provides, optionally, the expansion isentrope states of the reaction products as they expand adiabatically from the CJ point.

The basic assumptions made in these equilibrium thermochemical code calculations are that the chemical reactions reach their chemical equilibrium product composition at the sonic point of the plane, steady, unsupported detonation and that the composition does not change during the subsequent expansion. This assumption has two major drawbacks. First, it results in a detonation velocity which is independent of the drillhole diameter. This is obviously incorrect. We know, for example, that the detonation velocity of crushed prills ANFO in a 53 mm diameter drillhole in rock is no more than half of the calculated ideal value, which in turn is in good agreement with the measured detonation velocity in 300 mm diameter drillholes in rock (see, for example, Figure 3.1). Second, as the temperature in the reaction products decreases during expansion, it is apparent that the rate at which the chemical equilibrium is reached decreases so that reaction product compositions characteristic of a higher temperature are "frozen" at their high-pressure, high-temperature mix which, from then on, stays constant even at much lower temperatures.

To get around these difficulties, we will resort to computer calculations where we take into account the actual rate of energy release in the detonation reactions. Such calculations can be made using a variety of computer codes called hydrodynamic codes or hydrocodes. The term hydrocodes comes from their early use to model the detonation behavior and effects of military high explosives and warheads. There, the dynamic pressures in the materials involved in the high-velocity impact or detonation and in the shock waves generated are so much higher than the strength of the materials that it was first believed strength effects could be neglected. The flow of a solid material under these conditions could then be treated as a the hydrodynamic flow of a liquid. As has been found later, the strength effects in metals are seldom quite negligible. In most rock blasting calculations, the strength of the rock material does play a very important part. Thus, the use of the term *hydrocodes* for codes used to calculate acceleration of metal or rock blasting is not strictly correct at all, but the term is now so well-stablished that we may have to live with it.

In the hydrocode calculation, we will need to make assumptions and assume we know the answers to several questions which are really difficult to answer, such as:

1. While the explosive material reacts, or burns internally, is the so-far unreacted part of the material at the same temperature as that of the reaction products, or is the main part of the thermal energy mostly in the reaction products?
2. What is the equation of state, i.e., the relationship between pressure, density, and internal energy (or temperature) of the unreacted fraction of the reacting explosive?
3. What is the equation of state of the rock mass surrounding the charge?
4. What is the dynamic strength of the rock mass surrounding the charge under the complicated conditions of triaxial compressive stresses set up by the expansion of the high-pressure gas in the drillhole?
5. How do we take into account the fracture toughness of the rock mass, and the difference in strength between the joints and the blocks of rock defined by the joint structure?
6. At what point in time, or at what pressure or volume expansion ratio during the expansion of the reaction products from the detonation state does the effective rock fragmentation process end?

After having made assumptions in response to these questions, we can now calculate the detonation of an extended charge of given charge diameter in a given rock material. We can check these calculations against actual measurements of the detonation velocity of our chosen explosive in drillholes of the given diameter in the given rock material. The calculation will also provide the pressure as a function of specific volume of the reacting explosive as it expands while reacting. To distinguish this expansion pressure, specific volume curve from an ideal expansion isentrope where no reaction occurs, we may call it the reactive expansion P, v curve. From the expansion P, v curve, we can integrate the total work done by a unit mass of the explosive reaction products out to our assumed end point of effective rock blasting work. From this, we deduct the increase in the internal energy in the rock material in the near region surrounding the charge, which has been heated by the passage of the shock wave and subsequent deformation heating; this internal energy represents a loss energy, of no great use for the rock fragmentation. The resulting effective rock blasting energy or work is then divided by the effective rock blasting work of a standard explosive, preferably ANFO, detonating in a very large-diameter drillhole in a standard rock material. The resulting ratio we now call the *new rock blasting weight strength value s*, and we may use it in the same way we will describe in Chapter 6 for using the Langefors weight strength value s.

In the following, we will describe one detail in these calculations, namely the way in which the rate of chemical reaction in detonation influences the detonation velocity and pressure when the charge diameter is changed.

4.5.6 Rock Blasting Performance Value s from the Bdzil-Lee Detonation Shock Dynamics Evaluation

The main feature of any method for deriving a value of the rock blasting performance of an explosive is its ability to take into account the expansion of the reaction products during the ongoing chemical reaction of the explosive. The first to attempt such a treatment were Kirby and Leiper [1985] who applied the approximate detonation theory developed by Wood and Kirkwood [1954] to the steady-state detonation along the axis of a stick charge in order to calibrate a reaction rate model for industrial explosives. Their reaction rate model was based on an assumed physical model for the chemical reaction in detonating composite explosives. It contained three separate terms designed to represent, respectively, the initial hot spot reaction, the burning of a liquid component, and the burning of grains of a solid component surrounded by the liquid.

Later, Leiper and Cooper [1990] expanded the Kirby and Leiper reaction rate model to include three separate, additive terms in the burning rate equation: one term to describe a rapid reaction ignition phase, a second term to describe the slower burn of a liquid component, and a third term to describe the grain-burning of a solid granular component mixed into the liquid.

Recently, Kennedy and Jones [1993] carried out further refinements. Their new burning rate equation has the form

$$\frac{d\lambda}{dt} = (1 - \lambda)\left(\frac{P_h a_h}{\tau_h} + \frac{P a_i}{\tau_i} + \frac{P a_f}{\tau_f}\right) \tag{4.57}$$

where

$$P_h = \begin{cases} \dfrac{P}{4}\left(\dfrac{3P}{4P_c}\right)^3 & \text{for } P < \dfrac{4P_c}{3} \\ P - P_c & \text{for } P \geq \dfrac{4P_c}{3} \end{cases} \tag{4.58}$$

where λ is the extent of reaction, varying from 0 for the unreacted explosive to 1 for the detonation reaction products. P is the local pressure in the reacting explosive, the a factors in Equation 4.57 describe the assumed geometry of the burn front, controlling the burn of the hotspot, the liquid, and the granular component, and the τ are the characteristic reaction times for the three components. The a factors are Gaussian functions of λ. The subscripts are: h to indicate hotspot, i for liquid (intermediate reaction), and f for the solid (final reaction) component.

Kennedy and Jones, in an excellent way, have visualized how the burning rate varies with P and λ by plotting $d\lambda/dt$ as contour lines for different values of $d\lambda/dt$ in the λ vs. P plane, showing the reaction rate surface $\dfrac{d\lambda}{dt}(P,\lambda)$ which is characteristic for each explosives. Figure 4.22 shows the reaction rate surface for a proprietary blend of an AN/water-in-oil emulsion and ANFO, denoted heavy ANFO. The locus of the values of $d\lambda/dt$, λ, and P for the sonic points in detonations with different velocities (different charge diameters and different front pressures) were plotted in the diagram as a dot-dash line on the reaction rate surface. Also plotted are the two loci of the values of $d\lambda/dt$, λ, and P along a flow line (expansion line) through the detonation reaction zone for two charge diameters, a dotted line for a 51 mm diameter charge, and a dashed line for a 250 mm diameter charge.

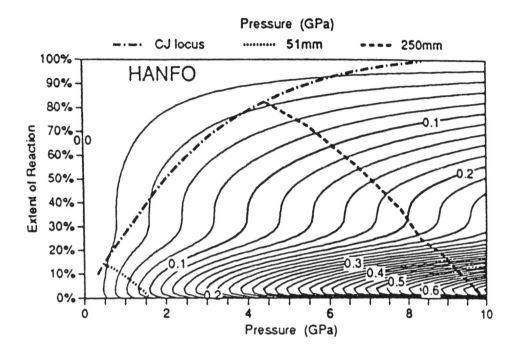

Figure 4.22. The reaction rate surface for a heavy ANFO explosive. The diagram shows contour lines of equal reaction rate as a function of the reaction extent λ and the local pressure p. The dot-dash line is the locus of sonic points of detonations with different velocity, the dotted line is the expansion line through the reaction zone on the axis of a 51 mm diameter stick charge, and the dashed line is the expansion line on the axis of a 250 mm diameter charge.

Kennedy showed that burning rate equations of the type shown in Equation 4.57, calibrated using experimental detonation data for steady state detonations in stick charges of different charge diameters, can be used successfully to model time-dependent (not steady-state) shock initiation and detonation states. This makes the burning rate equation extremely useful for predicting performance of explosives.

Similar results can be obtained by using the explicit Detonation Shock Dynamics (DSD) theory [Bdzil and Stewart 1989] for steady-state (constant axial velocity) detonation in a cylindrical charge, as described by Lee [1990] and by Lee, Bdzil, and Persson [1993]. The DSD theory replaces the Wood and Kirkwood detonation theory in the hydrodynamic code calculation in the above described technique. Lee [1990] used Bdzil's idea of the DSD theory and its assumption that there is a unique, single-valued relationship between the local detonation velocity D_n (in a direction normal to a curved detonation front) and the curvature κ of the front. With this asssumption, and assuming a specific form for the $D_n(\kappa)$ relationship, Lee could determine the parameters in that relationship for a given explosive using the recorded shape of the detonation front and the experimental axial detonation velocity, both as a function of charge diameter. He developed and used the relation:

$$\frac{D_n}{D_{CJ}} = 1 - \left(\alpha \kappa^{1/\nu} + \beta \kappa e^{\theta(D_{CJ}-D_n)}\right) \tag{4.59}$$

where D_{CJ} is the CJ detonation velocity, and α, β, and θ are parameters. From these same data, Lee could also determine the parameters in a general burning rate equation of the form

$$\frac{d\lambda}{dt} = k\,(1-\lambda)^\nu \exp\left(\frac{-A}{\sqrt{P/\rho^\tau}}\right). \tag{4.60}$$

In Equation 4.60, λ is the fraction burned, t is the time, P is the local pressure, ρ is the local density in the reacting material, and k, ν, A, and τ are parameters. For the emulsion explosive at density 1.25 g/cm^3, the parameters were found to have the following values:

In Equation 4.59	In Equation 4.60
$D_{CJ} = 6.788$ mm/μs	$k = 20 \times 10^6$ μsec^{-1}
$\alpha\ \ = 1.629$ mm$^{1/\nu}$	$A = 8.312$ (for P in GPa and ρ in g/cm^3)
$\nu\ \ = 1.832$	$\nu = 1.889$
$\beta\ \ = 0.0865$ mm	$\tau = 0.8418$
$\theta\ \ = 1.127$ μs/mm	

Please note that parameter values in above list are not in standard SI metric units.

(In the above descriptions of methods for evaluating the progressive reaction of the explosive in detonation, we have used the term "burning rate" for $d\lambda/dt$ and the term "fraction burned" for λ. In fact, the value of λ obtained from such evaluations really indicates how large a fraction of the calculated total explosion energy has been added to the internal energy of the mix of explosive and reaction products. The experiments used to determine the parameters in Equations 4.57 and 4.60 only allow us to find out how the internal energy of the mix changes as a result of the "burning" or chemical reaction of the explosive. They reveal nothing *a priori* of the actual chemical composition of the reacting explosive/reaction product mix. Questions as to how the explosive is being consumed are not answered: Is the explosive consumed in detonation by grain burning which would leave part of the explosive in its original form, or by a more or less homogeneous decomposition reaction which gradually changes the chemical composition of the greater part of the explosive from explosive to final reaction products? Such questions can be answered only gradually by extensive experimentation in which the explosive's composition and such properties as the ingredients' grain size are systematically changed. It would therefore be preferable to use the term *energy release rate* for $d\lambda/dt$ and the term *fraction of explosion energy released* for λ rather than the terms burning rate and fraction burned.)

Figure 4.23 shows for an ammonium nitrate-based emulsion explosive a comparison of the calculated detonation velocity as a function of inverse charge diameter obtained by Lee using Equations 4.59 and 4.60 and the experimental detonation velocities to which the parameter values in these equations were fitted. The turn-back of the detonation velocity at the lowest detonation velocities as indicated by the calculation is not expected to be real.

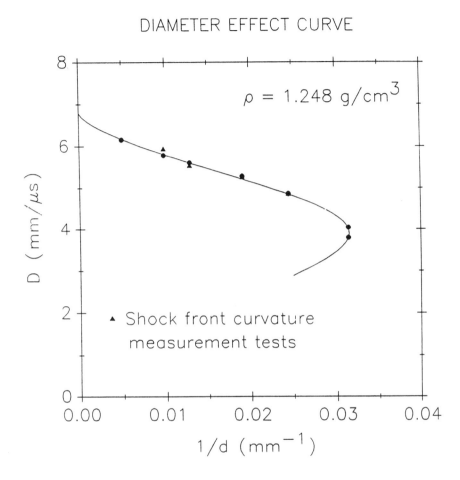

Figure 4.23. Fit between experimental and theoretical detonation velocities as a function of charge diameter for an ammonium nitrate-based emulsion explosive at density 1.25 g/cm^3 [Lee, 1990].

Figure 4.24 shows three calculated pressure profiles in the reaction zone for the ammonium nitrate based emulsion explosive at density 1.25 g/cm^3, corresponding to detonation velocities 6.7, 5.5, and 4 km/s. The dotted lines marked 25, 50, 75, and 90 % indicate the position of the points in the reaction zone where the corresponding percent of the chemical energy release reaction is completed. The dot-dash line is the locus of the sonic point. It is interesting to note that according to these calculations, the sonic point comes closer to the shock front the lower the detonation velocity is, while the point at which the chemical reaction is completed falls further and further behind.

Figure 4.25 shows, in the pressure vs specific volume plane, the shock Hugoniot for an unreacted ammonium nitrate based emulsion at density 1.14 g/cm^3, but otherwise similar to the emulsion in Figure 4.24, together with the Rayleigh line for the

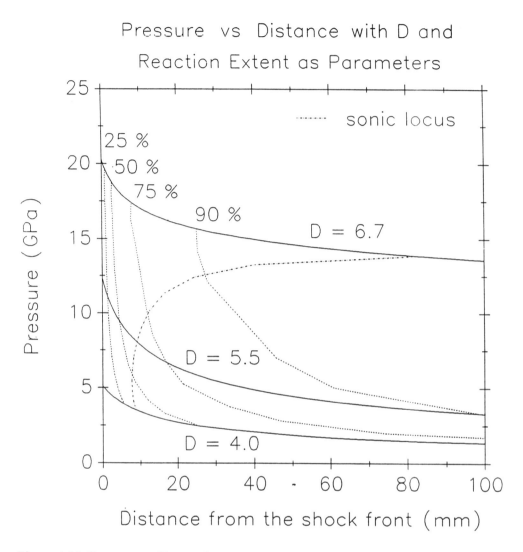

Figure 4.24. Pressure profiles on the charge axis within the reaction zone for three constant detonation velocities, 6.7, 5.5, and 4 km/s in stick charges of the ammonium nitrate-based emulsion explosive at density 1.25 g/cm³. The dotted lines marked 25, 50, 75, and 90% indicate the position of the points in the reaction zone where the corresponding percent of the chemical energy release reaction is completed. The dot-dash line is the locus of the sonic point [Lee, 1990].

Chapman-Jouguet (ideal, infinite charge diameter) detonation of velocity 6.2 km/s and the corresponding complete reaction product Hugoniot curve (full line). The dotted lines are the incompletely reacted Hugoniot curves for the three detonation velocities 6, 5, and 4 km/s. The crosses indicate the position of the sonic point on each expansion line. Recalling the discussion of the Chapman-Jouguet detonation (Figures 4.4 and 4.5), we can now see from Figure 4.25 how the front pressure of steady detonations in stick charges of different diameters decreases with decreasing detonation velocity as the charge diameter decreases.

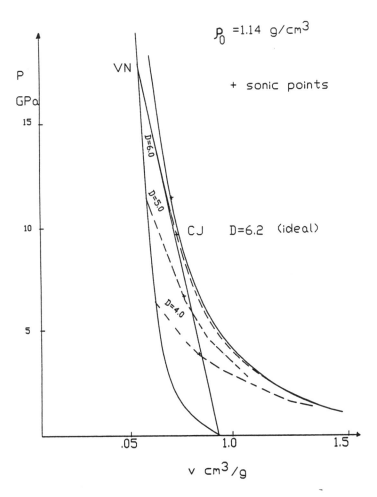

Figure 4.25. Unreacted and fully reacted ideal Hugoniot curves (full lines) for different constant velocity detonations in stick charges of an ammonium nitrate-based emulsion explosive at density 1.14 g/cm³. The ideal Chapman-Jouguet detonation velocity is 6.2 km/s. Also shown are three expansion lines for partially reacted material, corresponding to the three detonation velocities 6, 5, and 4 km/s. The sonic points on these Hugoniot curves are indicated with a cross.

We can couple these results with a model for the shock wave propagation and crack formation in different rock materials. Using experimental data for the detonation velocity of the explosive in drillholes of different diameters in different rock materials, we can utilize the locus of pressures and specific volumes during expansion (the *expansion line*) for the actual detonation velocity in a specific drillhole in a specific rock mass to determine the effective rock blasting energy (the integral under the expansion line out to the volume where effective rock fragmentation ends). We can express this energy as a fraction of the effective rock blasting energy of the standard explosive detonating ideally at its Chapman-Jouguet state. This gives us a value for the "new weight strength s" of the test explosive, detonating in a drillhole of specified diameter in the specific rock material. How this number will be used for charge calculations for practical rock blasting situations is described in Chapter 6.

Table 4.14. Properties of explosives from Nitro Nobel AB, Sweden.

Brand name of explosive	Density (kg/m^3)	Energy (MJ/kg)	Gas volume (m^3/kg)	Weight strength rel. ANFO	VOD (m/s)	Water resistance	Remarks
Composition B	1.7	4.8	0.78	1.14	8000	Excellent	60% RDX/40/%TNT, cast
Dynamex M	1.4	4.7	0.88	1.13	5000[1]	Very good	Flegmatized gel of nitroglycol and cellulose with AN (ammonium nitrate)
Emulet 20	0.22	2.4	1.12	0.73	1800[4]	Poor	ANFO/emulsion/styropore mixture graded in relation volume strength of 100% ANFO
Emulet 50	0.52	3.3	1.03	0.89	2600[4]	Poor	See above
Emulet 70	0.70	3.6	1.00	0.94	3100[4]	Poor	See above
Emulite 100	1.2	2.9	0.91	0.78	5600[3]	Excellent	Emulsion explosive with glass micro balloons as sensitizer
Emulite 150	1.15-1.25	4.1	0.84	1.01	4600-5100[1]	Excellent	Emulsion explosive with Al and glass micro balloons as sensitizer
Emulite 1200	1.25	2.7	1.02	0.77	6000[5]	Excellent	Primer sensitive bulk emulsion without Al for use above ground
Emulan 7000	1.25	2.9	1.02	0.81	5000[5]	Excellent	Pumpable, primer sensitive bulk emulsion/ANFO blend for use above ground
Gurit A	1.0	3.8	0.40	0.85	3000[2]	Fair[T]	NG/nitroglycol sensitized powder explosive with inert binder, not manufactured today
Gurit B	1.1	3.6	0.93	0.93	3000[2]	Fair[T]	AN-based explosive containing EGDN but not diatomaceous earth
Nabit A	1.04	4.2	0.92	1.04	3000[1]	Fair[T]	Powder explosive containing NG
Prillit A	0.85	3.9	0.975	1.00	2500[1]	Fair[T]	94.4% AN/5.6% fuel oil adsorbed into porous prills of 0.1-2 mm diameter

[1] 40 mm diameter without casing. [2] 17 mm diameter charge (without casing). [3] 63 mm diameter charge in plastic tubing. [4] 42 mm diameter charge in a steel tube. [5] 75 mm diameter charge in rock. [T] charge in plastic tubing. All explosives except Composition B are commercial. All brand names are trademarks of Nitro Nobel AB.

4.5.7 Explosive Properties of Some Commercial Explosives and Blasting Agents

The data presented in Tables 4.14 through 4.17 have been provided by several explosives manufacturers in different parts of the world. Since different manufacturers

Table 4.15. Properties of explosives from IRECO Incorporated, USA.

Explosive	Q total (MJ/kg)	Velocity[a] (m/sec)	Common charge diameter (mm)	Density (g/cm^3)	Gas vol. (l/kg)	Relative weight strength ANFO=1.00	Rheology type	Langefors weight strength rel. ANFO
Nitroglycerin Explosives:								
50% Extra Dynamite (Extra Dynamite)	3.912	3320[2]	>22	1.32	740	1.21		0.95
Gelaprime F (Gelatin Dynamite)	4.268	6000[2]	>50	1.43	672	1.16		1.00
40% Extra Gelatin (Extra Gelatin Dynamite)	3.996	4700[2]	>22	1.54	493	1.09		0.91
Unigel (Semi-Gelatin)	4.184	3850[2]	>50	1.27	829	1.14		1.02
Titan G Booster (Specialty Product)	4.265	6000[3]	>68	1.43	628	1.16		0.99
Non-Nitroglycerin Explosives:								
Iredyne 365 (Semi-Gelatin Dynamite)	3.787	3270[4]	>22	1.23	628	1.06		0.90
Irefo 403 (Specialty Product)	—	1450	—	1.35	—		Liquid propellant	
Permissible Explosives:								
Red HA (NG Ammonia Class)	3.285	2490[4]	>25	1.21	807	0.85		0.84
Irecogel B (Non-NG Ammonia Class)	3.305	5000[4]	>25	1.44	717	0.90		0.82
Irecoal E-1 (Emulsion)	3.305	4000[4]	>25	1.15	695	0.89		0.82
Bulk Explosives:								
Iregel 1.116 (Site Mixed)	2741	5200[1]	>150	1.25	964	0.78	Emulsion	0.77
Iregel RX (Repump)	2.678	5200	>100	1.20	941	0.76	Emulsion	0.75
Iregel RX Plus (Repump)	3.138	5200	>100	1.20	897	0.86	Aluminized emulsion	0.83
Iremex 560 (Heavy ANFO)	3.347	4000	>150	1.15	964	0.96	ANFO/ emulsion	0.89
Packaged Emulsions:								
Iremite 62 (Cap sensitive)	3.753	4900	>32	1.18	852	0.98	Firm	0.94
Iremite TX (Cap sensitive)	4.142	5000[2]	>50	1.24	785	1.12	Firm	1.01
Blastex 100	2.803	4900[2]	>38	1.20	829	0.97	Firm	0.75
Iregel 1135E (Shot bag)	3.305	5000[1]	>100	1.25	897	0.83	Soft	0.87
Iregel 1135P (Shot bag)	3.054	4500[1]	>100	1.25	919	0.83	Pliable	0.82
Slurry Explosives:								
Ireseis	3.222	4700[3]	>57	1.30	695	0.87	Gelled	0.80

[a] Velocities are measured in [1]150 mm, [2]50 mm, [3]63 mm, or [4]32 mm unconfined charges.

use different methods for determining explosive strength and other performance data, the tables should not be used for comparison of explosives from different manufacturers. These tables are only intended to provide an indication of the diversity of products offered, and to provide comparison between different explosives from each manufacturer.

Table 4.16. Properties of explosives from Dyno Explosives, Norway.

Product name	Density (g/cm^3)	Q (MJ/kg)	v (l/kg)	Weight strength rel. ANFO	D (m/s)	Consistency	Projectile impact[6] (m/s)
Dynex 205	1.17	4.00[1]	863[2]	1.00	5200[3]	Plastic	> 500
Dynex 300	1.17	3.06[1]	930[2]	0.82	4700[3]	Plastic	> 500
Dynamit	1.45	4.65[1]	875[2]	1.13	6000[4]	Plastic	150
Slurrit 50-10	1.25	3.10[1]	900[2]	0.82	43500[5]	Viscous	> 500
Slurrit 110	1.25	2.92[1]	834[2]	0.77	4800[5]	Fluid	900

[1] Total explosion energy with water in its gaseous phase. [2] Volume of detonation product gases at 25°C. [3] Detonation velocity, unconfined in 60 mm charge diameter. [4] Detonation velocity, unconfined in 50 mm charge diameter. [5] Detonation velocity, confined in 70 mm steel tube, wall thickness 5 mm [6] Critical impact velocity for initiation.

Table 4.17. Properties of Explosives from ICI Australia Ltd, Australia.

Product	Density (g/cm^{-3})	Available energy (MJ/kg)	VOD (m/s)	Gas volume at STP (l/kg)	Relative effective Energy (C.F. ANFO @ 0.8g/cm^{-3})	
					weight	bulk
"Powergel" 2131	1.15	3.286	5628	810	96	138
"Powergel" 3151	1.27	3.917	6350	801	120	190
"Powergel" 2880	1.25	3.679	6245	772	110	172
"Powergel" 2540	1.25	2.957	6078	850	98	153
"Energan" 2640	1.20	3.432	5942	921	109	163

"Powergel" and "Energan" are registered trademarks of ICI Australia Ltd.

Table 4.18. Properties of explosives from Kimit AB, Sweden.

Brand name of explosive	Density (kg/m^3)	Energy (MJ/kg)	Gas volume (m^3/kg)	Weight strength rel. ANFO	VOD (m/s)	Water resistance	Remarks
Kimit 80	1.1	4.1	0.74	1.06	4000	Excellent	Cap sensitive water gel, not manufactured today
Kimulux 82	1.18	4.5	0.81	1.13	4600	Excellent	Cap sensitive packaged emulsion with Al
Kimulux R	1.2	2.9	0.91	0.80	5500	Excellent	Primer sensitive bulk emulsion with Al
KP-primer	1.4	4.8	0.79	1.18	7500	Excellent	

Kimit, Kimulux, and KP-primer brand names are trademarks of Kimit AB, Sweden.

Chapter 5

Initiation Systems

Initiation systems (the common name for the systems needed to start the detonation in the main charges) include a great variety of specialty products, such as fuses, detonators, and booster charges. Initiation systems today are required to set off many charges in many separate drillholes in a predetermined timing delay pattern which is designed to provide optimal fragmentation and a minimum of ground vibration and flyrock.

5.1 Historical Note on Detonators

Before the days of Alfred Nobel, when rock blasting was done with black powder, initiation was started by the flame slowly propagating through a black powder fuse. The black powder fuse, still in use today in some mines, consists of a narrow column of black powder, lightly enclosed in a tarred hemp cover. It was ignited by holding a torch to the end of the fuse extending out of the drillhole, and the black powder, enclosed into the drillhole with a wad of hemp and sand stemming, was ignited to burn when the flame reached through the stemming to contact the black powder. The linear burning rate of black powder, like that of most other propellants, increases with pressure. The volume or mass burning rate is proportional to the burning surface; and by making the black powder grains of suitable size, the burning rate could be made high enough for the black powder (in the confinement of a drillhole stemmed with sand or crushed rock) to shatter the rock explosively.

The homogeneous liquid explosive nitroglycerin has a much smaller exposed burning surface than the granular black powder and requires a hotter flame to ignite. It could not reliably be ignited to self-accelerating burning by means of the flame from a black powder fuse, even in the confinement of a drillhole. Alfred Nobel tried unsuccessfully to ignite different mixtures of black powder and also neat, liquid nitroglycerin using black powder fuse. Freshly made, the mixture sometimes ignited; but with time, the nitroglycerin appeared to wet the black powder and reduce its ability to burn. Then in 1863, he found that, by enclosing a small charge of finely powdered black powder in a closed wooden capsule and leading the fuse through a tight-fitting hole in the capsule plug (Figure 5.1), he could reliably ignite nitroglycerin. Patents on this invention were applied for in 1864 and, with improvements, in 1865. In the tight confinement of the capsule, the burning rate of the black powder became sufficiently high for the capsule to explode and send what we now know is a shock wave into the surrounding nitroglycerin. (Initially, the thinking was probably to try to keep the black powder dry, away from the nitroglycerin, by enclosing the critical charge nearest to the end of the fuse in the capsule. Very likely, its shock initiating effect was understood only later.) Nobel soon realized that the shock from the exploding capsule was an important factor in the reliable initiation of nitroglycerin. He already knew nitroglycerin, with its high density and high explosion energy, had the potential to become a much more

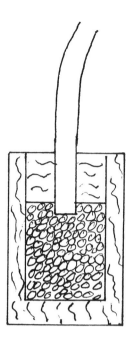

Figure 5.1. Alfred Nobel's first igniter cap for nitroglycerin consisted of an enclosed wooden capsule containing finely-powdered black powder, ignited by a black powder fuse.

effective rock blasting explosive than black powder. This was long before the concept of detonation had become clearly understood. The wooden capsule was an important first step toward the epoch-making invention of the detonator, and marked the breakthrough to the first use of detonating high explosives in rock blasting. Without the invention of the detonator, nitroglycerin was useless for rock blasting.

Further developments, described in a later (1865) Nobel patent for the detonator, made use of a small quantity of a primary explosive, mercury fulminate, pressed into a copper capsule which was crimped to the end of the black powder fuse. Of course, mercury fulminate, being a primary explosive, had the ability to burn to detonation very quickly. Figure 5.2 shows some present-day detonators that are now in large-scale use in rock blasting. In the following, we will outline the developments that led to these detonators, and also to newer developments that are still in the early experimental and testing stage.

About 1900 came the electrically initiated fuse-head. Connected to a short length of black powder fuse, in its turn connected to an ordinary detonator, this constituted the first electric delay detonator. Later, about 1920, came the instantaneous electric detonator in which a thin electrically-heated bridgewire set on fire a minute charge of a flame-producing compound. This in turn initiated a small quantity of primary explosive such as lead styphnate or lead azide. The primary explosive detonated, initiating detonation in the main charge — usually about half a gram of a relatively sensitive secondary explosive such as PETN, Tetryl, or RDX/TNT.

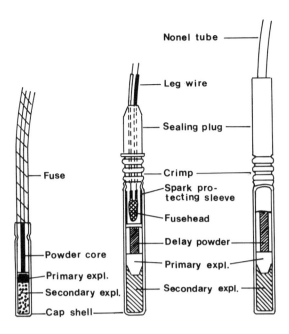

Figure 5.2. Detonators: (a) ordinary detonator for initiation by black powder fuse; (b) electric delay detonator; (c) NONEL delay detonator.

5.2 Delay Detonators without a Primary Explosive

Non-primary explosive detonators (NPED) have up to now been used mainly in military applications. By using a specially prepared secondary explosive, a new civilian detonator containing no primary explosive has been developed. When incorporated with the electric detonation of the NONEL system, this new detonator is expected to increase safety in the fields of civil engineering and mining.

5.2.1 Introduction

In the first Nobel patent for the detonator in 1865, a copper shell filled with a primary explosive, mercury fulminate, was crimped to the end of the black powder fuse. A primary explosive such as mercury fulminate has the ability to transform from burning to detonation extremely quickly.

At the beginning of the 20th century, a combination of primary and secondary explosives began to be used in detonators. This concept is still commonly used worldwide in detonators manufactured for civilian use.

During the last part of this century, significant progress in detonator technology has made it possible to transfer the ignition energy to the detonator in a variety of ways via electric wires and fuse heads, or via the NONEL tube. Pyrotechnic delay elements with a wide span of high-precision burning rates have been developed. Practical experience has guided the development toward the delay times most appropriate for different specific applications.

146 Chapter 5. Initiation Systems

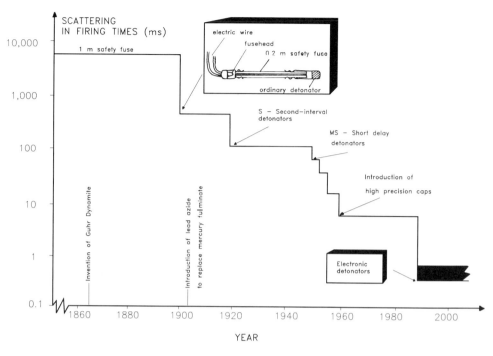

Figure 5.3. Historic review of the improvement in scatter for Swedish detonators.

In the 1950s, the introduction of short-delay or millisecond-delay blasting with 25 ms intervals was a major breakthrough for controlling the rock blasting process. It became possible to minimize ground vibrations and flyrock while simultaneously improving fragmentation. Drifting and tunneling demanded half-second interval times and today we can — with high precision — achieve up to 6 sec delays. The scatter with pyrotechnic delay charges can be held to within 1.5 to 2.5 % of the nominal delay time for short delays. Electronic detonators can be made with sub-millisecond delay scatter. Figure 5.3 shows how the scatter in firing times has decreased with the successive introduction of more and more refined detonator systems.

5.2.2 Present-Day Detonators

The detonators with pyrotechnic charges in use today are typically designed as shown in Figures 5.2 and 5.4. The electric detonator is equipped with a bridgewire which, when electrically heated, sets a fusehead on fire. This in turn ignites the delay charge which burns slowly, producing the required time delay.

When the delay charge is consumed, a small quantity of primary explosive, such as lead azide, is initiated. Finally, this detonation initiates the base charge which consists of a secondary explosive: for example, PETN, tetryl, or RDX.

Primary explosives generally consist of single molecules which permit them to decompose very quickly when initiated. They also have the ability, when ignited, to transit from burning to detonation at distances as small as a fraction of a millimeter, even under atmospheric pressure conditions. A few milligrams is enough to achive detonation. These properties of course make primary explosives very suitable for use in

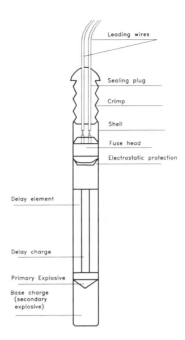

Figure 5.4. Example of a present-day pyrotechnic delay detonator using primary explosives.

the initiation process. But at the same time, however, their extreme sensitivity to heat, friction, and impact makes the explosives risky to handle.

Examples of primary explosives are mercury fulminate ($HgC_2N_2O_2$), lead styphnate ($PbC_6HO_2(NO_2)_3$), lead azide (PbN_6), and silver azide (AgN_3). They often have a high density — lead azide, for example, has a density of 4.71 g/cm^3.

A secondary explosive is much less sensitive to initiation than a primary explosive and can often, in a relatively small quantity and under atmospheric pressure conditions, burn without transiting to detonation.

For military applications, a number of instantaneous detonators based only on secondary explosive are available. One example is the exploding bridgewire detonator (EBW). The EBW detonator is detonated through a capacitor discharge into a low-inductance bridgewire circuit. Due to extremely rapid rise of the discharge current pulse (1000 A/μs) and to the high voltage applied across the bridgewire, the wire is heated to such high temperature that the metal evaporates and the bridgewire explodes. This shock initiates the secondary explosive pressed against the bridgewire. The time lapse from the burst of the bridgewire to the shock-wave break-out from the cap is of the order of one microsecond (1 μs), and it may have a scatter of 0.1 μs or less. The EBW detonators are used where extreme safety against accidental initiation is required. However, it has not been possible to develop this concept for the civilian market as the necessary delay times for rock blasting purposes cannot be achieved with a single firing set (blasting machine).

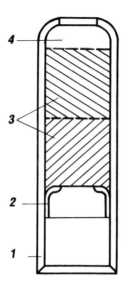

Figure 5.5. Initiation element.

5.2.3 The New Non-Primary Explosive Detonator

Black powder originated in China. The basic idea for Nitro Nobel's new detonator also originated in China. In 1984, this concept was procured by Nitro Nobel AB and patented (US Patent 4727808). In 1993, Nitro Nobel introduced the new detonator into the marketplace.

By 1995, the new non-primary explosive detonator had completely replaced the conventional (lead azide-containing) detonators in all electrical and NONEL detonator systems manufactured in Sweden by Nitro Nobel (such as NONEL UNIDET, NONEL LP, and NONEL MS).

Conceptual Design

The customers will not notice any difference when they examine the exterior of the new detonator. But the interior is totally different. The sensitive lead azide is replaced by an initiation element (Figure 5.5). This element consists of a steel shell (1), a sealing cup (2), PETN charges (3), and a delay charge (4). The PETN charges are characterized by different qualities and densities in order to achieve the stipulated function. By dividing the secondary explosive PETN into charges with various qualities, it is possible to control the transit from deflagration to detonation. The transition is termed *Deflagration-to-Detonation Transition* (DDT). An excellent recent article by McAfee, Asay, and Bdzil [1993] provides a lucid explanation of many of the complex phenomena which are parts of the DDT process.

5.2. Delay Detonators without a Primary Explosive

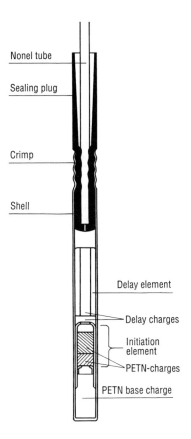

Figure 5.6. The new non-primary explosive detonator.

Safety

The conventional lead azide-based detonator has several disadvantages due to the primary explosive's sensitivity to heat, friction, and impact. When handling the detonator, any unintentional mechanical impact or bending of the detonator at the region of the lead azide may result in an explosion. Exposure to fire during transport or storage will also more easily result in unintended detonation due to the presence of a primary explosive. Several stages in the manufacturing process require handling the dangerous primary explosive, lead azide in which incidents leading to accidental explosions may occur, jeopardizing both people and process equipment.

The new non-primary explosive detonator (Figure 5.6) is much less sensitive to different stimuli than the conventional one. In reality, this means that a higher safety is introduced in all operations:

- Manufacturing
- Handling
- Transportation

An optimal overall increased safety is achieved when the new detonator is used together with the NONEL system, due to the properties of the non-electrical NONEL tube.

During the testing of the new detonator, mechanical impact from, for example, a falling rock has been simulated. The tests with the drop weight showed that safety from initiation by a falling rock is considerably improved with the NPED detonator compared to a regular NONEL or electric detonators.

Another benefit of the new detonator is the reduced risk for mass explosion during transportation. The flash-over distance is a way to measure the propagation from one detonator to another. The larger the flash-over distance, the more sensitive is the detonator. Flash-over tests show that the new detonators can be positioned at distances as small as 2 cm from each other in open air without any flash-over propagation. At this short distance, conventional detonators will always detonate due to flash-over.

Another benefit of the new detonator concerns the environment where the lead from the old detonators is an unwanted species. Due to the elimination of the lead azide, we have reduced the lead discharge by about 50 %. Some lead remains in the fusehead.

The development of the non-primary explosive detonator is another step toward safer rock blasting.

5.3 Electric Detonators

The instantaneous electric detonator was a step forward for safer ignition systems than the black powder fuse. The detonator was, however, restricted to single row rounds. Early efforts to use large multihole rounds generated tremendous ground vibrations and flyrock. It was obvious that great advantages could be achieved if a delay time could be introduced between the initiation of the different detonators. Experiments were carried out in Sweden in the 1940s where a mechanical-electric sequential blasting machine was used; but it was quickly found that this made the blasting operation unsafe. Misfires occurred as the electric wires leading to detonators intended for later initiate were cut off due to back break from the charges initiated earlier in the round. Misfires lead to increased risk for flyrock, and the recharging of the remaining rock containing undetonated explosive and detonators mixed into the partially broken rock is also a hazardous operation. There was a need for time delays placed in the hole itself if proper functioning was to be maintained.

By inserting a slow-burning pyrotechnic charge between the matchhead and the primary explosive in the detonator, a time delay could be introduced between the time of the initiating electric current pulse and the initiation of detonation. By making pyrotechnic charges of different length and with different compositions having different linear burning velocities, time delays ranging from a few milliseconds to several seconds could be built into detonators. The introduction of short-delay or millisecond-delay blasting was a major development that made modern large-scale industrial rock blasting possible. The rock blasting engineer now had a powerful tool for tailoring the blast to control fragmentation and environmental effects.

Multiple-row rounds were introduced which increased the productivity and cut the costs. A proper time delay made it possible for the blasting engineer to improve fragmentation and reduce environmental disadvantages such as ground vibrations, flyrock, and air blast. Several researchers have confirmed through experimental studies that an adequate delay time for good fragmentation in bench blasting is 6 to 15 ms per meter burden.

In drifting and tunneling, electric detonators with a delay time of up to 6000 ms are commonly used for the blast holes outside the cut. Detonators with delays of 0.5 second and up are usually named half-second (HS) delay detonators.

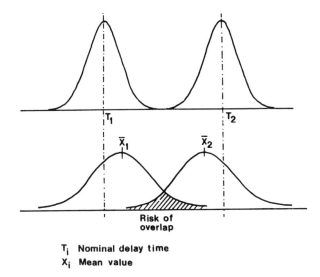

Figure 5.7. Distributions of firing times for good and bad precision detonators.

Mine safety regulations demand that any explosive materials for use in underground coal mines be tested to ensure that they do not ignite methane. Only those that pass the test are classified as permissible. This test requirement applies to detonators, as well as explosives. Due to the danger of igniting methane or coal dust, the cap shells must be made of such material that it does not reach high temperatures and burn when exposed to the air after the blasts. Therefore, regulations prohibit the use of aluminum shells in coal mining underground. Permissible detonators are made of commercial bronze or copper-plated steel and have copper or iron leg wires.

There is always an unavoidable scatter in the firing times of different detonators with the same nominal firing time. This can be due to small variations in the length, the packing density, and the composition of the pyrotechnic delay charges, or to changes in the burning rate occurring from ageing during storage. The burning rate may change due to slow oxidation reactions within the pyrotechnic charge itself, often accelerated by moisture. Because of this, it is always wise to store the detonators in such a way that earliest delivery is used first, and to avoid mixing in the same blast detonators with the same nominal delay time from lots of different age.

The effect on the blast of detonator time scatter can be of tremendous importance. Overlapping intervals can result in situations where holes are fired with a too large burden if the charge in the hole in front did not detonate properly. Coarse fragmentation, flyrock, and excessive ground vibrations result when delay precision is lost. An expected smooth blasted contour can also be spoiled if the time scatter of the detonation time of holes having detonators with the same nominal time delay becomes too large. Figure 5.7 shows firing time distributions for detonators of different precision.

Winzer et al. [1979] reported an investigation where the firing time and its standard deviation was measured for three electric MS cap series and one non-electric cap series. The results show that, with the quality of detonators used in their investigation, there existed a considerable potential for overlapping, especially for higher period numbers. Present-day so called "high precision" electric and NONEL detonators have greatly improved time accuracy and much less scatter is present.

The delay time scatter should be taken into account when calculating the probable maximum cooperating charge from several drillholes having detonators of the same nominal firing time. A certain degree of time scatter allows a larger total charge per nominal interval time to be used, since the probability that all charges would detonate at the same time is very low. The technique for calculating the maximum cooperating charge for estimating the ground vibrations generated from the round is described in detail in Chapter 13.

In using electric caps, the blasting engineer needs to connect the caps and include them in a firing circuit. Figure 5.8 shows some commonly used circuits. Detonators are usually connected in series when the number of detonators is low. Connecting the wires is easy in this case; one leg wire of one cap is connected to one leg wire from the next cap, and so on, until all the caps are connected into one big loop. The two free leg wires are connected to the firing line running to the power source. The whole circuit can easily be checked with a blasting galvanometer. The resistance of detonators, connecting wires, and firing line are given by the manufacturers. By applying Kirchhoff's law and Ohm's laws, the power requirements can be calculated.

When larger rounds are fired, it is recommended to connect the detonators in a parallel-series circuit to avoid having to use blasting machines with an excessively high voltage. The series may coincide with the rows as shown in Figure 5.8 but this is more often not the case. Important is that the series are balanced, i.e., that all of the parallel series should have nearly the same resistance to insure that the currents flowing through each series are approximately equal. If one series has many more detonators than the others, that series will have a lower firing current than the others and may not fire.

Good blasting practice must be maintained in all blasting operations. However, the use of electric caps usually requires more careful planning and control than if non-electric initiation systems are used. For safety reasons and to avoid misfires, the blasting engineer must be aware of the following factors.

Before bringing detonators into the blast site, a check should be made if *stray currents* exist in the area. These currents usually originate from electrically operated equipment or leaks in the electrical power distribution systems. In periods of high solar flare occurrence, additional stray currents are generated, particularly at northern latitudes. Stray currents can be a hazard if their magnitude is sufficiently high and if the current finds its way into the blasting circuit.

The blaster planning to use electric detonators must check if any radar, radio, or TV stations are operating in the neighborhood and he must find out the effect of these stations. The *radio frequency energy* can theoretically cause premature initiation if the blasting circuit acts as a receiving antenna. General rules, recommendations, or regulations exist which indicate the proper safety distances from transmitters of different types and different effects.

Table 5.1 outlines recommended safety distances for blasting with US manufactured electric detonators. Safety distances in these publications can also be found for electrical transformer stations and high-voltage power lines with their associated electric and magnetic fields. Through *induced currents* from AC transmission lines sufficient electrical energy can be introduced into a blasting circuit to cause an accidental detonation. The electric field of a high-voltage AC power line can charge the blasting circuit to a high voltage which, through *capacitive discharge*, can cause a spark from detonator to ground and thereby cause initiation.

The resistance of each cap should be checked with a galvanometer of a type recommended by the explosive manufacturer before priming and after the hole is loaded. Each

5.3. Electric Detonators

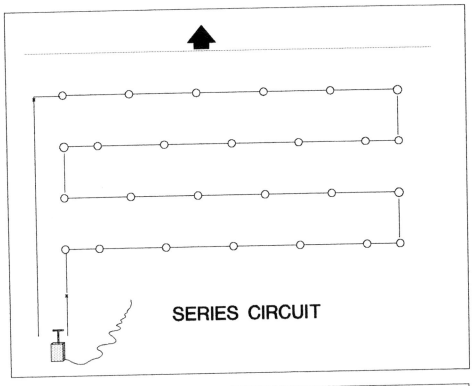

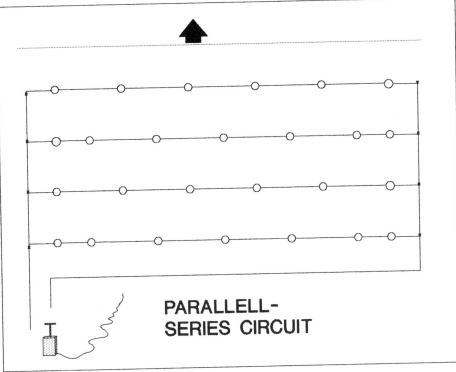

Figure 5.8. Examples of some commonly used circuits (arrows indicate direction of blasting toward the free surface).

Table 5.1. Safety distances for U.S. manufactured blasting caps [After IME Publication No. 20, Dec. 1988].

Recommended Distances for VHF TV and FM Broadcasting Transmitters	Effective Radiated Power (Watts)	Minumum Distance (Feet)		
		Channels 2 to 6	FM Radio	Channels 7 to 13
	Up to 1,000	1000	800	600
	10,000	1800	1400	1000
	100,000[a]	3200	2600	1900
	316,000[b]	4300	3400	2500
	1,000,000	5800	4600	3300
	10,000,000	10200	8100	5900

[a] Present maximum power channels 2 to 6 and FM is 100000 watts
[b] Present maximum power channels 7 to 13 is 316000 watts

Recommended Distances for Commercial AM Broadcasting Transmitters, 0.535 to 1.605 MHz	Transmitted power[c] (Watts)	Minumum distance (Feet)
	Up to 4000	800
	5,000	900
	10,000	1300
	25,000	2000
	50,000[d]	2900
	100,000	4100
	500,000	9100

[c] Power delivered to antenna
[d] 50,000 watts is the present maximum power of U.S. broadcast transmitters in this frequency range

Table 5.2. Nominal resistance of DuPont Electric Delay Blasting Caps (From DuPont's Blasting Handbook).

Length of legwire (ft)	4	6	8	10	12	14	16	20	24	30	40	50	60	80	100
Copper wire (Ohms)	1.16	1.24	1.32	1.40	1.48	1.57	1.65	1.81	1.97	2.21	2.06	2.32	2.59	2.61	3.01
Iron wire (Ohms)	2.00	2.49	2.99	3.49	3.99	4.48	4.98	5.98	—	—	—	—	—	—	—

Length of legwire (ft)	120	150	200	250	300	400
Copper wire (Ohms)	3.41	4.01	5.02	6.02	7.03	9.03
Iron wire (Ohms)	—	—	—	—	—	—

series of detonators should also be controlled to check that all the series in a *parallel-series* blast are balanced, having approximately the same resistance. Finally, firing lines and the *resistance of the blasting circuit* must be checked. The manufacturer should be able to supply data showing the nominal resistance of each detonator, the connecting wire, and firing line. Usually, detonators with different length of legwire have different nominal resistance (Table 5.2). Some manufacturers, however, provide electric caps which all have the same nominal resistance irrespective of the legwire length. This is achieved by using a different metal alloy for wires of different length. *Ground faults* (that is, contact between ground and the detonator wires which can occur when the leg-wire insulation has been damaged during loading) must be checked for any possible stray currents. This check can be made by using a high-resistance voltmeter recommended by the explosive manufacturer. The test for ground faults eliminates the need for testing to determine if any current leakage exists.

When using series or parallel-series circuits, *current leakage* always occurs to a certain extent. In particular, this phenomenon can arise when the ground is wet, when the ground is very conductive, or when conductive explosives such as ANFO are used. This current leakage can lead to misfires as part of the firing current is led through ground through a damaged insulation, bypassing several detonators, and returning to the detonator circuit at some other point of damaged insulation. Misfires can then occur by caps that are partially shorted out by current leakage and therefore do not receive enough firing current. Current leakage can be checked by measuring the resistance between one end of the circuit and a good ground (for example, a metal pole driven into wet ground).

The problems with current leakage are, to a large extent, related to insufficient or damaged legwire insulation. These problems can be avoided by the use of detonators having special heavy-duty insulation (OD detonators) for blasting underwater, in wet conditions, or in conductive rock (such as magnetite iron ore).

Thunderstorms can cause *lightning discharges* and electric field changes so rapid and of such magnitude that the induced currents in the detonator circuit can set off the blast. Therefore, all blasting operations using electric detonators must be suspended when thunderstorms approach and all personnel must proceed to safe areas outside the risk zone. Several manufacturers have developed lightning warning and forecasting systems which give information about approaching thunderstorms and any local high gradients of atmospheric static electricity.

If the firing voltage is too high or a high firing current has too long a duration, *arcing malfunctions* can prevent detonators from functioning in the intended way. Arcing can occur between legwires which have insufficient or damaged electrical insulation. Internal arcing between the inside cap shell and the bridge head can result in misfires as it might generate excessive heating inside the shell, resulting in premature shell ruptures which prevent the initiation of the pyrotechnic charge. Erratic timing can also occur, which can result in poor fragmentation, excessive flyrock, and large ground vibrations.

Each detonator manufacturer should be able to provide the user with a graph showing the recommended current-time combinations for avoiding these types of problems (Figure 5.9).

5.3.1 Formulae for Electric Circuits

The formulae needed for calculation of the detonator circuit resistance, the blasting circuit resistance, and power requirements of series are given below. Only the formulae for simple series and single parallel circuits are given. The reader should have little problem in combining them and applying them to combined series-parallel circuits.

Series Circuit

With reference to Figure 5.10, the blasting circuit resistance can be expressed by

$$R_o = \sum_{i}^{n} R_i + R_f + R_c \tag{5.1}$$

where R_i is the resistance of detonator i, R_c is the resistance of the connection wire, R_f is the resistance of the fire line. All resistances are in ohms.

Ohm's Law gives the voltage required from the power supply to provide the necessary current I if the blasting circuit resistance is known:

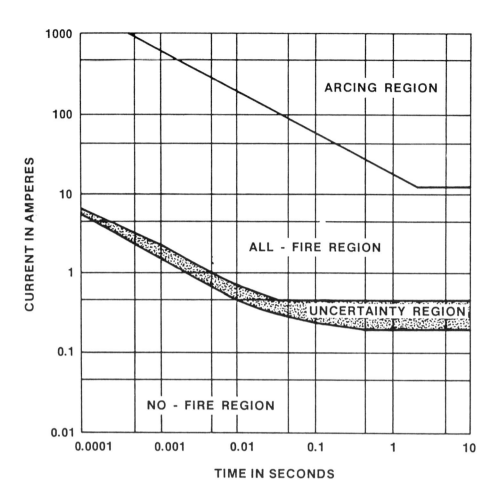

Figure 5.9. Electrical response characteristics of Atlas Electric Detonators (from Atlas Powder Company, 1987).

$$U_o = IR_o \tag{5.2}$$

where U_o is the voltage of the power supply in volts (V), and I is the current in amperes (A).

The power supply must have a sufficient capacity to feed the circuit with the required current and voltage. To compute the power needs (in watts, W) of the series circuit, use the equation

$$P = U_o I \tag{5.3}$$

or, when Ohm's Law is substituted,

$$P = \frac{U_o^2}{R_o} \tag{5.4}$$

or

$$P = I^2 R_o. \tag{5.5}$$

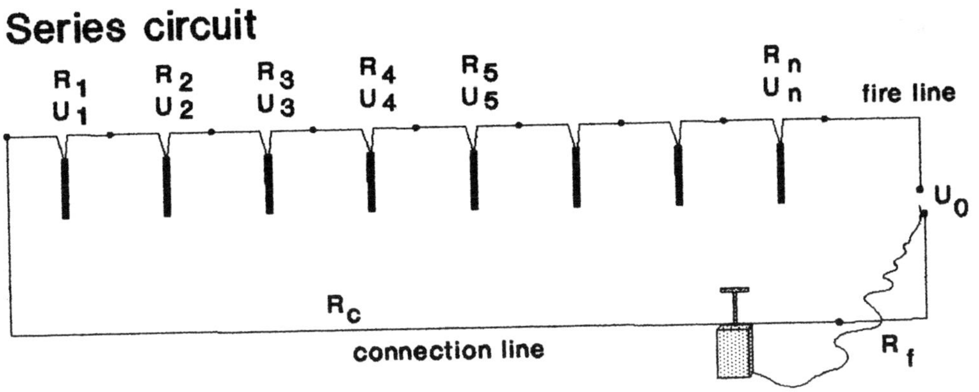

5.3. Electric Detonators

Series circuit

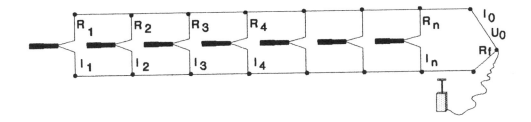

Figure 5.10. Series circuit.

Parallell circuit

Figure 5.11. Parallel circuit.

Parallel Circuit

For the parallel circuit (Figure 5.11), the blasting circuit resistance is

$$R_o = R_f + \frac{1}{\sum_i^n \frac{1}{R_i}} \qquad (5.6)$$

and the total current flowing is

$$I_o = \sum_i^n I_i . \qquad (5.7)$$

Series-Parallel Circuit

For blasts containing large numbers of drillholes, hybrid combinations of the series and parallel circuits are frequently used. By using such circuits, the firing voltage can be kept low (smaller than 1500–2000 volts) to reduce the risk of current leakage and discharge to earth from the detonator wires.

5.4 Semiconductor Bridge Ignition

Bickes and Schwarz of the Sandia National Laboratories (SNL), Albuquerque, NM, developed the semiconductor bridge in 1984–1985 and received US patent 4708060 in 1987. SCB Technologies, Inc., a New Mexico corporation obtained a license from SNL in 1989 for further development as well as production and commercialization of the SCB.

Most explosive devices use small bridgewires (sometimes called hot wires) to ignite explosives, pyrotechnics, or propellants pressed against the bridge. Passage of a current through the wire heats the wire and then ignites the explosive in a few milliseconds. The SCB method is different. It uses a heavily doped polysilicon bridge which is much smaller than conventional bridgewires. Passage of a current pulse with significantly less energy than that required or hot-wire ignition produces a plasma discharge in the SCB which ignites the explosive pressed against the bridge, producing an explosive output in a few microseconds.

Production of the bridges is a routine polysilicon-on-silicon wafer process. The finished wafers are diced into chips, and the chips are placed on a header holding the incoming electrical leads. Aluminum or gold wires, or tape-automated-bonding procedures, are then used to connect the bridge to the header leads. After assembly into a device appropriate for the particular use, application of a current pulse to the header pins transports the current through the bridge resulting in a plasma discharge that ignites the explosive or pyrotechnic material pressed against the SCB and produces an output in a few tens of microseconds.

Computer-generated masks are used to define the bridge; thus, the SCB design can be easily tailored for a particular application. Because of the intimate thermal contact of the bridge with the underlying substrate, excellent no-fire current levels are obtained (no-fire is the highest current which can be applied without igniting the explosive).

Direct comparisons of components built with SCBs substituted for bridgewires show that the energy for SCB ignition is at least 10 times less than that for bridgewires. Further, the function times for SCB devices are only a few tens of microseconds, approximately 100 times faster than hot-wire components. Table 5.3 shows a comparison of the ignition times and firing energies of two SCB igniters and a standard bridgewire igniter. SCB explosive safety, as determined from no-fire and electrostatic discharge (ESD) studies, therefore, is better than that of bridgewires.

SCB devices have a very wide range of applications. They are being developed for use in airbag ignition, in rock blasting detonators, and in military igniters and detonators. They are particularly attractive for applications where only a very low ignition energy is available.

The SCB technique also lends itself to being combined with other microelectronic circuits, such as those required in an electronic delay detonator for rock blasting. A recent patent (US Patent 5179248) granted to McCampbell and Hartman in 1993 features the use of a Zener diode incorporated on the SCB chip to provide protection of the semiconductor bridge against accidental overvoltages.

Table 5.3. Comparison of SCB and hot-wire actuators.

Test	Hot-wire	Type 3-2	Type 15
All-fire energy (mJ) (millijoules)	32.6 ± 1.02 (Ambient)	2.72 ± 0.48 (−65 °F)	1.33 ± 0.03 (−65 °F)
No-fire current (A)	1.1 (Ambient)	1.39 ± 0.03 (165 °F)	1.30 ± 0.12 (165 °F)
ESD test	Passed	Passed	Passed
Function time (μs)	3400 (Ambient)	60 (Ambient)	60 (Ambient)

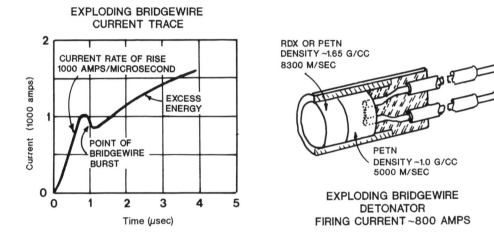

Figure 5.12. Principles of the EBW-cap (from Reynold Industries, Inc.).

5.5 High-Precision Electric Detonating Caps

For blasting operations or detonation research studies where extreme timing precision is required, several types of detonator can be used; the most frequently used is the exploding bridgewire detonator (EBW detonator).

The EBW detonator is detonated through a capacitor discharge into a low-inductance bridgewire circuit. Due to the short duration of the discharge current pulse and the high voltage applied across the bridgewire, ($\approx$1000 A during a fraction of a microsecond), the bridgewire is heated to such a high temperature that the metal evaporates and the bridgewire explodes, causing a combination of shock wave and thermal initiation of the secondary explosive pressed against the bridgewire. It is not necessary to have a primary explosive charge in an EBW detonator; therefore, a lower voltage which only heats and perhaps melts the bridgewire will not cause initiation. Also, the EBW does not detonate upon heating, such as in a fire. The EBW detonators are therefore also used where extreme safety against accidental initiation is required. EBW detonators are used to set off the conventional explosive charges in nuclear explosive devices, and are also the preferred way of setting off non-electric firing lines such as the NONEL system. The time lapse from the burst of the bridgewire to the shock-wave breakout from the cap is of the order of 1 μs, depending on the caps physical dimensions, and it may have a scatter of 0.1 μs or less (see Figure 5.12).

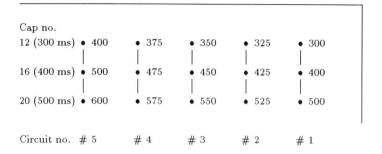

Figure 5.13. Nominal firing times with the sequential blasting machine.

5.6 Sequential Blasting Machine

Sequential blasting machines can be used to good advantage when combined with electric delay detonators. This type of sequential timing system permits the blasting operator to energize several independent firing circuits at a preset millisecond delay interval. The system can be used where there is a need to limit the maximum cooperating charge by distributing the explosive charges over more time delay intervals than are available in the pyrotechnic delay detonator series, allowing several identical series of pyrotechnic delay series to be fired one after the other in the same blast. We can illustrate the use of a sequential blasting machine with an example.

Assume that we have a three-row bench blast consisting of 15 holes and we only have three electric MS detonators #12, #16, and #20 with nominal firing times of 300, 400, and 500 ms, respectively.

Instead of firing row by row, we use a sequential blasting machine with five circuits and a preselected time interval of 25 ms. Figure 5.13 shows the nominal firing times.

By using a sequential blasting machine, numerous combinations of nominal firing times can be achieved. However, one must remember that this initiation method is a combination of high-precision electronic time delay circuits and pyrotechnic delay charges. The preselected electronic time interval between series should not be chosen too small because the scatter of two neighboring pyrotechnic delays with the same delay interval number may then result in overlapping.

5.7 Detonating Cord

Detonating cord is a narrow core of finely powdered PETN enclosed in a woven cover of polymer yarns and extruded plastics. Its water resistance and tensile strength can be varied through quality combinations of the yarn and the plastic cover. Usually initiated by an electric detonator, it propagates a violent detonation with a velocity of about 6500 m/s ($\approx$ 4 miles/s) — a detonation sufficiently strong to initiate dynamite or the special primer explosives used to initiate the insensitive blasting agents, and to cause considerable damage to ANFO through which it is led.

Detonating cord is used where the blaster prefers a tough non-electric initiation system because of the risks of premature initiation of detonation by electric detonators. It is often used in surface operations, like quarries and open pit mines where the frequency of occurrence of lightning is high, although it has been replaced more and

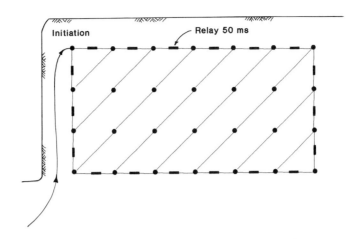

Figure 5.14. Multiple-row shot with detonating cord and with the rows delayed using 50 ms pyrotechnic relays. The blast also illustrates the use of redundancy in connecting the drillholes, in that each hole is reached by the firing impulse from two directions, thereby reducing the risk of detonation cutoff by a damaged detonating cord.

more by the NONEL system. The detonating cord can also preferably be used in underwater blasting operations where the environment is such that electrical connections are difficult to insulate. In perimeter blasting, like preshearing, it can be advantageous to utilize detonating cord for initiating the holes and thereby avoiding too large time delays.

The explosive strength of the cord is given in grams per meter (or grains of explosive per foot) of core load. The most commonly used strengths of detonating cords are 5 g/m (25 grains/foot) and 10 g/m (50 grains/foot). Higher-strength cords are available for special purposes. The 10 g/m cord is often used as the downline and the 5 g/m cord is used as the trunkline. These cords have an approximate initiating energy comparative to a No. 6 blasting cap and can initiate cap-sensitive products. Extra strength cord (called *seismic cord*) can also be used for special-purpose blasting in seismic prospecting.

It is possible to spread the firing of different holes or rows of holes in a sequence when using detonating cord for trunkline and downline by the use of detonating relays or MS connectors simply coupled into the trunkline between boreholes. Such relays consist of a pyrotechnic delay element having a primary explosive charge at each end, capable of igniting the delay element to burn when the detonation in the cord arrives, and at the other end of the delay charge initiate detonation again in the detonating cord continuing on. For example, DuPont MS connectors have delay intervals of 5, 9, 17, and 25 ms; Ensign-Bickford supplies MS connectors with delay intervals of 9, 17, 25, 35, and 65 ms; and AECI manufactures detonating relays with delay times of 12, 25, and 40 ms. The relays are bi-directional.

By using only surface delays, there is always a risk for cut-offs in later firing holes by ground movement initiated at a previous shot hole, even though detonating cords can accept a certain strain before detonation is cut off. When initiating a round with detonating cord, it is good practice to connect the charges in the round in such a way that every hole has the possibility to be initiated from two ways (Figure 5.14). This is called connecting a round with *redundancy*.

The air blast noise from the trunklines in a large surface blast is considerable, having an unpleasantly sharp intensity. Reduction of the noise can be made by covering the trunklines with sand. Another way to reduce noise is to initiate each downline with an electric or a NONEL blasting cap taped onto the cord, or to use the NONEL UNIDET system which avoids both the noise of the trunkline and the destructive effect of the detonating cord downline altogether, as described in the next section.

There are some other drawbacks with detonating cord that should be mentioned. The use of detonating cord for primer initiation can sometimes result in side initiation of the explosive, especially if the explosive is ANFO. The side-initiated underdriven radial detonation or deflagration has a low energy release, which can result in insufficient fragmentation and an excess of toxic fumes. When decking of ANFO in 4-in holes was tested in the Fabian Mine in Sweden blast performance control showed that the downline E-cord (5 g/m) initiated each deck when the cord detonated through instead of transmitting the detonation to the millisecond cap and primer.

These disadvantages have been partially removed by detonating cord with considerably lower explosive strength. DuPont manufactures a non-electric initiation system called "Detaline" which utilizes a low energy detonating cord. The explosive PETN core has a strength as low as 0.5 g/m (2.4 grains per foot). It can be used with conventional detonating cord downlines as well as with non-electric in-the-hole delay boosters. Six different millisecond surface delay periods (9, 17, 30, 42, 60, and 100 ms) exist and 19 in-the-hole delays from 25 to 1000 ms are available. Detonating cord is available at various tensile strength. It is a common practice to lower explosive cartridges into the hole by means of the cord.

5.7.1 NONEL

The usefulness of detonating cord is limited by the high cost of the cord, the unwanted noise associated with its use, and the risk for premature initiation by the downline passing through a sensitive explosive.

When using electric detonators, great care must be taken to ensure good electrical contact with the detonator wires. The risks of accidental premature initiation of a blasting round by stray currents, strong radio or radar signals, or lightning is always there. In mining and other underground blasting operations, more and more electrohydraulic percussion drilling rigs are introduced. In order to create a better working environment, developments of electric loading and hauling machines are being enhanced, and electric lighting of the workspace has become commonplace. As a consequence of the increasing use of electric equipment, the demand for non-electric ignition has also increased.

The NONEL system offers full immunity to electrical interference, creates no noise, and does not interfere even with the most sensitive dynamite explosive through which it leads the detonation impulse. It allows the use of in-the-hole pyrotechnic delays to the same extent as the electric detonators. Therefore, the NONEL system has largely replaced conventional electric detonators and also detonating cord for a large number of blasting operations.

The NONEL non-electric initiation system transmits the initiating signal in the form of a reaction-supported air shock wave in the air inside a narrow-gauge plastic tube which ends in a detonator. The inner wall of the tube is covered with a fine dust of explosive or reactive material which reacts chemically to support the air shock wave by heating and by the expansion of the gaseous reaction products.

The explosive load of the NONEL tube is less than 0.02 g per meter (20 mg/m) of tubing, sufficient to propagate the shock wave indefinitely through the tube, or until

Table 5.4. Delay intervals for NONEL MS and LP.

Period No.	Delay (ms)	Interval time (ms)	Standard wire length (m)
NONEL MS			
3-20	75-500	25	3.0, 4.8, 7.8, 15.0
NONEL LP			
0	25	—	
1-12	100-1200	100	
14,16,18,20	1400-2000	200	6.0, 7.8
25,30,35,40,45	2500-6000	500	
50,55,60			

Table 5.5. Delay numbers, delay times, intervals, and scatter for the NONEL LP system.

Delay number #	Delay time (ms)	Interval (+ms)	Scatter (±ms)
0	25	75	5
1-2	100-200	100	5
3-5	300-500	100	5
6-11	600-1100	100	8
12-20	1200-2000	200	50
25-60	2500-6000	500	150

it reaches the conventional cap delay element. The propagation velocity of the shock wave is about 2000 m/s and the shock wave travels through the tube without affecting the outside surface of the tube.

The NONEL system includes coupling blocks as a means of branching the tube and connecting several detonators into one network for multishot delay rounds.

As can be seen in Figure 5.2, the blasting cap and the delay system is essentially identical to that of an electric detonator. The flame and hot particles emitted from the end of the NONEL tube ignite the delay element in the same way as the flame and hot particles from the matchhead in an electric cap.

As is the case with a detonating cord blast, the only check needed to ensure reliable functioning is to follow the round to visually ensure that all connections are in place at the coupling points. There is no need for the elaborate checks required with the electric systems where a simple visual inspection does nothing to reveal a lack of electric contact in a connection.

The Swedish NONEL systems MS and LP are built up in a conventional way with pyrotechnic delays having steps of 25, 100, 200, or 500 ms in different delay time regions (Table 5.4 and Figure 5.15).

Every batch of NONEL detonators is test fired for delay time measurements. The sampling procedures and charts for inspection by variables for percent defective are based on international standard ISO 3951-1981 (E) and Swedish standard SS 02 01 40.

The precision requirements for the delay times are described in Swedish standard SS 499 07 07: "Initiation systems with non-electric signal conductors of low energy type — general requirements and testing". Table 5.5 shows the nominal delay times, the time interval between consecutive delay numbers, and the scatter for the NONEL LP system. Table 5.6 lists the nominal delays and the upper limit $\left(T_{upper}\right)$ and lower limit $\left(T_{lower}\right)$ of the delay times for each of the three NONEL systems: MS, LP, and UNIDET.

If the noise can be allowed, it is easy to use a 5 g/m detonating E-cord as a trunkline. Special connecting units are used to hook up the NONEL tube from each shothole to

Table 5.6. Delay times for NONEL detonators.

NONEL MS			
Period number	Nominal delay (ms)	T_{lower} (ms)	T_{upper} (ms)
3	75	62.5	87.5
4	100	87.5	112.5
5	125	112.5	137.5
6	150	137.5	162.5
7	175	162.5	187.5
8	200	187.5	212.5
9	225	212.5	237.5
10	250	237.5	262.5
11	275	262.5	287.5
12	300	287.5	312.5
13	325	312.5	337.5
14	350	337.5	362.5
15	375	362.5	387.5
16	400	387.5	412.5
17	425	412.5	437.5
18	450	437.5	462.5
19	475	462.5	487.5
20	500	487.5	550

NONEL LP			
Period number	Nominal delay (ms)	T_{lower} (ms)	T_{upper} (ms)
0	25	12.5	50
1	100	50	150
2	200	150	250
3	300	250	350
4	400	350	450
5	500	450	550
6	600	550	650
7	700	650	750
8	800	750	850
9	900	850	950
10	1000	950	1050
11	1110	1050	1170
12	1235	1170	1300
14	1400	1300	1500
16	1600	1500	1700
18	1800	1700	1900
20	2075	1900	2250
25	2500	2250	2750
30	3000	2750	3250
35	3500	3250	3750
40	4000	3750	4250
45	4500	4250	4750
50	5000	4750	5250
55	5500	5250	5750
60	6000	5750	6250

NONEL UNIDET			
Period number	Nominal delay (ms)	T_{lower} (ms)	T_{upper} (ms)
SL 0	0		
SL 17	17	14	20
SL 25	25	22	28
SL 42	42	37	47
SL 67	67	61	73
SL 109	109	102	116
SL 176	176	167	185
U 400	400	391	409
U 425	425	416	434
U 450	450	441	459
U 475	475	466	484
U 500	500	491	509

Quality inspection acceptance criteria:

$$\frac{T_{upper} - \bar{X}}{S} \geq 1.50$$

$$\frac{\bar{X} - T_{lower}}{S} \geq 1.50$$

where $\bar{X}$ = batch mean delay time, and S = standard deviation.

Substituting $\bar{X}$ with nominal delay gives

$$S_{max} = \frac{T_{upper} - T_{lower}}{3}.$$

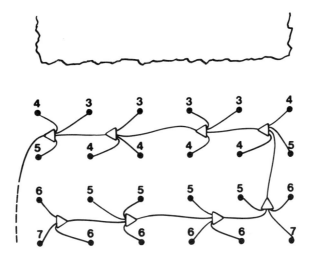

Figure 5.15. Example of coupling with connecting units and NONEL MS.

the cord. The hookup should sequence the initiation of charges in the drillholes in such a way that, if a failure occurs in the transmission in the NONEL tubing or in the E-cord trunkline, the initiation of all charges with intended delay times shorter than that at which the failure occurred will detonate. The detonators with shorter delays should be reached by the initiating shock or detonating wave before those with longer delays. In this way one can ensure that in the case of a transmission failure, or the omission of a connection, all charges detonating up to the point of failure have broken out their burden of rock, and the remaining charges are in the undisturbed rear part of the blast. It is very difficult and risky to re-drill or recharge a blast where some charges have detonated behind undetonated, charged drillholes.

Figure 5.16 demonstrates the easy hook-up procedure in a tunnel blast. The NONEL tubes from several holes are bundled together, and a simple clove hitch is formed around them with a detonating cord. Alternatively, a special bunch connector can be used.

Where suitable, combinations of detonating cord and non-electric shock-tube systems can be used. A description of a system provided by Ensign-Bickford, follows. The Primadet Noiseless Trunkline Delays (NTD) are used as surface delays; they are hooked up in such a way that they initiate detonating cord downlines (Figure 5.17). The direction NTD's are connected to the blastholes determines in what sequence the shotholes will detonate. The NTD's have delay times ranging from 5 to 200 ms.

The best solution is definitely to use an in-the-hole delay time considerably greater than the total surface-delay time. The NONEL UNIDET system (Figure 5.18) has an in-the-hole delay equal to 450, 475, or 500 ms which often is enough. However, Hansson [1985] reports about an extended NONEL UNIDET system where the in-the-hole delay was 2000 ms allowing a mass blast of 1713 holes in the top and bottom crown pillars at the Research Mine of Luossavaara, Sweden. The surface delays were 17, 25, and 42 ms and total surface delay was 1897 ms, i.e., 103 ms less than the in-the-hole delay. The delay times were accurate, with a low scattering and low probability for overlapping. This could be maintained as the delay charges were taken from the same batch.

The NONEL UNIDET simplifies the charging work as the same assembly is used in all the shotholes. The blasting foreman does not need to decide the delay times until

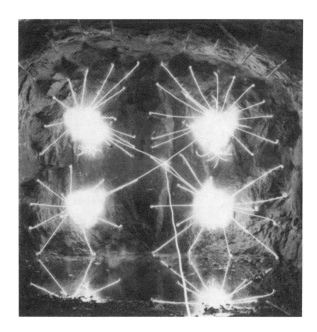

Figure 5.16. Connecting a NONEL tunnel round.

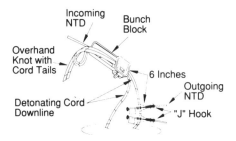

Figure 5.17. Example of a Noiseless Trunkline Delay system (NTD) (from Ensign Bickford).

The NONEL UNIDET simplifies the charging work as the same assembly is used in all the shotholes. The blasting foreman does not need to decide the delay times until after all holes are charged with explosive and detonators. Sometimes, conditions can change during explosive loading to influence the delay pattern. The UNIDET system also makes it possible to avoid storing a large assortment of delay detonators of different delay period numbers.

5.7. Detonating Cord

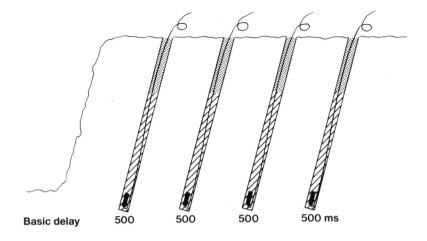

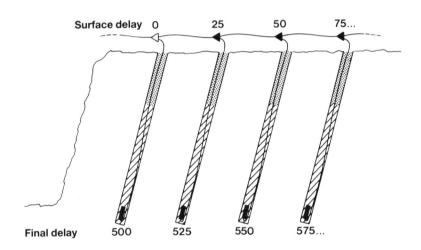

Figure 5.18. Principle of the NONEL UNIDET system.

5.7.2 Hercudet Initiation System

The Hercudet non-electric initiation system (Figure 5.19) is a non-electric detonator system which requires the same careful testing to ensure leak-free connections in the electric detonator systems. The system consists of a non-electric, pyrotechnic delay blasting cap connected to two plastic tubes (cap-leads) which must be filled by an explosive gas mixture which transmits the initiating signal by a gas detonation in the tubes. The outlet lead can be connected with a single tube trunkline leading to another hole and so on until the holes are hooked up in series or in several parallel series.

Some advantages with the Hercudet system are worth mentioning. First, the circuit can be checked with air and with a special pressure test module. The air flow will tell whether there is leakage or not. Second, the connecting tube system is completely inert

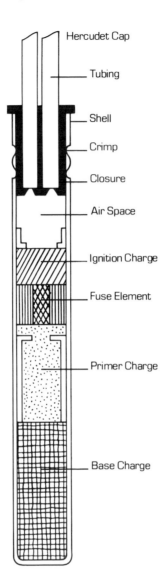

Figure 5.19. Hercudet detonator with 4 in cap-leads. 16- or 24-foot cap leads for underground applications are also available.

until charged with the proper fuel/oxidizer mixture.

However, the risk for damaging the tubes during charging might be somewhat larger for the Hercudet system than for the NONEL system since two plastic tubes are needed for each cap.

Initiation of the Hercudet system demands a blasting machine that first delivers a proper fuel/oxidizer mixture into the firing circuit and thereafter ignites the gaseous mixture by a spark plug. The fuel gas introduced into the tubes is a 50/50 mixture of methane and hydrogen with 6 ppm of an odorant added for leakage detection. Oxygen is used as the oxidizer; 60% oxygen and 40% of the fuel mixture are delivered into the circuit before ignition.

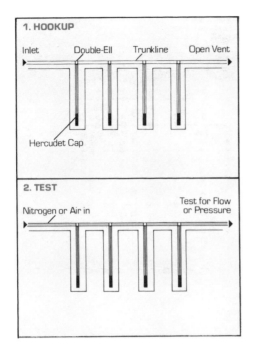
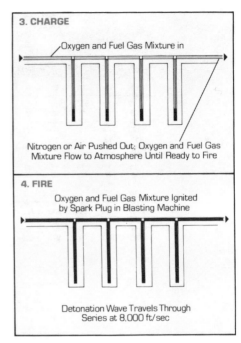

Figure 5.20. Principles for initiation with the Hercudet system.

The detonation wave travels through the tubes at 8000 ft/sec ($\approx$ 2400 m/s). Figure 5.20 shows the various sequences from hook-up to ignition.

As the combustible gas needs to flow through the entire circuit down each hole, through the cap, up the hole, to the next hole, and so on, one sometimes must consider the delay created by the finite detonation velocity in the gas. A hole depth of 20 m and a spacing of 5 m will correspond to an additional time delay of 19 ms $\left(\frac{20+20+5}{2400}\right)$ for each hole compared to its neighbor. This feature can be utilized to separate in time the detonation of several detonators of nominally the same delay time period for vibration control (Figure 5.21).

There are 16 short period delay intervals ranging from 50 to 850 ms. The long-period delay times range from 50 to 12.8 seconds in 17 intervals. Because of the greater simplicity of using the NONEL system, the Hercudet system has not reached widespread use; its production has recently been discontinued.

5.8 Electromagnetic Firing Methods

5.8.1 Nissan RCB System

An electromagnetic firing method reported by Nakano and Ueada [1983] was developed in Japan where underwater blasting operations were difficult to carry out by conventional methods due to rapid tidal current and deep water. The Nissan RCB System consists of three components: an oscillator, an exciting loop antenna, and firing elements equipped with receiver coils (Figure 5.22). The blaster set up (Figure 5.23) consists of a receiver coil, a firing condenser and an electron switch. By applying an alter-

170 Chapter 5. Initiation Systems

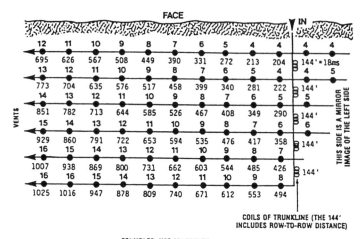

Figure 5.21. Nominal firing times for a pattern with 4.3 m (14 ft) row-to-row distance [from Hopler, 1980].

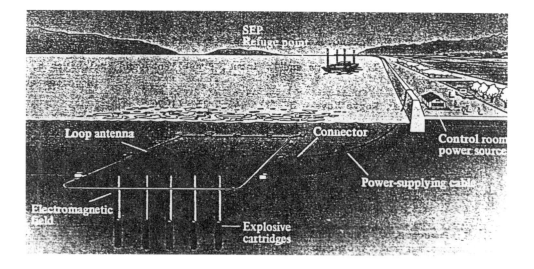

Figure 5.22. The NISSAN R.C.B. System for underwater blasting.

nating magnetic field with a frequency of 550 Hz using an alternating current through the loop antenna from a remote base, an alternating electric current is induced in the firing element. The alternating voltage is rectified to direct current via a diode and charged into the condenser. A firing signal of a different frequency triggers a separate circuit to discharge the capacitor charge through the bridgewires. The system shows promise of being simple and reliable to operate. A disadvantage is that it is not possible to use different delays for the different blastholes. Instead, one must divide the blasting operation into several rounds if ground vibration restrictions are present.

5.9 The Magnadet System

The Magnadet system developed by ICI Nobel's Explosives, Ltd., in the United Kingdom also employs a high frequency induced electromagnetic current to initiate electric detonators without having direct electric contact between the detonator and the blasting machine. The system is shown in Figure 5.24. Inside each key-shaped plastic housing is a toroid-shaped ferite magnetic core. The firing line carrying the alternating firing current is threaded through each toroid core. A transformer coupling is formed between the firing line and the leg wires which form a secondary winding around the magnet. For stray DC voltages, the bridgehead circuit is thus short-circuited by this winding providing a high degree of safety.

When the proper Magnadet delay cap has been inserted in the drillhole and the explosive loading is finished, a primary circuit is established by threading a connecting wire through the hole in the protective housing of the toroid of each detonator. The detonators are activated by an alternating current of 12 to 25 kHz sent out from a special blasting machine. Normal AC and DC power sources cannot activate the detonator, thereby making the Magnadet safe against stray currents. The Magnadet detonators employ pyrotechnic delays identical to those in regular electric detonators. A short and a long time delay series is available.

5.10 Electronic Detonators

In the past, all in-the-hole delays have employed a slow-burning pyrotechnic charge to create the time delay. The manufacturing of these delays demands a high level of quality control in order to keep the scatter sufficiently low that overlapping can be avoided. The scatter and the actual firing time can deviate somewhat from batch to batch in large-scale production, and the cap storage time and storage conditions may influence the burning time by various mechanisms of aging.

Over the last few decades, the demand for precision detonators has increased, partly because of a need for better blasting results, partly because of a need for better vibration control. The time scatter of present-day detonators is relatively high, and it increases with increasing delay time.

A number of authors in the past — Langefors and Kihlström [1963], Bergmann et al. [1974], Andrews [1980], Winzer et al. [1983], and Chiappetta and Borg [1983] — have reported small- and large-scale field experiments to determine the optimum delay time for best fragmentation. Langefors and Kihlström found the optimum delay time to be 3 to 5 ms/m burden. Others have reported the optimum delay to be in the range of 3 to 15 ms/m burden in bench blasting operations.

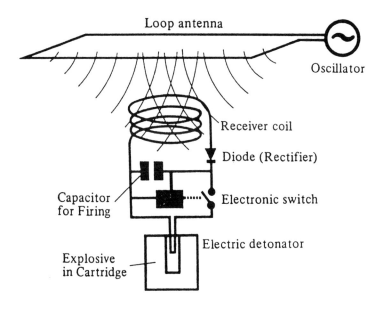

Figure 5.23. Principle for the electromagnetic firing method.

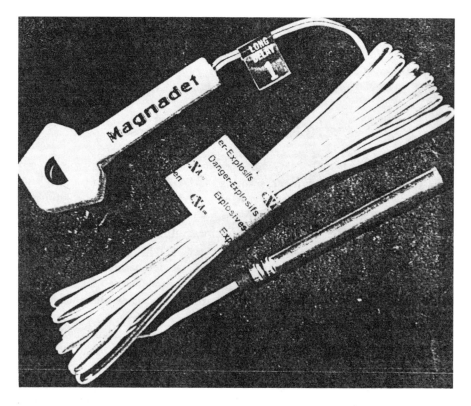

Figure 5.24. Magnadet initiation system (CXA Ltd.).

Studies by Beattie and Grant [1988], by König [1991], and by Norell [1985] indicate, however, that if more precise delay times are used it might be possible to achieve a better fragmentation.

The electric detonator with pyrotechnic delay is a superficially simple product. Increasing its timing precision would be a simple way of achieving the goals of high accuracy timing. In the following, we will discuss whether efforts in this direction are likely to succeed.

Most detonator manufacturers are now producing or developing high precision detonators with pyrotechnic delays, but there are certain limitations in both manufacturing and storage of the product. An electric detonator with pyrotechnic delay contains several elements in the chain that leads from ignition of the detonator to detonation of the charge in the drillhole. Each element involves a time delay, which is not exactly the same for all nominally equal detonators, i.e., each element has a certain amount of scatter in time. The scatter of the detonator's total delay time is influenced by all these elements.

The key elements are:

a. The fusehead (matchhead) composition should be such that it is ignited quickly by the bridgewire when the latter is heated by the electrical ignition current pulse

b. The fusehead ignites a limited quantity of an easily ignited pyrotechnic mixture, pressed on top of the delay element composition

c. The easily-ignited pyrotechnic mixture then must reliably ignite the delay element composition. The delay element must be manufactured with a minimum of variation in the burning rate, and it should burn stably at low temperature and low pressure

d. After it has burned through with a reproducible time delay, the delay composition ignites what is usually a primary explosive

e. The rapid burning of the primary explosive quickly transits into a detonation

f. Finally, the detonation of the primary explosive initiates detonation in the secondary explosive in the detonator

To keep within reproducible delay times, the raw materials in each batch of delay composition must be consistent and of even quality. Ingredients of low hygroscopicity should be used. Even so, the burning rate of the pyrotechnic composition may be affected by moisture during transport and storage if the temperature and relative humidity change. Many chemical ingredients in pyrotechnic delay compositions need to be handled with care because they are suspected of being a health hazard. Chromates, for example, formerly used extensively as oxidizing agents, are potentially carcinogenic. Authorities are also concerned about the possible health and environmental hazards of substances emanating from the detonators after the blast; for example, lead from the lead oxide Pb_3O_4 or lead dioxide PbO_2 used as alternative oxidizers in the pyrotechnic composition, or from primary explosives such as lead azide $Pb(N_3)_2$.

The practical difficulties of controlling the above mentioned key elements of the pyrotechnic delay detonator to achieve a radically increased timing precision appear to be insurmountable.

The electronic delay circuit offers such a radically increased timing precision. Thus, for a quantum step toward radically improved timing precision, most detonator manufacturers are looking toward replacing the pyrotechnic delay element with an electronic delay circuit. Using integrated circuits, timing precision will be measured in microseconds rather than in milliseconds as is the case for pyrotechnic delays. Figure 5.3 shows

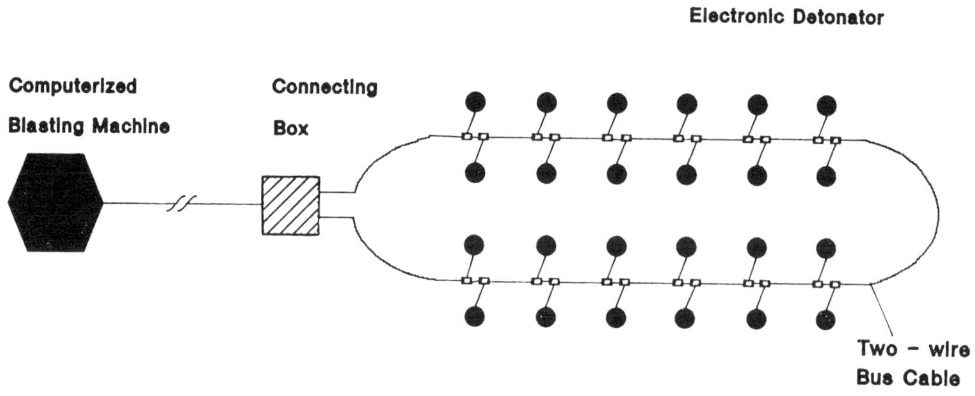

Figure 5.25. Scheme for connecting electronic detonators.

the historical development of the time precision of detonators, with the electronic delay detonator included.

Several authors, Svärd [1992], König [1991], Grant [1990], Hinzen et al. [1987], Niklasson and Keisu [1992], and Sakamoto et al. [1989] have reported encouraging blasting results with the use of extremely high-precision delay times.

Edwards and Northwood [1960], Siskind et al. [1980], Medearis [1977], Clark et al. [1983], Anderson et al. [1982], and Dowding [1985], among others, have indicated that blast damage to structures due to blast-induced ground vibrations is a function of the vibration frequency as well as its particle velocity. Several authors have concluded that, with higher precision firing times, it might be possible to deliberately modify the frequency spectra of the ground vibration wave by setting off the detonators in a round with precisely tuned intervals and thereby avoiding structure resonant frequencies.

High accuracy timing can help to develop new blasting methods where better fragmentation control can be achieved. It would also be possible to tailor the blast in such a way that, at a given distance, there will be minimal ground vibration disturbances. Additionally, massive blasts using a large number of shotholes could be fired since the system would allow a greater number of delay intervals.

Many explosives manufacturers are now developing their own electronic detonators and it appears likely that electronic detonators will be commercially available in quantity in the future. However, the cost of manufacturing the microchip is high compared to that of the simple pyrotechnic delay element, and it will be some time (from the writing of this book in 1993) before the price of an electronic detonator comes down to that of present-day electric detonators.

Even so, the benefits of using high-precision electronic detonators costing many times the price of electric detonators in many cases are such that the total cost for a blasting job requiring high precision may be lower using the expensive electronic delay detonators in place of ordinary electric detonators. The ongoing research into the use of detonators with exact timing shows that electronic detonators definitely have great potential in the following areas:

a. Improving contour blasting and decreasing the need for rock support
b. Controlling ground vibrations
c. Controlling rock fragmentation and heave

Worsey and Lawson [1983] have discussed the advantages with electronic detonators. Some of the features they mentioned are:

1. Safety features, including a unique fire control command which eliminates the majority of accidental electrical initiation
2. On-line programmability such that a single detonator may be programmed for any delay
3. A factory-programmed security code unique to the operator which will provide "ultimate security" and exclude unauthorized use
4. Interactive report-back facilities for checking the status of the detonator and making a circuit check before firing

As an example of the state of the art in the field of electronic detonators in 1993, the following is a description of the detonator designed by Nitro Nobel AB in Sweden.

5.10.1 Nitro Nobel's Electronic Detonator

Nitro Nobel's electronic detonator system has two closely interacting main components: the detonator and the blasting machine. Both are necessary for the proper functioning of the system.

An outline of a round using this system is shown in Figure 5.25. The detonators in the round are connected in parallel with arbitrary polarity. This is done by connecting each detonator to a two-wire bus cable via a terminal block, using special pliers. It is not necessary to strip the lead wires or the bus cable before connecting. If an error occurs, or if the detonator's safety function is triggered, this is automatically detected by the pliers. This measuring function can be repeated after completed connection. Finally, the bus cable is connected to the blasting machine via a terminal box and a firing cable.

Detonator

From the exterior, the electronic detonator looks exactly like a conventional electric detonator — it has the same dimensions and has two wires. The detonators are marked with delay period numbers between 1 and 250.

The overall interior design of the electronic detonator is shown in Figure 5.26. In principle, the detonator consists of an electronic delay unit in combination with an instantaneous detonator.

An integrated circuit on a microchip (4) constitutes the heart of the detonator. In addition, the detonator cap also contains a capacitor (5) for energy storage, and separate safety circuits (6) on the input side (toward the lead-in wires) in order to protect against various forms of electric overload. The chip itself also has internal safety circuits in the inputs. The fuse head (3) for initiating the primary charge (2) is specially developed to provide a short initiation time with a minimum of time scatter.

The most striking characteristic of the detonator is its flexibility. The period numbers, for example, do not state the delay time, but only the order in which the detonators will go off. Each detonator has its own time reference, but the final delay time is determined through interaction between the detonator and the blasting machine only immediately before initiation.

176 Chapter 5. Initiation Systems

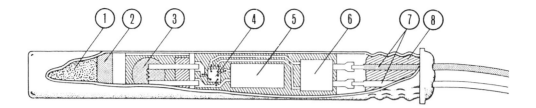

1 Base charge (PETN)
2 Primary explosive (Lead azide)
3 Matchhead with bridgewire
4 Integrated circuit chip
5 Capacitor
6 Overvoltage protection circuitry
7 Lead-in wires
8 Sealing plug

Figure 5.26. Electronic detonator.

Different electronic detonator systems can be selected based upon the customer's requirement for flexibility. Demands for a high degree of safety against accidental initiation lead to similarities in the (different) designs produced by different manufacturers. Typical characteristics for the electronic detonator include:

a. The detonator initially has no initiation energy of its own.

b. The detonator cannot be made to detonate without a unique activation code.

c. The detonator receives its initiation energy and activation code from the blasting machine.

d. The detonator is equipped with over-voltage protection. Low excess loads are dissipated via internal safety circuits. Higher voltages (>1000 V) are limited by means of a spark-gap. Large excess loads will burn a fuse in the detonator which incapacitates it, without making it detonate.

e. The initiation system operates with low voltages (<50 V), which is a great advantage considering the risk of current leakage from the lead-in wires.

Blasting Machine

The blasting machine constitutes the central unit of the initiation system. It supplies the detonators with energy and determines the delay time to be allocated to each period number. The unit is microcomputer controlled and its mode of operation can thus be altered with various control programs, while it can be uniformly designed from a hardware point of view. This gives part of the flexibility of the system.

The objective is that the operator at the blasting site should have a simple and well-known system to handle. Therefore, the controls for initiation of the round were designed to appear as conventional as possible in spite of the advanced internal design of the unit. A panel with lamps indicates what is happening and gives the go-ahead signal when the round is ready to be fired. If any errors are found, they are indicated on the panel and the blasting machine resets the system.

Delay time allocation to the detonators is carried out by uniquely coded signals to eliminate any possibility of error. The detonators do not react to any other code than the one from the Nitro Nobel blasting machine, and the risk for unintentional initiation because of spurious signals from other energy sources is thus eliminated. The blasting machine also performs an operation status control. This is done automatically by the machine. The ready signal for firing is given only after an approved result of this check.

System Characteristics

Typical system characteristics are:

a. The shortest delay time between two adjacent period numbers (equal to the shortest interval time) is 1 ms
b. The longest delay time is 6.25 sec
c. A detonator with a lower period number cannot be given a longer delay time than a detonator with a higher period number
d. Detonators with different period numbers cannot be closer to each other in delay time than the difference in their numbers. (For example, the interval time between No. 10 and No. 20 must be at least 10 ms)
e. The maximum number of detonators connected to each blasting machine is about 500

These rather unrestricted time allocation rules indicate that, to achieve the desired delay times for a particular round, many different number combination can be used. In practice, this means that the user, for most rounds, only needs a sufficient number of detonators with different period numbers in stock (not detonators with certain fixed time delays as is the case with present electric detonators). Figure 5.27 shows the regularity of the detonations in a round of 54 holes initiated with a series of 54 electronic detonators having a 12 ms interval time. It shows a ground vibration record from a round of 54 charges set off by 54 electronic detonators with 12 ms interval time. The vibration peak from each individual detonation can be clearly seen in the record in spite of the short interval time.

Mode of Operation

The preparatory work for a blasting operation includes determining the delay time for each blasthole in the round and charging the holes with detonators with suitably chosen period numbers. The blasting machine's time memory is then programmed with the necessary time information adapted to the period numbers chosen. This can be carried out with a computer or with a special programming unit connected to the blasting machine.

5.11 Safety Fuse

The safety fuse, or black powder fuse, consists of a black powder core wrapped in textiles and covered with waterproof materials, traditionally bitumen and wax. Normal unconfined burning rates for fuses are in the range 100 to 120 sec/m, corresponding to a propagation velocity of 8 to 10 cm/s, but safety fuses can be manufactured with much higher burning rates. Such cords are called *igniter cords*. Plant-manufactured safety fuse assemblies (Figure 5.2 (a)) are available in which the blasting cap and a length of safety fuse are connected together by means of a waterproof crimp. (Detonators in which initiation is by means of a safety fuse are called *plain or ordinary detonators*.)

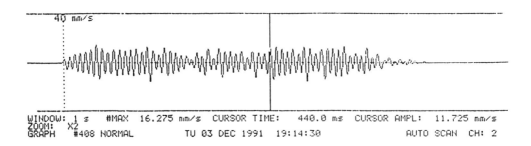

Figure 5.27. Ground vibration record from a round of 54 drillhole charges initiated by 54 electronic detonators with 12 ms interval time, showing clearly the vibration peak from each individual detonator. The figure is a good illustration to the high timing precision of the electronic detonators.

The igniter cord connectors are aluminum or brass devices for connecting the igniter cord to the safety fuse. Special thermalite connectors with an internal ignition charge that burn with an intense heat are also available.

In multihole shots, the capped fuses with their igniter cord connectors are cut to a uniform length and connected to an igniter cord at equal spacing. The burning rate of the igniter cord is chosen so that

a. The time lapse for the igniter cord to burn between successive cord connectors is longer than the scatter in burning times between fuses.
b. The igniter cord burning rate should preferably be high enough that all fuses are lighted and burning inside the shot holes before the first hole detonates. If this is not possible, one must be sure that the igniter cord flame front is sufficiently far away from the detonating holes so as not to be affected by them.

For large-scale application of multihole blasting with igniter cord, it is necessary to have a variety of cords with different burning rates. For example, in 1985, AECI manufactured seven different types of igniter cord with burning rates from about 3 to 40 sec/m.

Safety fuse is becoming less widely used for multiple shots as safer initiation systems are available with precise time delays. Still, in South African gold mines for example, the safety fuse is the predominant initiation system in use due to its low cost. In the beginning of the 1980s, 40% of the silver produced from the Coeur d'Alene mining district in the USA. was from cap and fuse blasting. This accounted for 20% of the entire US consumption of safety fuse.

5.12 Primer or Booster Charges

In large-diameter drillhole blasting, it is necessary to use a special primer charge, sometimes also called a *booster charge*, to initiate insensitive explosives and blasting agents such as ANFO and non-cap sensitive slurries, watergels, and emulsions. The primer explosive must be sensitive enough to be reliably initiated by the detonator, or by a detonating cord led through a hole in the primer, and the primer charge must be large enough to initiate the insensitive explosive. The choice of primer must take into consideration the diameter of the borehole, the explosive to be initiated, and the confinement.

5.12. Primer or Booster Charges

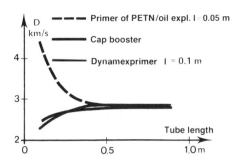

Figure 5.28. Detonation velocity of ANFO-K2Z (50 % prilled AN, 50 % crushed AN prills) in 52 mm steel tubes and initiated with different primers [from A. Persson, 1975].

The effectiveness of a primer is determined by its detonation pressure, diameter, its length. Although it is well known that the larger the diameter, the more effective is the booster charge, most commercially available booster charges have a length to diameter ratio of two or more. For use in wet or water-filled holes, the primer explosive also must be unaffected by being immersed in water, while waiting for the main charge explosive to be loaded.

To understand the requirements of a good primer, we need to understand the mechanism of shock initiation. When a shock wave of high amplitude propagates into an undisturbed explosive, the explosive will be heated due to compression and energy dissipation at the shock front. The shock wave will lose its strength and attenuate if no chemical reaction starts in the explosive to restore the mechanical energy expended. Local hot spots in heterogeneous explosives will be heated to a higher temperature than the surrounding explosive matrix. If the shock wave is strong enough, the hot spots will concentrate the energy so a chemical reaction is initiated and a decomposition takes place. If the heat production is higher than the heat losses to the surrounding explosive matrix, the reaction accelerates and a detonation wave is formed.

If the shock wave is weak, the chemical reaction, possibly started at the hot spot, will die out before the reaction has interacted with that at the neighboring hot spot.

Starting a detonation requires that detonation pressure and released energy (primer size) must exceed certain critical values, different for different explosives.

We learned in Chapter 4 that the Chapman-Jouguet detonation is represented by a steady-state detonation supported by the chemical energy release. If the shock at the detonation front is weakened due to, for example, local changes of energy content (bad mixing) or differences in salt particle sizes, etc., the shock wave will later return to its original stable condition when the explosive composition becomes normal again.

On the other hand, if the detonation wave transmitted from a primer to a blasting agent such as ANFO is stronger than the steady-state detonation wave of ANFO, an *overdriven* detonation occurs. This means that the released chemical energy is insufficient for maintaining the velocity of the detonation front, and the front will decelerate until the steady-state detonation velocity for ANFO is achieved (Figure 5.28).

Using a primer with a too low detonation pressure will cause an *underdriven* detonation in the ANFO charge. If the shock strength is of sufficient magnitude, the detonation velocity will increase asymptotically up to the steady-state detonation velocity; otherwise, it will fail.

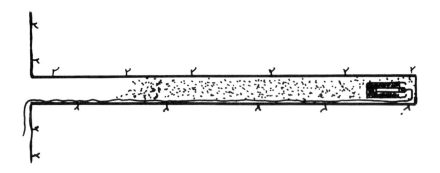

Figure 5.29. Bottom priming of ANFO with a correct cap direction.

Experiments carried out by Gert Bjarnholt of SveDeFo clearly show that the primer does not influence the detonation velocity of the primed explosive further away from the primer than 10 times the hole diameter.

The detonation pressure of the primer should preferably exceed the detonation pressure of the primed explosive. For estimation of the detonation pressure, one can use the formula

$$P = \frac{\rho_o D^2}{\gamma + 1} \qquad (5.8)$$

where P denotes pressure (N/m^2), D detonation velocity m/s, ρ_o initial density (kg/m^3), and $\gamma = -\left(\frac{\partial \ln P}{\partial \ln V}\right)_S$. We may use $\gamma = 3$ as an approximation at the detonation state for relatively high density explosives, which gives the widely used expression

$$P = \frac{\rho_o D^2}{4}. \qquad (5.9)$$

Primer manufacturers usually specify the detonation velocity, but they do not always state whether the detonation velocity was measured when the explosive was confined or unconfined or what the charge diameter was.

The primer should have sufficient length from the initiation point that there is time to reach a steady-state detonation velocity. A rule of thumb is that the length should be around 2 to 4 times the primer diameter, and this diameter should not be less than 0.5 times the hole diameter.

When the primer is initiated with a blasting cap, the cap should be directed with its bottom part toward the explosive to be primed (Figure 5.29).

Preferably, the shotholes should be bottom-primed for optimum breakage. When a hole is bottom primed, the gaseous detonation products are confined for a longer time period and perform better fragmentation work. Bottom initiation also usually produces less flyrock than when top initiation is employed.

For many purposes, dynamite is a sufficient primer; for others, the primer is made of pentolite (a cast mixture of PETN and TNT) or TNT with a central core of pentolite. Primers are often made with several axial holes for the detonator and detonator leg wires. Where in-the-hole delays are not needed, detonating cord can be used to directly initiate a primer by leading the detonating cord through the axial hole. A special type of sliding primer is used for so-called "deck-charging" in which the charge in the drillhole

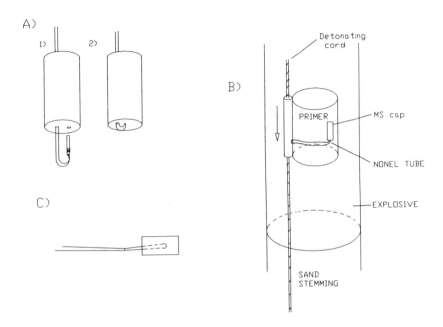

Figure 5.30. Primer charges: (a) cast pentolite primer 1 lb. with NONEL and MS cap, (b) sliding primer with detonating cord and NONEL delay detonator, (c) cap booster.

is separated in several "decks" by intermittent sand-stemming. Each deck is initiated by a special primer containing an individual delay detonator which is activated by the passing detonation front from the detonating cord. This stimulus is in turn conducted to the delay element in the initiating unit by means of a very low core load detonating cord 0.5 to 1 g/m (2.5 to 5 grains/foot) or by a NONEL shock tube. Figure 5.30 shows some examples of primer charges.

5.13 Invention of the NONEL and Other Shock Tube Based Non-Electric Detonator Systems

The non-electric fuse or shock tube which is the key element of the NONEL detonator system and many later non-electric detonator systems using the same technology was described by the inventor in his two base patents: P.-A. Persson [1967, 1968]. As described in these patents, the detonator system consists of a small diameter plastic tube having on its inner wall or on the surface of elongated elements inside the tube a light dusting of fine particulate explosive and/or reactive material. A shock wave introduced into one end of the tube was found to propagate at high velocity through the open channel inside the tube and the flame emanating from the other end of the tube was able to start burning in a pyrotechnic delay composition of the kind used in conventional electric delay detonators. The base patents also described means of initiating and branching the tubing to allow it to replace the electric wires or detonating cord in previously conventional blasting rounds. It may interest some readers to know, in the inventor's own words, how the invention came about:

"In 1965, I joined Nitro Nobel AB (then Nitroglycerin AB, Alfred Nobel's first company) as director of Nitro Nobel's Detonics Laboratory and the Swedish Detonics Research Laboratory, both at that time situated in Vinterviken in Stockholm, in the buildings that once had housed Alfred Nobel's first large scale plant for making detonators, nitroglycerin, and later dynamite. The former director of the laboratory, Ulf Langefors, had become the president of the company. Langefors was a visionary leader with a clear view of what would be demanded of future ignition systems for rock blasting. Langefors repeatedly voiced his conviction that a new non-electric blast initiation system was needed. The system he wanted was a fuse that involved the detonation of a minimal amount of explosive charge per unit length of the fuse. The detonation in such a fuse would pass through a drillhole filled with a sensitive dynamite without setting off or damaging the dynamite. Ideally, the fuse should end in a pyrotechnical delay detonator at the bottom of the drillhole where the initiation should then occur after the predetermined delay time had passed. It was Langefors' opinion that a person with a good grounding in detonation physics should be able to come up with such a system without much difficulty.

"At that time, the researchers at the Detonics Laboratory had completed a large program of research into the so-called channel effect. This was the effect of the air shock wave in the elongated air space or channel between a long stick charge and the walls of a drillhole with a diameter larger that of the charge. They had concluded that the air shock wave could de-sensitize explosives by compression ("dead-pressing") which could lead to detonation failures. They had also observed that the detonation could pick up again further down the channel. We speculated that the flow of hot, compressed air following the shock front might re-ignite burning which could transit to detonation in the explosive beyond the detonation failure point.

"To verify this theory, I devised an experiment (which was never performed) in which the stick of dynamite would be replaced by a stick of wood, the outside of which was smeared with Vaseline and dipped in PETN powder. I thought that the air blast along such a charge placed in a larger drillhole might initiate rapid burning of the PETN powder.

"Our laboratory librarian, Monica Petrini, a linguistics scholar, had become interested in the experimental work with explosives, and asked if she could perform an experiment, too. I suggested that she could take two 6 mm diameter paper bakelite tubes, smear Vaseline on the inside wall of one of them, and ask our explosives scientist to pour some PETN powder through it so that some PETN would stick to the inside tube wall. She should then insert a detonator into the end of each tube. After detonation, she should take careful note of the length of each of the remaining tubes. I thought that a longer portion of the tube with PETN might be destroyed near the detonator than of the tube without PETN. This would indicate that some of the PETN had been ignited by the shock wave. Monica came back and told me the experiment had not worked as I expected, because the entire tube with the PETN had disappeared! As Ulf Langefors had predicted, once such a fine lead had presented itself, it was indeed not difficult for me to conclude that a longer tube with a fine dusting of PETN or other reactive material on the inside wall might be used as just such a fuse system as he had wanted. In the production process, never patented but long kept secret, the reactive powder was added during extrusion of the hot Surlyn plastic tubing. Therefore, the more thermally stable fine HMX and aluminium powder mix later replaced PETN."

Chapter 6

Principles of Charge Calculation For Surface Blasting

6.1 Introduction

The aim in this chapter is to provide an understanding of the available techniques for selecting the type of explosive to use for a specific application, and for calculating the charge distribution within each drillhole, the pattern of drillholes within the blast, and the fragment size distribution of the broken rock.

In the previous chapters, we have reviewed and laid a foundation of specialized knowledge and understanding of the four major elements, or pillars, of the science of rock blasting:

The rock mass. An inhomogeneous structure of blocks of mostly hard, brittle rock materials, separated by weaker joints. The rock material's strength is low in tension but increases with confining pressure; the joints have an even lower tensile strength, a shear strength that increases with confining pressure, and often a much lower elastic modulus than the adjacent rock materials.

The mechanical processes of drilling and boring. Drilling, using hard tungsten carbide tipped drillbits and powerful rock drilling machines, makes it possible to place the explosive charge inside the rock mass where it can create mainly tensile as well as compressive stresses. Full-face boring, the technique of mechanical excavation of the entire tunnel without the use of explosives, provides the competition for rock blasting, a stimulus for further development of tunneling techniques.

The explosive material. A convenient, compact power and energy source has the ability to rapidly burn or detonate, thereby creating high pressures inside the drillholes, which sends shock waves and stress waves into the surrounding rock mass. A variety of different explosive materials provide a wide register of different detonation properties and rates of power release. These can be selected to suit the type of rock to be fragmented, within the confines of economic factors such as the selling price of the ores or minerals or ballast material excavated, the raw material cost of the explosive itself, the cost of fuel and other energy costs, and the cost of drilling and the alternative, mechanical excavation or tunnel boring.

The detonator and booster system. The means of triggering, or initiating, the explosive energy release in many drillholes cooperating in sequence, so that the explosive work is done in an orderly fashion, always working towards a free surface, to create a maximum of tensile stress and a minimum of compressive stress within the rock mass.

In the charge calculation, we tie these factors together and attempt to optimize the whole operation of rock blasting. The charge calculation is the engineering design of the blasting round. In its equations, we evaluate and weigh all the factors that influence the blasting fragmentation such as the type of rock mass and its strength and density, the available drillhole diameter, the depth of drilling, the burden (which we define as the distance between the drillhole and the nearest free surface) and the spacing (the distance between the drillholes in a row along the free surface), the properties of the particular type of explosive selected, and the time interval delay detonator system available. In this chapter, we will evaluate these variables which influence the charge calculation profoundly. The ultimate composite charge calculation equations become very complex when all the important factors are considered. For many purposes, and within limited ranges of the parameter values, simplified formulae and even rules of thumb based on experimental observations are sufficient. These are given later in this chapter.

In order to successively outline for the reader the importance of each factor upon the charge calculation, the variables will be isolated one by one, to clarify how the blasting result is affected as each variable changes. Finally, the variables will be connected to each other, resulting in the basic equations of rock blasting.

The transformation of the art of rock blasting into an engineering technique as presented in this chapter to a large extent is due to the pioneering work of Langefors and Kihlström which they described in their book, *The Modern Technique of Rock Blasting* [1963].

To get a proper understanding of rock blasting engineering, the reader should not immediately rush off to the final, simplified formulae, but first try to understand each single step, thereby arriving at the simplified formulae with a deeper understanding of the reality they represent.

6.2 Geometry Effects

The presence or absence of a free surface near the charge profoundly influences the ease with which the charge can break loose and fragment the surrounding rock. Without any nearby free surface, the charge can only damage the rock immediately surrounding the charge, hence no real fragmentation of the rock takes place. If a charge is placed in a drillhole at the center of a boulder, the entire boulder will be fragmented (if the charge is large enough). In the former case, we say that the charge is fixed. In the latter case, the rock can expand freely in all directions. In the this section, we will discuss a range of intermediate cases, and we will introduce the concept of the "degree of fixation" to put numbers to (quantify) the effect.

The size of the charge, and the linear scale of the rock mass to be fragmented by that charge, has an important influence on the rock blasting capacity of a given mass of explosive. This will also be discussed in detail below.

6.2.1 Degree of Fixation

Students of the science of rock mechanics are familiar with the fact that the strength of rock becomes higher as the surrounding stress field increases.

If the possible motion of the rock surrounding a drillhole is very constrained, such as when the distance between the drillhole and a free surface is large or when the number of available free surfaces is low, the explosive consumption for breaking loose the burden can be very high. Conversely, when there are free surfaces available in all directions from the drillhole, such as in boulder blasting, very little explosive is needed to break the rock. Figure 6.1 illustrates a number of different geometries and the corresponding relative charge weights needed for breaking loose the rock. The burden, i.e., the distance from the charge to the nearest free surface, has been kept constant in all these examples. The relative charge weights differ considerably depending on the geometry of the drillhole in relation to the free surface(s). For example, the charge weight needed to blast out the burden in crater blasting, where the drillhole is at right angles to the single free surface, a charge weight between 2 and 20 times the charge weight used in common bench blasting is needed. Conversely, in boulder blasting, where there is a free surface in all six directions, a charge size no greater than 25 % of that needed for common bench blasting is sufficient.

A comparison of the two examples A and C in Figure 6.1 shows that inclined holes break out the rock more easily than vertical holes. This advantage with inclined holes can be used to increase the burden and still keep the charge at the same level as for vertical holes. Thus, much more rock can be removed per unit weight of explosives and per unit length of drillhole. If the drilling equipment allows, it is often advantageous to use inclined holes to reduce costs or improve fragmentation.

For bench blasting with a fixed bottom, we can introduce a fixation factor f, which is defined as a function of the hole inclination n

$$f = \frac{3}{3+n} \qquad (0 \leq n \leq 1) \tag{6.1}$$

where the inclination n is given in m/m (horizontal meters per vertical meter) with $0 \leq n \leq 1$. n is thus the cotangent of the angle α between the axis of the drillhole and its projection on the free surface. ($\alpha = 90°$ is perpendicular to the free surface.) $n = 0$ indicates a drillhole at right angles ($\alpha = 90°$) to the horizontal free surface. Note that Equation 6.1 is only valid for holes inclined 45° to 90° relative to the free surface.

Figure 6.2 shows how the inclination affects the fixation factor. Empirically, it has been shown that the burden B is proportional to the square root of the inverse of the fixation factor f:

$$B \propto \frac{1}{\sqrt{f}} \tag{6.2}$$

Thus, for the same mass of explosive, the larger n is (i.e., the more inclined away from the vertical the holes are) the lower the fixation factor will be, and the larger the burden.

6.2.2 Size Relations

As early as 1725, Belidor [1725] showed that one part of the charge weight W_o used in blasting was proportional to the volume excavated, and another part was proportional to the surface area of that volume. This gives the relationship

$$W_o = k_2 B^2 + k_3 B^3 \tag{6.3}$$

where B is the burden and the k_i are constants to be determined.

Later, Langefors and Kihlström showed that the concentrated charge needed for breaking the rock can be expressed as a function of a power series by

$$W_o = \sum_{i=0}^{n} k_i B^i \tag{6.4}$$

Through calculations and experiments, they found that in ordinary hard gneissy bedrock with geometrical conditions as in Figure 6.3, the charge weight can be given as

$$W_o = a_2 B^2 + a_3 B^3 + a_4 B^4 \tag{6.5}$$

186   Chapter 6. Principles of Charge Calculation For Surface Blasting

Remarks.	No. of free surfaces available.	Relative charge size.
A. Common bench blasting	2	1
B. Inclined holes 0.5 m/m	2	0.85
C. Free bottom	2	0.75
D. Boulder blasting	6	0.25
E. Crater blasting	1	2 - 10

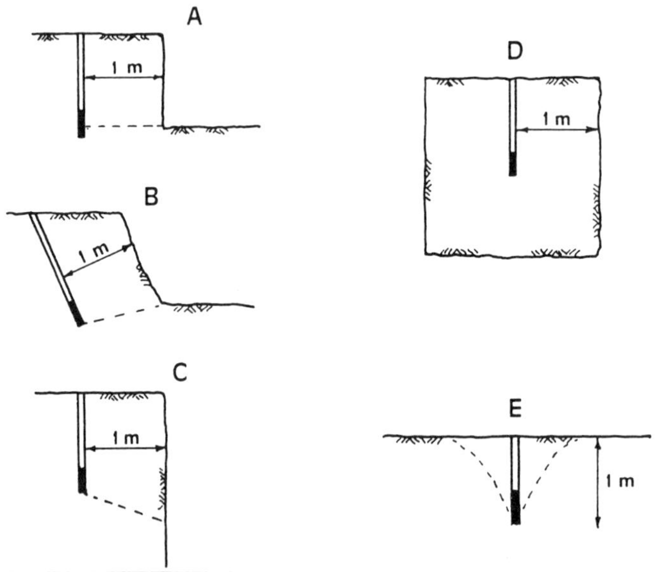

Figure 6.1. Influence of various geometries upon required charge weight for breaking loose a burden of 1 m.

Simply expressed, the first term $a_2 B^2$ represents energy losses that arise as new free surfaces are created, the second term $a_3 B^3$ represents the consumption of energy associated with the volume of rock removed, and the last term is the effect of gravity. The gravity term $a_4 B^4$ represents the energy needed to move or lift the center of gravity of the rock being removed against the force of gravity. This term is often called the *swelling* or *throwing* term and must be taken into account separately depending on the blast.

6.2. Geometry Effects

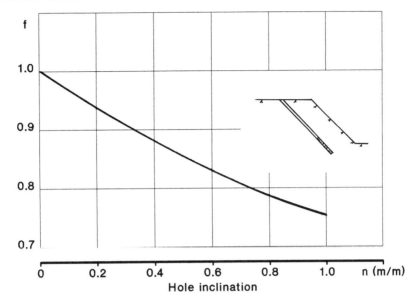

$n = 0$ (in m/m) for a vertically drilled hole.
$n = 1$ (in m/m) for a 45° slanted hole.

Figure 6.2. Fixation factor f as a function of the hole inclination n for bench blasting with a fixed bottom.

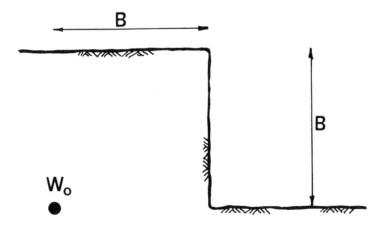

Figure 6.3. The simple case of a concentrated charge (W_o) in a geometry with equal burden and bench height (fixation factor = 1).

188   Chapter 6. Principles of Charge Calculation For Surface Blasting

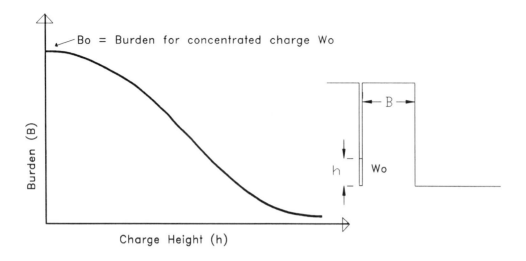

Figure 6.4. Effect of breaking capacity for elongated charges with equal weight.

In considering the feasibility of carrying out blasting operations on the moon or on Mars, the gravity factor needs to be adjusted to fit the local gravity field.

Practically all blasting operations are carried out with elongated charges instead of concentrated charges. It is clearly more efficient to utilize a sizeable portion of the length of the drillhole for charging with explosive, rather than have a concentrated charge at the bottom of an otherwise empty drillhole. Therefore, we need to understand what happens to the breaking capacity if the concentrated charge W_o is replaced by an elongated one of the same total weight.

Intuitively, one would expect the breaking capacity to decrease when the originally concentrated charge W_o is distributed over hole length h. This would happen if a given charge mass were used in a drillhole of smaller diameter and it does lead to a decrease in the burden. Figure 6.4 shows schematically how the breaking capacity of a charge of a given weight decreases with increasing charge length.

Instead, if we keep the hole diameter constant and increase the hole length, and at the same time we observe how large (relative to the elongated charge) the concentrated charge must be in order to perform the same work, the result shown in Figure 6.5 is obtained.

Figure 6.5 shows that, if the charge height increases from zero to about $0.3B$, there is no noticeable change in the breaking capacity relative to that of a concentrated charge with equal weight. Almost 100 % of the charge with length $0.3B$ is useful in obtaining breakage. When the charge height is increased to $0.5B$, about 90 % of that elongated charge weight is useful. At a charge height equal to the burden, only about 60 % of that elongated charge is useful.

Obviously then, it is not worthwhile to extend the charge to a height greater than $h = B$ because very little of the extra energy supplied will be of use in breaking out the rock at the toe.

However, it is possible to increase the breaking power by subdrilling under the pit floor, thereby extending the charge length into the subdrilling. A charge length of $0.3B$ placed below the pit floor is utilized almost to 100 %.

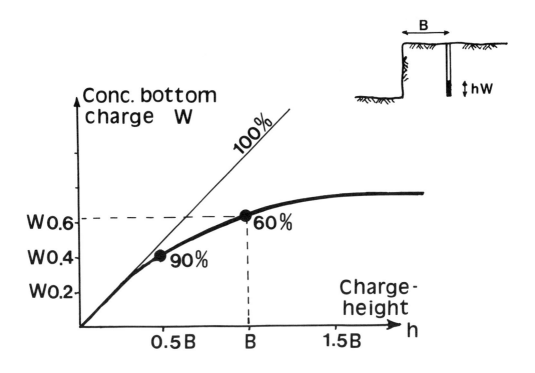

Figure 6.5. Breaking capacity at the toe of an elongated charge.

In the region of the drillhole away from the bottom, the top charge W corresponds to a concentrated charge equal to $0.6W$ and the subdrill charge $0.3W$ corresponds to a concentrated charge equal to $0.3W$. This implies that the total $1.3B$ long charge with a weight of $1.3W$ has the same breaking power as a concentrated charge with a weight of $0.9W$.

Thus, the elongated total bottom charge W_b is given by the charge height and the linear charge concentration ℓ_b (weight explosive per borehole length) as

$$W_b = 1.3 B \ell_b . \tag{6.6}$$

This charge corresponds to a concentrated charge W_o with the weight

$$W_o = (0.6 + 0.3) \ell_b B = 0.9 \ell_b B . \tag{6.7}$$

For calculation of the bottom charge for a single hole, the linear charge concentration ℓ_b is now given by combining Equations 6.5 and 6.7

$$0.9 \ell_b B = a_2 B^2 + a_3 B^3 + a_4 B^4 \tag{6.8}$$

or

$$\ell_b = 1.11 \left(a_2 B + a_3 B^2 + a_4 B^3 \right) . \tag{6.9}$$

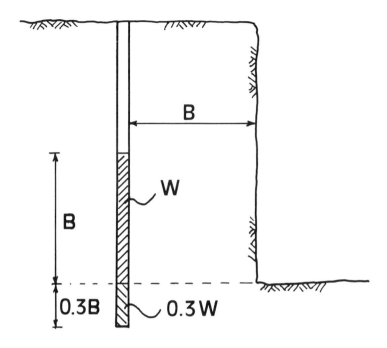

Figure 6.6. Favorable charge geometry for blasting with elongated bottom charges.

Equations 6.6 and 6.9 give *the bottom charge for a single hole* as

$$W_b = 1.3\, \ell_b B \approx 1.4 \left(a_2 B^2 + a_3 B^3 + a_4 B^4\right) . \tag{6.10}$$

Evaluation of test blasts has determined the coefficients a_2, a_3, and a_4 to be

$$a_2 = 0.07 \qquad a_3 = c \qquad a_4 = 0.004 \tag{6.11}$$

for B in meters and W in kilograms.

The constant c, corresponds to the amount of explosive (kg/m^3) needed for breaking loose the rock at the toe in a defined blasting geometry. Since c depends on the type of rock material in which the blasting is done, it was therefore called the *rock constant* by Langefors and Kihlström. c ranges from 0.2 to 0.6, with an typical value of 0.4.

Early experiments had shown that a concentrated charge can break the face up to a height above the pit floor equal to the burden B. The equivalent extended bottom charge of length $1.3B$ extending from a point $0.3B$ below the pit floor is able to break both the toe and a bench height up to $2B$ above the pit floor. Figure 6.6 shows such an arrangement.

For efficient utilization of the explosive charge, the drillhole above the charge should be *stemmed*, i.e., filled with coarse sand or crushed rock. Often the drill cuttings are used for this purpose, although a better stemming effect is obtained with coarser material.

For small hole diameters (≤ 2.5 inch), it is sufficient if the stemmed length h_s is equal to B. For large hole diameter bench blasting operations in hard rock, usually the stemmed or unloaded hole length has to be smaller than the burden if satisfactory fragmentation is to be achieved. A good rule of thumb, which gives satisfactory results

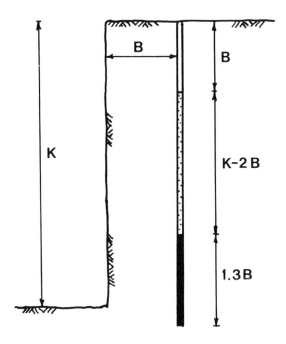

Figure 6.7. Charge distribution with full bottom charge, column charge, and unloaded hole length at the top equal to burden B.

for large and small drillhole diameters, is to use a stemming length equal to the square root of the maximum burden (expressed in meters) plus 1 meter, that is, $\sqrt{B+1}$.

If the bench height is higher than twice the maximum burden $2B$, an additional charge (the column charge) will be needed above the bottom charge to get satisfactory breakage along the face.

If the same type of explosive and the same linear charge concentration is used for the bottom charge and the column charge, the top part will be overcharged. Langefors and Kihlström confirmed, by test blasts in slabs of granite, that only 40% of the energy needed per meter length of drillhole in the bottom charge is needed as a column charge. This can be achieved by using smaller diameter cartridges or an explosive with less energy per unit volume for the column charge. Figure 6.7 shows such a charge distribution.

6.3 Specific Charge

When supervising rock blasting, it is always useful to note the amount of explosive needed to break the rock. Changes in explosive consumption per ton or per cubic meter of rock is always an indication of changed rock conditions. In mining, it is common to use explosive consumption per ton of ore as a measure of the blastability; in civil engineering, it is common to use explosive consumption per cubic meter rock. The ore is sold at a price which is given in $-per-unit-weight; therefore, it becomes natural to think in terms of explosives weight per mined-out ton of ore, material consumption per mined-out ton of ore, etc. In a civil engineering project, the contractor is required to excavate a certain volume of rock and, for him, it is natural to relate everything to the excavated cubic meters of solid rock, sometimes called *solid cubic meters* to distinguish it from the expanded volume of the broken rubble.

Both the mining engineer and the civil engineer will use the term *specific charge*, although both would be more justified to use the civil engineering concept. The rock mass density does not influence the energy consumption for breakage to a large extent. Significant differences when blasting in a high density or in a normal density rock with the same strength will normally only be seen from comparison of the throw or the heave of the fragmented material.

In this textbook, specific charge q is defined as kilograms explosive per cubic meter of broken rock:

$$q = \frac{W}{V} \tag{6.12}$$

where q is the specific charge in kg/m^3, W is the explosive weight, and V is the solid volume of the rock broken out.

From Equation 6.10, it can be shown that for burdens in the range of 1 to 10 m, almost the same specific charge will be needed to break the burden. This range is the interval where the law of conformity applies and where scaling will give reasonable results. The middle term $a_3 B^3$ thus dominates in this range. For much smaller scale blasting, the first term $a_2 B^2$ becomes important. For extremely large blasts, the third term $a_4 B^4$ dominates.

We will assume there is a single hole in a bench loaded with explosive to a height of B above the pit floor and stemmed to $0.4B$. We will also assume that the angle of breakage will be 90°. From Figure 6.8, the volume of broken rock V will be

$$V = \frac{\sqrt{2}B \sqrt{2}B}{2} \times 1.4B \tag{6.13}$$

and

$$V = 1.4B^3. \tag{6.14}$$

Equations 6.10, 6.12, and 6.14 give the specific charge q

$$q = \frac{a_2}{B} + a_3 + a_4 B \tag{6.15}$$

As an example, consider blasting in granite with a rock constant $c = 0.35$. Equation 6.15 becomes

$$q = \frac{0.07}{B} + 0.35 + 0.004B. \tag{6.16}$$

Figure 6.9 shows the specific charge q as a function of burden B according to Equation 6.16. It can be seen from this figure that the first term in Equation 6.16 will become more important for very small burdens. For large burdens, more energy will be needed per cubic meter of rock to heave the rock mass sufficiently, so the third term will dominate.

Linear charge concentration ℓ is often used in blasting terminology especially when dealing with any type of perimeter blasting. The term *loading density* (used in the USA) is equivalent to linear charge concentration. The expression simply gives the amount of explosive W used per loaded borehole meter:

$$\ell = \frac{\pi d^2 P}{4} \quad \text{kg/m} \tag{6.17}$$

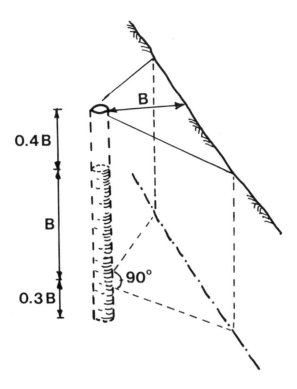

Figure 6.8. Single hole in a bench with burden B loaded to height B above the pit floor. Stemming length $= 0.4B$.

The angle of breakage 90° shown in the figure in reality varies with the burden. For small burdens, the angle is larger, approaching 180°. For large burdens near the maximum, the angle is often smaller. For simplicity, we use the value 90°, which gives the simple result that the volume of rock removed is B^2 times the hole length. In multi-hole blasting, the volume of rock removed by each hole is $B \cdot S$ times the hole length.

where d is the drilled hole diameter m, and P is the degree of packing (kg/m^3). If decoupled charges (i.e., charges in tubes that do not entirely fill the cross-sectional area of the drillhole) are used, d is the inner diameter of the tube. The degree of packing is given by mass of explosive per volume borehole.

A good rule of thumb for calculating the linear charge concentration is: *take the hole diameter in inches, raise it to the power of two, and divide by two*. This, surprisingly, gives the volume of the drillhole expressed in liters per meter drillhole length. Multiply this with the degree of packing P (expressed in kg/m^3), divide by 1000, and the result is the linear charge concentration:

$$\ell \approx \frac{d^2}{2} \frac{P}{1000} \quad \text{kg/m} \tag{6.18}$$

where d is in inches.

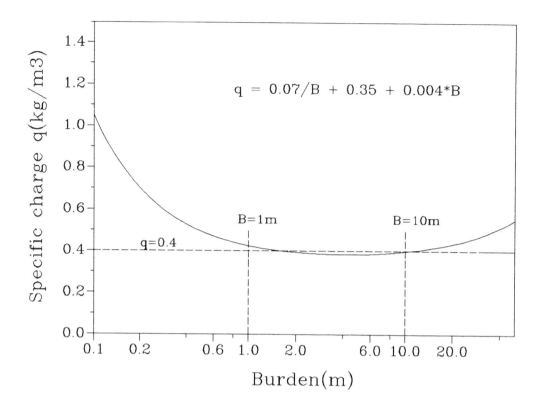

Figure 6.9. Specific charge as a function of burden.

6.4 Charge Calculations For Several Holes

When more than one hole is blasted in a round, several additional variables enter which all affect the blasting results — the spacing of the holes, the number of holes, and the delay time between initiation of adjacent holes. For simple blasting operations where the spacing S is equal to the burden B and where the delay time between adjacent holes is in the range of 6 to 15 ms/m burden, the charge in each hole can be reduced to about 80 % of the charge needed for a single hole.

Alternatively, if the spacing-to-burden ratio is 1.25 and, if this is considered together with Equation 6.9, the final equation for the linear charge concentration needed in a row of holes in a round will be

$$\ell_b = (0.8)(1.11)(1.25)\left(a_2 B + a_3 B^2 + a_4 B^3\right) . \tag{6.19}$$

Equation 6.19 can be written in the form

$$\ell_b = K \sum_{i=2}^{4} a_i B^i . \tag{6.20}$$

where $K = 1.11$.

Equations 6.19 and 6.20 are strictly valid for a row of vertical drillholes only if they are loaded with the explosive LFB dynamite which was used for the evaluation of the coefficients K and a_i. A correction term to take into account the strength of other explosives will be introduced later in this chapter.

The next two sections will describe the influence on the maximum burden of the three parameters: explosive performance, rock constant, and geometry.

6.4.1 Influence of Explosive Performance

To provide for the use of various explosives, it is necessary to have a basis for comparing the blasting "strength" of different explosives. Several methods have been developed to characterize the strength of an explosive. Examples of these are: the calculated explosion energy, the result of the ballistic mortar test, the Trauzl lead block test, the brisance test, the weight strength concept, and the underwater detonation test. However, these methods should be used with caution when stating the breaking capacity of an explosive in a rock material. The strength of the explosive is, in reality, dependent upon the type of blasting operation, and is different in different rock materials and in different blasting operations, for example, crater blasting, bench blasting, tunneling, etc. Just emerging are more sophisticated methods of ranking explosives for rock blasting by computer calculations, in which the burning rate of the explosive and the equation of state of its reaction products, determined experimentally, are matched with the experimentally determined dynamic properties of the rock mass. In effect, the entire rock blasting process is computer modeled. By using good judgement in the formulation of the equations used, and in the interpretation of the experiments, such modeling can give results in reasonable agreement with reality. (Computer modeling methods are described in Chapter 10.)

The best way to rank explosives, of course, would be to measure the rock-breaking capacity in different rock conditions with different blasting operations under different charging conditions. However, such an evaluation is prohibitively costly and time consuming. Instead, one is generally restricted to using one of the aforementioned methods for the comparison of strength.

In Sweden, the Langefors-Kihlström weight strength concept has been used extensively for the correlation of blastability of the rock mass with explosive strength.

The relation is written as

$$s = \frac{5}{6} \frac{Q_v}{Q_{v0}} + \frac{1}{6} \frac{V}{V_0} \qquad (6.21)$$

where s is the weight strength relative to a reference explosive (LFB-dynamite), Q_v is the explosion energy for 1 kg of the explosive used, V is the released gas volume at STP from 1 kg of the explosive used, $Q_{v0} = 5$ MJ, and $V_0 = 850$ liter.

Equation 6.21 is based on the fact that the expansion work performed by an explosive depends primarily on the heat of explosion and secondarily on the volume of the released gaseous reaction products. Table 6.1 shows the weight strength for some common explosives calculated using Equation 6.21.

The constants 5/6 and 1/6 in the formula were determined in field experiments where different explosives producing small or large gas volumes were used and compared to the standard LFB dynamite in bench blasting. In these experiments, a bench height of 7 m was used, and several rounds with 5 to 10 holes per round were blasted.

Table 6.1. Weight strength for some explosives.

Explosive	Q_v (MJ/kg)	V (m³/kg)	s_{LFB}	s_{ANFO}	Density (kg/m³)
LFB Dynamite	5.00	0.850	1.00	1.19	1450
Dynamex M	4.7	0.88	0.94	1.13	1400
ANFO	3.91	0.973	0.84	1.00	900
TNT	5.1	0.610	0.97	1.15	1640
PETN	6.38	0.717	1.20	1.43	1670
Nabit	4.42	0.904	0.91	1.08	1200
Gurit A	3.8	0.400	0.71	0.85	1000
NG	6.27	0.716	1.19	1.42	1590
Emulite 150	4.1	0.84	0.85	1.42	1200
Iremite 62	3.75	0.852	0.79	0.94	1180
Iregel RX	2.68	0.941	0.63	0.75	1200
Dynex 205	4.00	0.863	0.84	1.00	1170
"Powergel" 2131	3.29	0.810	0.71	0.84	1150
Kimit 80	4.10	0.74	0.89	1.06	1100
Emulet 20	2.4	1.12	0.61	0.73	220

The field experiments showed that the burden B was directly proportional to the square root of the weight strength s of the explosive, or

$$B \propto \sqrt{s}. \tag{6.22}$$

If the linear charge concentrations ℓ (expressed in kg LFB/m hole length to fit the parameter values $a_2 = 0.07$, $a_3 = c = 0.35$, and $a_4 = 0.004$) is given by the expression

$$\ell = \frac{\pi d^2 P s}{4} \tag{6.23}$$

then Equation 6.21 gives the weight strength s of an explosive in LFB dynamite equivalents, i.e., the number of kg LFB that give the same blasting effect as 1 kg of the explosive. For ANFO, $s = s_{ANFO} = 0.84$ (in kg LFB per kg ANFO). If we express the weight strength in ANFO equivalents, then

$$\ell = \frac{\pi d^2 P s_{ANFO} \cdot 0.84}{4} \quad \text{kg LFB/m hole length.} \tag{6.24}$$

Thus, for any explosive, the strength s_{ANFO} relative to ANFO is $s_{ANFO} = \frac{s}{0.84}$ (in kg ANFO per kg explosive).

The weight strength concept, defined by Equation 6.21, strongly overestimates the blasting strength of high-density, high-energy explosives such as aluminized TNT slurries and plastic dynamites with high nitroglycerin content. It also strongly underestimates the blasting strength of low flame temperature, low-density explosives such as ANFO and pure emulsion blasting agents.

Methods are now being developed for obtaining a better estimate of the blasting strength of these explosives, which also take into account the interrelationship between the blasting strength and the detonation velocity. The detonation velocity for most common rock blasting explosives decreases with decreasing hole diameter. It is also influenced by the density, compressibility, and strength of the surrounding rock material. The change in detonation velocity reflects profound changes in the rate of the chemical reaction. One method for estimating the blasting strength is described in Section 4.5.4.

In the absence of reliable weight strength data, the assumption can be made that most standard explosives give, to a first approximation, the same rock blasting effect per kilogram of explosive.

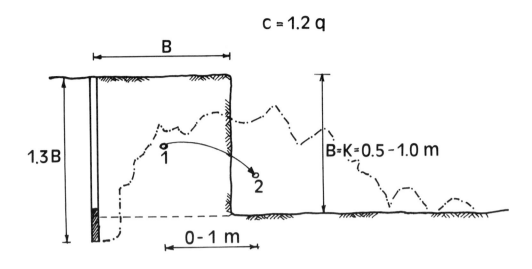

Figure 6.10. Blasting for determining the rock constant c.

6.4.2 Rock Constant

Several attempts have been made to evaluate what properties of a rock material are most important for determining (predicting) blastability. In the literature, correlations have been found between most of the material properties and the blastability for the special type of blast performed; however, there is no general concept that the blasting engineer can use when judging the blastability. Practical experience and skill are still major attributes for predicting the results of a blasting operation. It is obvious that structural geology plays a major role in breakage and fragmentation and often guides the necessary judgments prior to blasting. However, no one has yet managed to express this structural geology in a neat factor that can easily be used in charge calculations.

In this charge calculation concept, only one constant is used for describing the rock conditions — the rock constant. The rock constant c is an empirical measure of the amount of explosive needed for loosening 1 m^3 of rock.

When blasting in different Swedish rocks, it has been found that the value of c lies in the vicinity of 0.4. The c-value can be determined by trial blasting in a vertical drillhole with a hole diameter of about 32 mm. The vertical bench will be about 0.5 to 1 m high, the drillhole will have a depth of $1.3B$, and the burden will be equal to the bench height. In these tests, a start is made from a hypothetical c-value and the result is judged by the material thrown. In the event of a blown-out shot, a 30 to 50% larger charge is placed in a new hole having the same depth and burden. The process is repeated until the charge is large enough to break out the burden. If the center of gravity of the rock mass is thrown forward a distance of 0 to 1 m, the shot is regarded as being just sufficient (Figure 6.10). Longer throws of 2, 4, 6, and 8 m indicate excess charges of 10, 20, 30, and 40%, respectively. The c-value is then obtained by multiplying the amount of explosive used per cubic meter of rock by the factor 1.2 which was obtained by trial and error and from practical experience [Fraenkel, 1958].

Blasting in brittle crystalline granite gave a c-value equal to 0.2 kg/m^3. Blasting in rock with a strata perpendicular to the blast direction occasionally gave a c-value between 0.5 and 1.0 kg/m^3. In practice, all other normal fissured rock materials, from sandstone to granite, can be described by a c-value of about 0.4 kg/m^3.

6.5 Calculation of Burden

Taking explosive performance and hole inclination into account when calculating the burden, the linear charge concentration for the bottom charge ℓ_b can be written in two ways using Equations 6.19 and 6.24:

$$\ell_b = 1.11\,(f)\,\left(0.07B + cB^2 + 0.004B^3\right)$$
$$\ell_b = \frac{\pi d^2 (P_b\, s_b)}{4}. \tag{6.25}$$

The burden B can be evaluated by solving $f_n(B) = 0$ where

$$f_n(B) = \left(0.07B + cB^2 + 0.004B^3\right) - \left(\frac{\pi P_b d^2 s_b}{4.44\,f}\right). \tag{6.26}$$

Equation 6.26 (which can be evaluated using the Newton-Raphson iteration method) is valid when the hole length H

$$H > 1.3B + h \tag{6.27}$$

where B is the maximum burden, and h is the stemming length.

The Newton-Raphson iteration method can easily be used for the evaluation:

$$B_o = 40\,d$$
$$B_{n+1} = B_n - \frac{f_n(B_n)}{\frac{df_n(B_n)}{dB}} \tag{6.28}$$

If a small programmable calculator is not available, it is possible to use a simplified equation for evaluation of maximum burden. Equation 6.26 can be written as

$$B = \left[\frac{P_b \pi d^2 s_b}{4.44 f \left[\left(\frac{0.07}{B}\right) + c + 0.004B\right]}\right]^{1/2} \tag{6.29}$$

or

$$B \cong \frac{d}{2}\left[\frac{P_b \pi s_b}{1.11\,f\hat{c}}\right]^{1/2}$$

$$\hat{c} \approx c + 0.05 \tag{6.30}$$

$$f = \frac{3}{3+n}$$

Thus $\hat{c}$ is estimated to have a value of 0.05 higher than the rock constant c. This is valid for normal burdens in the interval 1 to 10 m (Figure 6.11).

If normal parameter values are inserted in Equation 6.30, an easy rule of thumb for calculating the maximum burden for a vertical hole in a bench of sufficient height will be given. Using, for example, $P_b = 1000$ kg/m^3, $s_b = 0.9$, and $c = 0.4$ results in

$$B \approx 40d. \tag{6.31}$$

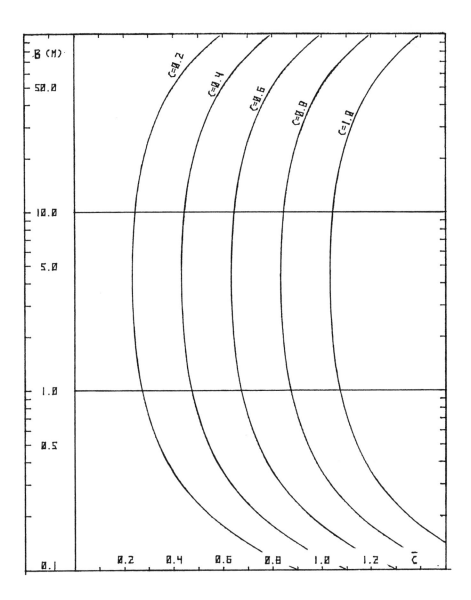

Figure 6.11. $\hat{c}$ as a function of the burden B with the rock constant c as a parameter.

6.5.1 Examples

Example 6.1. How much can the burden be increased if the hole inclination is changed from vertical to 70°?
From Equation 6.29,

$$B = \frac{d}{2}\left[\frac{P_b \pi s_b}{1.11\, f\hat{c}}\right]^{1/2}$$

$$f = \frac{3}{3+n}$$

<u>Vertical holes</u>
$\alpha = 90°$
$n_1 = 0$ m/m
$f_1 = 1$
$B_1 = k\left(\dfrac{1}{f_1}\right)^{1/2}$

<u>Inclined holes</u>
$\alpha = 70°$
$n_2 = 0.36$ m/m
$f_2 = 0.89$
$B_2 = k\left(\dfrac{1}{f_2}\right)^{1/2}$

$$B_2 = B_1 \left(\frac{f_1}{f_2}\right)^{1/2}$$

$$B_2 = B_1 (1.12)^{1/2}$$

$$B_2 \approx 1.06 B_1$$

The burden can be increased by 6%.

Example 6.2. What can be gained by changing explosive from ANFO to Watergel A?

$$\begin{array}{ll} \text{ANFO} & \text{Watergel A} \\ s_1 = 0.84 & s_2 = 0.77 \\ P_1 = 900 \text{ kg/m}^3 & P_2 = 1150 \text{ kg/m}^3 \\ B_1 = k(P_1 s_1)^{1/2} & B_2 = k(P_2 s_2)^{1/2} \end{array}$$

$$\frac{B_1}{B_2} = \left(\frac{P_1 s_1}{P_2 s_2}\right)^{1/2}$$

$$B_2 = \frac{B_1}{(P_2 s_2)^{1/2}}$$

$$B_2 \approx 1.08 B_1$$

The burden can be increased by 8% if a change from ANFO to Watergel A is made. From explosive and drilling costs, and fragmentation estimates, it is possible to judge for the quarry if this is an economical benefit or not.

If the bench height K is small compared to twice the burden for the hole diameter d chosen (for example in open pit mines where the bench height K needs to be kept low for safety (scaling) and mucking reasons, $K/d < 80$), Equation 6.26 will not be valid. Generally, when

$$H < 1.3B + h_s \tag{6.32}$$

where B is maximum burden, and h_s is the stemming height, there is no place for a full bottom charge. If a full bottom charge cannot be used, it is necessary to reduce the burden if satisfactory breakage at the toe is to be achieved. Figure 6.12 shows how the stemming height h_s, the charge length B, and subdrilling 0.3 B vary with hole diameter d. The line at $h = 1.3B + h_s$ is the line below which reduced charge and burden has to be adopted. It then becomes somewhat more complicated to calculate the burden, the subdrilling, and the explosive weight. Figure 6.13 shows the algorithm for this special case.

In Figure 6.13, W_{bmin} is the minimum charge weight required for breakage, W_b is the bottom charge weight, h_s is the stemming length, h_b is the height of the bottom charge, and P_b is the degree of packing for the bottom charge.

Figure 6.14 shows how the charge lengths vary for different hole diameters when drilled in a 10 m high bench.

6.6 Drilling Deviations

A blasting operation can be totally spoiled if the drilling precision is bad. Too large a burden in the bottom part of a round results in insufficient toe breakage and can result in floor protrusions (moguls) that have to be secondary blasted to satisfy a smooth floor. Too small a burden in the bottom part results in sometimes dangerous throw and flyrock. When the practical burden B_p is calculated, the drilling deviations must be considered.

The drilling deviation can be regarded as the sum of the three deviations:

   a. The collaring deviation α_1 (m) caused by mis-positioning the drill on the rock surface
   b. The angular deviation α_2 (m/m) caused by incorrect initial drilling direction
   c. The hole deviation α_3 (m) resulting from a curved drill hole

202 Chapter 6. Principles of Charge Calculation For Surface Blasting

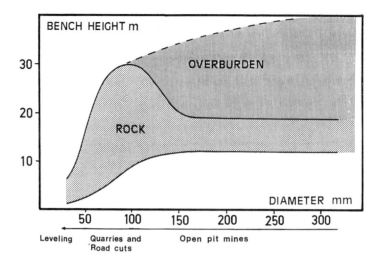

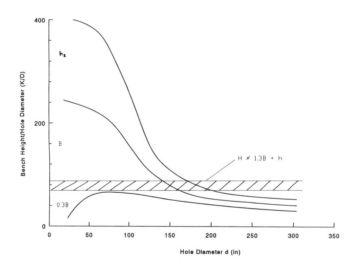

Figure 6.12. Some common relations between bench heights and hole diameters.

The collaring deviation depends on the precision of the marking of the hole on the rock surface prior to drilling and on the care taken by the drilling crew when the drill rig is positioned. A normal value for the collaring deviation is typically no more than three times the hole diameter.

The angular deviation depends upon how carefully the drilling crew can align the feeder and, of course, upon the stiffness of the drill rig.

The hole deviation depends on the ratio between hole length and hole diameter, and often increases with increasing feeder pressure.

Figure 6.15 shows the angular deviation for 347 holes measured during the drilling operations during construction of the Landvetter Airport outside Gothenburg, Sweden The standard deviation for the samplings was up to 2.5° or 0.0425 m/m drillhole depth. For a bench 10 m high, this means that 66.7% of all holes were drilled not more than 0.425 m away from the planned position, while the rest of the holes (33.3% of all holes) were drilled further away. If not more than 5% of the drilled holes are allowed to exceed

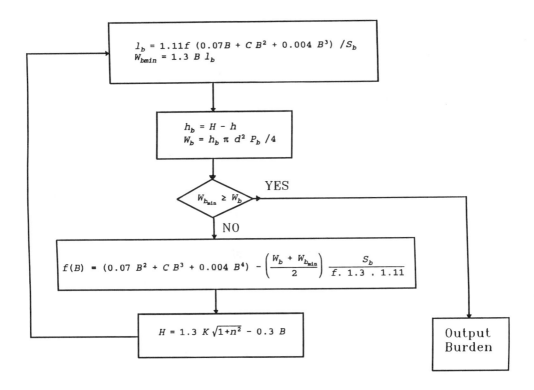

Figure 6.13. Flow chart for calculation of burden B when $H < 1.3B + h_s$.

the maximum burden, it can be shown statistically that one must calculate an angular deviation of 4.1° or 0.07 m/m when the practical burden is to be determined from the calculated maximum burden.

When automatic alignment equipment was installed at the drill rig, the drilling precision was very much improved (Figure 6.16).

The sampling showed a standard deviation of 1.1° or 0.019 m/m. If 5 % of the holes are allowed to exceed the maximum burden, we have to calculate an angular deviation $\alpha_2 = 1.85°$ or 0.32 m/m. This value is probably also the limit of achievement of manual alignment.

The hole deviation is usually of less concern when large-diameter, relatively shallow holes are drilled. For example, in open pit mining where 10 to 15 in diameter holes are drilled to a maximum depth of 18 m, there is no need to correct for the hole deviation.

However, in sublevel stoping where 6 in holes can be drilled up to 100 m, it is very important to map the holes in order to determine their curvature. Also, when small hole diameters (<100 mm) are used in high benches (10 to 30 m), it is very important to control the drilling deviation, especially in dipping formations with zones of rock with various hardness. If no stabilizers are used to guide the drilling, deflections can easily occur as the drill bit passes through a transition zone of rocks with different hardness. Long inclined holes tend to bend more and more upwards due to the bending of the drillrod which rests on the lower borehole wall during drilling. Long holes have a tendency to curve toward the ground surface.

204   Chapter 6. Principles of Charge Calculation For Surface Blasting

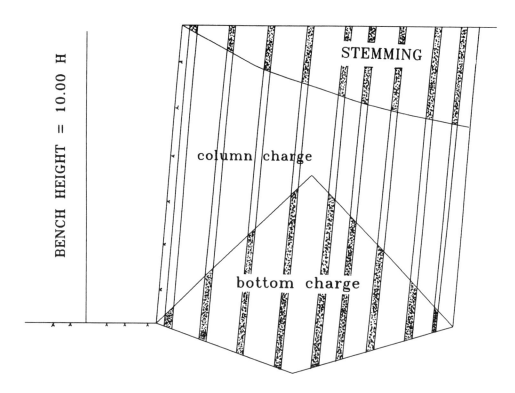

Figure 6.14. This figure shows the charge lengths for various hole diameters when a constant bench height of 10 m is used. Hole diameters are 0.01, 0.04, 0.07, 0.1, 0.15, 0.2, 0.3, 0.5, and 1 m.

Several different methods are commercially available for measuring drillhole deviations. In nonmagnetic zones, a camera together with a compass and an electromagnetic-activated pendulum can be used. Also used is a gyroscope together with a pendulum, or stress gauges fastened on a long rod lowered into the hole.

The REFLEX Maxibor is an instrument which measures the bending of a probe lowered into a borehole. Reflector rings are placed at intervals of 1.5 or 3 m. A liquid level sensor (see Figure 6.17), mounted in line with the reflector rings, defines the vertical plane. A light source is directed at and reflected by the rings. A CCD area sensor (from Video Camera Technology) receives the reflected image and passes it to a high-speed, real-time analyzer. After completion of a survey, information is transferred to a computer for further processing.

A more accurate drilling pattern can be calculated if all described drilling errors are known to some extent for the drilling equipment used.

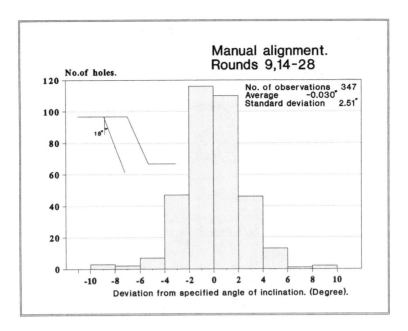

Figure 6.15. The distribution of angular deviation for 386 holes using manual alignment of the feeder.

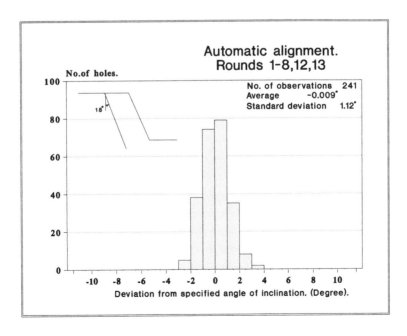

Figure 6.16. The distribution of angular deviation for 241 holes using automatic alignment.

206   Chapter 6. Principles of Charge Calculation For Surface Blasting

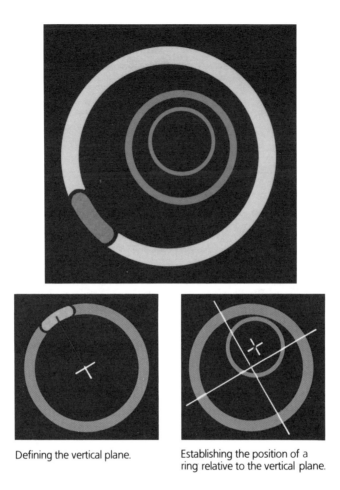

Defining the vertical plane.   Establishing the position of a
                                    ring relative to the vertical plane.

Figure 6.17. **REFLEX** Maxibor surveying equipment.

The total drilling deviation F for a vertical hole is given by

$$F = \alpha_1 + H\alpha_2 + \alpha_3 \tag{6.33}$$

where H denotes the hole length.

In the formulae used for bench blasting, we will only consider α_1 and α_2.

The burden must now be reduced according to

$$B_{P1} = B - (\alpha_1 + H\alpha_2) \tag{6.34}$$

6.6.1 Swelling

In multiple-row bench blasts, it is important to consider the fact that, the more rows that are blasted in the same round, the harder it will be to break out with the last rows of holes. This is due to the swelling of the rock mass upon fragmentation which increases the load of rock in front of each new row.

The swelling factor can be up to 40% for blasted material. Either one has to increase the charge successively in the rows or reduce the burden to achieve a controlled fragmentation and throw. For high benches, as in sublevel stoping, there also is a need for extra charge to compensate for gravitational forces at the toe. It has been shown that the easiest way to compensate for these effects is to reduce the burden.

In these calculations, the maximum burden B will be reduced by

$$B_{P2} = \frac{B}{1 + \left(\dfrac{N-1}{N+3}\right)\left(\dfrac{K}{33}\right)} \qquad (6.35)$$

where N is the number of rows, and K is the bench height in meters.

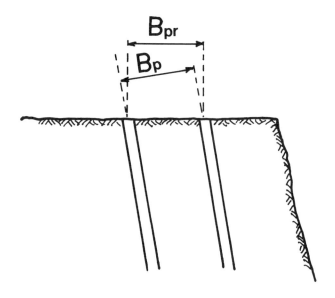

Figure 6.18. Practical and projected burden.

6.6.2 Practical Burden

The practical burden B_P, which is the right-angled distance between the holes at the surface (see Figure 6.18), is given by

$$B_P = \min\left(B_{P1}, B_{P2}\right). \tag{6.36}$$

In order to make it easier for the drilling crew, it is better to give them the burden projected on the horizontal B_{pr}

$$B_{pr} = \frac{B_p}{\cos\left[\arctan(n)\right]} \tag{6.37}$$

where n is the hole inclination given in m/m.

Chapter 7

Charge Calculations For Tunneling

7.1 Introduction

Tunnel excavation, or drifting as it is called in mining engineering, is a very important and central operation both in underground construction engineering and in underground mining. The growth and increasing population concentration in urban areas brings with it an increasing need for underground tunnels, for public transportation, for water and sewage conduits, and for electrical and telecommunication cable ways. In mining, it is not unusual for the percentage tunnel volume of rock excavated during the development of a new sublevel caving mine to be as much as 25 % of the total broken ore. The volume of rock then excavated on a continuing basis during the lifetime of an underground mine for transport, ventilation, and exploration drifts is also a considerable portion of the total volume excavated. The planning and excavation of drifts plays a major part in the total economy of the mine.

The hole diameters in use for drifting and tunneling have increased considerably. Not too long ago, a drift in Swedish bedrock with an area of 30 m^2 needed some 70 holes. The blastholes were 32 mm diameter and a parallel cut with empty holes of 76 or 102 mm were used. Today, at the LKAB iron ore mine, drifting is partly carried out with a hole diameter of 64 mm and often no empty large hole diameters were used in the cut. Rationalization and development of technology will make it possible to reduce the number of drill rigs in the future as the advance of rounds can be increased considerably. Figure 7.1 shows the result from a drifting round in Malmberget at the LKAB iron ore mine where 40 holes with 64 mm hole diameter were used for an area of 30 m^2. Drilling depth was almost 8 m and the advance was 7.5 m.

A continuing increase in the mechanization of mining and underground construction and the increasing transport volume in urban areas both demand larger tunnel areas. With modern, large, mechanized drill rigs, the hard work using hand-held pushers is gone, and a better working environment has been achieved. More rational methods for placing the drillholes and guiding the drill rod can be used, and the risk of personal injury has decreased radically. Unfortunately, a great deal of the hands-on experience and feeling for the rock material that the working man acquired when he was drilling close to the rock face and learned how to use the natural weak planes in the rock when placing the drillholes has been lost in the process. As a result of having separate shifts for drilling, loading, and hauling, more attention must be paid to developing standard, well-designed drilling patterns. The tunneling operation takes on the features of any mechanical engineering process — with drawings, records, and standards of quality. This chapter provides a formal procedure for charge calculation, applicable to most tunneling operations.

210 Chapter 7. Charge Calculations For Tunneling

Figure 7.1. A drift blasted with hole diameter 64 mm and an advance of 7.5 m (photo LKAB).

7.2 Development of Drilling Equipment

With mechanized drilling, the number of holes can be reduced to a certain degree because of the larger holes that can be produced. On the other hand, it becomes more difficult to use the larger holes because, if they are loaded full, they have too high a loading density and cause more damage to the remaining rock and more intense ground vibrations. On the other hand, the drilling precision has become very good with the use of parallel-guided booms and automatic devices for look-out angle setting. Figures 7.2 and 7.3 show what it could look like without the parallel-guided booms, i.e., using manual alignment.

Quite a few years have passed since the 1950s when the Swedish method of "One Man – One Machine" was introduced and made it possible, with the help of tungsten carbide bits, to achieve what was then regarded as amazingly increased production rates. Increasing labor costs later caused a demand for even higher degrees of mechanization in the 1960s, and penetration rates doubled with the development of twin hydraulic booms equipped with pneumatic rock drills. In the 1970s, a new generation of rock drills came on the market — the electrically powered hydraulic percussive rock drills, in

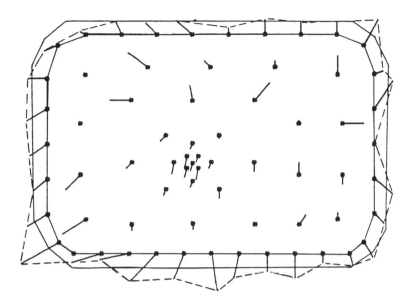

Figure 7.2. Without parallel-guided feed beams, the drilling precision can be very poor.

which a hydraulic fluid has taken over the function of the compressed air in transmitting the percussive action to the drill rod (Figure 7.4).

The introduction of hydraulic drilling rigs rapidly increased drilling rates by 50 to 100 % compared to that of the pneumatic percussive rock drill. It is not unusual today to achieve a rate of penetration of about 300 m/hr in hard bedrock (Figure 7.5.)

The working environment improved in many ways with the introduction of hydraulic rock drills. The noise level can be kept very low due to the absence of exhaust air. The oil mist has disappeared. The operator's view is clearer because water does not condense as it previously did when the high humidity exhaust air expanded and lowered the dew point.

For high-precision drilling, parallel-guided feed beams and automatic devices for precision setting of the direction of the drill can be provided. Automatic feedback control adjusts the feeding force and eliminates the problem of drill steels getting stuck when a joint is encountered during drilling. However, feeding forces have increased faster than the improvement of the drill rods and bits. This has created a trend toward larger drill hole diameters as the penetration rates have increased. Today, it is common to use 45 to 51 mm hole diameters for tunneling and drifting. However, during the last few years, research and development work have made it possible to manufacture small-size hydraulic rock drills with high penetration rates and still keep drill rod stresses low. Figure 7.6 shows the net penetration rates in Swedish bedrock for hydraulic machines used for drilling holes with a diameter of 45 mm.

The quality of button bits has steadily increased, and their durability is usually higher than for the insert bits. The quality improvement is visualized in Figure 7.7 where the drilling life of button bits (45 mm diameter) is plotted over the years 1976–1990.

212   Chapter 7. Charge Calculations For Tunneling

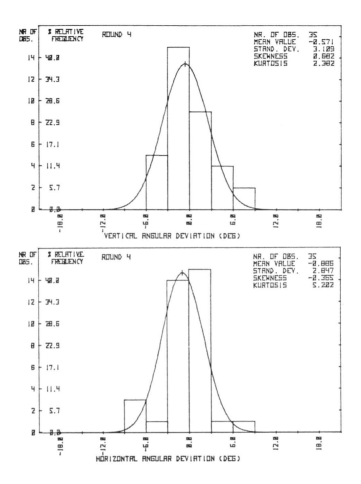

Figure 7.3. Angular deviation of holes shown in Figure 7.2.

The tests were carried out in one rock type in the Sandvik AB test mine. It is interesting to note that the durability has trebled since 1976, although the penetration rate has doubled during the same period. The scatter of drilling life has decreased continuously, indicating the development of a high-quality product.

Use of large mechanical drilling rigs in underground mines necessitates wider and higher drifts to allow room for the equipment. A larger arch of the roof in a drift will call for a smoother blasting procedure than a small one in order to prevent rock fall and give a sufficiently long stand-up time.

Mechanization of underground charging equipment has continued. Systems today are available for pneumatic loading of ANFO and for pumping emulsions. Today, explosives are very much tailor-made for the site-specific mine or tunnel. In dry conditions, it is common to use ANFO for the cut and the stoping holes, and a blowable mix of emulsion/ANFO/filler material with reduced strength in the contour. In wet conditions, emulsions are gaining ground due to their excellent water resistance. Figure 7.8 shows a highly mechanized underground charging rig. The operator can sit inside the cab and charge the holes with NONEL detonators and the necessary primers and explosives.

Figure 7.4. Atlas Copco all hydraulic drill rig, Boomer H127 for drifting and tunneling.

7.3 Charge Calculations

In order to clarify the conditions and principles for tunnel blasting, an analysis will be made in this chapter of the empirical relations which can be used for an economic and optimal blasting design.

The basic principles for the method of charge calculation are those developed by Langefors and Kihlström [1963].

Tunnel blasting is a much more complicated operation than bench blasting because the only free surface that initial breakage can take place toward is the tunnel face. Because of the high degree of constriction or fixation, larger charges will be required, leading to a considerably larger specific charge than in bench blasting. Figure 7.9 illustrates how the explosive consumption varies with the tunnel size.

Environmental factors must be considered when selecting the explosive in order to avoid exceeding the limits of allowable concentrations of toxic fumes. The small burdens used in the cut demand the use of an explosive agent which is insensitive enough that sympathetic detonation — premature initiation of charges from hole to hole — is impossible. Also, the explosive must have a sufficiently high detonation velocity to prevent occurrence of channel effects when the diameter of the charge is less than that of the drillhole (coupling ratio less than 1). The channel effects can occur if there is an air space between the charge and the borehole wall. Even with a high detonation velocity explosion, the expanding detonation gases can drive the air in the channel ahead of the detonation front in the form of a compressed air layer with high temperature and

214   Chapter 7. Charge Calculations For Tunneling

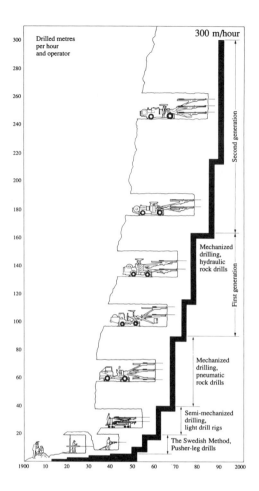

Figure 7.5. Progress of penetration rates [from Holdo, 1980].

high pressure. The high-pressure air surrounding the charge compresses the explosive in front of the detonation front. This may destroy the porous nature of the charge which provides the hot spots for initiation, or increase its density to a degree where the detonation could stop or the energy release might be reduced. Explosives used in the lifter holes at the bottom of the tunnel contour must also be able to withstand water. In the contour holes, special pipe charges with greatly reduced linear charge concentration should be used to minimize damage to the remaining rock.

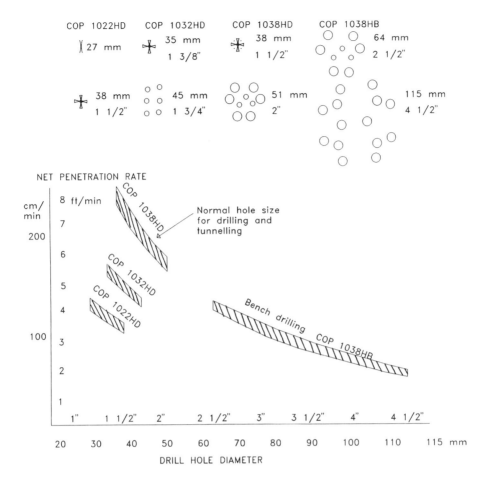

Figure 7.6. Net penetration rate in Swedish granite ($\sigma_c = 220$ MPa as a function of drill hole diameter for Atlas Copco hydraulic rock drills. Hole diameter is $\phi 45$ mm.

7.3.1 Dividing the Tunnel Face Area in Design Sections

To make the charge calculation easier, let us divide the tunnel face into five separate sections A–E (Figure 7.10). Each will be treated in its own special way during the calculation. Notation:

- A is the cut section
- B includes the stoping holes breaking horizontally and upwards
- C includes the stoping holes breaking downwards
- D includes contour holes
- E includes the lifter holes

The most important operation in the blasting procedure is to create an opening in the face in order to develop another free surface in the rock. This is the function of the cut holes. If this stage fails, the round can definitely not be considered a success.

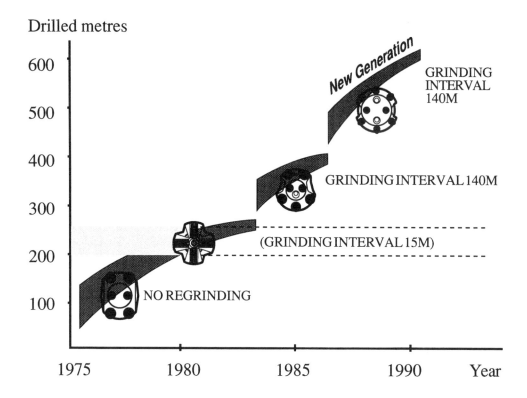

Figure 7.7. Test results from Bodås test mine. Service life development since the 1976 introduction of button bits for drifter drilling. Bit diameter is 45 mm. Drilling performed in granitized leptite (from Sandvik Rock Tools).

In the cut, the holes are arranged geometrically in such a way that firing the charges in sequence creates an opening which becomes wider and wider until the stoping holes can take over. The cut holes can be drilled to form a series of wedges (V-cut), to form a fan (fan cut), or in a parallel hole geometry; they may be drilled in a pattern close to and parallel with an empty, large central hole (parallel-hole-cut or parallel cut).

The choice of cut must be made with an eye to what drilling equipment is available, how narrow the tunnel is, and the desired advance. With V-cuts and fan cuts (where angled holes are drilled), the advance is strictly dependent upon the width of the tunnel. The parallel cut (four-section cut) with one or two centered large-diameter empty holes is being used extensively with large, mechanized drilling rigs. The advantages are obvious. In narrow tunnels, the large booms cannot be angled sufficiently to create the necessary V-cut angles; it is easier to maintain good directional accuracy in the drilling when all holes are parallel so there is no need to change the angle of the booms.

In the parallel cut, standard diameter holes are drilled with high precision around a larger hole (diameter 65 to 175 mm). The larger, empty hole provides a free surface for the smaller holes to work toward, and the opening is enlarged gradually until the stoping holes can take over the breakage.

The dominant type of parallel hole cuts is the four-section cut which is described and used in the calculations in the remainder of this chapter.

Figure 7.8. Rocmec 2000. A highly mechanized charging truck for drifting and tunneling (photo Nitro Nobel AB).

7.3.2 Advance

The advance is restricted by the diameter of the empty hole and by the hole deviations for the smaller diameter holes. For economy, the entire hole depth must be utilized. Drifting becomes very expensive if the advance is much less than 95% of the drilled hole depth. Figure 7.11 illustrates the required hole depth as a function of the empty hole diameter when a 95% advance is desired with a four-section cut.

The equation for hole depth H can be written

$$H = 0.15 + 34.1\phi - 39.4\phi^2 \tag{7.1}$$

where ϕ is the hole diameter expressed in meters and $0.05 \leq \phi \leq 0.25$ m.

The advance I for 95% advance then is

$$I = 0.95H \tag{7.2}$$

and the two equations above are valid for a drilling deviation not exceeding 2%.

Sometimes, there is an advantage in using two empty holes in the cut instead of one. This is, for example, the case when the drilling equipment cannot handle a larger diameter. Equation 7.1 is still valid if ϕ is recalculated using

$$\phi = d\sqrt{2} \tag{7.3}$$

where d denotes the hole diameters of the two empty holes.

The general geometry for the cut and cut spreader holes is outlined in Figure 7.12.

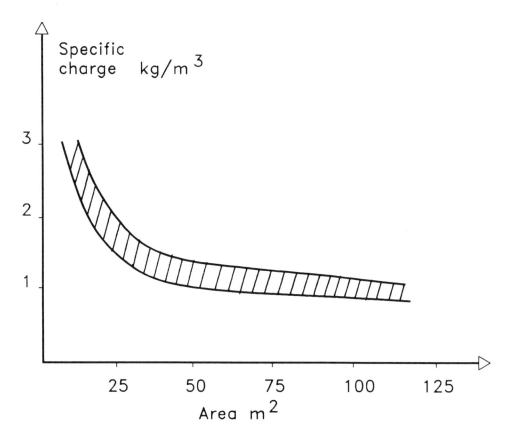

Figure 7.9. Specific charge as a function of the tunnel area.

7.3.3 Burden in the First Quadrangle

The distance between the empty hole and the drillholes in the first quadrangle should not exceed 1.7 times the diameter ϕ of the empty hole if a satisfactory breakage and cleaning of the resulting cavity is to take place. Breakage conditions differ very much depending upon the explosive type, structure of the rock, and distance between the charged hole and the empty hole.

As one can see in Figure 7.13, there is no advantage in using a burden greater than 2ϕ because the aperture angle will then be too small for the heavy charge. Plastic deformation would be the only effect of the blast. Even if the distance is smaller than 2ϕ, too high a charge concentration could cause a malfunction of the cut due to the rock impact and sintering which prevents the swelling that is necessary. If the maximum acceptable hole deviation is of the magnitude 0.5 to 1%, then the practical burden B_1 for the cut spreader holes must be less than the maximum burden B of 1.7ϕ.

We use

$$B_1 = 1.5\phi. \tag{7.4}$$

When the deviation exceeds 1%, B_1 must be reduced even further. We then use the formula

$$B_1 = \left[1.7 - (\alpha_2 H + \alpha_1)\right] \phi \tag{7.5}$$

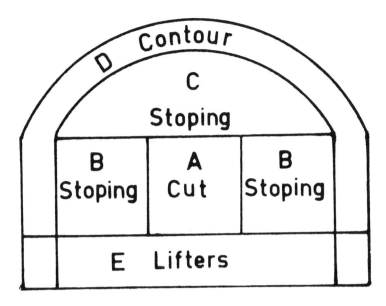

Figure 7.10. Sections A–E represent types of holes with different blasting conditions.

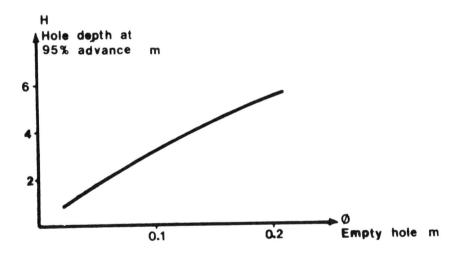

Figure 7.11. Hole depth as a function of empty hole diameter for a four-section cut.

where the term $(\alpha_2 H + \alpha_1)$ represents the maximum drill deviations, α_2 is the angular deviation in m/m, H is the hole depth in meters, and α_1 denotes the collaring deviation in meters.

In practice, the drilling precision is good enough to allow the use of Equation 7.4.

220 Chapter 7. Charge Calculations For Tunneling

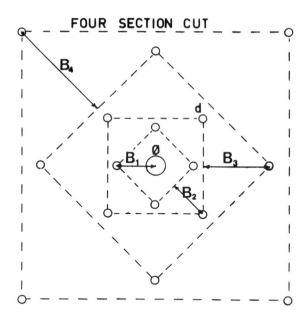

Figure 7.12. Four-section cut: V_i represents the practical burden for quadrangle number i.

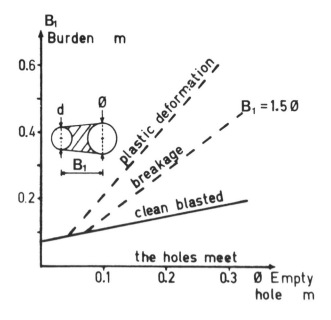

Figure 7.13. Blasting result for different relations between the practical burden and the empty hole diameter. Hole deviation is less than 1 %. Adapted from Langefors-Kihlström.

7.3.4 Charge Concentrations in the First Quadrangle

Langefors and Kihlström [1963] verified the following relation between the linear charge concentration ℓ (expressed in kg/m), the maximum distance between the holes B, and the diameter of the empty hole ϕ, for a borehole with a diameter of 0.032 m:

$$\ell = 1.5 \left(\frac{B}{\phi}\right)^{1.5} \left(B - \frac{\phi}{2}\right). \tag{7.6}$$

In order to utilize the explosive in the most effective way, we should use a burden of $B_1 = 1.5\phi$ for a deviation of 0.5 to 1 %.

Remember that Equation 7.6 is only valid for a drillhole diameter of 0.032 m. If larger holes are to be used in the round, an increasing charge concentration per meter drillhole must be used. To keep the breakage at the same level, it is necessary to increase the concentration appropriately with the drillhole diameter. Thus, if a drillhole diameter of d_2 is used instead of $d_1 = 0.032$ m, the charge concentration is determined by

$$\ell_2 = \frac{d_2 \ell_1}{d_1}. \tag{7.7}$$

Obviously, when the diameter is increased, the coupling ratio and the borehole pressure decrease. It is important to carefully select the proper explosive in order to minimize the risk for channel effects and incomplete detonations.

Taking the rock material and type of explosive into consideration, Equation 7.6 now can be rewritten for a general hole diameter d (in meters)

$$\ell = 55 \, d \frac{\left(\frac{B}{\phi}\right)^{1.5} \left(B - \frac{\phi}{2}\right) \left(\frac{c}{0.4}\right)}{s_{ANFO}} \tag{7.8}$$

where s_{ANFO} denotes the weight strength relative to ANFO, and c is defined as the rock constant.

Often, the charge concentration cannot be changed in so many ways. Due to the limited assortment of explosives and charge diameters available from the explosives manufacturer, the charge concentration is given beforehand and we need to calculate the burden from Equation 7.8 instead. This can easily be solved numerically by using a pocket calculator.

7.3.5 The Second Quadrangle

After the first quadrangle is calculated, we encounter a new geometrical problem when solving the burdens for the following quadrangles. Blasting toward a circular hole demands a higher charge concentration than blasting toward a rectangular opening due to the higher constriction (fixation) and a less-effective stress wave reflection.

If, as shown in Figure 7.14, there is a rectangular opening with side length A and the burden B is known, the charge concentration ℓ relative to ANFO is given by

$$\ell = \frac{32.3 \, d \, c \, B}{s_{ANFO} \left[\sin\left[\arctan\left(\frac{A}{2B}\right)\right]\right]^{1.5}}. \tag{7.9}$$

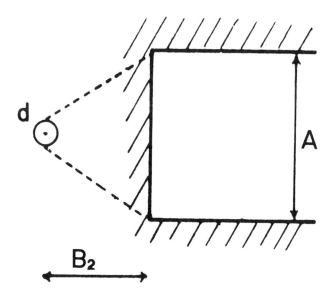

Figure 7.14. Geometry for blasting toward a rectangular opening.

If we instead start from the assumption that we know the charge concentration for the actual explosive and the rectangular opening A, we can express the burden B with good accuracy explicitly as a function of A and ℓ.

$$B = 8.8 \times 10^{-2} \left[\frac{A \, \ell \, s_{ANFO}}{d\,c} \right]^{1/2}. \tag{7.10}$$

In calculating the burden for the new quadrangle, we should remember to consider the faulty drilling F as defined in Equation 7.5. We have to treat the holes in the first quadrangle as if they were placed at the most favorable location.

From Figure 7.15, we can see that the free surface B to be used in Equation 7.10 is given from the calculation of the burden B_1 in the first quadrangle

$$A = \sqrt{2}\,(B_1 - F). \tag{7.11}$$

Finally, insertion gives us the burden for the new quadrant

$$B = 10.5 \times 10^{-2} \left[\frac{(B_1 - F)\, \ell \, s_{ANFO}}{d\,c} \right]^{1/2}. \tag{7.12}$$

Of course, this value must be reduced by the drilling deviation to obtain the practical burden

$$B_2 = B - F. \tag{7.13}$$

There are a few restrictions that must be put on B_2. It must fulfill the condition

$$B_2 \leq 2A, \tag{7.14}$$

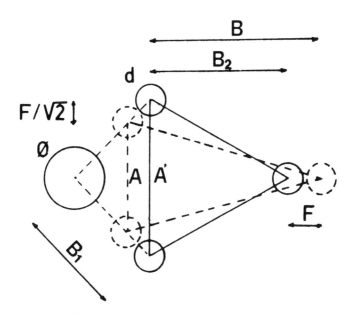

Figure 7.15. Influence of the faulty drilling.

if plastic deformation is not to occur. If no deformation occurs, the charge concentration should, according to Equations 7.9 and 7.14, be reduced to

$$\ell = \frac{32.3\, d\, c\, (2A)}{s_{ANFO} \left[\sin\left[\arctan\left(\frac{1}{4}\right)\right]\right]^{1.5}} \quad \text{(in kg/m)} \tag{7.15}$$

or

$$\ell = \frac{540\, d\, c\, A}{s_{ANFO}}. \tag{7.16}$$

If the restriction for plastic deformation cannot be fulfilled, it is usually better to choose an explosive with a lower weight strength in order to optimize the breakage.

The aperture angle should also fall below 90°. If not, the cut will lose its character as a four-section cut. This means

$$B_2 > 0.5A. \tag{7.17}$$

Gustafsson [1973] suggests the burden for each quadrangle to be $B_2 = 0.7A$.

A rule of thumb for the number of quadrangles in the cut is that the side length of the last quadrangle A should not be less than the square root of the advance. The algorithm for calculation of the rest of the necessary quadrangles is the same as that described above for the second quadrangle.

Holes in the quadrangles should be loaded so that a hole length h_s of 10 times the hole diameter is left unloaded

$$h_s = 10\, d. \tag{7.18}$$

224 Chapter 7. Charge Calculations For Tunneling

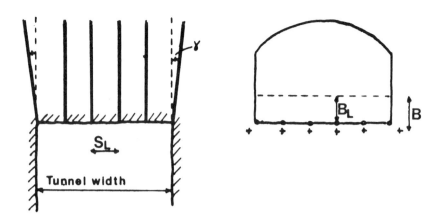

Figure 7.16. Blasting geometry for lifters.

7.3.6 Lifters

The burdens for the lifters in a round are in principle calculated with the same formula as for bench blasting. The bench height is just exchanged for the advance and a higher fixation factor is used due to the gravitational effect and to a greater time interval between the holes.

The maximum burden B can be solved from

$$B = 0.9 \left[\frac{\ell \, s_{ANFO}}{\bar{c} f} \right]^{1/2} \tag{7.19}$$

where f is the fixation factor and $\bar{c}$ is the corrected rock constant calculated as

$$\begin{aligned} \bar{c} &= c + 0.05 & \text{if } B \geq 1.4 \text{ m.} \\ \bar{c} &= c + \frac{0.07}{B} & \text{if } B < 1.4 \text{ m.} \end{aligned} \tag{7.20}$$

Equation 7.19 is valid when the lifter holes are drilled with a square pattern, i.e., $S/B = 1$ where S is the hole spacing. If the spacing is made larger, the burden should be reduced proportionately. Typically for lifters, $f = 1.45$.

In positioning the lifters, one must remember to consider the look-out angle γ (Figure 7.16). The angle is dependent upon available drilling equipment and hole depth. For an advance of about 3 m, a look-out angle equal to $3°$ (corresponding to 5 cm/m) should be enough to give space for drilling the next round.

Hole spacing should be equal to B, but we need to check if there is room for the holes along the tunnel width. If there is not, the spacing must be changed.

The largest number of lifters N which can be fitted across the tunnel width is

$$N = \text{nearest higher integer of } \left[\frac{(\text{Tunnel width} + 2H \sin \gamma)}{B} + 2 \right]. \tag{7.21}$$

The spacing S_L for the holes (except the corner holes) is evaluated by

$$S_L = \frac{(\text{Tunnel width} + 2H \sin \gamma)}{N - 1}. \tag{7.22}$$

The practical spacing S'_L for the corner holes is equal to

$$S'_L = S_L - H \sin \gamma. \tag{7.23}$$

The practical burden B_L should be reduced by the look-out angle in the bottom and the drilling deviation

$$B_L = B - H \sin \gamma - F. \tag{7.24}$$

The length of the bottom charge h_b needed for loosening the toe is

$$h_b = 1.25 B_L. \tag{7.25}$$

Finally, the length of the column charge h_c is given by

$$h_c = H - h_b - 10d \tag{7.26}$$

and the concentration of this charge can be reduced to 40 % of the concentration in the bottom charge. This is, however, not always common since it is time-consuming work. Usually the same concentration is used both in the bottom and in the column.

For lifters, an unloaded hole length of $10d$ is usually used.

A condition that must be fulfilled if Equation 7.19 is to be used is

$$B \leq 0.6H. \tag{7.27}$$

If this is not the case, the maximum burden must be successively reduced by lowering the charge concentration. Thereafter, the practical spacing S_L and burden B_L can be evaluated.

7.3.7 Stoping Holes

The method for the calculation of the stoping holes in the sections B and C (Figure 7.9) does not differ much from the calculation of the lifters. For stoping holes breaking horizontally and upward in section B, the fixation factor f is 1.45 and the S/B ratio is equal to 1.25. The fixation factor for stoping holes breaking downwards is reduced to 1.2 since gravity is working to our advantage. The S/B ratio should be 1.25.

The linear charge concentration ℓ in the column for both types of stoping holes should be equal to 50 % of the charge concentration for the bottom charge.

7.3.8 Contour Holes

If smooth blasting were not to be used, the burden and spacing for the contour holes would be calculated according to what was said above regarding stoping holes breaking downward. However, blast-damaged roof and walls in a drift often need an extensive amount of support, which is expensive. For example, a 3 m long, 50 mm diameter drillhole completely filled with ANFO ($\ell = 1.5$ kg/m) is capable of producing a damaged zone of 1.3 to 1.2 m radius in competent rock. In almost any type of rock mass, compensating for such extensive damage by any kind of support is extremely expensive.

With smooth blasting, this damage zone is reduced to a minimum. Even in low-strength rock, a long stand-up time can usually be achieved by more careful contour blasting. Our experience shows that the optimum spacing is a linear function of hole diameter

$$S = kd \tag{7.28}$$

226 *Chapter 7. Charge Calculations For Tunneling*

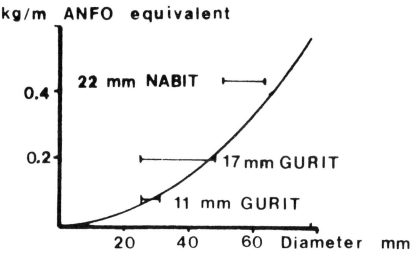

Figure 7.17. Minimum required charge concentration for smooth blasting and recommended practical hole diameter for Nabit and Gurit charges.

where the constant k is in the range of 15 to 16. An S/B ratio of 0.8 should be used. For a 51 mm hole diameter, the spacing will be about 0.8 m and the burden about 0.6 m.

The charge concentration (expressed in kg per m borehole length) is also a function of the hole diameter. For hole diameters up to 0.15 m, we have the relation

$$\ell = 90\,d^2 \tag{7.29}$$

where d is expressed in meters.

In smooth blasting, the total hole length must be charged to avoid the collar being left unbroken.

In Figure 7.17, ℓ is plotted as a function of d.

7.4 Sample Charge Calculation

In this section, we will list the input and results of a charge calculation for a tunnel round applying the methods described above. The resulting calculated drilling hole pattern is shown in Figure 7.18.

7.4.1 Input Conditions

Hole diameter = 45 mm
Empty hole, $\phi = 102$ mm
Tunnel width = 4.5 m
Abutment height = 4.0 m
Height of arch = 0.5 m
Smooth blasting in the roof
Lookout for contour holes $\gamma = 3°$ (0.05 rad)
Angular deviation $\alpha_2 = 0.01$ m/m
Collar deviation $\alpha_1 = 0.02$ m

Explosive: a water gel explosive is used with three cartridge dimensions:
$\phi 25 \times 600$ mm, $\phi 32 \times 600$ mm, and $\phi 38 \times 600$ mm
Heat of explosion = 4.5 MJ/kg
Gas volume at STP = 0.85 m^3/kg
Density = 1200 kg/m^3
Rock constant $c = 0.4$

7.4.2 Calculation

Weight strength relative to LFB (Equation 6.21).

$$s_{LFB} = \frac{5 \times 4.5}{6 \times 5.0} + \frac{1 \times 0.85}{6 \times 0.85} = 0.92$$

and

$$s_{ANFO} = 0.92/0.84 = 1.09$$

Charge concentration	ϕ (mm)	ℓ (kg/m)
	25	0.59
	32	0.97
	38	1.36

7.4.3 Advance

Using an empty hole diameter $\phi = 102$ mm, Equation 7.1 results in a hole depth of 3.2 m and the advance is 3.0 m.

7.4.4 Cut

First Quadrangle

Maximum burden $B = 1.7\phi = 0.17$ m
Practical burden $B_1 = 0.12$ m
Charge concentration $\ell = 0.58$ kg/m
ℓ for the smallest cartridge is 0.59 kg/m which is
  sufficient for clean blasting the opening
Unloaded hole length $= 10d = 0.45$ m
Hole distance in quadrangle $A' = \sqrt{2}B_1 = 0.17$ m
No. of 25×600 cartridges $= (3.2 - 0.45)/0.6 = 4.5$

Second Quadrangle

The rectangular opening to blast toward is
$A = \sqrt{2}\,(0.12 - 0.05) = 0.10$ m
Maximum burden for $\phi 25$ cartridges $B = 0.17$ m
Maximum burden for $\phi 32$ cartridges $B = 0.21$ m
Maximum burden for $\phi 38$ cartridges $B = 0.25$ m

Equation 7.14 says the practical burden must not exceed $2A$. This implies that the $\phi 32 \times 600$ cartridges are the most suitable ones in this quadrangle.

>Practical burden $B_2 = 0.16$ m
>Unloaded hole length $h_s = 0.45$ m
>Hole distance in quadrangle $A' = \sqrt{2}(0.16 + 0.17/2) = 0.35$ m
>No. of $\phi 32 \times 600$ cartridges $= 4.5$

Third Quadrangle

>$A = \sqrt{2}(0.16 + 0.17/2 - 0.05) = 0.28$ m
>Use 38×600 cartridges with charge concentration $\ell = 1.36$ kg/m
>Maximum burden $B = 0.42$ m
>Practical burden $B_3 = 0.37$ m
>Unloaded hole length $h = 0.45$ m
>Hole distance in quadrangle $A' = \sqrt{2}(0.37 + 0.35/2) = 0.77$ m
>No. of $\phi 38 \times 600$ cartridges $= 4.5$

Fourth Quadrangle

>$A = \sqrt{2}(0.37 + 0.35/2 - 0.05) = 0.70$ m
>Maximum burden $B = 0.67$ m
>Practical burden $B_4 = 0.62$ m
>Unloaded hole length $h_s = 0.45$ m
>$A' = \sqrt{2}(0.62 + 0.77/2) = 1.42$ m
>No. of $\phi 38 \times 600$ cartridges $= 4.5$

The side length of this quadrangle is 1.42 m which is numerically comparable to the square root of the advance expressed in meters. Therefore, there is no need to add more quadrangles.

7.4.5 Lifters

Use $\phi 38 \times 600$ cartridges with a charge concentration of $\ell = 1.36$ kg/m.

>Maximum burden $B = 1.36$ m.
>No. of lifters $N = 5$.
>Spacing $S_L = 1.21$ m.
>Spacing, corner holes $S'_L = 1.04$ m.
>Practical burden $B_{PL} = 1.14$ m.
>Length of bottom charge $h_b = 1.43$ m.
>Length of column charge $h_c = 1.32$ m.

This charge concentration will be 70 % of the bottom charge concentration; $0.70 \times 1.36 = 0.95$ kg/m.

Use 2.5 cartridges $\phi 38 \times 600$ as the bottom charge and 2 cartridges $\phi 32 \times 600$ as the column charge.

7.4.6 Contour Holes, Roof

Smooth blasting with 25×600 cartridges is specified.
Spacing $S = 0.68$ m
Burden $B = S/0.8 = 0.84$ m
Due to lookout and deviation, the practical burden becomes

$$B_{PR} = 0.84 - 3.2 \sin 3° - 0.05 = 0.62 \text{ m}$$

The minimum charge concentration for this smooth blasting is

$$\ell = 90d^2 = 0.18 \text{ kg/m} \tag{7.29}$$

The charge concentration for the $\phi 25 \times 600$ cartridges is 0.59 kg/m, which is considerably more than what is really needed.

No. of holes; integer of $(4.7/0.68 + 2) = 8$
5 cartridges per hole are used

7.4.7 Contour Holes, Wall

The abutment height is 4.0 m and from the calculation it is known that the lifters should have a burden of 1.14 m, and the roof holes should have a burden of 0.62 m. This implies that there are $4.0 - 1.14 - 0.62 = 2.24$ m left in the contour along which to position the wall holes.

By using a fixation factor $f = 1.2$, and an $S/B = 1.25$, Equation 7.19 results in a maximum burden of $B = 1.33$ m.

Practical burden $B_{PW} = 1.33 - 3.2 \sin 3° - 0.05 = 1.12$ m
No. of holes = integer of $(2.24/(1.33 \times 1.25) + 2) = 3$
Spacing $= 2.24/2 = 1.12$ m
Length of bottom charge $h_b = 1.40$ m
Length of column charge $h_c = 1.35$ m

2.5 cartridges $\phi 38 \times 600$ are used as the bottom charge, and 2 cartridges $\phi 32 \times 600$ are used in the column.

7.4.8 Stoping

The side of the fourth quadrangle in the cut is 1.42 m, and the practical burden B_{PW} for the wall holes was determined to be 1.12 m. As the tunnel width is 4.5 m, a distance of $4.5 - 1.42 - (2 \times 1.12) = 0.84$ m is available for placing horizontal stoping holes.

Maximum burden $(f = 1.45)$ $B = 1.21$ m.
Practical burden $B_{PH} = 1.21 - 0.05 = 1.16$ m.
Instead the burden $B_{PH} = 0.85$ m is used, due to the tunnel geometry.

The height of the fourth quadrangle was 1.42 m, and this will of course determine the spacing for the two holes, which becomes 1.42 m.

For stoping downward:

Maximum burden $B = 1.33$ m.
Practical burden $B_{PD} = 1.28$ m.

230   Chapter 7. Charge Calculations For Tunneling

The maximum height of the tunnel is specified to be 4.5 m. If we subtract the height of the fourth quadrangle (1.42 m), the burdens for the lifters (1.14 m) and the roof holes (0.62 m), there is 1.32 m left for a stoping hole. This is just a little more than the practical burden, but if the stoping holes are placed at 1.28 m above the cut, the remaining 0.04 m will in all probability be removed by the overcharged contour. Furthermore, the formulae used in the calculation have a safety margin that can tolerate small deviations.

Three holes for stoping downward are positioned above the fourth quadrangle (Figure 7.18). The charge distribution for the stoping holes is the same as for the wall holes.

A summary of explosive consumption for the round in our example is given in Table 7.1.

7.4.9 Initiation Sequence

From the above, it is obvious that the charges in a tunnel blast must be initiated in sequence so that the opening produced by a previous hole can be utilized by the following holes. For the blast in our example, the best sequence is:

1. Cut (in the order: first quadrangle, second quadrangle, third quadrangle, fourth quadrangle)
2. Stoping (stoping from the side and stoping downwards)
3. Lifters
4. Contour Holes, Roof
5. Contour Holes, Wall

The cut is fired using millisecond caps. Since the rock removed between each hole in the cut and the central empty hole must be blown out to provide expansion room for the rock removed by the next charge, a long enough time interval between these holes is needed — in this example, we have used 75 ms.

The rest of the round is fired using half-second interval caps.

The entire firing time sequence is shown in Figure 7.18.

Table 7.1. Summary of explosive consumption.

Hole type	Number of holes	Number of cartridges			Charge per hole	Total
		$\phi 25$ mm	$\phi 32$ mm	$\phi 38$ mm	(kg)	(kg)
1st quad.	4	4.5			1.59	6.37
2nd quad.	4		4.5		2.62	10.48
3rd-4th quad.	8			4.5	3.76	29.36
Lifters	5		2.0	2.5	3.20	16.00
Roof	8	5.0			1.77	14.16
Wall	6		2.0	2.5	3.20	19.20
Stoping	5		2.0	2.5	3.20	16.00

Total charge weight = 111.6 kg
Cross-sectional area = 19.5 m²
Advance = 3.0 m
Specific charge = 1.9 kg/m³
Total no. of holes = 40
Hole depth = 3.2 m
Specific drilling = 2.2 m/m³

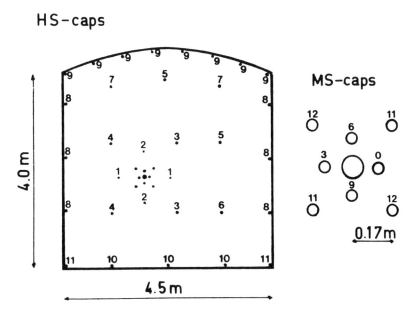

Figure 7.18. Calculated drilling pattern; MS stands for msec caps (4 no. = 100 msec) and HS stands for half-sec caps (1 no. = 0.5 sec).

Chapter 8

Stress Waves in Rock, Rock Mass Damage, and Fragmentation

In primitive rock blasting, the purpose may be simply to remove rock without respect for the size and shape of the broken rock. This type of rock blasting occurs only in small-scale blasting in very sparsely populated areas, and even there is becoming less frequent. Growing concerns for the environment and a growing understanding of the economic usefulness of crushed rock for a ballast material, for road building, or for building construction generally dictate that the broken rock be cleared away to be used as fill material, or preferably as a raw material for building purposes. Even if the broken rock simply has to be transported away, there is always a need for making the fragment size small enough to suit the equipment available for loading the broken rock on to a truck.

Very often, the end purpose of rock blasting is to produce input material for a crusher. This is the case in most mining and construction rock blasting. Fragments produced by blasting should then not only be small enough for the loading equipment at hand, but they should also be small enough to pass easily into the crusher opening. The economic success of a mining operation often hinges on this, since the primary crusher otherwise becomes a bottleneck choking the flow of ore into the benefication plant (the plant where the ore is concentrated).

Sometimes, as in dam construction or when building a wave breaker, it is necessary to produce quantities of large and regularly shaped boulders.

Even where the main purpose of excavation is to create an empty space surrounded by rock walls, such as a tunnel or an underground machine hall, the fragment size is important because it influences the speed of loading or mucking and the capacity of the transport equipment. In urban areas, the broken and crushed rock is a valuable commodity.

In all these cases, the degree of fragmentation influences the economy of the excavation job. Thus, predicting fragmentation is an important art to learn. Three factors control the fragment size distribution: the quantity of explosive, its distribution within the rock mass, and the rock structure. In this survey of the fragmentation process, we will first consider the purely theoretical case of blasting in an infinite, homogeneous, brittle material. We will do this only in order to understand the mechanism of crack formation around the borehole. We will then introduce free surfaces and see how they influence the process of fragmentation. Then, we will see how the structure of joints and fissures in real rock influences the process. We will summarize the existing empirical knowledge relating rock fragment size (boulder size), burden, spacing, and specific charge. Finally, we will explore how the calculated vibration velocity around an extended charge can be used as a measure of the degree of rock damage and fragmentation caused by the charge and as an instrument to predict rock damage.

234   *Chapter 8. Stress Waves in Rock, Rock Mass Damage, and Fragmentation*

8.1 Fracture Initiation

When the charge detonates in the borehole, the detonation wave will propagate along the hole with a velocity of 3000 to 6000 m/sec depending on the type of explosive and the charge diameter. At the front of the detonation wave, the pressure is between 0.5 and 20 GPa, or more normally between 5 and 10 GPa for a hole filled with a high explosive. If the borehole is packed full of the explosive, a pressure about half the front pressure will initially act on the borehole wall. The pressure on the wall is lower than the front pressure because the wall material is being accelerated outward. If the charge does not fill out the hole, the gas will first expand radially and upon equilibration within the larger hole volume will exert a much lower pressure on the wall.

In either case, the pressure will propagate out from the borehole into the rock as a shock wave with a conical front coaxial with the hole. As a consequence of the simultaneous axial and radial expansion of the reaction products, the wall pressure will initially drop quite quickly. The front pressure of the shock wave therefore will decrease at a rate even higher than that determined only by the radial expansion of the front. The shock front cone angle, determined by the relative magnitudes of the shock and detonation velocities, will therefore decrease with increasing radius because of the gradually decreasing front pressure. When blasting hard rock with a low detonation velocity blasting agent, the shock waves in the rock may have a velocity higher than the detonation velocity. The detonation front may even move at subsonic velocity with respect to the surrounding rock. In these two latter cases, a relatively weak stress wave will precede the detonation front along the drillhole, prestressing the rock material. The details of the mechanics in these two cases are still not very well explained and understood. They represent an interesting area for future research.

Due to the radial outflow of material accompanying the shock wave, the tangential pressure will decrease more rapidly than the radial and axial pressure and will ultimately cause radial cracks to appear. Up to that time, the material between the borehole and the shock front is compressed and deforms elastically or plastically depending on the pressure and the strength of the rock. Figure 8.1 is a high-speed camera photograph of a detonating charge placed in a 5.2 mm diameter drillhole in a block of plexiglas. We can clearly see the expanding drillhole and the shock front. The material behind the front is still completely transparent, indicating that the material is deforming plastically without cracking. At a certain distance behind the detonation front, the material becomes opaque because of the radial cracks. These do not initiate at the borehole wall. Instead, they start at a point one or two hole radii further out, and then propagate radially outward and inward. This picture of the initial expansion process is supported by computer calculations. These indicate that the tangential stress reaches the tensile fracture value about two hole radii out from the borehole wall, while all stresses are still positive both at the hole wall and further out. Poncelet [1961] pointed out the possibility of different types of fracture in different radial regions in the rock outside the borehole wall.

These regions are clearly seen in Figure 8.2 (top), which is a slice cut out from a plexiglas cube after detonation of a low-density PETN charge in a 3 mm hole. The hole has returned to its original size after having expanded to 2–3 times this size. The region near the hole has probably yielded without cracking. The intermediate zone has fractured by shear cracks, and the outer region by radial cracks.

Initially, the number of radial cracks is quite large, but only a few of these cracks propagate very far because of the stress relaxation spreading from the longest among them. In the absence of a free surface, a few cracks become much longer than the others.

8.1. Fracture Initiation 235

Figure 8.1. Detonating PETN charge in a 5.2 mm borehole in plexiglas.

The number of long cracks is often around five. The radial crack propagation velocity is initially of the order of 1000 m/s, gradually decreasing.

As the shock wave velocity in hard rock is of the order of 4000–5000 m/s, the radial crack length by the time the shock wave reaches the free surface parallel to the drillhole is less than 25 % of the distance to the free surface. The compressive shock wave is then reflected back from the free surface as a tensile wave. If the specific charge is of the order of 5 kg/m^3 or more, the tensile wave in granite is strong enough to cause spalling fractures with the fracture surfaces parallel to the free surface. Smaller charge weights such as 0.5 to 1.0 kg/m^3 (which are normal in bench blasting) do not cause spalling fractures in granite at the free surface.

Experiments by Field and Ladegaard-Pedersen [1969] showed the importance of the interaction between the expanding radial crack system and the reflected tensile wave. This interaction gives a greater propagation velocity to those cracks which are nearly parallel to the (reflected) tensile wave front (Figure 8.3.)

In the meantime, another mechanism has come into play — the longest cracks have extended inward and reached the borehole wall. This is supported by two observations. First, in plexiglas experiments, reaction product gases are seen to emerge from cracks reaching the free surface at an early stage of the process. Second, when an explosive

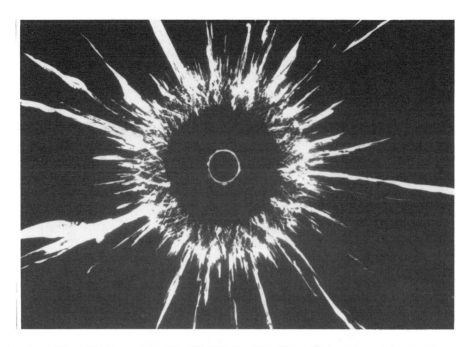

Figure 8.2. A crack pattern (top) in plexiglas around a charge detonated in a 3 mm diameter drillhole. Cracking (bottom) around a $d = 38$ mm drillhole in granite ($\sigma_C = 200$ Mpa, $\sigma_t = 12$ Mpa) filled with a $d = 22$ mm diameter plastic tube containing an emulsion explosive, Kimulux 42. The maximum crack length is 45 cm (photo by SveDeFo).

with an excess of carbon in the reaction products is used, carbon is deposited on the surface of the largest cracks near the borehole (within a distance of about 10 to 20 hole radii from the hole). Thus, the gas pressure will extend into the cracks, acting on the crack walls. The gas flow velocity into the cracks may or may not be sufficiently high for the gas to reach the crack tip. The flow in the narrow crack is associated with an appreciable pressure and heat loss and is initially not very much greater than the crack velocity. Even if the gas pressure in the crack decreases rapidly outward, the lever or wedge effect of the large gas pressure on the crack surface nearer to the borehole will make an appreciable contribution to the tension at the crack tip. This gas pressure contribution will also be greater for a long crack than for a short crack.

All these factors would appear to preferentially aid those cracks singled out by the interaction with the returning tensile wave and thus to extend a long-enough distance to reach the free surface. There is very little overall displacement of the rock prior to this time. This was very clearly shown by the experiments in granite by Norén [1956] and also by the plexiglas experiments by Persson and Ladegaard-Pedersen [1968]. After the cracks have reached the free surface, the rock is suddenly accelerated by the remaining gas pressure, which acts upon the forward half of the drillhole and at least part of the crack surface during a short initial period.

8.2 Interaction Between Radial Cracks and the Returning Tensile Wave

Field and Ladegaard-Pedersen [1969] investigated, in a series of model scale tests in plexiglas, the interaction between the growing radial crack system and the tensile wave originating at the free surface and caused by the reflection of the initial shock wave. The tensile wave increases the tensile stress at the tip of those cracks which are parallel to the curved, returning tensile wave front (Figure 8.3). Therefore, those cracks traveling in a direction at an angle of 40 to 80° to the normal of the free surface are given a greater propagation velocity. They will travel ahead of the intermediate cracks and by relaxation of the surrounding material will reduce the velocity of these. The theory thus developed gives a plausible explanation of earlier observations by P.-A. Persson, Ladegaard-Pedersen, and Kihlström [1969] who found that the break-out angle (the top angle at the drillhole of the wedge of the rock by the charge) increases with increasing burden and with increasing height of bench or length of the charge (Table 8.1).

8.3 Stress Waves in Rock

The detonation of an explosive charge in a borehole in rock gives rise to a strong initial shock wave which then decays into stress waves in the surrounding rock. For a borehole fully charged with a strong explosive, the initial shock wave pressure exceeds the strength of rock, and we get a very complicated shear deformation pattern which ultimately leads to crushing of the rock around the borehole. Figure 8.4 shows the results of a computer calculation of the stress waves in the near region around the borehole. Further out, the conditions are favorable for the formation of radial cracks, such as those shown in Figure 8.2 of the crack pattern in plexiglas.

As the wave moves radially out from the borehole, the amplitude (pressure) decreases and the wave becomes a purely elastic compressive wave, the P-wave.

Additional waves, the S-wave and the Rayleigh wave, are formed as a result of the interaction of the P-wave with the free surface (Figure 8.5).

Table 8.1. Increase of break-out angle with burden.

Semi-angle for break-out	Burden (m)	Bench height (m)	Stemming height (m)	Total charge weight (g)
75°	0.35	1.65	0.58	110.4
77.5°	0.5	1.8	0.58	110.3
85°	0.6	1.9	0.58	110.6
90°	0.8	2.3	0.58	110.0

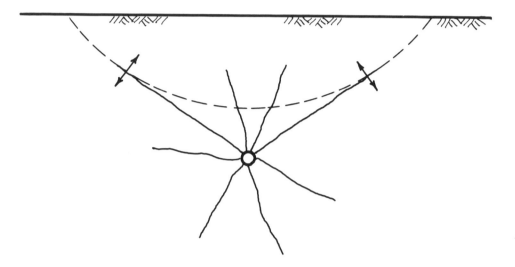

Figure 8.3. Interaction between reflected, tensile wave and growing crack system.

When we are discussing wave strength in this far-field region, it becomes useful to use the peak particle velocity as a measure. Figure 8.6 shows the approximate decrease of the peak particle velocity with distance away from a charge in a 250 mm diameter borehole. Finally, at very long distances, compared to which the charge dimension is small, the peak particle velocity follows approximately the relation

$$v = K \frac{W^\alpha}{R^\beta} \qquad (8.1)$$

where W is the charge weight; R is the distance; and K, α, and β are constants. (For hard bedrock, $K = 0.7$, $\alpha = 0.7$, and $\beta = 1.5$ if W is measured in kg, R in m, and v in m/sec.) Figure 8.7 shows the considerable scatter, as large as ±1 order of magnitude, due to rock structure-related differences in wave transmission of different rock masses at many different locations in different types of rock. For measurements taken at one location with blasts set off in the same general area, the scatter is always much less than that shown in Figure 8.7.

The stress waves move with different velocities:

$$c_P \approx 5000 \text{ m/sec}, \qquad c_S \approx 3500 \text{ m/sec}, \qquad c_R \approx 2500 \text{ m/sec}.$$

Depending upon the wave type, we can get an estimate of the stress σ or strain ϵ in the rock if we consider the motion as a simple harmonic oscillation (extension or bending),

$$\epsilon = \frac{\sigma}{E} \approx \frac{v}{c}. \qquad (8.2)$$

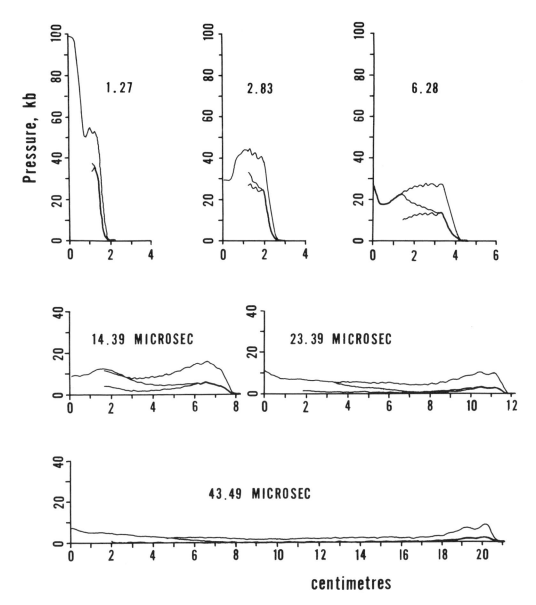

Figure 8.4. Calculated reaction product pressure in a 25 mm diameter drillhole and principal stresses in granite as a function of radius. Initial pressure in borehole 100 kbar. Upper curves, radial, intermediate curves, axial, and lower curves, tangential stresses. (a) 1.27 μs, (b) 2.83 μs, (c) 6.28 μs, (d) 14.39 μs, (e) 23.39 μs, and (f) 43.49 μs after start of motion.

Solid, unfractured granite may be expected to fail in tension at a stress of perhaps 30 MPa, corresponding to a strain of 0.1 to 0.2%, that is, a particle velocity between 2 and 10 m/s depending on the wave type. But normal fissured rock will undoubtedly show damage at lower stress levels, such as between 0.5 and 1 m/s, and the orientation of the fissures in relation to the wave propagation direction may also be important.

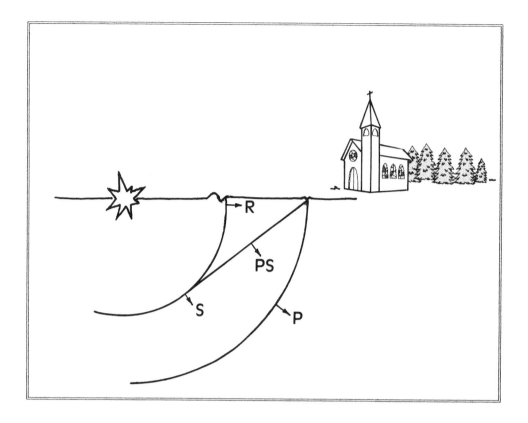

Figure 8.5. Far field stress waves (ground vibrations).

8.4 Rock Mass Damage and Fragmentation by Blasting

The tensile strength across a joint is normally by far the weakest link in the general strength of a rock mass. As we have seen, the stress wave generated by a detonating charge gives rise to transient deformations in a rock mass. Part of these deformations are tensile, and where they occur across a joint, it may crack open, even at a stress level much lower than the tensile strength of the solid rock material surrounding the joint.

Because of the short duration of the tensile transient, the crack may open only a few microns or a few tens or hundreds of microns. However, the damage is irreversible, a crack once opened will not heal, and the result is a slight permanent swelling of the rock mass and a lowering of the rock mass strength. The more intense the stress wave, the greater the degree of swelling and the greater the loss in strength.

In a typical rock mass, there are several, often intersecting joint systems of different strength. As shown schematically in Figure 8.8, the charge size will influence not only the extent of damage, but also the fragment size distribution. Close to the charge, gross joint damage will lead to complete separation of fragments of rock along the surfaces of all the intersecting joint planes. In an intermediate region, the two weakest joint planes will separate. Further away from the charge, only the weakest of the joint systems will get damaged. Outside of the largest circle, essentially no damage will occur. In this way, we may understand in a qualitative way how the joint structure, the charge size, and the drillhole separation will influence the shape and size of fragments in blasting.

8.4. Rock Mass Damage and Fragmentation by Blasting

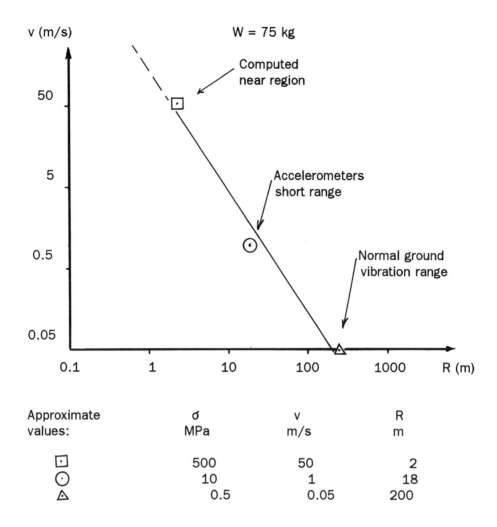

Figure 8.6. Stress wave particle velocities from 250 mm charge, 15 m long.

Obviously, the large-diameter drillholes lead to a much greater region of damaged rock at the same linear charge density than the small-diameter holes.

Certainly, very close to the charge, the stresses are high enough to cause cracking and crushing of the homogeneous rock material between joints. However, in most competent rock, this affects only a very small fraction of the rock volume fragmented.

The maximum tensile or shear stress in homogeneous rock increases approximately as the square root of the charge diameter and decreases approximately linearly with increasing distance from the charge. The stress level is also influenced by the presence of nearby free surfaces. In a jointed rock mass, the local tensile stress resulting from the detonation of a given charge a given distance away from the blasthole is greatly influenced by the joint structure, particularly by any open cracks or joints filled with soft material. Also, the fracturing occurs along the joints rather than along directions of maximum shear stress because the joints are so much weaker than the solid rock between the joints.

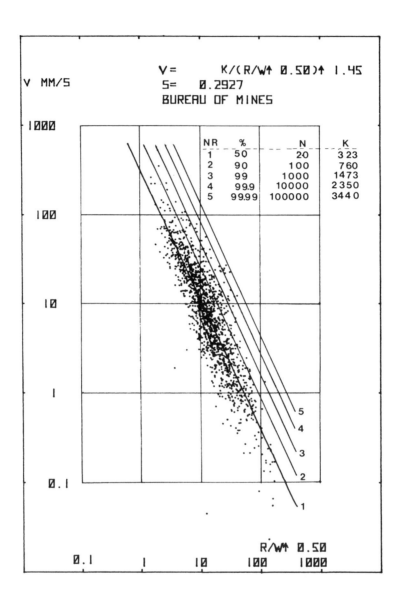

Figure 8.7. Peak particle velocity v as a function of $R/W^{0.5}$ for measured data according to the US Bureau of Mines at different confidence levels and different number of rounds.

An exact calculation and prediction of the joint damage and fragmentation of a real rock mass is out of reach, even with present-day computer capabilities.

However, by applying experience gained from the study of ground vibrations and the associated rock damage in real life rock blasting, we have been able to make surprisingly reliable predictions, not only of where incipient rock damage will occur some distance away from a given charge, but also of the degree of fragmentation in the near region.

We will now outline this technique. We estimate the vibration particle velocities in the near region around an extended charge, taking into account the influence of nearby free surfaces. We then use these velocity values to estimate tensile stresses and rock

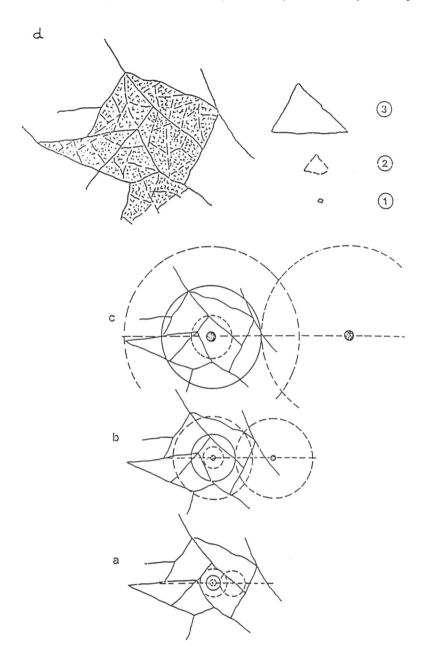

Figure 8.8. Schematic of the influence of charge size on the fragmentation of jointed rock [Persson, 1973]. In (d), 1, 2, and 3 are three different systems of joints, of increasing strength, $(\sigma_1 > \sigma_2 > \sigma_3)$. In (a), (b), and (c) are shown circles around drillholes of three different diameters, small, medium, and large, superimposed on the same joint system. The innermost of the three circles (dashed) indicates the greatest distance away from each drillhole where the stress exceeds or equals σ_1, the intermediate circle (full line) indicates the greatest distance from the hole where the stress exceeds σ_2, and the outer circle (dashed) indicates the greatest distance from the drillhole where the stress exceeds σ_3. Inside the innermost circle, then, all three joint systems fracture, in the intermediate ring area only the two weakest, and in the outer ring area, only the weakest joint system fractures.

8.5 Estimation of Near Region Vibration Particle Velocities

Let us first assume that we have an extended charge of length H with linear charge density ℓ in a drillhole in rock (Figure 8.9). To find the resulting stress at point P at a perpendicular distance r from the charge axis, we will also make an additional assumption. We will assume that at any one point away from the charge, the vibration peak particle velocity resulting from the detonation of each part of the charge is numerically additive. We will then take the resulting vibration peak particle velocity to be representative of the stress caused by the vibration.

Many authors have attempted to integrate the wave packages coming from each part of the charge as they arrive at the point of observation, taking into account the different times of arrival due to the distance difference. However, the peak particle velocity, and thus the peak stress in the rock, does not occur when the front of the wave reaches the point of observation. Instead, it occurs when the entire mass of rock between the drillhole and the free surface is set in motion as a result of the presence of the free surface, which happens after the reflected tensile wave from the free surface passes by the drillhole. On that time scale, the small differences in shock front arrival time is negligible. Therefore, it is allowable to integrate the vibration contributions from the elements of an extended charge without consideration to the small differences in timing of the arrival of the direct wave front as is done below.

The vibration velocity v resulting from a unit charge W at a distance R is given by Equation 8.1 which is repeated here as Equation 8.3 for completeness of the discussion in this section.

$$v = K \frac{W^\alpha}{R^\beta} \tag{8.3}$$

where the constants K, α, and β may have the values 0.7 m/s, 0.7, and 1.5, respectively.

Let us now integrate the transformed function w, which we may regard as the vibration intensity. (w is a quantity linear in W, and is therefore easier to integrate over the charge length.)

$$w = \left(\frac{v}{K}\right)^{1/\alpha} = \frac{W}{R^{\beta/\alpha}}. \tag{8.4}$$

For a very small charge dW, the vibration intensity dw is given by

$$dw = \frac{1}{R^{\beta/\alpha}} dW. \tag{8.5}$$

We can now integrate Equation 8.5 using

$$dW = \ell \, dx$$

and

$$R = \left[r_0^2 + (x - x_0)^2\right]^{1/2} \tag{8.6}$$

to get

$$w = \ell \int_{x_s}^{x_s + H} \frac{dx}{\left[r_0^2 + (x - x_0)^2\right]^{\beta/2\alpha}} \tag{8.7}$$

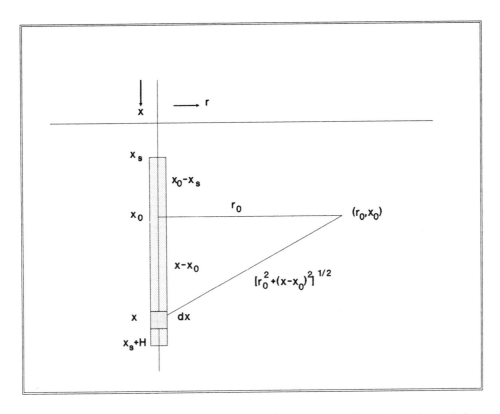

Figure 8.9. Integration of the surface wave effect in the near region of an extended charge.

which together with Equation 8.4 gives

$$v = K \left[\ell \int_{x_s}^{x_s+H} \frac{dx}{[r_0^2 + (x - x_0)^2]^{\beta/2\alpha}} \right]^{\alpha}. \qquad (8.8)$$

For $\beta = 2\alpha$, Equation 8.8 can be integrated to give

$$v = K \left(\frac{\ell}{r_0}\right)^{\alpha} \left[\arctan\left(\frac{H + x_s - x_0}{r_o}\right) + \arctan\frac{(x_0 - x_s)}{r_0} \right]^{\alpha}. \qquad (8.9).$$

Figure 8.10 shows curves of the resulting vibration velocity v for different linear charge concentrations ℓ as a function of distance r to the charge axis (a) at points level with one end of a 15 m long large-diameter charge, and (b) at points level with the center of a 3 m long smaller diameter charge.

8.6 Influence of Nearby Free Surfaces

In the above treatment, the presence of the original nearby free surfaces, such as the top of the bench, is not specifically taken into account. In a general way of course, the free surface toward which the rock vibrates is taken into account. The experimental data points from which the parameter values in Equation 8.3 were originally determined are the measured vibration peak particle velocities at the ground surface resulting from

246 Chapter 8. Stress Waves in Rock, Rock Mass Damage, and Fragmentation

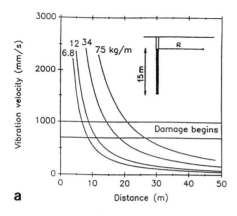
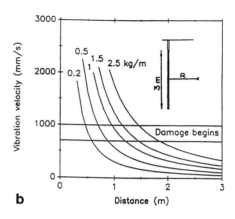

Figure 8.10. Calculated peak vibration velocity as a function of the distance to (a) one end of a 15 m long, large-diameter charge, and (b) the center of a 3 m long, smaller-diameter charge, with the linear charge concentration as a parameter. The charge arrangement is typical for bench blasting with large-diameter holes.

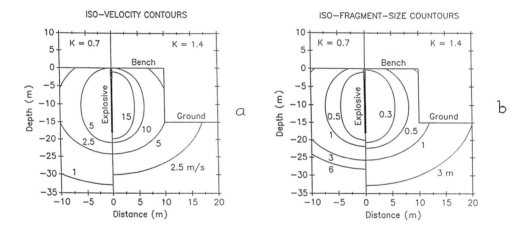

Figure 8.11. (a) Contours of equal vibration velocity corrected for the presence of additional free surfaces; and (b) the corresponding contours of equal fragment size distribution.

the detonation of a concentrated charge in rock a relatively long distance away. Values of vibration peak particle velocity measured by accelerometers in drillholes in rock — placed 5 m below the top free surface of the bench on a level with the top of the 15 m long charge — were in good agreement with the predictions of Figure 8.10a. This indicates that the approach is valid. The free surface created by the blast itself is another, equally important nearby free surface, especially when we are discussing damage effects in the remaining rock near the drillholes in smooth blasting, such as those at the damage limit for small linear charge concentrations in Figure 8.10b.

It appears likely that the presence of an additional free surface will increase the vibration velocity more at a point close to the free surface than at a point further away from it. However, in the absence of experimental evidence for the magnitude and

Table 8.2. Damage and fragmentation effects in hard Scandinavian bedrock resulting from vibrations with different values of the peak vibration particle velocity. Density $\rho_0 = 2600$ kg/m^3, sound velocity $c = 4900$ m/s, Young's Modulus $E = 60,000$ MPa.

Peak particle velocity (m/s)	Tensile stress (MPa)	Strain Energy (J/kg)	Typical effect in hard Scandinavian bedrock
0.7	7	0.65	Incipient swelling
1	10	1.33	Incipient damage
2.5	25	8.3	Fragmentation
5	50	33	Good fragmentation
15	150	300	Crushing

variation with distance of this effect, we will simply assume that for a given charge the vibration velocity resulting from a given charge in the material to be broken loose is twice that in the material left standing behind the charge (at points equidistant from the charge). We have done this by increasing the value of the parameter K in Equations 8.3 to 8.9. Figure 8.11 shows the iso-velocity contours (contours of equal velocity) obtained based on this assumption, together with the resulting iso-fragment size contours (contours of equal fragment size). This technique, even though crude, makes it possible to calculate both the fragment size distribution in the rock broken loose and the extent of damage to the remaining rock, each as a function of the distance to the charge.

8.7 Rock Mass Damage and Fragmentation

We are now ready to use these estimated vibration peak particle velocities to predict the associated tensile stresses. Based upon experiment, we have found that in hard Scandinavian granitic or gneissy bedrock, a vibration peak particle velocity in the region 0.7 to 1 m/sec begins to give measurable damage, in the form of slight swelling and slightly decreased shear strength. Velocities in the region of 2.5 m/sec are characteristic of the range where fragmentation begins to become marginal, at a distance equal to the critical burden in hard bedrock. Velocities around 5 m/sec are characteristic of very good fragmentation, still mainly along planes of weakness; whereas velocities in the neighborhood of 15 m/sec are needed to crush the solid granite. By using the simple equations for the strain ϵ, stress σ, and strain energy e_s in a stress wave in a bar

$$\epsilon = \frac{v}{c}, \qquad \sigma = \frac{v}{c} E, \qquad e_s = \frac{1}{2}\frac{\epsilon \sigma}{\rho_0} = \frac{1}{2}v^2, \qquad (8.11)$$

and by assuming the wave is propagating with velocity $c = 4900$ m/sec in a rock mass with elastic modulus $E = 50,000$ MPa, we find that the vibration peak particle velocities in the range 0.7 to 15 m/s are equivalent to stress levels from 7 to 150 MPa, as shown in Table 8.2. Table 8.2 also contains values for the strain energy, as defined in Equation 8.11. Traditionally, strain energy has been used as a measure of the fragmentation achieved in mechanical crushing of rock. The strain energy imposed on a given mass of rock just before it fractures has been found to correlate very well with the fragment size distribution of the fragments obtained after fracture. By using the ranges of strain energy indicated in Table 8.2, we can begin to apply the concept of strain energy as a measure of damage to the remaining rock, fragmentation, and crushing. Obviously, the levels of strain energy at which these typical effects occur depend on the type of rock, and great caution should be taken to establish the limiting values for any new type of rock mass before using data such as from Table 8.2 for engineering design purposes.

248 Chapter 8. Stress Waves in Rock, Rock Mass Damage, and Fragmentation

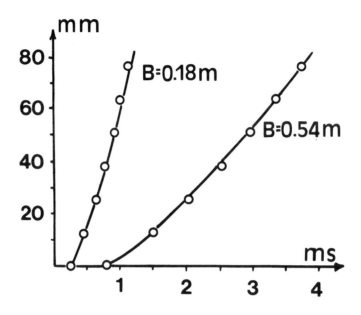

Figure 8.12. Distance-time diagram for the front of the detached rock [after Norén, 1956].

In weaker rock materials, these effects may occur at lower vibration peak particle velocities. Observe, though, that a rock material that has a low strength as measured in laboratory samples may still, as part of a rock mass, have joint strengths comparable to those of granite and may therefore be no easier to fragment. It may also, if not unduly damaged by careless blasting, leave an equally good and competent rock surface in a tunnel or on a slope as does the hard granite.

8.8 The Rock Acceleration Process

Measurements of the acceleration and velocity of the free surface in front of a borehole show quite clearly that the movement of the free surface caused by the arrival of the shock wave is negligible in normal bench blasting.

The time interval between detonation and the start of motion of the free surface is between 5 and 10 times the shock travel time from the borehole to the free surface.

After this time interval, there is a rapid acceleration to the final break-out velocity. Norén [1956] first observed this phenomenon in full-scale experiments in granite, using a 1830 mm (6 foot) long borehole of 32 mm ($1\frac{1}{2}$ in) diameter loaded with 38 mm ($1\frac{1}{4}$ in) diameter dynamite cartridges. The measurements were made with an electromechanical device in which the motion of the rock face was recorded by the successive breaking of a series of thin copper wires by an insulated knife edge moving with the rock. Norén's results are shown in Figure 8.12.

The motion of the rock face caused by the arrival of the shock front was not measurable in these experiments, but we may see quite clearly that the start of real motion is quite close to the time of breakage of the first wire. In Table 8.3, the time interval between detonation and start of motion of the free surface is compared with the time of travel of the shock wave (with an assumed velocity of 5000 m/sec) from the borehole to the rock face. The time interval is about 7 times the shock travel time.

Table 8.3. Comparison of time interval between detonation and start of motion of the free surface and the time of travel of the shock wave from the borehole to the rock face.

Burden (m)	Time interval between detonation and start of motion of free surface (ms)	Shock travel time from detonation to free surface (ms)	Ratio between time interval and shock travel time
0.18	0.25	0.036	6.94
0.23	0.31	0.046	6.74
0.28	0.42	0.056	7.50
0.54	0.83	0.108	7.68

8.9 Fragment Size Distribution

Langefors and Kihlström [1963] in an empirical way described the fragmentation in blasting. Through field experiments, they studied the influence of burden and specific charge on the fragment size distribution. Careful observations were carried out in quarries and other types of blasting operations where bench blasting was utilized. Based on these observations and measurements, they developed a relationship between the average boulder size, the burden, the type of rock, and the specific charge. The average boulder size L was defined as the approximate sieve size through which 90 to 95 % of the fragments from a round could pass.

Figure 8.13 shows the relation between the average boulder size and the specific charge of explosive in kg/m^3 for different burdens. From the figure, we can see that the average boulder size can be reduced if we keep the specific charge constant and decrease the burden. We can also conclude that it is cheaper to achieve good fragmentation by reducing the burden rather than only increasing the specific charge. This is especially true when a high specific charge is used. It is common experience that a switch to blasting with a larger hole diameter under otherwise unchanged conditions must be accompanied by an increase in the specific charge. This behavior is caused mainly by the rock structure itself. When the burden and spacing is increased, the average joint spacing or distance between weakness planes will play an increasingly important role in fragmentation.

In examining the relative magnitudes of the costs for explosives and drilling (Figure 8.14), we note that the cost for explosive can be as much as five times the cost for drilling when a large hole diameter is used. Taking into account both the influence of burden and specific charge upon the average boulder size (Figure 8.13) and the relative magnitudes of explosives and drilling costs, it is often a good decision to reduce both the burden and the specific charge if a better fragmentation is wanted. The lower explosive cost will pay a part of the cost for the extra drilling if the same hole diameter is used. This method, which assumes that decoupled charges are used, results in a lower borehole pressure and a minimization of the fraction of very finely fragmented material, called *fines*, in the fragmentation distribution.

Larsson [1974] defined the average fragment size k_{50} as the fragment with a side length equal to the sieve size through which 50 % of the blasted rock would pass. Sieve analysis from several rounds indicated that fragmentation distribution could be determined with sufficient accuracy if the average fragment size k_{50} was known. This fragment size could be calculated if the burden, the specific charge, and the rock constant were known. It can be shown that there exists a fairly good correlation between the average boulder size L and the average fragment size k_{50}

$$L \approx 2.6\, k_{50}. \tag{8.12}$$

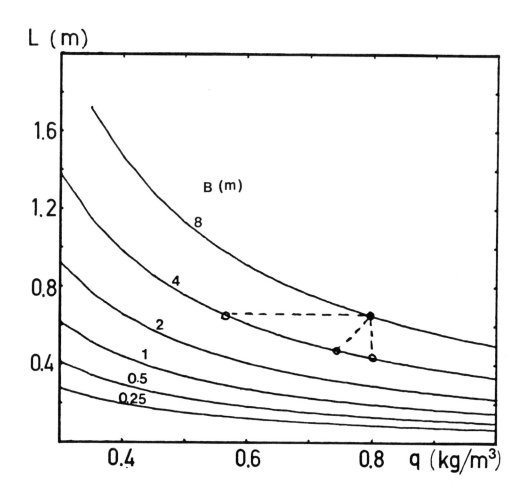

Figure 8.13. Average boulder size L as a function of specific charge q for various burdens V.

From Larsson's empirical fragment distribution, Holmberg [1981] derived the mathematical distribution function for the fragment size by using a Weibull distribution

$$y = 1 - \exp\left[-\left(\frac{0.76x}{k_{50}}\right)^{1.35}\right] \tag{8.13}$$

where x is the sieve size (m), y is the weight percent passing through the sieve size x, and k_{50} is the average fragment size.

Figure 8.15 shows the calculated percent passing as a function of sieve size with k_{50} as a parameter. Equation 8.12 also determines the average boulder size L.

From these mathematical formulae, it is possible to calculate with a good accuracy the average fragment size k_{50} and thereby describe the fragment distribution on the basis of rock type, specific charge, and burden. The relationship also determines the average boulder size L, shown as a function of specific charge with hole diameter as a parameter in Figure 8.13.

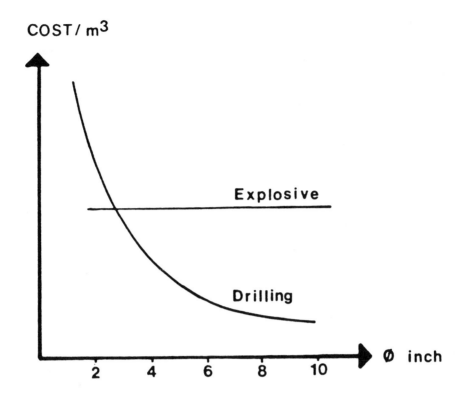

Figure 8.14. Cost per cubic meter of solid rock for explosives and for drilling as a function of hole diameter (inch) at constant specific charge.

8.9.1 Short Delay Multiple-Row Rounds

Short delay blasting was introduced as a means of reducing ground vibrations blasting. Before long, however, it was found that the fragmentation was also much improved by the introduction of multiple-row rounds. With multiple-row blasting, the specific charge could be increased without hazardous throw. On one hand, the short delay time between firing of consecutive rows is long enough to allow the rock of one row to move away, and the free face of the following row to be uncovered, prior to the detonation of its charges. On the other hand, the short delay time is short enough for the rock from the previous rows to still be hanging in the air at the detonation of the second row, creating an effective curtain to stop rock fragments from the second row from moving with a speed above the average. In addition, collisions between blocks moving with different velocities will considerably increase the fragmentation.

The optimal delay time τ, i.e., the delay time at which the best fragmentation is achieved, is a function of the burden B. According to Langefors and Kihlström [1963], for burden between 0.5 and 8 m,

$$\tau = kB \tag{8.14}$$

where k has a value in the range from 3 to 5 ms/m.

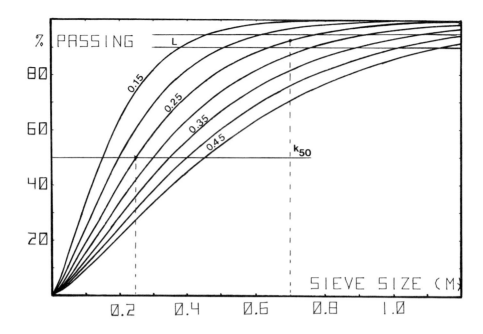

Figure 8.15. Fragment distributions from rounds with an average boulder size L or an average fragment size k_{50}.

8.9.2 The Wide-Spacing Blasting Method

If the hole spacings S in the rows are increased while keeping the product of spacing and burden $S \cdot B$ constant, the fragmentation in short delay multiple-row rounds will become improved. This method, proposed by Langefors and investigated in model-scale [1965], was worked out and tested in full-scale by Persson and Ladegaard-Pedersen [1970]. The ratio of spacing S to burden B was previously kept in the range 1 to 1.3. Through systematic investigations using model tests keeping the product $S \cdot B$ constant, the fragmentation was found to increase with S/B up to $S/B = 8$. The causal connection is principally simple. Figure 8.16 shows schematically one-row rounds with S/B from 0.25 to 4. At $S/B = 0.25$, the burden is a layer, the thickness of which is four times the spacing or distance between the holes. It is easily seen that this layer will be thrown out as a whole — divided into a few large blocks only. At $S/B = 1$, the course of events is similar, but the cracks from each charge assert themselves more and the fragmentation will improve. At $S/B = 4$, there will be little cooperation between the charges in the one-row round. Using multiple-row rounds, however, the sections between the holes in each row will be loosened and fragmented by the charges in the following rows.

Table 8.4 gives some comparative results for granite (Umeå) and limestone (Stora Vika). At both places, the bench height was 20 m and the hole diameter 78 mm (3 in). According to Table 8.4, the total number of boulders in limestone in blasts with $S/B = 6$ was 47 per 1000 m^3 of rock, compared to 121 in a blast with $S/B = 1.25$. In granite, the number of boulders decreased to 90–112 in 5 rounds with $S/B = 4$, as compared with 171–219 for $S/B = 1.25$. The mucking capacity increased from 1500 to 2000 ton/day, and the total costs for drilling, blasting, and mucking dropped significantly.

8.9. Fragment Size Distribution

Table 8.4. Comparison between conventional rounds using $S/B = 1.25$ and $S/B = 1.5$ with wide-spacing rounds using S/B values between 4 and 8 [after Persson and Ladegaard-Pedersen, 1970].

	Granite (Umeå, Sweden)		Limestone (Stora Vika, Sweden)			
S/B	1.25	4.0	1.5	4.3	6	6
Rounds	4	5	8	1	1	1
Holes/round	10–44	41–68	39-69	67	66	45
Rock volume, 1000 m³/round	1.3–5.9	4.0–7.4	4.7-8.3	11.1	7.9	5.8
B, m	2.0–2.3	1.3	2	1.4	1	0.9
SB, m²	5-7	6.7	6	8.4	6	6.5
Hole diam., mm	76	76	76	76	76	76
Spec. charge, kg/m³	0.43	0.67	0.53	0.44	0.51	0.49
Spec. drilling, m/m³	0.15	0.19	0.17	0.12	0.17	0.16
Boulders/1000 m³	171–219	90–112	121	57	47	86
Mucking, ton/day	1500	2000				
Boulder size	—	—				
Costs in US/m³ (1970)						
Drilling	0.20	0.26				
Blasting	0.27	0.35				
Boulders	0.35	0.17				
Mucking	0.65	0.50				
Total	1.45	1.26				

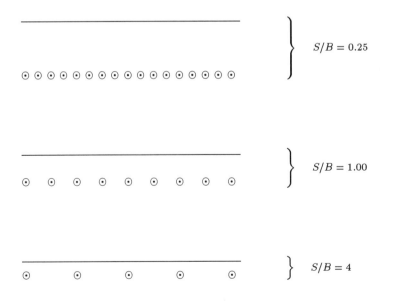

Figure 8.16. One-row rounds.

254   Chapter 8. Stress Waves in Rock, Rock Mass Damage, and Fragmentation

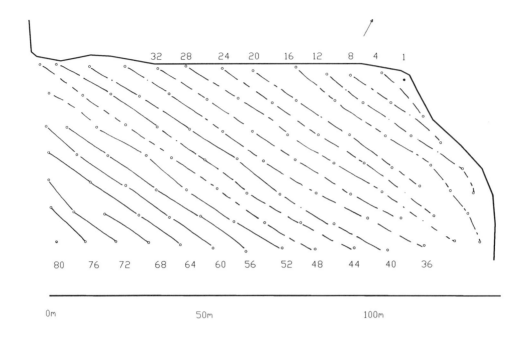

Figure 8.17. How to use the short interval delay pattern to produce a blast with wide hole spacing from a nearly square drill pattern. Numbers indicate the cap number used in each row of holes connected by a dot-dash line. Time interval between rows in 100 msec.

The superior fragmentation associated with the wide-space blasting method may also be obtained with a "square" drilling pattern by suitably choosing the ignition delay pattern. Figure 8.17 shows the drilling and ignition delay pattern in which the rock is thrown at an angle of 25 to 30° to the rows of boreholes. This method is particularly attractive where the blasting and drilling operations are carried out separately at widely different times.

8.10 Bench Blasting with Auxiliary Holes

In bench blasting with large borehole diameters, a large portion of the upper part of the holes must be left unloaded in order to prevent flyrock from the hole openings. However, this often results in the production of large boulders from the upper, unloaded part of the bench. By the addition of short auxiliary holes of smaller diameter between the main holes, it is possible to achieve a more satisfactory result. Figure 8.18 gives an example of a bench round where the maximum throw forward had to be kept within 15 m (50 ft). The holes were inclined. The main holes of 50 mm diameter (2 in) were drilled to full depth and charged only to induce breakage and swelling, including the necessary technical margin. The auxiliary holes, of 32 mm diameter ($1\frac{1}{4}$ in) were drilled to about half the depth and charged near the limit for full breakage of rock in front of them, but with no throw. With this drilling pattern, a fragmentation corresponding to the small burden of the small holes was obtained without drilling all the holes to the bottom. Because of the practical complications of operating two types of drilling equipment, this method has not been used very extensively.

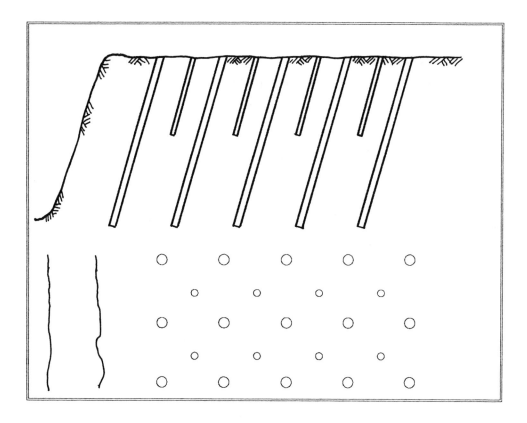

Figure 8.18. A method of improving fragmentation in a blast using large borehole diameters.

8.11 Coarse Fragmentation

If a coarse fragmentation is wanted, there are some useful rules of thumb. Use:

1. A small spacing to burden ratio
2. A small number of rows
3. An instantaneous ignition of each row
4. A small column charge concentration

A small number of rows should be used, as this makes it possible to choose a burden close to the critical burden. If some holes should not break the toe burden, the costs for secondary blasting for leveling the bottom would not be too high.

The crushing due to collisions between blocks blasted in subsequent rows will also be reduced if a small number of rows is used. The column charge should preferably be reduced in order to lower the borehole pressure. This will reduce the crushing at the borehole vicinity and decrease the dynamic load.

8.12 Influence of Geological Discontinuities upon Fragmentation

Blasting in a homogeneous isotropic medium naturally does not result in the same fragmentation pattern as when the medium is permeated with discontinuities. In most rock materials, fissures occur, thus reducing the explosive-induced stresses due to shock wave reflections. Radial cracks from the charge are effectively arrested at the fissures when the stress concentration factor becomes too low and the gas penetrates through already existing fissures. Bedding planes in sedimentary rocks and foliation in metamorphic rocks can result in the rock material properties which are dependent upon loading direction. The previous stress-time history in apparently homogeneous sedimentary rocks and the differences between the principal stresses can very much change the explosive-induced fracture pattern and thereby affect the fragmentation.

Anisotropy and the directions of discontinuities must be considered when an optimal blasting direction is to be determined. These features also introduce a scale effect into the fragmentation prediction and may influence the choice of blast hole diameter. For a given charge concentration per rock mass, a better rock fragmentation is achieved if the explosive is well distributed within the rock mass as is the case when using small-diameter holes. However, it can sometimes be more economical to utilize cheap (lower cost per borehole volume), large-diameter holes, which can be drilled to a longer hole epth with better precision than small-diameter holes and thus decrease the expensive development work. If large-diameter holes are introduced, the amount of explosive must usually be increased if the same fragmentation is to be achieved. However, this is not always necessary. Sometimes the size of the geological discontinuities is such as to provide a small enough fragment size when large-diameter holes are used even without increasing the total mass of explosive.

A number of authors have studied the influence of the rock structure on the blasting result. The degree of fragmentation achieved was found to be highly influenced by the character of the fissures and joints in the rock mass. Blasting techniques should be designed to take advantage of the direction and the dip of existing weakness planes in order to improve the blasting results.

Let us review some experiments carried out in fissured materials in order to establish a background for understanding the fragmentation process.

Seinov and Chevkin [1968] reported a model experiment where they studied the influence of rock structure upon fragmentation. They simulated joints and filled fissures by using three glass plates ($6 \times 30 \times 30$ mm) separated by materials with different acoustic impedance and width (Figure 8.19). Air, water, kaolin, and concrete were used.

The models were blasted in a metal frame in which the specimens were held between rubber packing to prevent the fragments from flying apart. Fissure widths and filling material were changed during the experiments. The first plate was always completely crushed to fine particles, thus, the fragmentation quality was only recorded for the second and the third glass plates.

Results from these experiments indicated that, in a medium with open fissures, an increased amount of explosive may lead to improved fragmentation (depending upon the width of the fissure). When the fissures were filled with kaolin, energy was absorbed from the explosion and fragmentation did not significantly improve when the specific charge was increased. When the acoustic impedance between the plates was increased (by using a cement mixture), an increase in the fragmentation was achieved. Fragmentation was

8.12. Influence of Geological Discontinuities upon Fragmentation

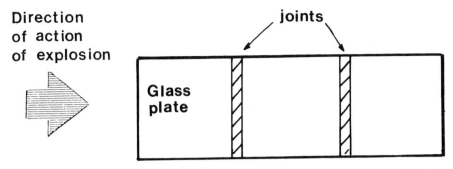

Figure 8.19. Model experiments in glass plates performed by Seinov and Chevkin.

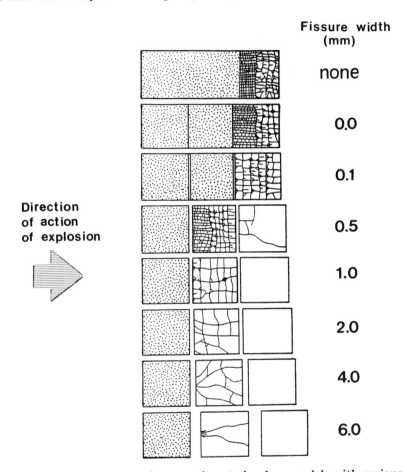

Figure 8.20. Results of fragmentation experiments in glass models with various fissure width. Dotted regions represent fragments smaller than 0.8–1.0 mm.

slightly greater when water was used instead of air in the fissures. The authors also claim that even with a very careful fitting of the plates to one another, the transmitted shock wave energy was strongly reduced due to the wave interference at the interface. It must be pointed out that, in the experiments described above, the released gas pressure could not be utilized due to the experimental geometry. The main energy utilized in these experiments is the shock wave energy. Figure 8.20 shows some of the results obtained when different fissure widths were used.

Bhandari [1974] describes a cratering test that was performed in a limestone quarry to observe the effects of bedding and joints. On the basis of these experiments and the knowledge of breakage principles in homogeneous rock, he pointed out the following: (a) in a non-homogeneous rock, the shock wave suffers greater attenuation and dispersion than in a homogeneous rock; (b) the joint width and the filler material determine this attenuation and the further propagation of the stress wave through the existing joints; (c) leakage of gas into joints significantly reduces the available borehole pressure and changes the normal fragmentation process; and (d) large joints parallel to the borehole mostly terminate the crater process and decrease the broken rock volume per charge weight.

Holmberg [1983] described the same effects, but here, the conclusions were based upon drifting in Precambrian rock with a schistose and gneissic structure containing about 10 different joint sets. Very cautious blasting had to be done to reduce damage to the remaining rock. When the perimeter hole distance was too large, the length of the radial crack created was determined by the distance from the closest joint to the borehole. The detonation gas penetrated through the radial cracks into the joints and lifted or just loosened the intermediate block defined by the joints between two contour holes. As a result of this phenomenon, the floor became rough and the digging efficiency became very low due to the "teeth" developed. This effect was later suppressed by decreasing the charge weight and the hole spacing to such a degree that each block between the joints sets was intercepted by a contour hole. During the excavation, it was found that, for effective breakage, the hole distance should be smaller than the average joint spacing.

T. From [1976] made model-scale experiments in PMMA (plexiglas) where he studied the influence of discontinuities upon fragmentation. He glued thin PMMA plates together and varied the strength of the glue. He observed that, when he placed the hole axis parallel to the weakness planes, coarser fragmentation was achieved when a high-strength glue was used. The normal fracture history observed from high speed camera photographs was the following: bending occurred in the plates after the explosive detonated and a crack was developed at the surface perpendicular to the hole axis due to the tensile stresses caused by the bending; when the crack had reached the maximum propagation velocity and bending increased, branching occurred and the crack developed into several cracks which increased fragmentation. The reason that coarser fragmentation was achieved when high strength glue was used probably depended upon the facts that the higher acoustic impedance made it easier for radial cracks to develop through the plates and the strength between the plates was such that the plates did not break along the weakness planes. For a low-strength glue, the radial cracks only developed to the first weakness plane and the fragmentation was based mainly upon the bending which also separated the plates, and smaller fragmentation occurred. When the angle between the hole axis and the weakness planes was changed from 0 to 90°, the fragmentation became coarser (Figure 8.21).

Barker et al. [1979] made a photoelastic investigation in polymeric models of the initiation of the fragmentation mechanism at the borehole wall. They found that, if the reflected stress wave was absent, no circumferential cracking occurred. It was also found that without the influence of the free face (and the reflected stress wave in a flaw-free material), the final crack pattern consisted of 8 to 12 dominant radial cracks that emerged from a very dense radial network surrounding the borehole. The final fragmentation consisted only of large sector-shaped fragments. In models containing artificial flaws, Dally and Fourney [1977] found a marked difference in the fragmentation. Barker et al. [1979], induced flaws in Homolite 100 to study how flaws enhance fragmentation. They reported that flaws in the immediate vicinity of the borehole wall usually were

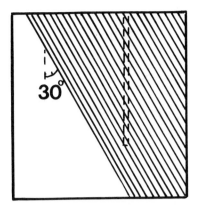 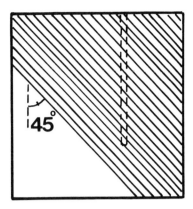

Figure 8.21. Fragmentation became coarser when the angle between the weakness planes and the hole axis increased.

initiated by both the S- and P-wave tensile tails. Flaws outside the borehole vicinity were initiated by the P-wave tensile tail. The maximum distance that the P-wave can initiate flaws is dependent upon the flaw geometry and the P-wave attenuation. Flaws close to a free surface were found to be more sensitive to the reflected P-wave, and cracks developed if a critical stress intensity factor was achieved at the flaw site. These flaws generally did not propagate in the radial direction. The strongest flaw-initiating mechanism is the combined action of the PP- and PS-wave near a free surface.

8.13 Influence of Void Volume upon Fragmentation

Jarlenfors and Holmberg [1980] performed model-scale tests at the Swedish Detonic Research Foundation to study how the void volume influenced the rock fragmentation. A study was also performed to see what influence the ignition interval between the rows had upon fragmentation when blasting was carried out with a small void volume. The experiments were conducted with PMMA and a mixture of epoxy and quartzsand. A steel sheet metal box was used to contain the models to be blasted (Figure 8.22). PETN was used as the explosive and was loaded in holes having diameters of 1.5 and 2.0 mm. Eight rows were fired in every test. The epoxy-quartzsand mixture was earlier used as a model-scale material by Naarttijärvi et al. [1980] for cautious blasting in simulated underground driftings.

Three different void volumes were used: 11.1, 16.7, and 50 %. These experiments showed that the best fragmentation was achieved for a void volume of 50 %. When the void volume was decreased, an increase in the fragmentation size occurred (Figure 8.23).

They concluded that, if blasting took place with one row at a time, the fragmentation became coarser as the void volume decreased. It was also found that, if blasting was performed toward a compacted, fragmented rock mass, the fragmentation was increased when several rows were fired with a proper initiation time delay compared with single-row blasting. The explanation for this phenomena must be that when row by row blasting takes place with a proper sequence, the movement of the burden produces a free surface, permitting the reflected stress wave to be utilized for fragmentation.

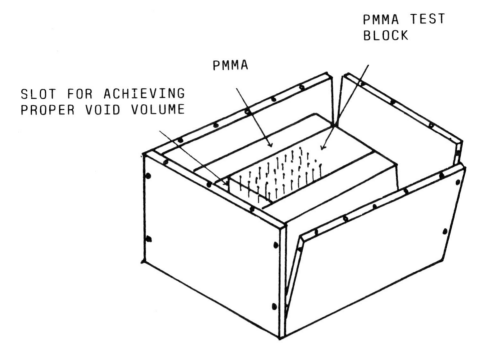

Figure 8.22. Blasting geometry for fragmentation experiments with various void volumes.

In order to determine the fragmentation for different shapes and sizes of void volumes, Federenko and Kovtun [1977] carried out an investigation on a scale of 1:100. A mixture of martite ore and cement was used for the 50 to 100 kg models. The mixture had the following properties:

Density	3060 kg/m^3
Porosity	31 %
Compressive strength	14 MPa
Tensile strength	1.2 MPa
Young's Modulus E (dynamic)	17 GPa

By changing the shape of the void volume, the influence on the fragmentation could be described. The different shapes of the voids used in the model-scale experiments are shown in Figure 8.24.

Various void volumes in the range 3 to 50 % were tested. A 3 % void volume resulted in a small number of developed cracks because of the confined holes. Fractures were mainly developed because of the compressive stress and very little because of the tensile stresses. A void volume of 5 to 6 % developed a greater number of cracks; and when the void volume was 7 to 8 %, a higher fragmentation degree was achieved, and the rock was broken and expanded into the void.

An increase in the swell to about 35 % (void volume 26 %) produced a decrease in the yield of oversize fragments. A further increase in void volume, however, led to an increase of the oversize yield. The curves shown in Figure 8.25 reveal how the shape and the size of the void volume affected the ore crushing quality.

The authors have shown that the formation of cracks which give rise to fragmentation of the blasted layer occur at a minimum void volume of 5 to 6 %. A horizontal, rectangular void (where the rock broke simultaneously from opposite sides) gave the best fragmentation. This happened when the swell was 30 to 34 % (void volume 23 to 25 %).

8.13. Influence of Void Volume upon Fragmentation

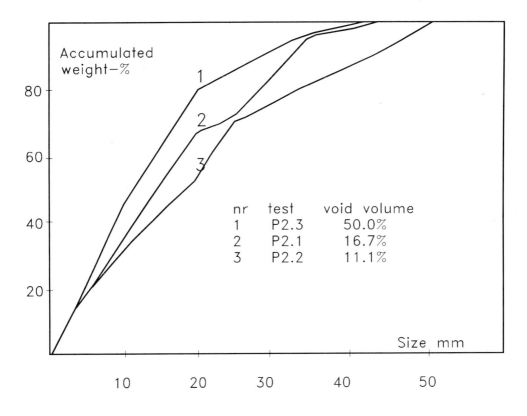

Figure 8.23. Achieved fragmentation distribution for blasting in PMMA with 11.1, 16.7, and 50% void volumes.

The authors described full-scale tests which were performed in an iron ore mine. For blasting from both sides toward a slot, the optimum size of the void volume was 21 to 23%.

Another Russian scale-model test (120 × 100 × 200 mm in size) with a 2:1 sand-cement mixture was performed by Volchenko [1977]. He studied the fragmentation of the mixture after it was blasted toward a compressible bed of crushed iron ore.

The crushed ore which had been placed in front of the concrete block had a void volume of 23 to 38% before the blast. The depth of compression due to the blast was determined by means of the movements of metal plates fixed in the compressible crushed ore (Figure 8.26).

Instantaneous one-row blasting with different spacing-to-burden ratios was tested. Three holes per row were drilled and 1.3 g PETN per hole was used.

Sixty-five experimental blasts were performed. They found that the best fragmentation occurred when the blasting took place toward crushed ore with a swelling equal to 40% (void volume equal to 29%) (Figure 8.27.)

The authors later confirmed the model-scale experiment results through pilot-scale blasts of 3.5 M tons of ore in a mine.

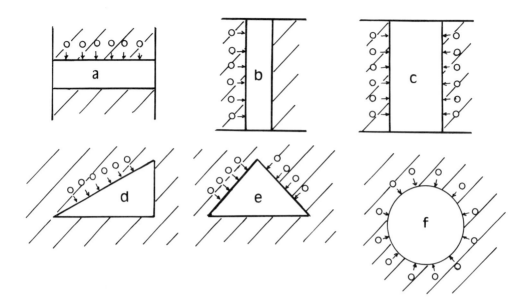

Figure 8.24. Shapes of voids.

During the years 1982–1986, half-scale experiments were carried out by Olsson [1987] in LKAB's mine in Malmberget, Sweden. By extremely cautious blasting, a small bench was established in the magnetite ore. Thereafter, a concrete wall was grouted in place at such a distance from the bench face that the wanted void volume was achieved. Before each blast, a heavy steel plate was bolted on top of the bench and the concrete wall. This made the blast confined so the only direction the fragmented material could go was into the void volume.

Data for the blasts using 3 holes/row and 5 rows were:

Hole diameter	11 mm	Bench height	0.7 m
Burden	0.28 m	Bench area	1.4×0.7 m^2
Spacing	0.7 m	Specific charge	1.4 kg/m^3

The results showed that much improved fragmentation was achieved in confined volume blasting compared to ordinary blasting. The best fragmentation occurred when the void volume was 30 to 33 % of the rock volume to be fragmented. It is obvious that when the fragmented material collides with the concrete wall or with the previously fragmented material, additional fragmentation takes place. Part of the kinetic energy is used for this additional fragmentation.

Especially in underground mining, it should be possible to utilize the positive effect with a reduced void volume to improve fragmentation.

8.13. Influence of Void Volume upon Fragmentation

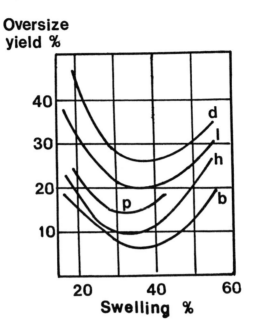

b: rectangular space, opposed ore breaking.
h: rectangular space, one-sided breakage.
p: triangular space, opposed breaking.
l: cylindrical space.
d: triangular space with one-sided breakage.

Figure 8.25. Influence of shape and compensation space on ore crushing quality.

264 Chapter 8. Stress Waves in Rock, Rock Mass Damage, and Fragmentation

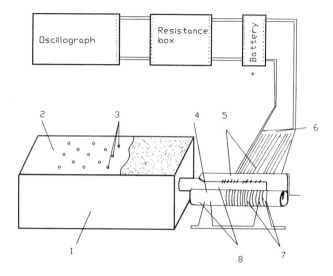

Figure 8.26. Blasting box used for fragmentation experiments by Volchenko.

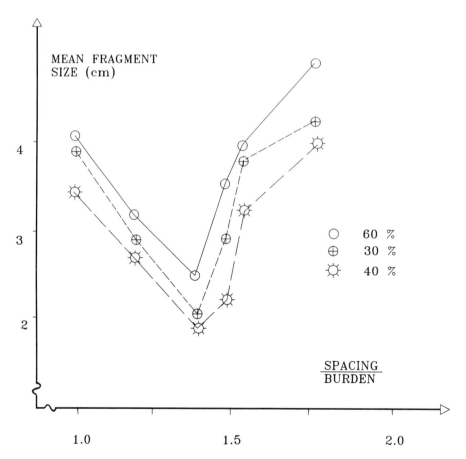

Figure 8.27. Best fragmentation occurred when the swelling was 40 % and the spacing-to-burden ration was 1.4.

Chapter 9

Contour Blasting

9.1 Introduction

After a rock blasting operation is completed, the remaining rock mass itself is very often a part of a structure that is expected to remain stable for a long time. The required stand-up time, i.e., the time the rock mass remains stable, depends on the purpose for which the structure is intended. The strength and stability of the rock mass surrounding large, permanent, underground storage rooms is of major importance to prevent rock fall that could badly damage people and equipment. Road and railway cuts along steep mountain sides and tunnels through mountains or underground beneath cities or under water must be stable enough for a long, safe lifetime. Abutments for large bridges and dams must maintain their integrity and strength without failure for a very long stand-up time. The failure of a major dam could mean a major catastrophe for a large region including villages and cities. In large open pit mines, the amount of extra rock that needs to be removed for even a small increase in the slope angle of the hanging-wall is astronomically large — leading to a demand for high steep slopes which must remain stable to avoid failures and costly scaling.

In all these cases, a blasting technique must be used which minimizes the damage done in the remaining rock. The less competent the rock mass itself is, the more care has to be taken in avoiding damage.

Several steps may be taken to reduce the damage caused by blasting. Where control of the perimeter is desired, methods lowering the charge concentration adjacent to the final perimeter are usually employed.

This chapter reviews some of those methods and discusses the rock mass damage from a rock mechanics point of view.

9.2 Smooth Blasting and Presplitting

All methods of cautious blasting have one common objective: to better distribute the explosive energy which is transmitted into the rock mass by the action of the detonation reaction product gas pressure on the drillhole wall, in order to reduce the dynamic stress, fracturing, and back break of the remaining rock. Most of these methods have been developed in the field, mainly by trial and error. We will describe one method for calculating the rock damage which is based on the intensity of blast vibration in the adjacent rock mass.

266 Chapter 9. Contour Blasting

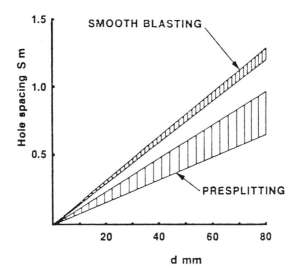

Figure 9.1. Recommended ranges of hole spacing as a function of hole diameter for smooth blasting and presplitting.

Smooth blasting is a method where the row of holes adjacent to the planned contour is fired at the end of the round, with a light charge per hole, with a small spacing, and usually with an S/B ratio of 0.8.

In presplitting, the contour charges are fired before the rest of the charges, most often in a separate round. The crack running from hole to hole then has to be accommodated by elastic deformation of the rock on both sides — because no rock is broken loose. Therefore, presplitting needs a closer spacing of contour holes, about 50 to 75 % of that for smooth blasting. Thus, presplitting becomes more expensive than smooth blasting (Figure 9.1). Because of this, and also because it is often difficult to fit in an extra blasting operation in the tight shift cycle, presplitting is very seldom used underground in Sweden. Smooth blasting is the main method, particularly in tunneling.

Empirically, it has been shown that the minimum required linear charge concentration for smooth blasting and presplitting is a function of the hole diameter

$$\ell = 90\, d^2 \tag{9.1}$$

where ℓ is the charge concentration in kilograms per meter borehole, and d is the hole diameter in meters.

For best results in smooth blasting, the charges in the contour should be initiated simultaneously so that they can cooperate fully. However, because the extension of the crack between the holes involves the relatively slow dynamic motion of the considerable mass of material within the burden, the latitude for time scatter in the times of detonation is larger than might be expected. Also, the simultaneous firing of a large number of contour holes in a tunnel blast would create very large ground vibrations. Therefore, the contour holes are often distributed over several time intervals; and because the contour rows are fired at the end of a long-duration tunnel blast with many time delays, the scatter in timing of these long delay detonators is considerable.

9.3 Rock Damage in Blasting

When an explosive charge detonates in a borehole, the high-density gaseous reaction products exert a high pressure on the drillhole walls which sets the drillhole walls in motion outward, creating a dynamic stress field in the surrounding rock. The initial effect in the nearby rock is a high-intensity, short-duration shock wave, which quickly decays. The continued gas expansion leads to further motion and sets up an expanding stress field in the rock mass. Where the free surface is close enough to the borehole, the rock breaks loose. In other directions, the motion spreads further in the form of the well-known ground vibration waves. These are a complicated combination of elastic waves in which the rock reverberates in the compressive, shear, and surface wave modes. Each mode or wave type (P-, S-, and R-waves) has a characteristic propagation velocity c which is some fraction of the sonic velocity that is a material property of the rock mass. Each particle in the rock mass runs through a complicated, approximately elliptical motion in several cycles with varying amplitude. The highest velocity is attained during this motion, and the peak particle velocity v decreases with the distance from the charge. Damage is a result of the induced strain ϵ which, for an elastic medium in the sine-wave approximation, is given by the equation

$$\epsilon = \frac{v}{c}. \tag{9.2}$$

In the region close to the charge, permanent damage begins to occur at a given critical level of particle velocity, which is different for different rock masses. Whether the damage affects the stand-up time of the rock contour depends on the character of the damage, the rock structure, the groundwater flow, and last but not least, on the orientation of the damaged planes in relation to the contour and to the direction and magnitude of the static load.

The extent of rock damage can be approximately correlated with the peak particle velocity, which is proportional to strain as a measure of the damage potential of the wave motion. Of course, the surrounding rock mass contains a number of potential weak planes, each of which is able to withstand a different level of peak particle velocity. In experiments, the extent of the damage zone has been determined by a comparison of the crack frequency before and after the blast by using core logging or with a borehole periscope. In addition, the total expansion of many small cracks or joints due to the incipient blast damage can be evaluated by extensometers fastened in drillholes at different distances from the blast.

From the extensive studies made of structural damage to buildings and constructions due to the detonation in a drillhole of a single charge, we know that reliable predictions of damage can be made if we know the peak particle velocity. The peak particle velocity can be predicted using the empirical equation:

$$v = K \frac{W^\alpha}{R^\beta} \tag{9.3}$$

where v is the peak particle velocity in mm/sec, W is the charge weight in kg, and R is distance in m.

The constants K, α, and β depend on the structural and elastic properties of the rock mass and vary with each particular blasting site. Typical values for hard rock masses are: $K = 0.7$ m/sec, $\alpha = 0.7$, and $\beta = 1.4$.

Equation 9.3 is valid only for distances that are long in comparison with the charge length, i.e., where the charge can be treated as concentrated.

In Section 8.5, we described a method for determining the ground vibration velocity and rock damage effects in the region adjacent to an extended (long) charge. The resulting peak vibration velocity was described by Equation 8.8, for arbitrary values of α and β, or by Equation 8.9, for the special case where $\beta = 2\alpha$. The resulting peak vibration velocity was shown in Figure 8.11 as a function of the perpendicular distance to the charge with the linear charge concentration ℓ as a parameter. These figures are reproduced here as Figure 9.2. In the next section, we will describe how these diagrams can be used to compute the damage zone in the remaining rock adjacent to a row of contour holes.

9.4 Computed Damage Zones

Figure 9.2a shows a diagram of the peak particle velocity v as a function of the perpendicular distance to the extended charge R with the linear charge density as a parameter, for a charge 3 m long. For a rock mass which gives an incipient fracture at a vibration particle velocity within the range 700–1000 mm/sec, the radius of the zone of incipient fracture is about 0.25–0.35 m around a 45 mm hole of length 3 m, charged with the special 17 mm Gurit charge that is used for smooth blasting in Sweden. These charges have a linear charge density of about 0.2 kg/m, giving a damage zone of approximately 0.7 m. By contrast, a 45 mm drill hole fully loaded with ANFO has a linear charge density of 1.5 kg/m and produces a zone of incipient damage in the same rock of about 1.5 m radius. This is the typical drillhole and charge arrangement used in the perimeter holes in a tunnel round.

Figure 9.2b shows a similar calculated diagram for a charge 15 m long in the linear charge density range typical of large open-pit mining. A 250 mm borehole fully loaded with a heavy slurry explosive giving a linear charge density of about 75 kg/m produces damage in a zone of radius 25 m or more. This is the typical drillhole and charge arrangement used in a production hole in an open pit mining operation.

Diagrams such as these (and calculations for other charge lengths, linear charge densities, and charge arrangements relative to the free surface) have shown themselves to be very useful for the design of blasting rounds to specified criteria for controlling blast damage to the remaining rock.

9.5 Blast Planning

It is not unusual for tunnel blasters to use great care in selecting a small enough charge for the contour holes, but to fail to consider the damaging effects of the charges in the adjacent row of holes. Charging the adjacent rows with a heavy charge results in cracks spreading further into the remaining rock than from the smooth blasted row (Figure 9.3). It is better to optimize the charge calculations such that the damage zone from any hole in the round will not exceed the damage zone from the contour holes. This can easily be done by use of Figure 9.2(a), where the extent of the damage zone is given for different charge concentrations.

Table 9.1 contains recommended burdens for some earlier Swedish explosives for use in smooth blasting. For a hole diameter of 48 mm and 17 mm Gurit pipe charges, a burden of 0.8 m is normal. From Figure 9.2(a), we find that this charge will result in a damage zone of about 0.3 m. Choosing a fully charged hole of ANFO ($\ell = 1.6$ kg/m)

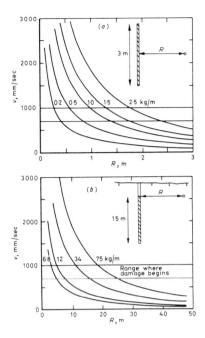

Figure 9.2. Estimated peak vibration velocity as a function of distance for different linear charge densities, kg/meter borehole length and distance ranges typical of (a) small-diameter hole tunnel blasting and (b) large-diameter hole open-pit blasting.

in the next row with a damage zone of 1.5 m will unfortunately result in a damage zone that extends 0.4 m further into the rock $(1.5 - 0.8 - 0.3 = 0.4$ m$)$ than the damage zone from the contour-hole charges. Instead, the charge concentration should be limited to one that results in a damage zone equal to that caused by the contour-holes plus their burden, i.e., $0.4 + 0.8 = 1.2$ m (Figure 9.4) — such a charge should have ℓ in the range 0.8 to 1.2 kg/m. In fact, even if this is done, the major damage done to the remaining rock more than 0.4 m away from the contour is due to the second row of holes. The vibration velocity due to these holes drops off less rapidly in that region than that from the contour holes.

It is apparent from this example that a reduction of the damage zone can be obtained by a reduction of the charge concentration per meter of borehole. This obviously results in increased costs for drill and blast operation, but these are balanced by the advantage of a safer roof and decreased costs for grouting and maintenance. Smooth blasting also results in less overbreak which results in reduced cost for finishing the tunnel wall, whether by lining, by shotcreting, or by steel supports.

Table 9.2 lists a number of explosives used for smooth blasting and presplitting from different manufacturers.

Table 9.1. Some common explosives and recommended burdens used for tunneling in Sweden.

Explosive	Dynamex B			ANFO	Nabit	Gurit
Hole diameter, mm	45	45	45	45	45	45
Charge diameter, mm	45	40	32	45	22	17
Explosive density, kg/l	1.15	1.45	1.45	0.95	1.1	1.2
Weight strength	1.14	1.14	1.14	1.0	1.09	0.83
Charge concentration						
real, kg/m	2.0	1.6	1.12	1.5	0.40	0.24
kg ANFO/m	2.3	1.82	1.28	1.5	0.44	0.20
Borehole pressure						
kbar	55	32	13	21	3.3	0.9
GPa	5	3.2	1.3	2.0	0.3	0.09

Drillhole diameter (mm)	Charge concentration (kg ANFO/m)	Charge type and size	Burden (m)	Hole spacing (m)
25–32	0.08	11-mm Gurit	0.20–0.45	0.25–0.35
25–48	0.20	17-mm Gurit	0.70–0.90	0.50–0.70
51–64	0.44	22-mm Nabit	1.00–1.10	0.80–0.90

Table 9.2. Explosives for smooth blasting and presplitting.

Explosive name	Manufacturer	Cartridge size (mm)	Linear charge concentration (kg/m)	(lb/ft)	Type of explosive
Iremite Presplit	IRECO	ϕ 22	0.48	0.32	Emulsion
IRESPLIT D	IRECO	ϕ 32	0.75	0.50	NG-based
Smoothex	AECI	ϕ 18	0.19	0.13	NG-based
F-pipe charge	Forcit	ϕ 11	0.11	0.07	NG-explosive
F-pipe charge	Forcit	ϕ 17	0.22	0.15	NG-explosive
Profile charge	SEI	ϕ 19	0.33	0.22	NG-explosive
Gurit	Nitro Nobel	ϕ 11	0.11	0.07	NG-explosive
Gurit	Nitro Nobel	ϕ 17	0.24	0.16	NG-explosive
Emuline	Austin	ϕ 22	0.45	0.30	Non NG high explosive + cord
Red-E-split A	Austin	ϕ 22	0.45	0.30	Semi-gelatin dynamite
Red-E-split B	Austin	ϕ 29	0.84	0.56	Ammonia dynamite
Red-E-split C	Austin	ϕ 22	0.34	0.23	Ammonia dynamite
Red-E-split D	Austin	ϕ 29	0.63	0.42	Ammonia dynamite
Power split	Atlas Powder	ϕ 22	0.47	0.31	Emulsion
Power split	Atlas Powder	ϕ 25	0.62	0.41	Emulsion
Power split	Atlas Powder	ϕ 51	2.45	1.63	Emulsion
Larvikit	Dyno Norway	ϕ 17	0.26	0.17	NG-powder
Emulite 100 Gurit	DWL	ϕ 17	0.25	0.17	Emulsion
Isanol 25/75	—	bulk loaded ϕ 45	0.34	0.23	ANFO/polystyrene
Emulet 20	Nitro Nobel	bulk loaded ϕ 38	0.22	0.15	Emulsion/ANFO/polystyrene
Emulet 50	Nitro Nobel	bulk loaded ϕ 38	0.50	0.33	Emulsion/ANFO/polystyrene

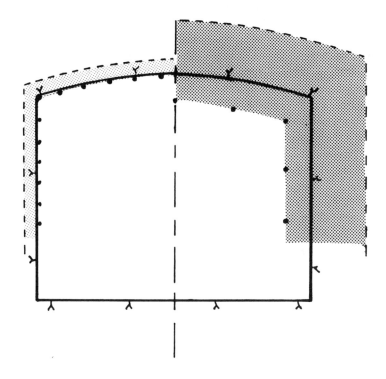

Figure 9.3. Stoping holes may cause more damage than the contour holes!

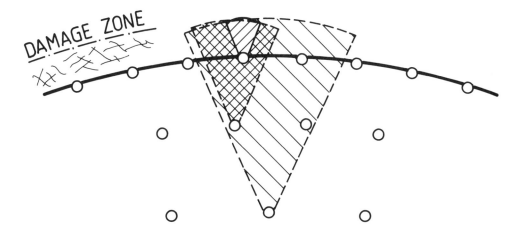

Figure 9.4. Care must also be taken to the damage zone caused by adjacent rows to the contour.

9.6 Experimental Observation of Rock Damage

It is, of course, necessary to correlate the computed or measured vibration velocities with observed damage to the rock mass. The most direct method — but also the most expensive — for this is core drilling.

In a production blast with 250 mm boreholes, a diamond-drilled core sample of what was to become the remaining rock was taken out before loading the blastholes. The core hole was drilled horizontally from the bench face through the entire bench to be blasted into the rock behind. After the blast, new core samples were taken from new horizontal diamond drill holes, drilled very close by and parallel with the old ones. In this way, the increase in crack frequency due to the blast could be determined. The results obtained can only be presented in a probabilistic manner. We found that for the blast described, there was a 50% probability of damage at a distance of 22.5 m from the nearest hole and a 5% probability for damage at a 32 m distance. Damage was considered to have occurred when the number of cracks after the shot is measurably greater than that before the shot.

Rock mechanic tests carried out at Swedish Detonic Research Foundation [Persson and Mäki, 1973] indicated that damage was caused mainly in the rock structural discontinuities such as joints, cracks, or other weakness planes in the rock mass. Carefully controlled tensile tests indicated that a permanent decrease in rock shear strength occurred already when a closed crack or joint was opened by an amount of a few microns (μm). The damage was invisible to the naked eye, but in the rock mass would be measurable in the form of a slight swelling. The best way of measuring such small swelling damage was by an extensometer fitted into a hole drilled through the adjacent rock before the blast.

9.7 Cautious Blasting in Open Pit Mines

At the planning stage of an open pit mine where the excavation is considered to be optimized, the total slope angle is of primary concern. From an economic point of view, the steeper the slope, the less waste there is to excavate. Dramatically, it can be said that the slope should be stable until the last shovel is hauled — then failure should occur.

The overall maximum slope angle for an open pit mine is normally pre-determined by probability calculations based on the known rock mass strength properties. These can be determined from diamond core drilling or from observation of the structural discontinuities in a bench wall of the rock mass adjacent to the future pit wall during the early stages of the excavation. But the boundary conditions of course require the use of a proper cautious blasting method. Blasting without concern will, in the long run, successively lead the way to an unstable slope due to excessive back break, to undercutting of the toe, and to other disturbances in the rock mass continuity. Deterioration due to groundwater flow and weathering due to temperature fluctuations (particularly through the freezing point) would also help to contribute to potential slope failures.

Figure 9.5 shows the influence of back break on the overall slope angle. Assume the blasting method is such that the back break flattens the bench slope angle to 55°. It can be seen that for a 15 m bench height and a specified safety berm width of 7 m, the maximum overall angle cannot exceed 42°. A 30 m height final bench would increase the maximum overall slope to 46°. However, if more attention is paid to the blasting method, and the back break is reduced to allow a bench face sloping (for example) 80°,

9.7. Cautious Blasting in Open Pit Mines 273

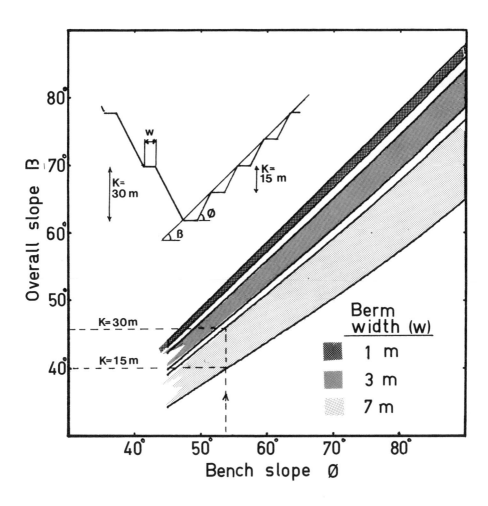

Figure 9.5. Maximum overall slope angle for different bench slopes, bench heights, and berm widths.

Figure 9.5 shows that at least a 55° overall slope angle can be maintained. This is, of course, only valid if the undisturbed rock has sufficient strength.

For a 300 m deep open pit mine, the cost saving by increasing the overall slope only 1° is on the order of $1 million dollars just for blasting, mucking, and hauling the waste rock. Cautious blasting also guarantees higher safety margins for miners and equipment.

When perimeter blasting in a large open pit mine using large diameter boreholes, it is very seldom that smooth blasting can be used to its full potential because of the cost aspect. The drilling costs would be enormous if thousands of square-meter pit slope should be perimeter blasted with small-diameter holes and special, expensive small-diameter charges. Large capital investments for large hole drilling equipment has already been made in an open pit mine and, most often, there is an overcapacity for drilling. Thus, the same hole diameter used in the production rounds can often be used also for perimeter blasting. By using deck charging, decoupled charges, or explosives with very low density, it is possible to lower the charge weight per hole in the first, second, and sometimes also the third row adjacent to the slope for preventing pit wall damage.

It is also important to mention the damage caused to pit slopes by the ground vibration from the total round. It is not uncommon today to blast single rounds totaling a fragmented rock volume of 100,000 m^3. Field experiments carried out by the Swedish Detonic Research Foundation indicate acceleration levels on the order of 0.1 g at distances 300 to 900 m away from such a round. The frequency spectra of these vibrations show that most of the energy in the Rayleigh wave is transmitted in the 10 Hz region and at a wave velocity of 2000 m/s. The wave length is 200 m.

Parts of the pit wall, for example one bench face, can be treated as a relatively stiff body when affected by the ground vibration because the bench height is small compared to the wavelength. Figure 9.6 depicts a simple safety factor calculation for plane failure of a 6 m high crest which is affected by different magnitudes of acceleration. The direction of the acceleration is described along the x-axis. The normal stress and the shear stress along the potential failure plane is affected by the acceleration, and thus the safety factor is changed. As seen from Figure 9.6, a safety factor of one (1) will be reached if the acceleration is about 0.25 g. The safest way to avoid long-range vibration damage is to reduce the number of blastholes per round. It is probably not enough just to rearrange the delays. The ground vibrations interact at longer distances and the number of available delays to choose from is limited. Instantaneous row-by-row initiation should definitely not be used. The vibration has a cycle time which is about the same as the detonator time delay — this could lead to resonance build-up of more violent vibrations.

9.8 Controlled Fracture Growth

To avoid damage in the remaining rock and leave a strong, smooth surface after the blasting operation, it would be advantageous to control the fractures. For example, in smooth blasting, the ideal result is to have major cracks traveling in the directions between the perimeter holes instead of going backward into the remaining rock. One way of achieving this is to introduce notches into the surface of the borehole wall since this results in a very high stress concentration at the crack tip when the gas pressure acts on the drillhole wall.

Field and Ladegaard-Pedersen [1972] showed the existence of this effect by model blasting in PMMA. The crack pattern was obtained by cutting two slits into a brass tube that was lowered into a borehole and filled with explosive (Figure 9.7). The brass tubes protected the holes at all points except where fractures were required. An initial blast resulted in two short cracks at each hole. After recharging, it appeared that most of the available energy went into the extension of the two cracks formed by shot 1, and no unwanted radial cracks were produced.

Full-scale experiments to implement fracture control procedures in a tunnel project have been carried out by Thompson [1979] in the USA. Using a special tool to notch the drillholes along most of its length, it was possible to control the cracks. The cutting edges were designed to cut notches 6.4 mm deep.

The perimeter holes in each experimental round were loaded with a specially designed string of explosives consisting of a concentrated bottom change of two sticks of 40 % extra gelatin and a distributed column charge. The column charge generally consisted of a 1.5 m length of 0.09 kg/m (400 grain/ft) primacord which was supported at the center of the drillhole by a specially designed spider tube (Figure 9.8).

The result was reported to be 10 to 30 % better than conventional contour blasting methods. However, the average cycle time was longer due to the extra time required to notch the perimeter holes.

9.9. Shaped Charges for Boulder and Contour Blasting

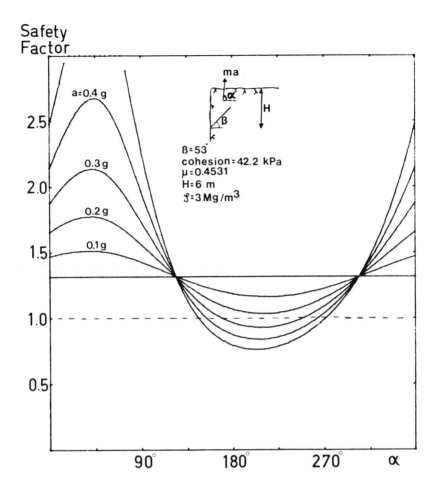

Figure 9.6. Influence of the acceleration upon the safety factor for a 6 m high bench crest with a potential failure plane dipping 53°.

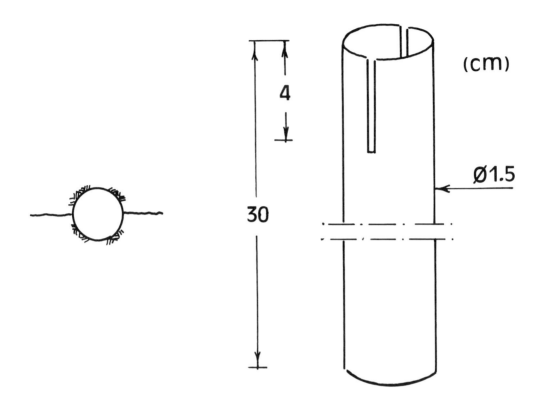

Figure 9.7. Method to control the fracture growth.

9.9 Shaped Charges for Boulder and Contour Blasting

The discovery of the shaped charge effect has been credited to C.E. Munroe at the Naval Torpedo Station in Newport, RI. In 1888, Munroe observed that, if cast high explosives — with letters indented into the explosive's surface — were detonated close to a steel plate, the letters would be engraved into the plate surface.

In the 1930s, the idea of concentrating the effect of the explosion in one direction by the use of hollow charges (charges with a countersunk depression) was introduced for military purposes. The details of the mechanics of jet formation and target penetration of such charges are given in Chapter 15. After World War II, a large quantity of surplus military shaped charges became available at very low prices. Entrepeneurs in the mining business tried to find a practical use for these charges. It was found favorable to use the low-cost shaped charges for secondary blasting of boulders.

Linear shaped charges have also been tested for contour blasting where the notching is made at the same time as the blasting. This was reported by Naarttijärvi, Rustan, Öqvist, and Ludvig [1980c] and by Bjarnholt, Holmberg, and Ouchterlony [1982].

9.9. *Shaped Charges for Boulder and Contour Blasting* 277

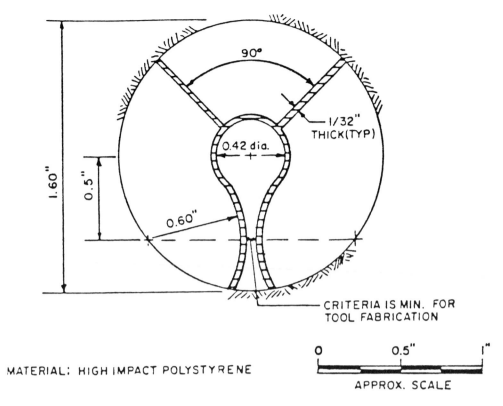

Figure 9.8. Spider tube used to center the Primacord.

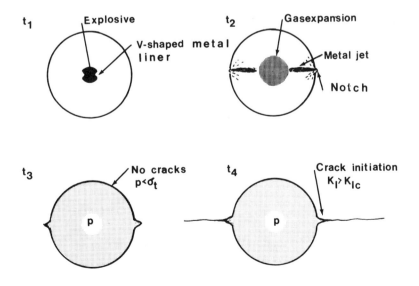

Figure 9.9. Principle for a linear shaped charge used in contour blasting.

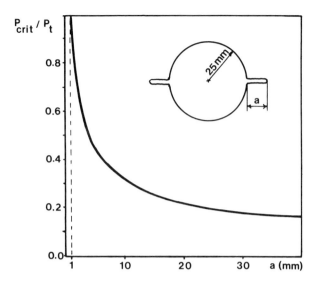

Figure 9.10. Ratio of borehole pressure necessary for crack initiation P_{crit} and needed borehole pressure when no notches are available P_t as a function of the length of the notch a in a $d = 50$ mm borehole in Bohus granite. $K_{IC} = 2$ MN/m$^{3/2}$ and $P_t = 16$ MPa.

9.9.1 Crack Initiation

In order to facilitate the crack growth in a given direction and to suppress unwanted crack initiation around the borehole, it is possible to utilize side notches extending the length of the borehole while adjusting the borehole pressure to an adequate magnitude. These ideas were described by Field and Ladegaard-Pedersen [1969]. The notches will reduce the borehole pressure required for crack initiation and at the same time make it more difficult for crack initiation in other parts of the borehole wall (Figure 9.10). The notches do not have to be very deep to achieve the desired effect.

Based on earlier work by Ouchterlony [1972], the required depth a of the notches can be estimated using the equation

$$K_I = 2 \cdot 1.12 \cdot p\sqrt{\pi a} \qquad (9.4)$$

where K_I is the stress intensity factor and p is the borehole pressure. To initiate cracks only at the notches, K_I must exceed the fracture toughness K_{IC} in the material and p must be less than the necessary crack initiation pressure P_t for a borehole where only the natural micro defects of the length a_t are available. From Equation 9.4, a_t can be solved:

$$a_t = \frac{1}{5\pi}\left(\frac{K_{IC}}{P_t}\right)^2. \qquad (9.5)$$

P_t will probably, due to loading times and geometrical effects, be of the order of 2 to 4 times higher than the uniaxial tensile strength σ_t.

Using data for granite ($K_{IC} = 2$ MN/m$^{3/2}$ and $P_t = 16$ MPa), we get $a_t = 1$ mm. The granite has, according to this calculation, about 1 mm deep natural micro defects along the borehole wall. By using notches with a depth $a = 5$ mm, the required borehole pressure P_{crit} can be reduced by at least 50 % for a 50 mm diameter borehole.

9.9.2 Design Parameters

The charge consists of an extended charge equipped with a V-shaped metal liner. The charge is initiated at the end. The detonation wave propagates along the charge and forms a metal jet with a velocity of several thousand meters per second.

Only a small part of the metal liner is accelerated to this high velocity jet. The main part of the metal material forms a slug with considerably lower velocity, not contributing to penetration of the target material. The depth of penetration d in the target material is dependent on liner material, explosive performance, and charge geometry.

Interesting parameters for linear shaped charges are the influence of:

- a. Material and thickness of the V-shaped liner
- b. Angle between the legs in the V-shaped liner
- c. Distance (stand off) between the charge and the target material
- d. Detonation pressure for the explosive

SveDeFo has used different types of charges (see Figure 9.11) to study these parameters. The charges were designed using an aluminum U-profile filled with explosive and a V-shaped aluminum liner. Steel SIS 1311 was used as a target material.

9.9.3 Liner Material

Aluminum is often used in an explosive composition as a fuel to increase the explosion energy. Aluminum oxide Al_2O_3 is formed in the detonation products and contributes to the energy release through its very high heat of formation.

By using aluminum as a liner in a linear shaped charge for contour blasting, it is possible to burn the aluminum in the hot reaction products after the jet has notched the borehole wall. This buffers the borehole pressure, especially during the later part of the crack propagation phase. The cracks can thus be propagated longer which means that the hole spacing can be increased, leading to decreased drilling costs.

By choosing a ductile metal with a higher density (e.g. copper), the depth of the notches can be somewhat increased.

Angle Between the Legs of the V-shaped Liner

The properties of the jet are influenced by the angle between the legs in the V-shaped liner. The results from the tests carried out with angles 75°, 90°, and 105° show that the largest depth is achieved for an angle of 90°. With an angle of 75°, the irregular notch in the target material shows that the jet is unstable. With an angle of 105°, a more shallow groove was achieved (Figure 9.11).

Stand Off Between the Linear Charge and the Target Material

The depth of the notch will first increase with the stand off s because the effective length of the jet increases with increasing stand off. For a very long stand off, the scatter in jet velocity and direction is increased, resulting in reduced penetration.

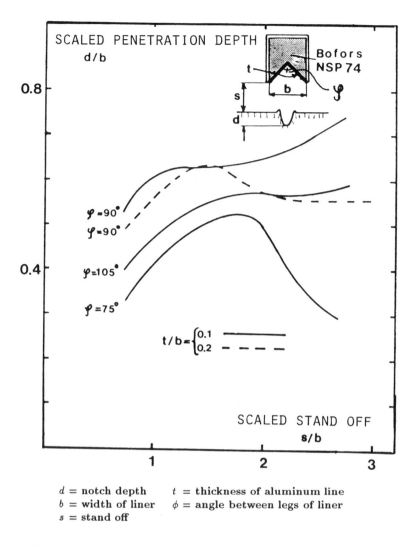

Figure 9.11. Scaled notch depth as a function of scaled stand off between charge and a target material, steel SIS 1311.

d = notch depth t = thickness of aluminum line
b = width of liner ϕ = angle between legs of liner
s = stand off

Explosive

Important parameters for the explosive are:
1. A high degree of homogeneity in the charge
2. A high detonation pressure

A high detonation pressure gives a high jet velocity important for good notching performance.

In the main part of the tests, a plastic PETN-based explosive (Bofors NSP-74) was chosen. This explosive has a fairly high detonation pressure and good homogeneity. NSP 74 has the following properties:

9.9. Shaped Charges for Boulder and Contour Blasting

Charging density	$\rho_0 = 1.50$ g/cm^3
Detonation velocity	$d = 7200$ to 7400 m/s
Critical diameter (unstemmed)	$d_{cr} = 2$ mm
Detonation pressure (CJ)	$P_d = 20$ GPa

Borehole Pressure for a Decoupled Charge

When an explosive charge detonates in a borehole, the gas pressure acting at the borehole wall first fluctuates due to the dynamic overpressure close to the detonation front. After a short while, the pressure will stabilize at a quasistatic value which is called the *borehole pressure*.

When the charge is decoupled by using a charge diameter much smaller than the diameter of the borehole, we have chosen to approximate the pressure initially acting at the borehole wall with the quasistatic borehole pressure. For the low borehole pressure achieved with decoupled charges, we can regard the borehole wall as stiff, and approximate the quasistatic borehole pressure with the constant volume explosion pressure P_e for the explosive. P_e can be calculated from thermochemical calculations. Figure 9.12 shows results from calculations carried out with SveDeFo's computer program NITRO-DYNE, [Holmberg, 1977].

Equation 9.1 gives the recommended linear charge concentration for smooth blasting and presplitting. This equation really defines a charge weight of 114.6 kg/m^3 of borehole volume, which in turn specifies a borehole pressure, which for ANFO is about 130 MPa.

$$\text{Charge weight per unit hole volume} = \frac{\ell}{\frac{\pi d^2}{4}} = \frac{90\,d^2}{\frac{\pi d^2}{4}} = \frac{360}{\pi} = 114.6 \quad (9.6)$$

This value for ANFO has been marked in Figure 9.12.

By calculating the borehole pressures for some commonly used contour blasting agents, it is seen that the pressure interval for contour blasting is about 70–150 MPa.

In Figure 9.10, it was shown that, for contour blasting with a notched borehole, the borehole pressure can be reduced to about 1/3 of the pressure needed when no notches are grooved into the borehole wall, i.e., to about 25–50 MPa. According to Figure 9.12 with NSP 74 explosive, one should then use a charge concentration of about 30 kg/m^3 borehole volume pressure.

When using a linear shaped charge to create the notches and blast the rock, one should increase the charge concentration by about 30% to compensate for the energy required for the notching.

Penetration in Granite

From the experimental work and theoretical calculations, a linear shaped charge was manufactured for use in boreholes with a diameter of about 50 mm. A charge was designed with a linear charge concentration of 90 g Bofors NSP-74 per meter with two aluminum liners. Experimental tests showed that, with this type of charge, it was possible to create two notches with a depth of 3 mm in steel and a depth of 5 mm in granite. The distance between the charge and the target was 22 mm.

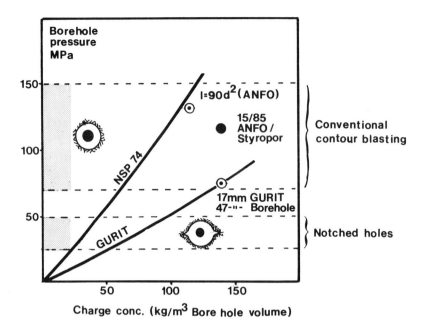

Figure 9.12. Borehole pressure as a function of the charge concentration for some explosives. Gurit is the commonly used Swedish explosive for contour blasting, with charge diameters 11 mm or 17 mm.

9.9.4 Comparison of Methods for Controlled Fracture Growth

At the Nordkross AB Hakunge quarry, a field test was performed to compare three methods for controlling fractures. The methods were

1. Linear shaped charges
2. Mechanical notching
3. High-pressure water jet

The rock was a competent granite gneiss with bands of amphibolite. Uniaxial compressive strength was 120 to 220 MPa, the predominant vertical joint spacing was 2.2 m, and the horizontal was 1.8 m.

In the test, various spacings were used to investigate also if it was possible to lower the drilling cost by separating the perimeter holes further than normal for conventional smooth blasting. [See Bjarnholt et al., 1988, and Holloway et al., 1987]. Boreholes with a diameter of 89 mm were drilled in a 5 m high 20 m wide bench. The burden was 1.6 m for all experiments, but the spacing was varied between 1.3–1.8 m. By using a detonating cord trunkline at the surface and NONEL downlines, it was possible to keep an interhole delay time of 1.2 ± 0.1 ms. The geometry of the bench is shown in Figure 9.13.

A mechanical notching tool with two tungsten carbide inserts was used to breach the holes after drilling. The effective notching time was about 1.5 min per 5 m borehole. The notch depth was about 12 mm. The tool we used (see Figure 9.14) was a modified version of a research tool supplied by Kennametal Corp.

High-pressure water jets with a pump pressure of 135 MPa were used to cut notches in the borehole walls after the holes were drilled. The two diametrically opposed nozzles

9.9. Shaped Charges for Boulder and Contour Blasting   283

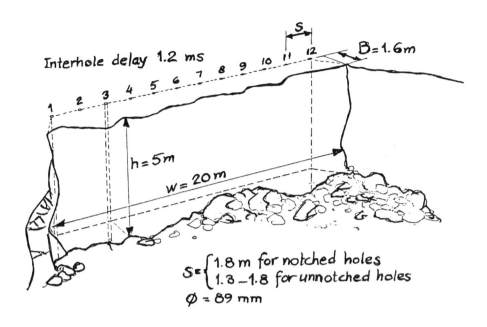

Figure 9.13. Bench blasting geometry [from Bjarnholt et al., 1988].

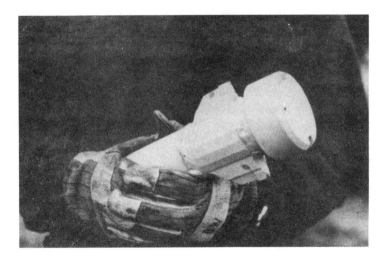

Figure 9.14. Mechanical notching tool with two tungsten carbide inserts.

had a diameter of 1.3 mm. They were incorporated into a traversing carriage powered by the hydraulic feed system of the drill rig. Three up and down passes in each hole at a rate of 7.8 m/min were required to produce notch depths which varied from 5 to 20 mm. The effective notching time per hole was about 3 minutes. Water consumption was about 1 liter per second. The high pressure water jet notching system was supplied by Atlas Copco MCT.

284   Chapter 9. Contour Blasting

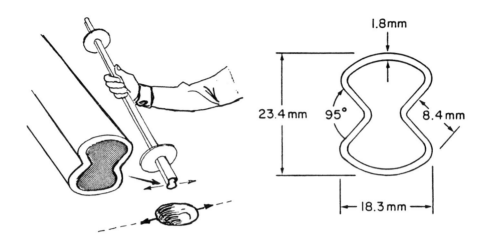

Figure 9.15. Linear shaped charge for notching [from Bjarnholt et al., 1988].

A linear shaped charge was also designed to create the notches. This charge created two notches 15 to 20 mm deep and removed the rock in one operation. It was charged with 0.35 kg of cast Composition B (60% RDX, 40% TNT) explosive per meter length. Figure 9.15 shows the charge.

When the blasts took place, carbon resistors were used as piezoresistive transducers to provide a record of the pressure-time history in the borehole. The sensors were placed just beneath the stemming. In the field test, 0.85 kg Dynamex M or 1.3 kg Emulite 150 (emulsion with 5% Al) were used as the bottom charge. In the column charge, Gurit 22 mm diameter, Nabit 25 mm diameter, or the shaped charge was used.

Figure 9.16 reflects the pressures built up when the charge detonated. The low pressure of 20 MPa from the Gurit column charge with its low linear density is overruled by the pressure pulse of about 75 MPa from the concentrated bottom charge of 1.3 kg Emulite 150 after about 3.1 ms.

The authors state that, with notched boreholes and about half the normal column explosive charge weight, they could increase the hole spacing by 40% compared to conventional smooth wall blasting with 89 mm diameter holes. They used the surface roughness achieved as a quality measure to judge the results. Surface roughness was measured with a laser-based distance meter equipped with an angle indicator. The shaped charges were easy to handle and gave a good result with 15 to 20 mm deep notches.

The other two methods had some drawbacks. Mechanical grooving with the tool was difficult since the tungsten carbide inserts tended to break when the hole was not sufficiently straight. It was also a difficult problem to maintain the preferred direction of the notches, which were about 12 mm deep. High-speed water jet notching was reported to be selective in the wrong way, that is, in softer granite gneiss, a much deeper notch (20 mm) was achieved than in the tougher amphibolite (5 mm). The tough material requires deeper notches than the soft material. Cold-climate water jet cutting also formed too much ice in the notches and in the boreholes.

9.9. Shaped Charges for Boulder and Contour Blasting 285

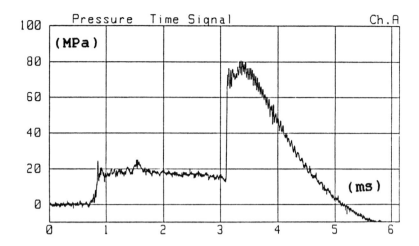

Figure 9.16. Typical pressure time-history for a Hakunge blast hole.

Chapter 10

Computer Calculations for Rock Blasting

10.1 Introduction

The use of a computer now makes it possible to solve from the outset many multi-variable problems in the optimization of complex operations involving rock blasting that could previously be solved only by trial and error. Many variables must be taken into account when optimizing the capacity and economy in the design and planning of a surface mining operation or an underground excavation project. Overall economic considerations influence the optimum value of each variable selected in the search for the best available excavation technique. For example, the use of large drillhole diameters generally leads to decreased costs because fewer pieces of equipment can be used, and fewer workers are required, compared to when blasting is done with smaller-diameter holes. However, large hole diameters cannot be used everywhere. Sometimes the large investment required for the drilling equipment is prohibitive. Sometimes, such as when blasting near urban areas, environmental or safety requirements may set limits on the noise, vibration levels, or flyrock distances that limit the drillhole diameter, or to the charge weight of explosive. Crushing and mucking machine size must be compatible with the drilling equipment, the rock fragment size, and the required production capacity. All these factors can be taken into account by the use of computer programs that can handle the large amount of calculation work involved.

In recent years, the widespread use of personal computers has drastically increased the demand for rock blasting computer programs. As a result, there are now a great many computer programs commercially available for a variety of tasks which in the past were beyond the reach of the blasting engineer equipped with only a slide-rule or a hand-held calculator. Some of these currently available programs are listed in the next Section. All of these are proprietary. Some are restricted with respect to who is allowed to access the programs. All of these programs represent a considerable investment in programming time and are therefore quite expensive for the individual, such as a student, to purchase. Most programs currently in use will soon be out of date and will be replaced by others, even more powerful and more elegant as the power of the personal computer reaches new heights. Therefore, to illustrate the fundamentals of rock blasting computer programs, we have included in this chapter two of the first programs developed in the 1960's, one for bench blasting calculations, the other for tunneling calculations. These programs are now outdated and outperformed by more modern programs. Yet for the student, they illustrate the principles. Later, a more modern program can be acquired.

10.2 Some Commercial Rock Blasting Computer Codes

Without pretense of complete coverage of all available computer codes, a short selection of more modern computer codes for rock blasting calculations is given below. In most cases, these codes are in a state of continuing, rapid development and modification. For additional up to date information, please contact the indicated organization directly.

Program Name	Organization
BLASTEC	Nitro Nobel AB, Sweden
"3 x 3"	JKMRC, Australia
BLASTCALC	Noka Software Systems, Canada
BLAST DESIGNER	Precision Blasting Services, USA
CARE	Atlas Copco, Sweden
DYNOVIEW	Dyno Explosives Group, Norway
SABREX	ICI Explosives, Scotland
SAROBLAST	University of Luleå, Sweden
SWEBENCH	SveDeFo, Sweden
TIGERWIN (TIGER for WINDOWS)	BAI Inc., USA

10.3 The Fundamentals of Two Rock Blasting Codes

Langefors and Kihlström [1963] structured and formulated the basic engineering principles for the calculation of many different rock blasting situations. These principles and the fundamental equations for engineering calculations of rock blasting are described in Chapters 6 and 7. They are based on extensive experimentation and study of the detailed functioning of explosive charges in rock blasting in different geometries. These equations and tabulated data were used in writing the two sample computer programs that will be described in this chapter. One is for bench blasting, the other for tunnel blasting.

The interested reader is encouraged to use his or her own imagination to develop their own codes for other rock blasting operations, following the general principles of charge calculation. This book contains the basic sets of equations for a number of such codes. The reader who does not want to or who perhaps cannot afford to buy a commercial program can make his own. Each reader will have his or her own set of economic parameters to add, in the form of local wages and costs for raw materials, power, fuel, and other supplies of different kinds, as well as the data for the local types of rock encountered. Programs such as these make it possible to save time and reduce costs in the planning and production stages of a rock blasting operation. With these programs, one can easily carry out a number of calculations to study the influence of hole diameter, type of explosive, ratio of spacing to burden, bench height or advance, etc., on the drilling, charging and explosive costs, and on the cost of the entire operation.

The tunneling program described below is the result of a cooperative development effort in which Atlas Copco AB and Nitro Consult AB have participated. The bench blasting program was originally developed by Nils Lundborg of the Swedish Detonic Research Foundation and has been further improved by other members of the Foundation staff.

10.4 Computer Program for Tunneling

The cross-sectional shape of any tunnel can be specified simply by tabulating the coordinates of the tunnel contour in the input data list. For example, a mining access tunnel may be given an asymmetric cross-section to make room for ventilation tubes or water pumps for drainage while minimizing the cross-sectional area. For most common regular tunnel sections, however, it is enough to specify the abutment height, or the area and the radius of the contour of the roof. The list of input data providing the necessary information required for the calculation includes

1. Geometry of the tunnel
2. Type of cut
3. Smooth blasting or normal blasting of the contour
4. Hole diameter
5. Collaring deviation
6. Angular deviation
7. Planned advance or hole depth
8. Types of blasting agents and degree of packing
9. Weight strength of the explosives
10. The toughness of the rock described by a rock constant
11. Cost of the drilling equipment including depreciation
12. Costs of explosives and detonators
13. Number of workers, wages and an efficiency factor
14. Ventilation costs

A large number of different cuts have been developed and used during the last few decades. A cut is a special arrangement of the first few drillholes intended to develop additional free surfaces toward which the rest of the holes can break the rock. The type of the cut, the advance, and the number of holes depend on the rock strength, the blasting agents, and the size of the heading, and upon the available drilling equipment. Angled cuts, such as the fan cut and the plow cut normally permit an advance per round equal to about 40 to 50% of the tunnel width. The advance is determined by the geometry of the tunnel. In narrow tunnels it is often more economical to use parallel cuts to pull long rounds. The program gives a choice of five different cuts:

1. Plow (or plough) cut or the V-cut
2. Fan cut
3. Parallel cut with one empty hole
4. Parallel cut with two empty holes
5. Parallel cut with three empty holes

To avoid contour damage, smooth blasting can be selected in the program. Tunnels to be used for a long period often become cheaper in the long run if smooth blasting is used since the cost for maintenance of the tunnel roof is decreased. Smooth blasting can greatly reduce the damage and crack formation in the surrounding rock. Smooth blasting also decreases the risk of water penetration, decreases the amount of concrete needed for grouting and for support. It also retains a greater part of the rock mass strength, which decreases the risk of rock fall from the roof.

Several parameters used in the code are assigned default values if an input value is not specified. In order to reach the results predicted in the calculations, it is important to carefully follow up the implementation in the field, that is, to check and recheck that the drilling accuracy and the amount of explosives loaded into each hole agrees with the computer input data. Drilling accuracy is of paramount importance in achieving good

Chapter 10. Computer Calculations for Rock Blasting

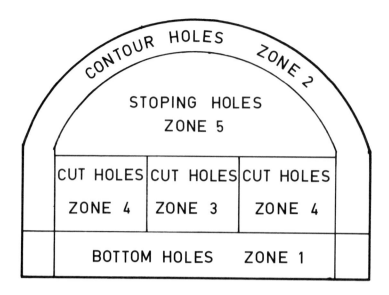

Figure 10.1. Dividing the tunnel sections into five zones.

blasting results. Having a good grasp of the drilling accuracy in real life will result in more realistic calculations and a better economy when planning a new drilling pattern. Drillholes should be charged with no more explosive than prescribed by the computer output.

Figure 10.1 shows the basic principle for a calculation of the different holes. The program divides the tunnel section into five zones containing the following four types of holes:

1. Bottom holes
2. Contour holes
3. Cut holes
4. Stoping holes

The computer calculation proceeds in the opposite time sequence to that of the real blast. For each zone, the limited number of drillholes are positioned with even separation, and the resulting charge in each hole is calculated. The first area to be calculated is zone 1 where the bottom holes are situated. These holes demand the largest specific charge because they have to lift the mass of rock toward the free face created by the cut. Then the contour holes in zone 2 are calculated. These have a very low specific charge, determined by the need for minimum damage to the remaining rock. The cut holes in zones 3 and 4 are calculated according to the methods detailed in Chapter 6, and finally the position and charging of the stoping holes in zone 5 is determined. These holes do not need as large a specific charge as do the holes in the bottom section because the force of gravity helps in removing the rock.

Figures 10.2, 10.3, and 10.4 show the printouts of three different sample calculations, and a segment of the output list for one of them is shown in Figure 10.5.

A follow-up of blasting operations in most countries in Europe and in Zaire, Hong Kong, Chile, and India where the tunnel program has been used shows that very reliable results can be obtained by designing the "blast" in the computer using the methods for charge calculation developed by Langefors and Kihlström.

Figure 10.6 shows a comparison between the results obtained in practice at four differently blasting sites before and after the calculation was made. In each case, the drillhole pattern calculated by the program produced a better result than that developed by the experience derived from the random process of field trial and error.

10.5 Computer Program for Bench Blasting

The bench blasting program covers the calculation of bench blasting and single hole blasting with free or inclined bottoms. Bench blasting using a free bottom can, for example, be used where a fault is situated at the toe of the bench or in the blasting work of underground storage rooms when a "center section method" is used for removing the rock between two headings placed above each other. For blasting in open pits and quarries where throw can be tolerated, it is also possible to calculate crater blasting where the unloaded hole length agrees with the optimal charging depth according to Livingstone's crater theory. The program also calculates the maximum throw length of flyrock for each row of holes, following principles explained in Chapter 12. This is very important information where safety areas have to be determined. The required input for an ordinary bench blasting operation is:

1. Diameter of drill hole
2. Ratio of spacing to burden
3. Planned surface area of the round
4. Face height
5. The toughness of the rock given as a rock constant
6. Degree of packing of the bottom and the column charge
7. Weight strength of the blasting agents
8. Hole inclination
9. Angular deviation
10. Collaring deviation

The Wide Space Blasting Method (see Chapter 8) was developed in the period 1965-72 and shown to produce considerable benefits in terms of better fragmentation, less secondary blasting, and consequently greater throughput of material for equal drilling and explosives consumption in large-scale blasting operations. The beneficial effect of using this method on the average fragment size has also been incorporated in the program.

The program which is based on empirical results from follow up and controlled experiments is very reliable and is routinely used by the staff of Swedish Detonic Research Foundation. A similar program is used by several construction companies in Sweden.

The bench blasting program was also used to judge the economic value of introducing automatic control systems mounted on the drill rigs to reduce the angle of deviation of the drillhole. The angle of deviation strongly affects the drilling pattern when small drillholes diameters are used together with high arches. By using the program to calculate the results of many different blasts with varying drillhole deviations, an accurate assessment of the new automatic drill rig control system could be obtained.

Figures 10.7 and 10.8 show sample printer and plotter outputs from calculations for blasts with hole diameters 76 mm (3 in) and 191 mm (7.5 in). Figure 10.9 lists a small bench blasting program.

292 Chapter 10. Computer Calculations for Rock Blasting

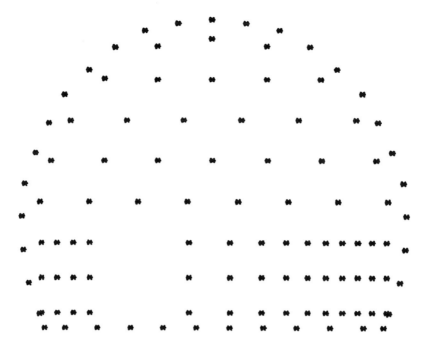

AREA	50.35	M²
ADVANCE	2.7	M
DRILL	31.0	MM
HOLES	108	
KG/CU.M.	1.0	
DM/CU.M.	2.3	

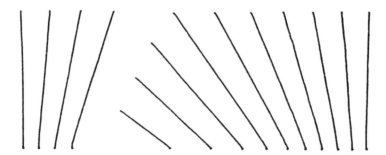

Figure 10.2. Fan cut.

10.5. Computer Program for Bench Blasting

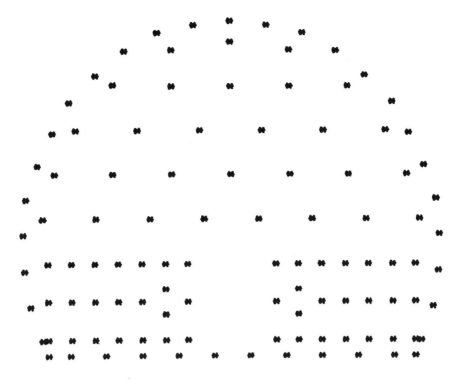

AREA	50.35	M²
ADVANCE	2.80	M
DRILL	31.0	MM
HOLES	110	
KG/CU.M.	1.0	
DM/CU.M.	2.3	

Figure 10.3. Plow (or plough) cut.

294 Chapter 10. Computer Calculations for Rock Blasting

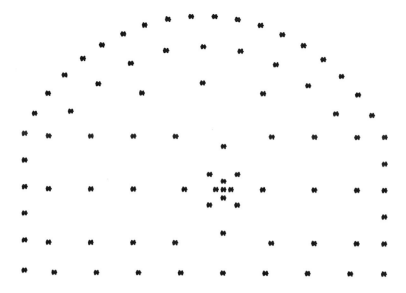

AREA	80	M²
ADVANCE	4.6	M
DRILL	51.0	MM
HOLES	82	
KG/CU.M.	1.1	
DM/CU.M.	1.1	
EMPTY HOLE	150.0	MM

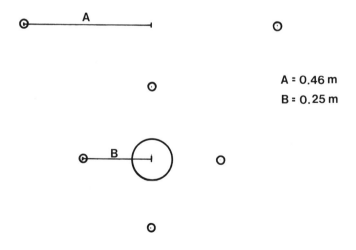

A = 0.46 m
B = 0.25 m

Figure 10.4. Parallel hole cut.

10.5. Computer Program for Bench Blasting

PARALLEL HOLE CUT WITH ONE EMPTY HOLE
FOLLOWING CONDITIONS ARE VALID
****

PROFILE GIVEN BY ARCH-RADIUS		ROCK DRILL BIT	51.00 MM	NO. OF WORKERS	0.
CONTOUR SMOOTH		SPEC. DRILL BIT	150.00 MM	HOUR-WAGE	-0.00 KR
		ROCK CONSTANT	0.40	EFFICIENCY	-0.00 PERC.
SECTION AREA	80.00 M2				
WIDTH	12.00 M	EXPLOSIVE STRENGTH		PRICE	
ARCH-RADIUS	6.80 M	BOTTOM CHARGE	100.00 PERC.	BOTTOM CHARGE	-0.00 KR/KG
X-MIN CUT-CENTER	-0.00 M	COL.CONTOUR	55.40 PERC.	COL.CONTOUR	-0.00 KR/KG
X-MAX CUT-CENTER	-0.00 M	COL.BOTTOM	100.00 PERC.	COL.BOTTOM	-0.00 KR/KG
Y-MIN CUT-CENTER	-0.00 M	COL.REMAIN	88.60 PERC.	COL.REMAIN	-0.00 KR/KG
Y-MAX CUT-CENTER	-0.00 M			DETONATORS/HOLE	-0.00 KR
ADVANCE/ROUND	4.60 M	DEGREE OF PACKING	1.20 KG/DM3		
				DEPRECIATION/ROUND	-0.00 KR
				DRILLING COSTS/M	-0.00 KR
				EMPTY HOLE COST/PC	-0.00 KR

THE CALCULATIONS GIVE THE FOLLOWING RESULTS
****

HOLES DRILLED/ROUND	82	HOLES	COSTS		
DRILLED METERS/ROUND	390.13	M	DRILLING/ROUND	0.00	KR
DRILLED METERS/CU.M.ROCK	1.06	M/M3	DRILLING/CU.M.ROCK	0.00	KR/M3
CONSUMPTION OF EXPLOSIVES					
BOTTOM CHARGES	234.76	KG	EXPLOSIVES/ROUND	0.00	KR
COLUMN CHARGES	51.89	KG	DETONATORS/ROUND	0.00	KR
COLUMN BOTTOM	40.42	KG	TIME-DEPENDING/ROUND	0.00	KR
COLUMN REMAIN.	62.23	KG	REMAINING/ROUND	0.00	KR
/ROUND	388.81	KG	BLASTING/ROUND	0.00	KR
/CU.M.ROCK	1.06	KG/M3	BLASTING/CU.M.ROCK	0.00	KR/M3
MAX CO-OPERATING CHARGE IN DYNAMEX-EQUIVALENTS			TOTAL COST/ROUND	0.00	KR
CUT	9.34	KG	TOTAL COST/CU.M.ROCK	0.00	KR
BOTTOM-HOLE	6.81	KG/HOLE			
STOPING-HOLE	6.85	KG/HOLE	TIME CHARGING AND BLASTING MAN-HOURS		
CONTOUR-HOLE	1.24	KG/HOLE	/ROUND 0.00 HOURS /CU.M. 0.00 HOURS/M3		

TABLE OVER HOLE-DATA

HOLE NO.	X-COORD COLLAR-ING	T-COORD COLLAR-ING	DIRECT. OF HOLE	DEPTH OF HOLE	BOTTOM CHARGE	COLUMN CHARGE	HEIGHT BOTTOM CHARGE	HEIGHT COLUMN CHARGE	DIAMETER OF HOLE	TYPE HOLE
	M	M	DEGREES	M	KG	KG	M	M	MM	
1	0:00	0.00	0.00	4.87	2.77	4.04	1.07	3.80	51.00	BOTTOM
2	1:10	0.00	0.00	4.87	2.77	4.04	1.07	3.80	51.00	BOTTOM
3	2:50	0.00	0.00	4.87	2.77	4.04	1.07	3.80	51.00	BOTTOM
4	3:90	0.00	0.00	4.87	2.77	4.04	1.07	3.80	51.00	BOTTOM
5	5:30	0.00	0.00	4.87	2.77	4.04	1.07	3.80	51.00	BOTTOM
6	6:70	0.00	0.00	4.87	2.77	4.04	1.07	3.80	51.00	BOTTOM
7	8:10	0.00	0.00	4.87	2.77	4.04	1.07	3.80	51.00	BOTTOM
8	10:90	0.00	0.00	4.87	2.77	4.04	1.07	3.80	51.00	BOTTOM
9	12:00	0.00	0.00	4.87	2.77	4.04	1.07	3.80	51.00	BOTTOM
10	0:00	0.00	0.00	4.87	2.77	4.04	1.07	3.80	51.00	BOTTOM
11	0:00	.91	2.49	4.69	.22	1.84	.09	4.60	51.00	CONTOUR
12	0:00	1.71	2.49	4.69	.22	1.84	.09	4.60	51.00	CONTOUR
13	0:00	2.50	2.49	4.69	.22	1.84	.09	4.60	51.00	CONTOUR
14	0:00	3.30	2.49	4.69	.22	1.84	.09	4.60	51.00	CONTOUR
15	0:00	4.10	2.49	4.69	.22	1.84	.09	4.60	51.00	CONTOUR
16	:41	4.76	2.49	4.69	.22	1.84	.09	4.60	51.00	CONTOUR
17	:88	5.37	2.49	4.69	.22	1.84	.09	4.60	51.00	CONTOUR
18	1:42	5.92	2.49	4.69	.22	1.84	.09	4.60	51.00	CONTOUR
19	2:02	6.41	2.49	4.69	.22	1.84	.09	4.60	51.00	CONTOUR
20	2:67	6.82	2.49	4.69	.22	1.84	.09	4.60	51.00	CONTOUR
21	3:37	7.16	2.49	4.69	.22	1.84	.09	4.60	51.00	CONTOUR
22	4:09	7.42	2.49	4.69	.22	1.84	.09	4.60	51.00	CONTOUR
23	4:85	7.60	2.49	4.69	.22	1.84	.09	4.60	51.00	CONTOUR
24	5:61	7.68	2.49	4.69	.22	1.84	.09	4.60	51.00	CONTOUR
25	6:39	7.68	2.49	4.69	.22	1.84	.09	4.60	51.00	CONTOUR
26	7:15	7.60	2.49	4.69	.22	1.84	.09	4.60	51.00	CONTOUR
27	7:91	7.43	2.49	4.69	.22	1.84	.09	4.60	51.00	CONTOUR
28	8:63	7.16	2.49	4.69	.22	1.84	.09	4.60	51.00	CONTOUR
29	9:33	6.83	2.49	4.69	.22	1.84	.09	4.60	51.00	CONTOUR
30	9:98	6.41	2.49	4.69	.22	1.84	.09	4.60	51.00	CONTOUR
31	10:58	5.92	2.49	4.69	.22	1.84	.09	4.60	51.00	CONTOUR
32	11:12	5.37	2.49	4.69	.22	1.84	.09	4.60	51.00	CONTOUR
33	11:58	4.75	2.49	4.69	.22	1.84	.09	4.60	51.00	CONTOUR
34	12:00	4.10	2.49	4.69	.22	1.84	.09	4.60	51.00	CONTOUR
35	12:00	3.30	2.49	4.69	.22	1.84	.09	4.60	51.00	CONTOUR

Figure 10.5. A segment of the output list.

296   *Chapter 10. Computer Calculations for Rock Blasting*

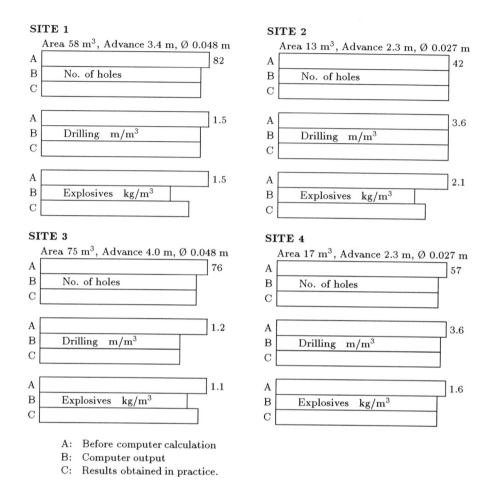

Figure 10.6. A comparison from four differently situated sites between the results obtained in practice before and after the calculation was made.

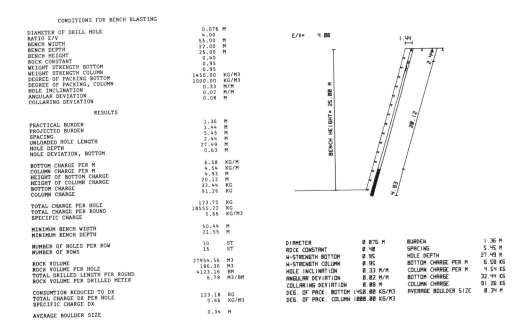

Figure 10.7. Bench blasting calculation for $d = 3$ inches.

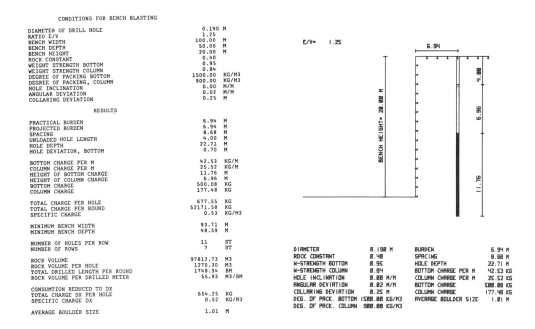

Figure 10.8 Bench blasting calculation for $d = 7.5$ inches.

Chapter 10. Computer Calculations for Rock Blasting

```
10 FIXED 2
20 REM N=INCLINATION M/M, D=DIAMETER M, P1=DEGREE OF PACKING KG/M↑3
30 REM K=BENCH HEIGHT M, S1=WEIGHT STRENGTH, C=ROCK CONSTANT
40 REM H0=STEMMING
50 N=0
60 D=0.25
70 P1=1400
80 H0=3
90 K=10
100 S1=1
110 C=0.4
120 F=3/(3+N)
130 REM INITIAL GUESS
140 B=40*D
150 H=K+SQR(1+N↑2)+0.3*B
160 IF H<(2.3*B) THEN 230
200 Y=P1*D↑2*S1*P1/F/4.44
210 J=0
220 GOTO 300
230 H=1.3*K*SQR(1+N↑2)-0.3*B
231 L1=1.11*F*(0.07*B+C*B↑2+0.004*B↑3)/S1
240 Q0=1.3*B*L1
250 H1=H-B
260 Q1=H1*P1*D↑2*P1/4
270 IF ABS(Q0-Q1)<0.0001*Q1 THEN 420
280 Y=(Q0+Q1)/2*S1/F/1.3/1.11
290 J=1
300 X=B
310 X1=(0.07*X↑2+C*X↑3+0.004*X↑4)/X-Y*((J=1)/X+(J=0))
320 X2=(2*0.07*X+3*C*X↑2+4*0.004*X↑3)/X
330 X3=X-X1/X2
340 IF ABS(X3-X)<1E-03 THEN 370
350 X-X3
360 GOTO 310
370 B=X3
380 WRITE (15,390)L1,Q0,H1,Q1,B,H
390 FORMAT 5X,6F8.2
400 IF J=0 THEN 411
410 GOTO 150
411 H1=1.3*B
412 H2=K*SQR(1+N↑2)+0.3*B
413 IF ABS(H2-H) >= 0.001 THEN 150
420 PRINT "MAXIMUM BURDEN           "B
430 PRINT "LENGTH OF BOTTOM CHARGE  "H1
440 PRINT "HOLE LENGTH              "H
450 PRINT "LENGTH OF COLUMN CHARGE  "H-H1-H0
460 END
```

49.62	645.02	0.00	0.00	7.97	10.00	
31.11	322.51	2.63	160.93	7.35	10.61	
26.34	251.72	3.44	236.55	7.28	10.79	$k = 10\,m$
25.80	244.13	3.54	243.12	7.27	10.82	
25.76	243.63	3.54	243.56	7.27	10.82	

MAXIMUM BURDEN   7.27
LENGTH OF BOTTOM CHARGE   3.54
HOLE LENGTH   10.82
LENGTH OF COLUMN CHARGE   4.27

25.76	243.59	3.54	243.59	11.69	28.00	$k = 25\,m$
25.76	243.59	15.20	243.59	11.69	28.51	

MAXIMUM BURDEN   11.69
LENGTH OF BOTTOM CHARGE   15.20
HOLE LENGTH   28.51
LENGTH OF COLUMN CHARGE   10.31

Figure 10.9. Listing of a small program for calculating the maximum burden.

Chapter 11

Blast Performance Control

In the initial stages of testing new explosives, initiation methods, or mining and excavation methods, good planning of the work is only one of many requirements for success. It is also necessary to follow the different phases in the field operation when the methods are finally tested on a production scale. A large number of things can go wrong during the implementation of a field test and, when something malfunctions (which it sometimes does), it is tremendously difficult to draw the right conclusions from the results. There are numerous parameters to consider, and one should be sure to identify the parameter that has to be changed to eliminate the effect and consequences of a malfunction.

Therefore, we strongly recommend that a blast performance control element be incorporated in the initial stage of any test (Figure 11.1).

The blast performance control should cover the following parameters:

    1. Blast geometry   4. Breakage performance
    2. Initiation performance   5. Fragmentation
    3. Detonation performance   6. Blast damage

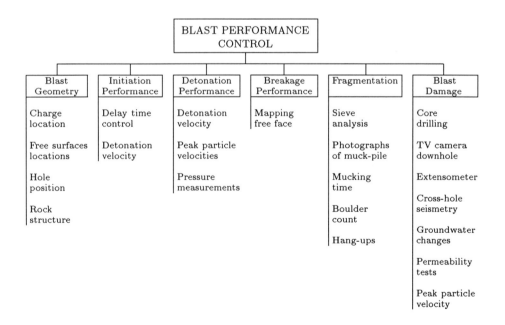

Figure 11.1. Important parameters to study in blast performance control.

1. Blast geometry involves the knowledge of the exact position of the charges and the free surfaces. Through hole logging, the hole positions and rock structure at actual depths relative to the free surface can be known. This enhances the calculation and positioning of each charge to achieve optimum breakage. It also makes it possible to tailor blasting charges close to weak rock structures such as a hanging wall contact zone which must not be damaged to avoid ore dilution. It is also much safer to compare empirical blast data from previous blasts with future blasting operations if one knows that the data has been carefully and correctly gathered and documented.

2. Initiation performance can be checked by detonation velocity measurements or by ground vibration measurements with accelerometers positioned close to the round. The results can be used to answer questions regarding the delay intervals and actual firing times, specifically to determine if any stress wave initiation between adjacent primers occurred, or if the initiation was unsatisfactorily causing an underdriven detonation in the bottom charge. By knowing the exact initiation time, it is also possible to record the P-wave travel time between explosive and gauges.

3. Detonation performance can be monitored by detonation velocity measurements or by the use of accelerometers. When ground vibration measurements are carried out, the pulse shape and the signal magnitude will help in the analysis of whether or not the entire charge functioned properly. Shock wave desensitization or "dead pressing" of the charge by ground movement or gas penetration through fractures connecting two holes can influence neighboring charges. Control of moisture in charges and correlation with laboratory data can help to explain inferior detonation performance for ANFO-type explosives. Microballoon sensitized explosives might lose their sensitivity due to dynamic loading from adjacent drillholes, especially if any fracture system intersects several drillholes. Pressure measurements in water-filled drillholes can be used to evaluate this effect.

4. Breakage performance should be checked after each round. This information, together with detonation performance data and descriptions of rock structure orientation and strength, may enable immediate changes of the blast design for the next round to further optimize existing practice.

5. Fragmentation should be observed after each round, if at all possible. Not all mining methods allow an immediate fragmentation study, but, with good planning and time coordination of the mining sequences, it is quite often possible to empty the draw points and perform the study. The size distribution and the shape of the fragmented rock adds more information to the evaluation of the test blast. The fragmentation study is best and most correctly done by sieve analysis; however, this method is labor-intensive and costly. Photographs of the muck pile or the draw points where reference frames are laid out give significant information to the fragmentation characterization. Mucking time and boulder counting are other useful parameters. Recent real-time imaging techniques allow continuous monitoring of the fragment size distribution as the broken rock flows by on a conveyer belt, or passes by on top of a truck load of rock.

6. Blast damage observations of the ground surrounding the mining area are fundamental in evaluating whether the blast design and linear charge concentrations selected are optimal and match the particular requirements.

Pre- and post-blast core drilling with crack frequency counting, extensometer measurements, drillhole inspection with TV cameras, permeability measurements, cross-hole seismic measurements, and peak particle velocity measurements are some of the usual methods for characterizing disturbances in the rock mass continuity.

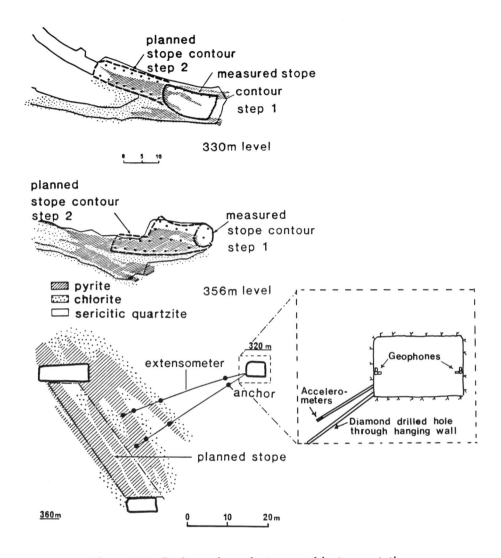

Figure 11.2. Geology, planned, stope, and instrumentation.

Figure 11.2 shows the instrumentation used when SveDeFo was involved in testing open stoping in a narrow ore body at the Kristineberg mine of Boliden Mineral AB, Sweden. This was the first time $6\frac{1}{2}$ in hole diameters were used in underground Swedish mining.

Various aspects of the Kristineberg test were reported by Krauland [1985] and Falkdal [1985]. The major concern was to avoid any caving of the weak hanging wall. The test showed that — by using careful control, accurate measurement, and sufficient expertise — the blasting method could be refined to match the particular requirements of a potentially very difficult site. No caving whatsoever occurred during or after the excavation phase. The mining sequence and hanging wall deformation is shown in Figure 11.3.

Figure 11.4 shows a blast performance test carried out at the LKAB mine Kirunavaara. The test had two aims: (1) to determine if the fan worked properly without detonation failures due to wet ANFO and (2) to determine if any simultaneous initiation

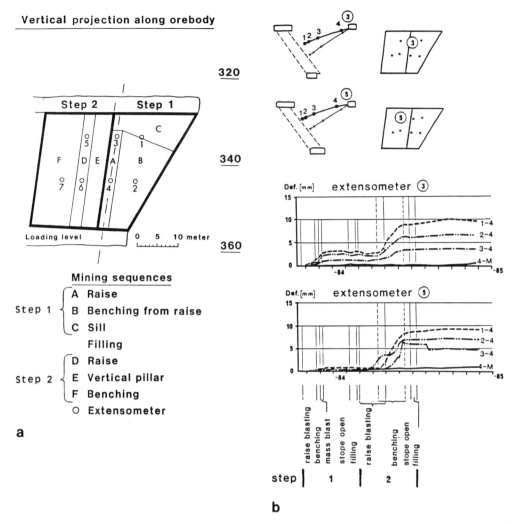

Figure 11.3. (a) Mining sequences and (b) Hanging wall deformation [from Krauland, 1985].

of neighboring shotholes could occur because of the small distances between the hole distances and thereby the small distances between neighboring primers. The results of the measurement (Figure 11.4) show that, for this particular fan, everything functioned well. Alternate intervals show a higher signal amplitude because these charges were closer to the accelerometer gauges than the other charges.

11.1 The Vertical Crater Retreat (VCR) Blasting Technique

The introduction of large-diameter drillholes in mining has made it possible to reduce development work because drilling has become more precise, and longer holes can be drilled. In sublevel stoping, holes up to 100 m long can be drilled from the drill level to the undercut. However, a horizontal stoping method needs a slot to be developed for the blasting operation.

Lang [1976] outlined a concept for VCR where vertical stoping takes place with concentrated "nearly spherical" charges in vertical drillholes blasting down toward a hori-

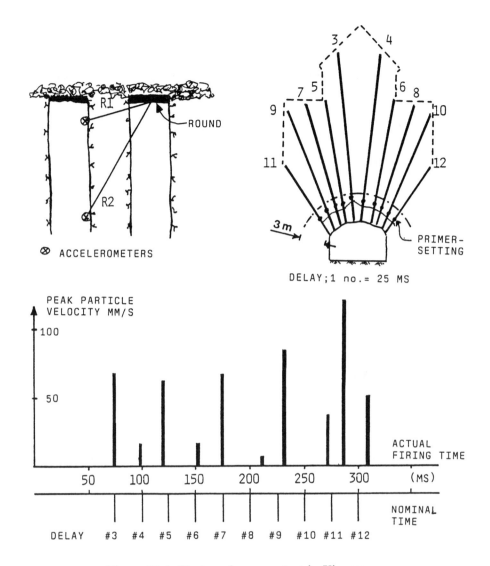

Figure 11.4. Blast performance test in Kirunavaara.

zontal free surface. The concentrated charges usually have a length equal to six times the hole diameter. These charges can break a considerable amount of rock if the charges are placed at their optimum distance from the free surface.

The method is being applied frequently to primary stopes and pillar recovery where it can also eliminate raise boring, slot cutting. Dilution of ore by backfill is greatly reduced which is a major advantage with the method. VCR has also been used extensively for producing vertical shafts by a technique termed *drop raising*, where vertical holes are drilled the entire length from the upper level to an undercut or lower tunnel, and a series of VCR blasts break the rock down toward the lower free surface. If vertical (or inclined) large-diameter holes are drilled on a designed pattern from a cut over a stope or pillar to bottom in the back of the undercut, and concentrated (short) charges of explosives are placed within these holes at a calculated optimum distance from the lower free surface toward the undercut and detonated (Figure 11.5), a vertical thickness of ore will be blasted downward into the previously mined area. As this

Figure 11.5. Vertical Crater Retreat Method.

loading and blasting procedure is repeated, mining of the stope or pillar retreats in the form of horizontal slices in a vertical upward direction until the top sill is blasted away and the shaft or stope is completed.

The application of VCR to mining has been reported by many authors since the pioneering work by Lang [1976, 1977, 1978, 1981]. Contributions to the present technique have been made by Almgren *et al.* [1978], Crocker [1979], Monahan [1979], Mitchell [1980], Miller [1979], Rowlandson [1980], Holmberg *et al.* [1980], and Niklasson [1979, 1984].

True VCR mining is really carried out only when the charge is blasted toward a free surface which is perpendicular to the hole axis. Sometimes, the term VCR is used although the mining method is a stoping method with endslicing or inverted benching.

Some mines are currently using a modified method where the charges are longer than six times the hole diameter. The term *vertical retreat mining* (VRM) has sometimes been used to indicate this modification.

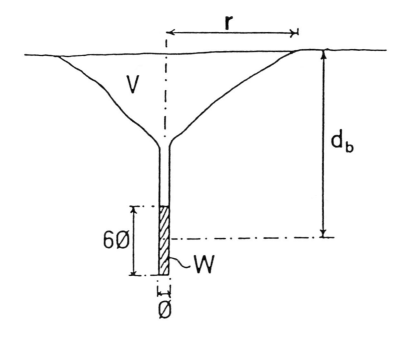

Figure 11.6. Blast geometry in cratering.

11.2 The Cratering Theory

The concept of crater blasting with the charge applied in a drillhole at right angles to the free surface and its development into a useful blasting method is generally attributed to Livingston [1962]. Additional extensive studies of crater blasting have been reported by Lang [1983]. Crater blasting is also a versatile tool for studying the blasting phenomenon, the performance of explosives, and the blastability of different rock materials and rock masses. The VCR blasting method mentioned above has grown out of studies of crater blasting.

A crater blast is a blast in which a spherical or near-spherical short charge (1:6 diameter-to-length ratio) is detonated beneath a surface that extends laterally in all directions beyond the point where the surrounding material would be affected by the blast.

The following notation and terminology is used (Figure 11.6) in VCR:

ϕ Hole diameter
6ϕ Charge length (standardized to 6 times the charge diameter)
d Depth of burial — distance from surface to center of charge
d_o Optimum depth of burial — the depth of burial at which the greatest volume of rock is broken
N Critical distance — the depth of burial at which the effects of a cratering charge are just noticeable on the surface
r Radius of crater
r_o Radius of crater formed at optimum depth of burial
V Crater volume
W Charge weight

Figure 11.7. Principle for the VCR blasting method [from Lang, 1981a].

It has been found that there is a definite relation between the energy of the explosive and the volume of the material that is affected by the blast. This relationship is significantly affected by the placement of the charge.

Livingston has determined that a strain-energy relation exists, expressed by an empirical equation

$$N = E_s W^{1/3} \qquad (11.1)$$

where N is the critical distance at which breakage of the surface above the spherical charge does not exceed a specified limit, E_s is the Strain Energy Factor (a constant for one given explosive-rock combination), and W is the weight of the explosive charge.

Equation 11.1 may be written into a different form as

$$d = \Delta E_s W^{1/3} \qquad (11.2)$$

where d is the distance from the surface to the center of gravity of the charge, i.e., the depth of burial, and Δ equals d/N which is a dimensionless number expressing the ratio of the depth of burial to the critical distance.

When d is such that the maximum volume of rock is broken to the fragment size required, this burial depth is called the optimum distance d_o (Figure 11.7).

Tests with a dynamite (DxB-dynamite) in a Swedish granite gave the results:

$$E_s = 1.5$$

$$d_o = \frac{N}{\sqrt{3}}.$$

Using this result together with Equation 11.2, we can express the optimum depth d_o for DxB-dynamite in granite and for various degrees of packing P and hole diameters ϕ as

$$d_o = \frac{E_s \phi}{\sqrt{3}} \left(\frac{3P\pi}{2}\right)^{1/3} \tag{11.3}$$

where $E_s = 1.5 \text{ m·kg}^{-1/3}$.

The curve can be plotted to visualize the optimum depth and the charge weight as a function of the hole diameter (Figure 11.8).

11.3 Choosing the Best Explosive for VCR Mining

When the rock material to be blasted is the same, but several different explosives are considered, the cratering theory may be used to determine the best suitable explosive through the application of Livingston's Breakage Process Equation

$$V = A B_m C W (E_s)^3 \tag{11.4}$$

where V is the crater volume, W is the charge weight, E_s is the strain energy factor, A is the energy utilization number, B_m is the material's behavior index, and C is the stress distribution number.

While V, W, and E_s can be measured with certainty from test blasts, it remains for the observer to determine the variables of A, B_m, and C.

The energy utilization number A, the ratio of the volume of the crater within the limits of complete rupture at any depth to the volume at optimum depth (where the maximum proportion of the energy of the explosion is used in the failure process), is

$$A = \frac{V}{V_o}. \tag{11.5}$$

At the optimum depth where fracturing is the most efficient, A equals 1.0. Accordingly, the numerical values of A are less than 1.0 at other charge depths.

The Material's Behavior Index B_m is a constant for a given type of explosive and weight of charge in a given material. B_m is measured at optimum depth, and

$$B_m = \frac{V_o}{N^3}. \tag{11.6}$$

Equation 11.6 has been derived from

$$V_o = B_m \left(W E_s^3 A C\right) \tag{11.7}$$

where $A = 1$ at optimum depth d_o, and $C = 1$ if the charge is spherical.

We may conclude that both A and B_m describe the effect of the explosive upon the failure process in blasting. The Energy Utilization Number A best describes effects due to variation in energy density with distance; B_m best describes effects due to the variation in energy density accompanying changes in stress-strain relations as measured at a given reference energy level.

We now give an example which will demonstrate the application of the Breakage Process Equation (Equation 11.4) for the comparison of the performance of the explosives in the same rock.

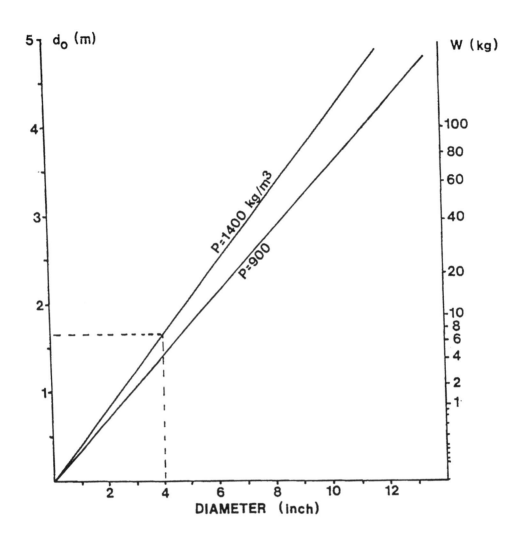

Figure 11.8. Optimum depth of burial and charge weight as functions of charge diameter for VCR blasting.

Basic cratering research was conducted in a hard cherty magnetic iron formation with types Slurry 1 [Selleck, 1962], and Slurry 2 [Lang, 1962]. The curves of V/W vs. Δ for the two experiments are plotted in Figure 11.9. The optimum depth ratio was found to be the same for both explosives ($\Delta_o = 0.58$), but E_s and N were different.

The values of A were calculated for each crater and the results were plotted against depth ratio (Figure 11.10). The two curves are similar to those of V/W vs. Δ. This diagram clearly indicates that in the case of Slurry 2, more energy is being used in the

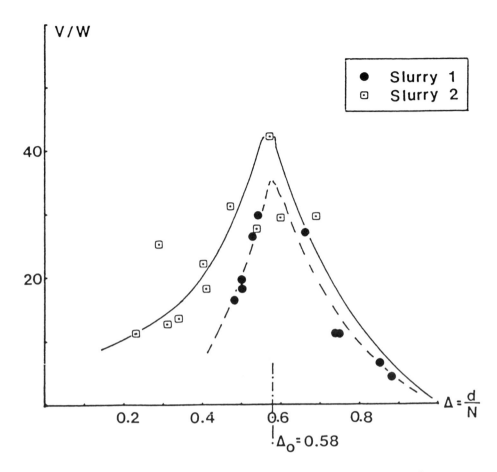

Figure 11.9. V/W (Crater volume/charge weight) vs. Δ (depth of burial d/critical distance N).

secondary fragmentation range and in the flyrock range than in the case of Slurry 1. This is responsible for better fragmentation and more gas energy. The results from production scale blasts confirmed the results of cratering experiments.

The Material's Behavior Index for both explosives was calculated at optimum depth and found to be:

Slurry 1: $B_m = 0.42$
Slurry 2: $B_m = 0.33$

Higher values of B_m are characteristic of brittle type failure. Experiments show that B_m decreases as the material becomes more plastic-acting, a fact that is true in this experiment too. Slurry 1 had a high detonation velocity; thus, the material was acting in the brittle manner. Due to the 10% Al content, Slurry 2 had a lower velocity of detonation, and the load was a slower and more sustained type. Hence, the same rock behaved rather in the plastic manner.

The Stress Distribution Number C was 1 because both experiments employed short, nearly-spherical charges.

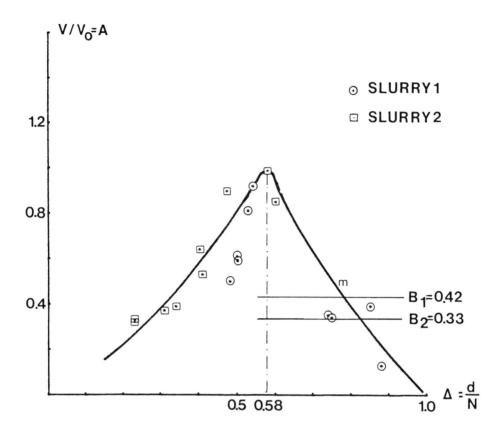

Figure 11.10. The energy utilization number A plotted against depth ratio for Slurry 1 and Slurry 2.

From this, we may conclude that when comparing different explosives in the same rock material, the comparison must be made while keeping the geometry of the blast constant. Otherwise, the results will be misleading. The three-step sequence of events in a proper comparison of different explosives in the same rock material is thus the following:

1. Separate cratering experiments should be conducted with the different explosives in the same material.
2. Determine N and Δ_o for each experiment.
3. If this information is for the design of VCR type blasting, then d_o and optimum spacing should be calculated for each explosive and ore combination. These criteria should be used in each respective stope.

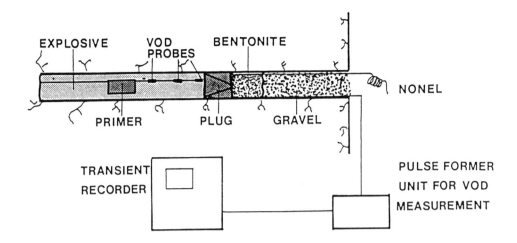

Figure 11.11. Small scale crater test experimental setup.

11.4 Small-Scale Cratering Tests

Location of Test

The aim of small-scale crater tests is to obtain results which are helpful in making qualified predictions of the blasting results to be expected in the full production stope. It is necessary to do the cratering tests as close as possible to the stope where the VCR method will be used. Different rock properties and structural geology may cause over- or under-estimation of the depth of burial for the production blasts.

If the depth of burial is less than optimum, this will still result in a satisfactory breakage — but the cost for drilling and explosives may be too high. If the depth of burial is larger than the optimum one, bells or unsatisfactory fragmentation may occur.

During the development work for a new stope, it is sometimes possible to carry out the test in the stope (in the undercut). If this cannot be done, then one must carefully observe the rock structure through field mapping procedures to evaluate whether the results achieved at the test area are the same as they would be in the production stope. Mäki [1982] reports an investigation where rock structure was mapped. He strongly emphasizes the influence of rock structure upon the cratering results, especially for small-diameter charges.

Crater tests can be carried out by drilling holes of various depths into the side (rib) of a drift. Hole diameters not smaller than 4 inches should be used. Preferably, the velocity of detonation should be registered, thus providing blast performance data that can be used in the evaluation of the test results, the initiation system, and the explosive function. Figure 11.11 shows a crater test experimental arrangement.

It is easier to carry out the tests in the rib, but if the structural geology is layered or significantly nonisotropic, the test holes should preferably be oriented in the same direction as the shotholes in the production stope.

After firing, the structural weakness planes which may have influenced the size or shape of the crater are mapped and photographs should be taken. Crater depths, radial fractures, and crater volume should also to be measured.

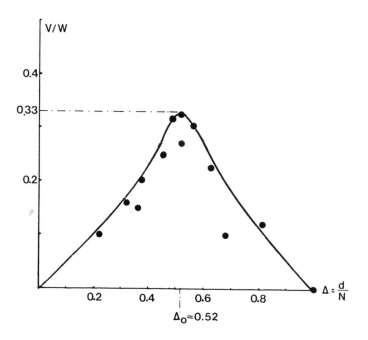

Figure 11.12. Scaled volume as a function of scaled depth.

11.5 Evaluation of Crater Results and Scaling

In the following, we describe a fictitious test result where a 0.6 m long TNT-slurry charge was detonated in 102 mm (4 in) diameter holes. The density of the slurry is 1400 kg/m^3, which gives a charge weight of 6.8 kg. The critical distance N was determined to be 2.5 m, and the scaled volume is plotted against the scaled depth in Figure 11.12.

Input data:

Hole diameter	ϕ	$= 0.1$ m (4 in)
Charge length	6ϕ	$= 0.6$ m
Charge weight	W	$= 6.8$ kg
Critical distance	N	$= 2.5$ m

Calculation of the Strain Energy Factor gives us

$$E_s = \frac{N}{W^{1/3}} = \frac{2.5}{(6.8)^{1/3}} \approx 1.32 \text{ m kg}^{-1/3}.$$

The optimum depth ratio, $\Delta_o = 0.6$, can be evaluated from the fictitious test results plotted in Figure 11.12. The calculated optimum depth of burial, d_o, is

$$d_o = \Delta_o N = (0.6)(2.5) = 1.5 \text{ m}.$$

Figure 11.12 gives us the value of V/W corresponding to the optimum depth ratio

$$\frac{V}{W} = 0.4 \text{ m}^3\text{kg}^{-1}.$$

11.9. Evaluation of Crater Results and Scaling

Hence,
$$V = (0.4)(6.8) = 2.72 \text{ m}^3.$$

As the crater was conical-shaped, one may calculate the radius r_o

$$V = \frac{\left[\pi r_o^2 \left(d_o + \frac{\text{(charge length)}}{2}\right)\right]}{3} \tag{11.8}$$

$$r_o = \left[\frac{(3)(2.72)}{\pi \left(1.5 + \frac{0.6}{2}\right)}\right]^{1/2} \approx 1.4 \text{ m}.$$

The following basic data is now available:

$W = 6.8$ kg $\quad\quad \Delta_o = 0.6$
$N = 2.5$ m $\quad\quad d_0 = 1.5$ m
$E_s = 1.32$ m kg$^{-1/3}$ $\quad r_0 = 1.4$ m

These data will now be used when we scale up the test result in order to apply it to our production stope where the same explosive will be used, but the hole diameter is 0.165 m (6.5 in). The charge length is $6\phi = 1.0$ m with a weight of 30 kg.

Following the Livingston theory, the corresponding values may be calculated while $E_s = 1.32$ remains constant.

Critical distance for the 30 kg charge weight is given by

$$N_1 = E_s W_1^{1/3} = (1.32)\left(30^{1/3}\right) \approx 4.1 \text{ m}.$$

The center of this charge should be at the optimum distance d_{o1} from the back of the stope

$$d_{o1} = \Delta_o N_1 = (0.6)(4.1) = 2.46 \text{ m}$$

or as the scaling factor F can be expressed as

$$F = \left(\frac{W_1}{W}\right)^{1/3} = \left(\frac{30}{6.8}\right)^{1/3} \approx 1.64$$

$$d_{o1} = d_o F = 2.46 \text{ m}.$$

Use $d_{o1} = 2.4$ m for the design; and the radius r_{o1} becomes

$$r_{o1} = r_o F = (1.4)(1.64) \approx 2.3 \text{ m}.$$

It is important to ensure complete breakage of the rock between two adjacent holes in the stope by designing an optimum spacing between holes. The recommended hole spacing, S_{o1}, should be in the range

$$1.2 r_{o1} \leq S_{o1} \leq 1.6 r_{o1} \tag{11.9}$$

so

$$(S_{o1})_{min} = 2.76 \text{ m}.$$
$$(S_{o1})_{max} = 3.68 \text{ m}.$$

It is more prudent to design the first stope using the min-value, and then, if satisfactory fragmentation and breakage permit this spacing, to increase it gradually in further stopes. However, $S = 3.0$ m would probably not cause any problem with the achieved test curve shape shown in Figure 11.12.

The advance I will be

$$I = d_{o1} + \frac{\text{(charge length)}}{2} = 2.4 + 0.5 = 2.9 \text{ m}$$

and the specific charge q becomes

$$q = \frac{W}{IS^2} = \frac{30}{(2.9)(3^2)} \approx 1.15 \text{ kg/m}^3 .$$

11.6 Application of the VCR Concept to Production Stopes

Full-scale VCR in the production stopes must be planned and followed up precisely if satisfactory results are to be achieved. The VCR method does not allow anyone to observe the blasting results as there is seldom entry to the mined-out room. Therefore, all logging of hole depths and estimation of the breakage must be carried out through the production holes.

Before each shot, the following planned, calculated, or computed data should be available for each shot hole:

1. Hole location
2. Hole length
3. Position of plug
4. Position of lower stemming
4. Depth of burial of charge
6. Position of upper stemming

The upper stemming should have such length that it enhances maximum confinement, but it must not freeze.

Figure 11.13 shows the flow sheet which was used by Niklasson of the Swedish Detonic Research Foundation when the D-stope was crater blasted in the Swedish Research Mine at Luossavaara. The additional information achieved from the fragmentation at the drawpoints contributed to the blast design and the excellent results achieved.

The computer database was of great help as one could quickly plot the contour of the stope crown after each blast. This visual information of the exact contour (Figure 11.14) gave information as to whether any caving had occurred which might change the blast design. Rock structure can sometimes influence the unwanted caving that produces large boulders to such an extent that the surface of the stope crown should not be kept horizontal, but instead follow an inclined weakness structure. This type of plot is of great help in any mining operation where the position of the charges and their delay intervals are to be determined.

Figure 11.15 shows two loading configurations used in the VCR stope in Luossavaara. Wooden plugs have been used for sealing the holes. These plugs work nicely if the "bells" have been developed, and they are easily positioned after they have been lowered down to exact depth by simply dropping a small piece of rock through the hole. (Emulite is the Nitro Nobel AB trade name for one of their emulsions, and Reolit is a TNT-slurry.)

The specification for the explosives used in the VCR stope are given in Table 11.1. The VCR results are summarized in Table 11.2.

11.7. Detonation Performance Check Using Tracer Elements

Table 11.1. The three different explosives used in the test.

Explosive	Density (kg/m^3)	Velocity of deton. (m/s)	Borehole diameter (inches)	Strength per vol. (rel. to ANFO)	Energy content (MJ/kg)	Gas volume (STP) (m^3/kg)	Detonation pressure (estimated) (GPA)
ANFO, prilled	825	3900	4 in	100%	3.9	0.97	3
Emulite VCR	1280	5900	6.5 in	157%	4.2	0.83	11
Reolit K	1450	4800	4 in	161%	3.7	0.76	9

Table 11.2. Summary of the most important results from the VCR stope.

	ANFO	Reolit	Emulite	Emulite
Geometry	2×2 m	3×3 m	2.5×2.5 m	3×3 m
Tons of ore blasted	4090	5302	7441	3950
Boulder, %	5.6	5.0	2.2	0.5
No. of hang ups	0	5	0	0
Reolit, kg	654	1880		
ANFO, kg	1724	—		
Emulite	—	—		
Specific charge, kg/ton	0.58	0.36	0.30	0.43
Specific drilling, m/ton	0.065	0.030	0.035	0.040
Average advance, m	2.4	3.1	3.0	2.6
Average width of stope, m	12	11	12	8
k_{50}, m	0.18	0.15	0.10	0.15
Costs SEK/ton	12.69	8.44	7.91	10.27
Overbreak, % of theoretical advance	32	25	21	4

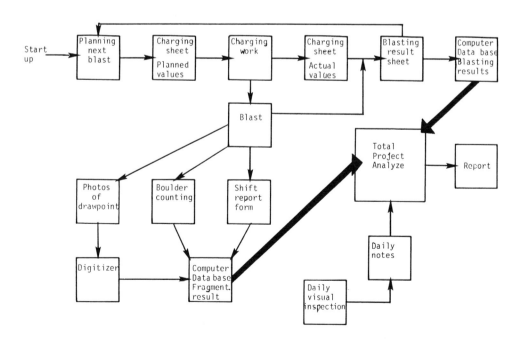

Figure 11.13. Flowsheet of VCR blasting in D-stope [from Niklasson, 1984].

316 *Chapter 11. Blast Performance Control*

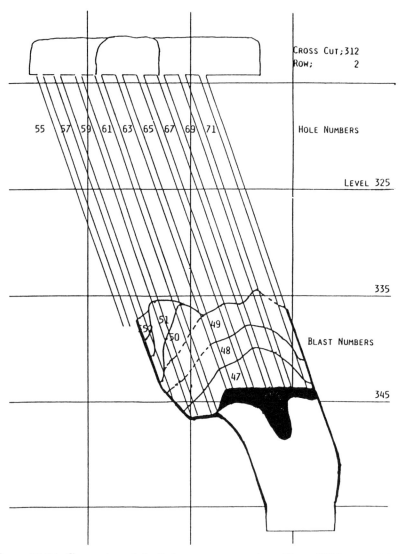

Figure 11.14. Computer plot of stope crown contour [from Niklasson, 1983].

11.7 Detonation Performance Check Using Tracer Elements

In VCR blasting and in other applications where the result of the blast is not immediately accessible for inspection, the question as to whether a specific charge detonated properly or not is sometimes raised.

Persson and Höglund [1991] report successful trials of detecting detonation performance with tracer substances underground in sublevel caving operations. They placed caps containing tracer substances (such as xylen, toluen, etc.) in those parts of the explosive to be controlled. This methods works very well, especially in sublevel caving where the detonator was positioned in the collar part.

When the detonation takes place, the tracer caps are demolished and the substance leaks out. The presence of the tracer substances in the explosive fumes indicates that

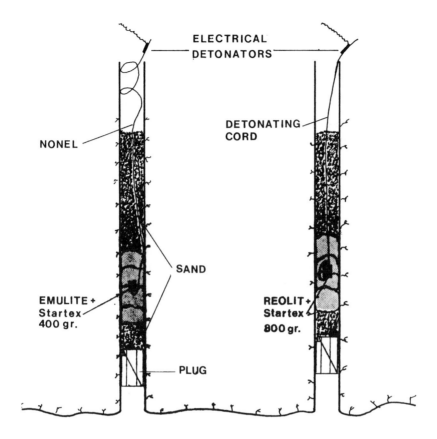

Figure 11.15. Loading configurations for VCR [from Niklasson, 1983].

the detonation wave reached the tracer cap; 10 measuring points can be detected in the same round using 10 different tracer substances and the method works faultlessly if the passages for the ventilated air are known. The method is very useful in underground blasting, especially where the results of the blasting cannot be inspected visually.

11.8 Cast Blasting

In most rock blasting operations the broken rock should be thrown only far enough to allow room for expansion of the broken rock. The cost of the entire blasting operation including that of mucking is generally minimized if the rock is not thrown too far away.

However, in certain blasting operations the controlled throw of the blasted rock can be a very cost effective method of moving large quantities of rock. An early example of such a blasting operation was the formation of a large dam across the river valley near Alma Ata, Kazakhstan, by the explosive throw of the adjacent mountain side. The dam still effectively protects the downstream towns from flood damage. In studies carried out in preparation for the blast, Lavrentiev [1974] showed theoretically, and verified by model blast experiments, that in order to throw a given volume of rock material held together without change in shape, the rock volume should ideally be totally surrounded by the explosive charges, with the greatest charge concentration below and behind the rock mass, seen from the direction of throw.

The greatest present-day application of explosive casting is in strip mining where large volumes of broken overburden rock have to be moved across the wide ditch at the bottom of which the coal seam will be exposed for loading. Huge dragline machines having bucket volumes as large as 220 cu yd are used for this purpose. The productivity of the very expensive dragline can be greatly increased if much of the overburden rock is thrown, or *cast*, across the ditch by the explosive employed in the round, so that only a small fraction of the overburden must be moved by the dragline. Typically, between 40 and 80 % of the original mass of overburden material can be moved across the ditch by the action of the explosive, leaving from 60 to 20 % of the mass to be moved by the dragline. The explosive is generally charged into drillholes which are evenly distributed through the rock mass in a staggered pattern. In the blast, the rock mass is moved row by row, with the time delay between rows increasing to several hundred milliseconds towards the rear of the blast. This allows the material thrown by the previous rows time to move freely out of the way of the material thrown by the following rows. The technique for achieving the cost-effective explosive casting or throw of the overburden with only a moderate increase in specific charge and with control over flyrock is called *cast blasting* or sometimes *explosive casting* .

Cast blasting increases the overburden-to-coal ratio which can be mined economically, and deeper coal seams can be mined effectively with smaller, less expensive equipment which was formerly limited to the mining of shallow formations.

Chapter 12

Flyrock

12.1 Introduction

Basic to all rock blasting is a careful optimization of the charge concentration needed for loosening the rock. In the 1950s through their experimental investigations, Langefors and Kihlström developed empirical formulae for calculating the charge weight (charge concentration) and charge distribution required for different blasting geometries. They also showed that the muck pile distribution depends solely on the excess charge used in the blast. Thus, the charge required to just break the rock but not move it can be calculated. In production blasting, some excess charge has to be added to facilitate easy loading of the broken rock and reduce wear on the loading equipment. Some excess charge is also needed to compensate for drilling deviations and for lifting the rock to accommodate the volume expansion (swelling) of the rock mass as it is fragmented. About 16 % of the energy of the total excess charge is used for the throw. The rock throw obviously increases with increasing total charge weight. In normal bench blasting, the center of gravity of the broken rock is moved a distance equivalent to a few bench heights only.

Under special conditions, however, individual rocks or even boulders can be thrown remarkable distances. Figure 12.1 shows a boulder of iron ore with a weight of about 3 tons that was thrown to land on a bench 300 m away from and 40 m above a standard bench blast in the LKAB Svappavaara open pit mine. The drillhole diameters used were 194 mm ($7\frac{5}{8}$ in) and the explosive was a heavy TNT-slurry. The average specific charge in the round was 1.1 kg/m^3.

Flyrock in rock blasting has been a serious problem since blasting began several hundred years ago. Men have been killed; buildings, equipment, and materials destroyed. These hazards are greatest in urban areas; but as the blasting rounds have become ever greater with larger-diameter drillholes, it is now a serious problem even outside urban areas, for example in the neighborhood of large open pit mines. It is not unusual for flyrock to travel a kilometer or more from a blast site using large-diameter drillholes when no precautions have been exercised to reduce the flyrock distance. Apart from the obvious risk for damage to buildings and production equipment, this creates an unacceptable threat of third-party property damage, injury, and even loss of life.

This chapter lays a foundation of understanding of the flyrock problem — how and why the flyrock occurs and how precautions can be taken to reduce or avoid them. Experiments carried out by the Swedish Detonic Research Foundation will be used to illustrate and describe the mechanism of flyrock generation.

320   Chapter 12. Flyrock

Figure 12.1. Flyrock can be thrown remarkable distances. This 3 ton boulder was thrown more than 300 m from the blast.

12.2 Influence of the Specific Charge

Model-scale studies of rock blasting in plexiglas (polymethylmethacrylate, PMMA) have been of great help in understanding and predicting what happens when flyrock is generated in rock blasting. PMMA has the advantage of being transparent, which makes it possible to observe by high-speed photography the development of cracks and fragments.

Ladegaard-Pedersen and Persson [1973] used model experiments in PMMA for rock throw studies. Figure 12.2a shows a PMMA block with dimensions 100 × 60 × 60 mm. This model corresponds to a bench with one drillhole. Three blocks with different hole diameters are illustrated. The charge weights of PETN used were varied by almost a factor of two in the different pictures (Figure 12.2b, c, d). The charge weights were 140, 227, and 550 mg, respectively. Using an exposure time of 0.1 μs, a Kerr-cell camera took pictures 210 μs after ignition. The pictures show that, if the specific charge is increased, not only does the fragmentation become better but the velocity of the broken material increases. The throw velocity as a function of specific charge is plotted in Figure 12.13.

Half-scale blasting experiments in granite indicated that, even if a very small charge was placed in the bottom part of the drillhole, flyrocks from the collar could achieve a considerable velocity. In these and other PMMA experiments, it was first noticed that, as the gaseous detonation products ventilated through the collar, fracture planes perpendicular to the drillhole axis were initiated. The high-pressure gas penetrates into those cracks, increases the stress concentration factor, and breaks pieces surrounding the collar (Figure 12.4). These pieces are accelerated by the ventilating gas and thrown considerable distances.

12.2. Influence of the Specific Charge

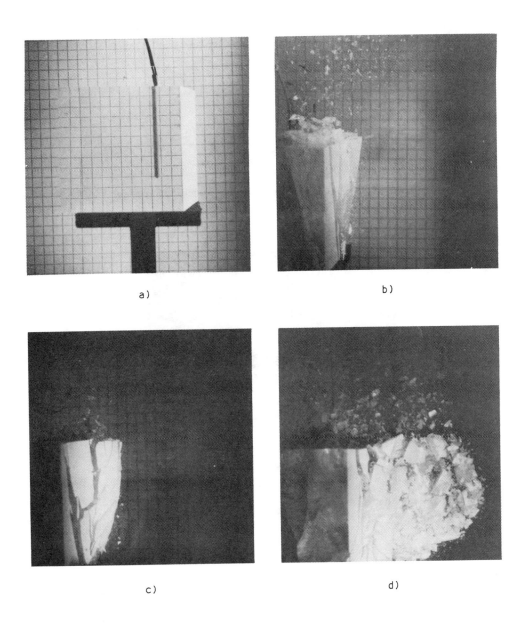

Figure 12.2. PMMA block with dimensions 100×600×60 mm. Burden is 15 mm and the borehole diameter is 2.0, 2.8, and 4.0 mm in the pictures b, c, and d, respectively. Time after ignition is 210 µs.

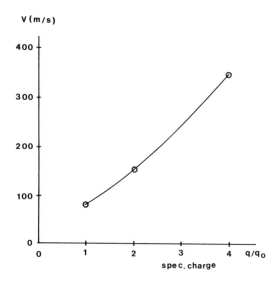

Figure 12.3. Relationship between throw velocity and specific charge for PMMA experiments.

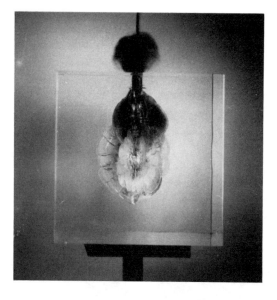

Figure 12.4. Radial cracks caused by ventilating gas can create high-velocity flyrocks.

Holmberg and G. Persson [1976] studied the flyrocks in field experiments with high-speed cameras. They showed that most of the collar flyrocks are thrown in a direction following the drillhole axis. Figure 12.5 shows collar flyrocks from hole diameters of 25, 38, and 76 mm. These experiments also confirmed that the scatter of the angle of throw increases as the unloaded hole length decreases.

Larger-scale field experiments based on the model studies were carried out to investigate how the throw length is affected by the specific charge in rock blasting.

12.2. Influence of the Specific Charge

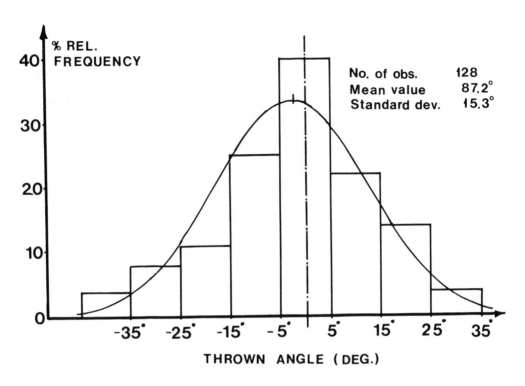

Figure 12.5. Histogram describing the thrown angle of flyrocks relative to the borehole axis.

Ladegaard-Pedersen and A. Persson [1973] conducted a series of bench blast tests with a single hole in rock boulders. The drillhole diameter was 25 mm (1 in). High-speed photography was used to study the movement of the rock fragments. The experimental arrangement is shown in Figure 12.6. After each shot, the distance from the shot hole and the angles of the flyrocks were determined. The maximum throw as determined in these tests is plotted against the specific charge in Figure 12.7. As can be seen from the diagram, it is possible to avoid flyrock if the charge concentration is made smaller than a critical value (in this case, 0.2 kg/m^3). However, it will hardly be possible to carry out practical rock blasting at such a low specific charge since there will always be different rock strengths, variations in burden, faults in the rock, and so on, which cannot be predicted and which would undoubtedly increase the risk of a missed round (no breakage). Therefore, the specific charge must be increased beyond the critical charge size in order to avoid missed rounds or unacceptably large rock fragment size.

Experiments in model-scale and larger field experiments reveal that top-initiated holes give greater crater depths. Model experiments in PMMA indicate that the crack lengths close to the bottom part of the charge are longer if bottom initiation is used. Also, better use is made of the expansion energy for breaking the toe when bottom initiation is used, as a longer duration pressure pulse is achieved.

Figure 12.6. Experimental arrangement to study flyrocks.

12.3 Theoretical Relations Between Flyrock and Drillhole Diameter

The theory for predicting flyrocks from blasting operations in hard competent rock has been established in an excellent way by Lundborg [1974].

Figure 12.8 shows a number of boulders of different size surrounding a spherical charge. The impulse density i delivered from a spherical charge may be written

$$\int p\,dt = i \sim \frac{W^{2/3}}{R} \qquad (12.1)$$

where p is the pressure, W is the charge weight, and R the distance from the charge.

For a constant value of the scaled distance $r = R/W^{1/3}$, we get

$$i \sim W^{1/3} \sim d \qquad (12.2)$$

where d is the charge diameter.

Equation 12.2 shows that the impulse density is proportional to the charge diameter when R is constant. It may easily be shown that this also holds for a cylindrical charge.

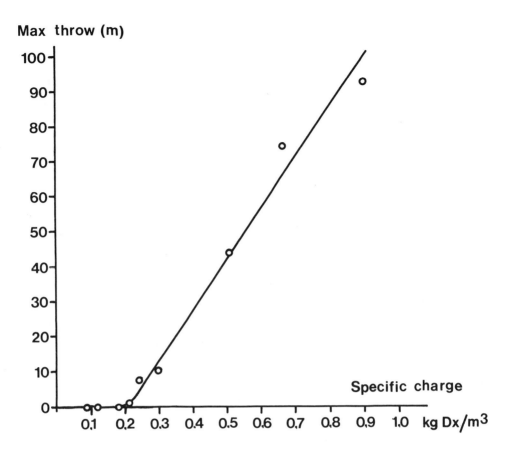

Figure 12.7. Maximum throw length as a function of specific charge.

The force F on the boulders is proportional to the impulse density and the projected area of the boulders; that is,

$$F \sim i\phi^2 \qquad (12.3)$$

where ϕ is the diameter of a boulder.

The total impulse I of a boulder then is

$$I = \frac{\pi\phi^2 i}{4}. \qquad (12.4)$$

It is well known that the impulse may also be written

$$I = mv \qquad (12.5)$$

where m is the mass and v the velocity of the boulder. This yields

$$I = \frac{\pi\phi^2 i}{4} = \frac{4\pi\rho\phi^3 v}{3 \times 8} \qquad (12.6)$$

where ρ is the density of the rock, or

$$i = \frac{2\phi\rho v}{3} \qquad (12.7)$$

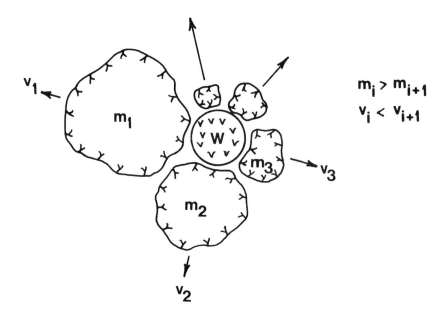

Figure 12.8. Spherical charge, W, surrounded by stones of varying size.

and, from Equation 12.2,
$$d = k\phi\rho v \qquad (12.8)$$
where k is a constant.

To determine the constant k, measurements of $\phi\rho v$ were made for different d values. By doing so, the relation
$$\frac{\phi\rho v}{2600} = 10\,d \qquad (12.9)$$
was obtained where ρ is the density of the rock in kg/m^3 (2600 kg/m^3 is the average density of granite), ϕ is the boulder diameter in meters, v is the velocity in m/s, and d is the drillhole diameter in inches.

12.4 Measurements of Flyrock Velocity and Maximum Throw

To investigate the validity of Equation 12.9 and to determine the factor of proportionality, several blasted rounds were photographed and the flyrock velocities measured. In a number of these rounds, the maximum distance of throw and the diameter of each flyrock boulder were also measured.

In most open pit mining, the trend nowadays is to use large drillhole diameters of $d = 178$–381 mm (7–15 in) and a bench height of 10 to 15 m. These conditions result in burdens B in the range 7 to 10 m with the hole spacing $S = 1.25B$. In bench blasting with hole diameters up to 63.5 mm (2.5 in), the unloaded hole length can be kept equal to the practical burden, which is about 35 times the drillhole diameter. In blasting situations where the ratio between the hole depth and the hole diameter is low (such as in large hole diameter, low bench open pit mining), it is not possible — from an economic point of view — to keep the stemming length equal to the burden. This would cause unsatisfactory fragmentation and raise the costs because of secondary blasting. This is

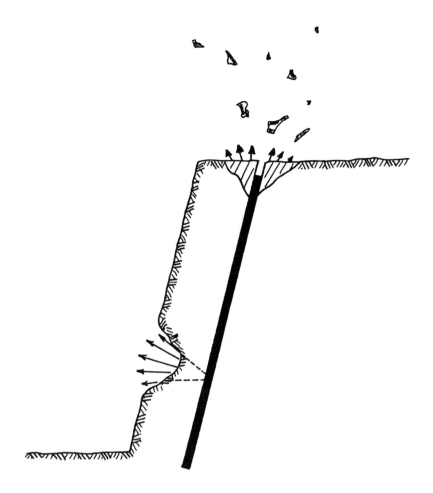

Figure 12.9. Crater effects that could cause flyrocks in bench blasting.

unfortunate since a charge placed too near a free surface results in flyrock throw from the upper end of the drillhole. In granite, under no circumstances should the distance from the charge to the free surface be less than 50% of the maximum burden if the cratering effect is to be suppressed. One also had to remember that if several benches have been blasted on top of each other, the bottom part of the previously blasted bench forms a layer of heavily fragmented rock at the top of the bench underneath.

A good rule of thumb is to keep the stemmed hole length h_s as follows:

$$h_s \geq B \qquad \text{when } d \leq 2.5 \text{ in.} \qquad (12.10)$$
$$h_s \geq B^{1/2} + 1 \qquad \text{when } d > 2.5 \text{ in.} \qquad (12.11)$$

In a weak fractured rock it is not unusual to see stemming lengths on the order of 5 to 8 m for hole diameters of 254 to 330 mm (10 to 13 in).

Most of the flyrock comes from the surface, but it is also possible to obtain crater effects from the front of a round if the burden is reduced because of inhomogeneities. Figure 12.9 shows examples of how crater effects could occur in bench blasting.

From these values, the flyrock velocities were calculated assuming initial throw angles of 45° as giving a maximum throw.

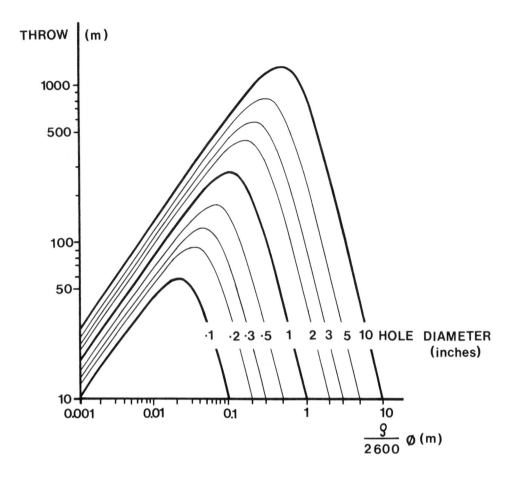

Figure 12.10. Calculated maximum throw vs. boulder size with borehole diameter as a parameter.

12.4.1 Calculation of the Throw

From the previous section, it is known that $\phi\rho v$ is a function of d. As the density of the rock is known, we may choose a value of ρ which yields a given velocity, v. Using a computer, we can then estimate the air drag and calculate maximum throw as a function of stone diameter with d as a parameter. These results are given in Figure 12.10. From the calculations, the maximum throw is

$$R_{max} = 260\, d^{2/3}. \tag{12.12}$$

The diameter of these boulders is

$$\phi = 0.1 d^{2/3} \tag{12.13}$$

where ϕ is in meters and d in inches.

The formulae are valid for the specific charge used when blasting in hard competent rock like gneiss, granite, pegmatite, crystalline limestone, hard iron ore, etc.

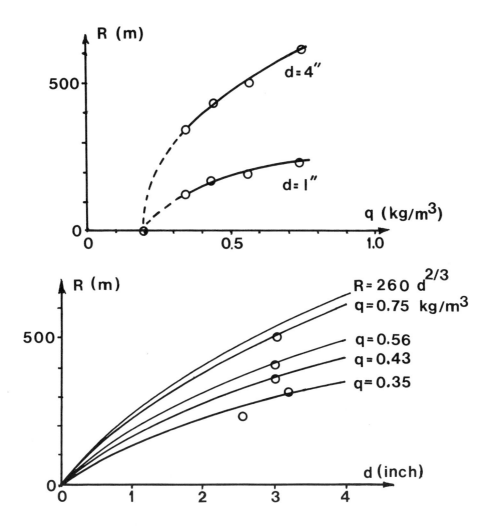

Figure 12.11. Throw length as a function of specific charge (a) and as a function of hole diameter (b).

Lundborg [1981] indicates that for weaker rocks where a lower specific charge is used, the maximum thrown length is reduced (Figure 12.11a and b).

From Figure 12.10 and Equations 12.12 and 12.13, it is shown that some boulders are thrown quite far. For example, with 254 mm (10 in) drillholes, the maximum throw is about 1200 m with the stone diameter of half a meter. This means that very large areas must be evacuated to avoid accidents. People within this area must be protected against flyrock, no matter what the cost. Since the enclosed area is great, the probability of hitting a single spot is small, but this cannot be used as an excuse not to protect people in the area. An alternative is to reduce the flyrock throw distance.

One way to reduce the throw distance is to cover the rounds. This is routinely done for small drillhole diameter rounds when blasting in city center conditions. The recommended practice is that the weight of the cover should be equal to the weight of the round blasted. This is clearly impossible for large blasts using large drillhole diameters.

12.4.2 Precautions to Avoid Flyrock

For small hole diameter blasting in densely populated areas or in such places where flyrock can cause damage to people or other objects, special precautions have to be considered.

12.4.3 Effect of Rock Structure

Even if the geology is similar within a single blasting site (e.g., a quarry), the rock structure may vary considerably. Fissures, joints, and weakness planes are never the same from point to point. Due to weathering, the surface rock is usually more nonisotropic than the deeper rock mass. Therefore, it is extremely important to map fissures and joints that can cause problems during drilling and blasting. Before any drilling from the ground surface takes place, the surface must be cleaned to make inspection more reliable. The results of the inspection must be kept in mind when the drilling and ignition pattern are designed. Previous excavation can give significant information about the rock structure; for example, horizontal joints, which cannot be visually detected from the surface, may be traced from the remaining rock face of a previous round.

12.4.4 Drilling

When planning the blast, the blasting engineer must remember that a larger hole diameter results in larger charges and often a higher specific charge in the round. As previously shown, this increases the risk of flyrock. To minimize the risk of flyrock, drillholes must be set out as accurately as possible, and special instruments for alignment of the holes should be used to prevent over-confined holes. It is easy to overestimate the burden if the terrain is hilly. The burden the distance between the free face and the drillhole axis (measured *perpendicular* to the drillhole axis). Trench blasting and blasting where the bench height is small compared to the burden are commonly regarded as the most difficult blasts when it comes to preventing flyrock. The confinement and the specific charge is high, and the charge is sometimes placed close to the surface.

12.4.5 Charging and Initiation

It is important that the staff carrying out the charging work follow exactly the instructions given for charging each drillhole. The bottom charge must be of correct length and the column charge must have the correct charge concentration. Stemming should be used, and the stemming should be of proper length.

In field experiments carried out by Holmberg and Persson [1976] and Holmberg [1978], the effect of the stemmed and unstemmed length was investigated. Intuition says that, if the stemming length or the unloaded hole length is increased, the throw would be reduced. This has also been assumed by other authors, e.g., Roth [1979]. However, it was surprising to learn from the field experiments that, at first, the throw length R increased with increasing stemming length or unloaded hole length h_s. A maximum was achieved for $h_s/d \approx 10$; and then the throw length decreased until a critical depth was achieved where no flyrock at all was developed (Figure 12.12).

There is a simple explanation for this phenomenon: when the explosive column is close to the surface, the detonation and its shock wave are sufficient to fragment the

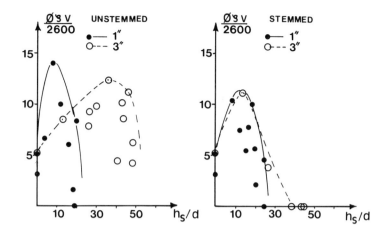

Figure 12.12. The figure shows the relation between $\phi \rho d$ which is proportional to the thrown length and the unloaded length (a); stemmed hole (b).

rock at the collar. If the rock is very finely fragmented, the small fragments will not be thrown very far because they have a very low mass and are effectively stopped by the air drag.

If the charge is very far away from the surface, larger pieces of rock will form the probable flyrocks. These have a large mass and, for the same impulse, their velocity and length of throw will consequently be short.

Somewhere between these two extremes, the charge is at the distance from the surface where flyrocks formed at the collar have just the right size to be thrown the longest distance.

Expressing the critical depth where no flyrocks are developed as h_s/d, one would perhaps expect, as Langefors and Kihlström [1963] and Livingston [1962] have pointed out, that the ratio h_s/d should be unaffected by the hole diameter d. However, field experiments showed that, for over-confined holes where the burden does not break, the ratio h_s/d increased with increasing hole diameter (Figure 12.13).

Figure 12.14 shows photographs taken before and after one of the field shots. Table 12.1 lists data for the 3 inch holes that were shot.

If bulk explosives are used, the charge weight pumped or poured into each hole must be checked to avoid filling too much explosive into cavities, open cracks, and other subsurface openings in the rock, since this could radically increase the charge concentration. If such cavities or openings in the rock do exist, the drillhole should be lined with a rigid liner. Alternatively liquid explosives should be packaged in strong plastic bags.

In designing the initiation pattern, one should avoid using interval times higher than 100 ms between adjacent drillholes. For longer time intervals, blasted rock from an adjacent, previously-ignited drillhole will have moved too far to cover the throw of rock from the latest ignited hole. A shorter time interval will also decrease the strain in the cover material and prevent the cover from moving away before the adjacent hole is ignited.

Top initiating should be avoided. Field observations confirm that the probability for collar flyrock is greater if top initiation is used than if bottom initiation is used. The initiation pattern must also be designed in a proper manner so that each hole only has

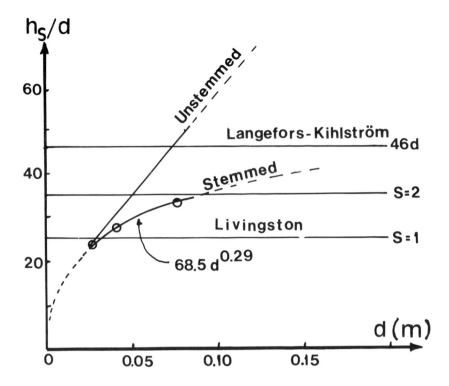

Figure 12.13. Critical depths where no flyrocks occurred relative to the hole diameter according to some authors.

its own designated burden to work against. Precautions must be taken to prevent over-confined holes. Partial ignition failures, where a hole in a rear row is ignited without the rows in front being ignited, invariably results in greatly increased flyrock problems.

12.4.6 Covering

Covering of the round must be carried out carefully and correctly. There are two types of covering in common use:

1. Fragment protective covering
2. Heavy covering

The fragment protective covering must be able to allow the detonation product gas to pass through it and is only intended to prevent small stones and loose material laying on top of the round from becoming airborne. The heavy covering is intended to hold the round together so that no part of the loose rock can escape when the round is fired. Generally, the lightest covering is placed uppermost and the heaviest beneath.

When heavy covering mats are used, one should begin covering the round behind the back of the round and work forward, overlapping the mats (Figure 12.15). When the round is fired, there is a rippling of the mats, and they therefore do not follow the round forward — this would occur if the cover were placed in the opposite order. If inclined holes are used, it can be advantageous to secure the mats so they cannot slide away.

Figure 12.14. The rock surface before and after a shot in which the charge was overconfined (the burden did not break out) where flyrock was produced for a large h_s/d value ($d = 38$ mm, unstemmed hole length 0.4 m).

Table 12.1. Data for 3" test shots. A dynamite Dynamex B was used in the field experiment.

Test Shot	Hole depth (m)	Unstemmed length (m) h_s	Stemmed length (m) h_s	h_s/d	Charge weight (kg) W	$\rho\phi v/2600$	Degree of packing (kg/m³)
2-1-74	4.83	2.28		30.0	16.2	9.8	1400
2-2-74	4.79	3.05		40.1	10.8	4.3	1368
2-3-74	4.80	2.75		36.1	12.6	12.2	1355
2-4-74	4.80	3.30		43.4	9.0	10.1	1323
2-5-75	4.80	3.52		46.3	7.2	11.2	1240
2-6-74	4.82	3.66		48.2	7.2	4.2	1368
2-7-74	4.70	3.06		48.3	9.0	6.2	1210
2-8-75	4.80	3.31		43.6	7.2	8.7	1065
2-9-75	4.80		3.36	44.2	7.2	0.0	1102
2-10-75	4.80		3.30	43.4	7.2	0.0	1058
2-11-75	4.80		2.95	38.2	9.0	0.0	1072
2-1-76	4.86	2.00		26.3	14.4	7.9	1110
2-2-76	4.86		2.00	26.3	14.4	3.8	1110
2-3-76	4.86	1.00		13.2	18.5	8.5	1054
2-4-76	4.86		1.00	13.2	19.8	11.2	1131
2-5-76	4.86	2.00		26.3	14.4	9.3	1110
2-6-76	4.86	0.00	0.00	0.0	25.2	5.1	1143

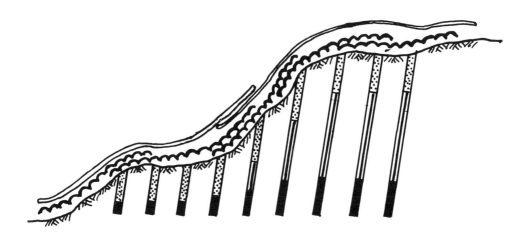

Figure 12.15. A proper covering should be laid from the back of the round and forward.

The material used for anchoring the mats should have a smaller strength than that of the covering mats in order to avoid damage to the mats.

Suitable heavy covering material include:
- Blasting mats made from tire rubber
- Blasting mats made from logs

Blasting mats made from tire rubber, fabricated from strips of used tires held together by steel wires, are used increasingly and perform in an excellent way (Figure 12.16). The mats, each of which cover a surface area of about 10 m², have a considerable weight; a front-end loader or a digger must be used for covering and uncovering the round.

Figure 12.16. Bench blasting in the city of Stockholm. A covering of rubber blasting mats was used.

Suitable splinter protective material include:

- Industrial felt mats
- Tarpaulins
- Closed-meshed mats

In bench blasting, a remaining muck pile with a height equal to the bottom charge can be left at the toe to act as a cover for the bottom charge. The hole deviation is greatest at the bottom of the holes; therefore, the risk for flyrock is greatest from the bottom part of the bench front.

For tunneling, a curtain made of wire, logs, and industrial felt is most suitable for covering the entrance during the blast.

Experiments have been carried out by the Swedish Detonic Research Foundation (SveDeFo) to investigate the possibility of using a light cover which can be placed and anchored without the use of any machines. A high-strength Kevlar canvas with a net which could be anchored to the ground by nylon lines was used. However, high-velocity, sharp-edged rock fragments were found to be able to rip the canvas and escaped with a considerable velocity, forming dangerous flyrock.

Chapter 13

Ground Vibrations

13.1 Introduction

Ground vibrations are an integral part of the process of rock blasting. The sudden acceleration of the rock by the detonation gas pressure acting on the drillhole walls induces dynamic stresses in the surrounding rock mass. This sets up a wave motion in the ground much like the motion in a bowl of jelly when disturbed by the action of a spoon. The wave motion spreads concentrically from the blasting site, particularly along the ground surface, and is therefore attenuated, since its fixed energy is spread over a greater and greater mass of material as it moves away from its origin. Even though it attenuates with distance, the motion from a large blast can be perceived from far away. Humans and animals alike react to the faint motion with reflexes of alert and fear, which have their origin in the distant past, when the need to react by flight to the approach of a bigger animal was a condition for survival.

In the near region, the ground vibration can cause damage to buildings and other man-made structures by causing dynamic stresses that exceed the building material's strength, in much the same way as it happens in the fracturing of the rock material itself. A building, being much less rigid than the solid rock mass, can be damaged even a long distance away from a carelessly designed blast.

The ground vibration effects of blasting operations on building structures and human beings need to be predicted, monitored, and controlled by the blasting engineer as part of optimizing the job. Doing this right is crucial for the economy of most rock blasting operations. Being too conservative about the ground vibration levels in planning the blasting work can increase the blasting cost to such an extent that the contract is lost, and the owner chooses another contractor for the job. Being too liberal might result in damaging nearby buildings and the claims and legal expenses for this might change the profit balance from positive to negative.

This chapter will introduce the basic concepts about ground vibrations needed for the successful planning of blasting operations. The blasting engineer or student who requires more detail should read specialized papers on the topic of ground vibrations (such as those by Richart *et al.* [1970], Bollinger [1980], and Dowding [1985]) each of which cover this subject in its own excellent way.

The material presented in this chapter is taken, in part, from a publication by Holmberg *et al.* [1984].

13.2 Definitions and Basic Terms

Vibration and shock can be described mathematically as the way a body or particle moves in time. The time-dependent motion can be given alternatively in terms of displacement, particle velocity, or acceleration. Which alternative to select is dictated not only by requirements of the problem at hand, but also by the type of vibration measurement equipment used. Conversion between displacement, particle velocity, and acceleration can be done by integration or differentiation with respect to time; however, differentiation twice of a poor displacement record is a very difficult and risky undertaking.

13.2.1 Types of Vibrations

The simplest form of vibration is the simple harmonic movement, often called a *sine* (sinusoidal) vibration.

$$x(t) = A\sin(\omega t + \phi) \tag{13.1}$$

where A is the displacement amplitude, ω is the angular frequency, t is the time, and ϕ is the phase angle.

For many practical applications, where the vibration can be approximated by a single sine wave, the phase angle has little significance. The vibration can then be characterized by two parameters: the amplitude and the frequency. In dealing with wave interference, however, the phase angles of the interacting waves of course are of vital importance, and for many other applications, it is important to know the vibration velocities in the three principal directions.

Referring to Figure 13.1, it can be shown that for the sinusoidal wave, the conversion between velocity and displacement (amplitude) by integration on the one hand, and between velocity and acceleration by derivation on the other, is equivalent to the division or multiplication of the velocity by ω, respectively, and the displacement in time by a phase angle equal to $\pi/2$.

13.2.2 Parameters Describing Vibrations

There is a great variety of terms used to describe the intensity of the vibration and its variation in time. Different terms are used for different types of vibration depending on their nature (e.g., if they are continuous vibrations or transient vibrations). In many cases, a satisfactory description of a particular vibration can be given by several different sets of parameters. An important part in the treatment of many vibration problems is to find a set of parameters suitable to the problem in question. It is often a matter of finding the sine function that is equivalent to the actual vibration. The effect of the sinusoidal vibration on building elements or human beings can be determined by calculation or by experiment. The characterization can be done in the time domain (description of lapse) or in the frequency domain (frequency spectrum). The latter is perhaps the more common.

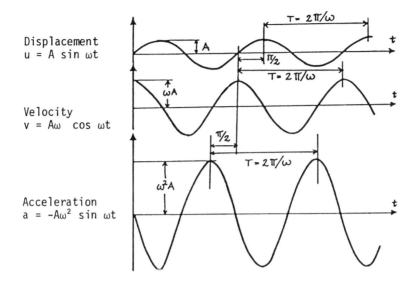

A = displacement amplitude (m)
T = period (s)
F = frequency $\left(=\dfrac{1}{T}=\dfrac{\omega}{2\pi}\right)$ (s^{-1})
ω = angular frequency (rad/s)

Figure 13.1. Harmonic oscillation.

13.2.3 Parameters

For a signal with a given particle velocity, acceleration, or displacement, we use the following definitions:

The peak value $\hat{x}$ is the maximum absolute value of the vibration during a time interval. Additionally, the sign, indicating whether the peak values are positive or negative needs to be taken into account.

The R.M.S.-value x_{eff} is defined as

$$x_{\text{eff}} = \left[\int_{t_1}^{t_2} \frac{x^2 dt}{t_2 - t_1}\right]^{1/2}. \tag{13.2}$$

For a pure sine wave, the $x_{\text{eff}} = \dfrac{\hat{x}}{\sqrt{2}}$.

The *mean value* $\bar{x}$ over the time interval t_1 to t_2 is defined as

$$\bar{x} = \int_{t_1}^{t_2} \frac{x \, dt}{t_2 - t_1}. \tag{13.3}$$

The mean value for vibrations is often defined to be 0, that is, $\bar{x} = 0$.

The *standard deviation* σ for the time interval t_1 to t_2 is defined as

$$\sigma = \left[\int_{t_1}^{t_2} \frac{(x - \bar{x})^2 dt}{t_2 - t_1}\right]^{1/2}. \tag{13.4}$$

If the mean value $\bar{x} = 0$, then the standard deviation is equivalent to the R.M.S. The *undirected mean value* $|\bar{x}|$ over the time interval t_1 to t_2 is defined as

$$|\bar{x}| = \int_{t_1}^{t_2} \frac{|x| dt}{t_2 - t_1}. \tag{13.5}$$

13.2.4 Different Types of Wave Motion

The best-known wave motion may be the wave motion in water. If an object is thrown into water, the waves propagate symmetrically outward, and from experience we know that the further away from their origin these waves propagate, the less the water is disturbed — the wave motion is dampened. The waves we can see on the water surface represent a transmission of energy from one point in the water to another. The main difference between water waves and ground vibration waves is that the water waves are gravity waves where the driving force is gravity, whereas the ground waves are elastic stress waves where the driving force is the elastic stress set up by deformations within the strength limits of the rock material. In construction work, in addition to the ground vibrations generated by the blasting, activities such as driving, piling, and trucks or trains traveling nearby also generate ground vibrations of similar amplitudes.

To understand and predict what will happen to structures at a given distance from an excitation source, there must be a fundamental knowledge of existing types of waves, how the wave motion is propagated, and how the structure in question is affected.

In gases and liquids that cannot support shear stress, only one type of stress wave motion is possible: the bulk compressional wave. This type of wave propagation can also arise in ground and other solid materials, where it is often called a *shock wave* (P-wave). The compressional wave is a longitudinal wave, e.g., the particle motion is parallel to the direction of the wave propagation.

In solid elastic materials, *shear waves* (S-waves) also exist, e.g., a wave motion of distortional movement. Shear waves are generally transversal waves, e.g., the particles move transversely to the direction of the wave propagation. P- and S-waves are both volume waves since they propagate in a three-dimensional space.

At interfaces between different media (for instance, at interfaces between ground and air, between ground and water, or between layers of ground of very different elastic characteristics), different types of surface waves are developed. In rock blasting, the most important surface wave is the Rayleigh-wave or R-wave (Figure 13.2). In this wave, in contrast to water waves, the particles at the surface describe a retrograde elliptical motion. The vertical component of the particle motion has its maximum just below the surface, but thereafter diminishes relatively rapidly with the depth.

The different wave types are characterized among others by different wave propagation velocities; the P-wave has the highest velocity, the S-wave is often considerably slower, and the R-wave in turn is still a little slower than the S-wave. The wave velocity is a function of the elastic properties of the material. At long distances from the

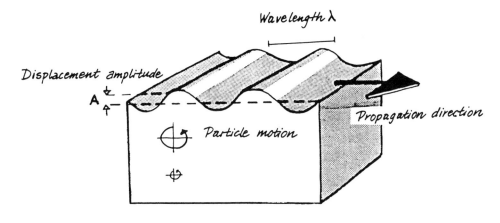

Figure 13.2. The Rayleigh-wave.

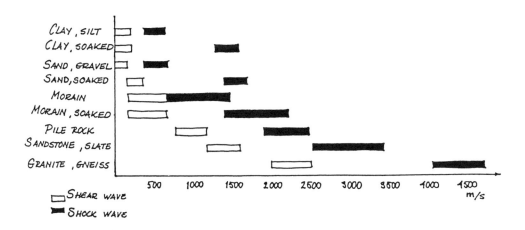

Figure 13.3. Typical wave velocities for P- and S-waves [IVA, 1979].

source, the R-wave is often the only distinguishable wave due to its much lower attenuation characteristics. Some typical wave velocities for P- and S-waves are shown in Figure 13.3.

The wavelength for a periodic wave is the distance measured transversely to the wave front between two successive points with the same motion phase (cf. period). For a wave with frequency f and propagation velocity c, the wavelength λ is

$$\lambda = \frac{c}{f}. \qquad (13.6)$$

The wavelength is a parameter of great significance for the influence of vibrations on different structures. Unfortunately, in many practically important blast vibration situations, the wavelength is of the order of 12 to 30 m, i.e., similar to the dimension of many buildings. This is a disadvantageous wavelength range. If the wavelength were much shorter than the building, the displacements would be too small to cause damage, unless the vibration velocity is very high. If it were much longer than the building, the building would essentially be displaced up and down with the wave without damage, at least for moderate vibration velocities. When the wavelength matches the length of the building, the building has to flex and bend with the undulation of the ground, which is liable to cause damage already at moderately high vibration velocities. The reason why this wavelength range dominates is simple: a vibration source does not generally generate vibrations with wavelengths greater than its own dimensions, and the characteristic size of many blasts is in the above-mentioned size range. Waves with shorter wavelengths are generated by the detonation of the charge, but these attenuate rapidly with increasing distance.

At some distance from the source just mentioned, this wavelength range will dominate — whether it is generated by blasting, other construction activities, or by road or rail traffic. At very long distances, the Rayleigh wave in rock can have a wavelength longer than 100 m.

13.2.5 Wave Propagation Velocities

The following equations are valid for the type of wave indicated:

P-waves:
$$c_P = \left[\frac{E(1-\nu)}{\rho(1-2\nu)(1+\nu)}\right]^{1/2} \quad (\text{m/s}) \tag{13.7}$$

S-waves:
$$c_S = \left(\frac{G}{\rho}\right)^{1/2} = \left[\frac{E}{2\rho(1+\nu)}\right]^{1/2} \quad (\text{m/s}) \tag{13.8}$$

Rayleigh-waves:
$$c_R \approx c_S \frac{0.86 + 1.14\nu}{1+\nu} \quad (\text{m/s}) \tag{13.9}$$

where E is the modulus of elasticity (Pa), G is the shear modulus of the material (Pa), ρ is density of the material (kg/m^3), and ν is Poisson's ratio.

13.3 Damage to Buildings and the Reasons for Annoyance

13.3.1 Reactions to Vibrations

During thousands of years, man has developed a perception system well adapted to earlier environments. To be warned in time for landslides, the approach of large animals, and the like, the ability to perceive minute vibrations was of vital importance. As a result, the sensitivity of the human perceptive system to vibration is very high, similar to the extremely high sensitivity of the ear to sound waves. What was in former times an advantage is nowadays perhaps a source of problems, particularly to the blasting engineer. Vibrations are still to a certain extent interpreted as warning signals, even though it may be obvious as soon as the source is identified that no danger exists. Vibrations from sources which are difficult to identify, such as from blasts some distance away in a city, are unconsciously registered by humans as very disturbing — especially if there has been no warning beforehand (Figure 13.4).

Figure 13.4. Human response to vibrations.

Ground vibrations often occur together with other disturbances, such as noise and infrasonics, which can intensify or mask their effects. The human's response depends on both genetic conditions and education. A certain dull sound together with vibrations can remind of an earthquake and cause great fear within certain individuals. Noise and vibrations from a passing bus can often be rapidly identified for what they are, lessening their effect when compared with similar vibrations from an undetected source.

Identification of the excitation source often produces a reassuring effect, but there are certain exceptions. If the source is not acceptable as a reasonable or beneficial activity in the neighborhood, considerable irritation may occur; for instance, over the sound from a motor bike in a residential area where such noises are typically forbidden. Each time the motor bike is heard, the disturbance grows (the irritation spiral). The long-term effects of disturbances thus depend as much on the acceptability of the source as on the actual measurable magnitude of the disturbance. A vibration can be given a low threshold value by a person, even if it does not directly negatively influence the person. He might be afraid that the vibrations might cause cracking in his house, wake up his sleeping child, ... (*the expectation effect*). It is therefore important to provide information about how vibrations affect buildings and other structures, in order to break the irritation spiral and decrease the expectation of damage.

Conversely, when the vibration is known to be the result of an activity beneficial to the beholder, he or she often has a surprisingly high tolerance level to such vibrations. This is a well-known phenomenon, easily observable in many small mining communities, the economic well-being of which may be entirely dependent on the mine as the one main source of steady employment in town.

A sleeping person is also influenced by earlier learning. A person who is used to environmental vibrations sleeps much better in a disturbed environment than one who is used to a silent environment. However, even if a person does not actually wake up, or remember the next morning that there was a vibration, it is possible that the vibrations have an unconscious effect. It is not very well known as yet at which vibration level persons accustomed to disturbance in their environment become disturbed.

Very heavy vibrations can of course be directly harmful to the human body, but such levels seldom occur in buildings. The greatest problems facing the blasting engineer are, therefore, the disturbance effect and the expectation effect; the latter especially with regard to buildings belonging to the person perceiving the disturbance.

344   Chapter 13. Ground Vibrations

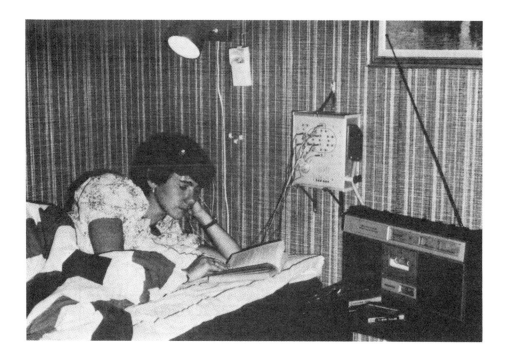

Figure 13.5. Interior of a bedroom where bed is connected to a shake table.

Reiher and Meiser [1931] investigated the influence of the intensity of vibrations on the degree of human annoyance. However, the vibrations they used were not of a transient nature like the one emitted from blasting operations. Instead, vertical sinusoidal vibrations with a duration of 500 s were used. The test group of 15 persons reported that the vibrations were "barely noticeable" when the peak particle velocity was $\approx$ 0.02 mm/s ($f = 3$–25 Hz), and "uncomfortable" when the velocity was 0.5 mm/s at $f = 30$ Hz and 50 mm/s at $f = 5$ Hz.

In another study, Wiss and Parmelee [1974] used vertical transient vibrations (damped sinusoidal) with a duration of 5 s. The mean value for "strongly perceptible" was reported to be near 25 mm/s for the frequencies 2.5 to 25 Hz.

It is obvious from these two tests, and others, that the perceptiveness of vibrations is reduced when the exposure time decreases.

Arnberg [1983] carried out sleep studies with the help of a vibration table upon which a bed was mounted. As the test persons slept in the bedroom, vibration recordings digitized earlier controlled the behavior of the vibration table. Various traffic intensity and vibration magnitudes were simulated and the human response was measured through EEG monitoring, and morning interviews of how the person in question felt : if he or she was tired, etc. The average test subject reported that she felt disturbed and tired at a particle velocity of 5 mm/sec. Figure 13.5 shows the interior of the bedroom. The bed itself was mounted on a shake table.

Most likely, it will be impossible to establish a vibration level where nobody will complain. There is always some percentage of persons in a population who will complain no matter how small the disturbance is. People are much more tolerant if the exposure time is small, if the duration time of the particular job (construction work, mining activity, ...) is short, or if they are properly informed of why they must be exposed to the vibrations how the vibrations will affect them and their property and they are informed of the exact time when they will be exposed to the vibration.

International standards are being prepared where the levels for acceptable vibrations lie immediately above the threshold at which they are possible to perceive. Some of the threshold values are so low that if the standard comes through, they will strongly affect the mining and construction industry. However, the future will show to what extent these international standards will influence blasting in practice.

13.3.2 The Origin of Damage to Buildings

The connection between vibration and damage to buildings is a complicated one for many reasons. Buildings are constructed in many different ways: some are more solidly built than others, and they have different dimensions, materials, methods of construction, and types of foundation. Moreover, the intensity, type, frequency range, and wavelength of the vibrations and the direction of incidence of the wave front relative to the main axis of the building structure all play an important part in the origin of damage.

However, not even all these factors are sufficient to judge the risk for vibration damage. A most important factor is the static stress influencing the building elements, and how much the original values have been increased by ground settlement, moisture variations, and temperature variations. In extreme cases very small vibrations can be the triggering cause of significant damage, almost entirely due to a static stress condition that existed long before the vibration arrived.

To systematize the problem area to some extent, building damage can be divided into three general categories:

1. Direct vibration damage
2. Accelerated ageing
3. Indirect vibration damage

Direct vibration damage is generated directly by vibrations in a construction previously undamaged and unexposed to prior abnormal states of stress.

Buildings not founded on solid rock are more or less damaged over time by settlement in the foundations. Dynamic stresses from vibrations can accelerate the development of such damage. This is often called *accelerated ageing*. It should be remembered that even normal building use gives rise to stresses as people walk, run, dance, move furniture, and slam doors. Wind, snow loads, and temperature variations are perhaps the greatest environmental factors contributing to the stress in buildings.

In special cases, vibrations can cause settlement that can later result in damage to buildings. This is an example of indirect vibration damage in buildings.

13.3.3 Primary Damage Criteria

The relation between applied stress and strain, and the strength of the building elements, determines whether a building will remain undamaged. If the stress induced by the vibration exceeds the ultimate strength, damage will occur.

When dealing with vibrations, it must be remembered that the stresses are composed of a static part and a dynamic part and that their sum is important. The dynamic stress is initiated directly by the vibrations, while the static stress is determined by applied loads and prior deformations of the construction, most often acting over a long period of time.

For practical use in connection with traffic and construction activities where a large number of neighboring sites may be exposed to disturbances of short duration, methods for calculating stresses involving strain measurements or vibration measurements are generally too complicated and expensive. Instead, standard values for permitted vibrations based on experience are used. These values take into account the type and condition of the building, the state of the foundation, and the types of vibrations. The standard values used today cannot guarantee that new damages will not occur (even if the vibration level never exceeds the standard value); yet based on a great deal of experience, damage will occur only in a limited number of cases.

For many types of blasting vibrations, a set of reliable standard values has been established. The standard values refer only to direct vibration damage. Accelerated ageing can occur even at very small vibration levels, and it is impossible in practice to use standard values which would exclude this type of damage. To attempt this would be equivalent to stating that no disturbance would be acceptable [Holmberg et al., 1981].

For other types of vibration sources, there are few, generally accepted standard values because the number of well-documented damage cases is very limited. Vibrations from traffic, pile driving, sheet piling, and soil compaction seldom reach levels where they can cause direct vibration damage. A reasonable starting point then is to relate the intensity of stress caused by the vibrations from an external source to the inevitable stresses that a building withstands due to the indoors environment and to the effects of weather and variations in the climate.

Stagg et al. [1984] compared the strains produced by blast vibrations with strains produced by daily temperature and humidity changes and strains produced by the normal indoor activities. The experiment showed that the household activities resulted in strains corresponding to blast vibration levels of 22.4 mm/s and the daily environmental changes gave a corresponding blast level equal to 76 mm/s.

13.4 Limiting Vibration Levels for Buildings and Installations

Over the last 30 years, blasting in residential areas has mainly been done in accordance with a certain vibration level less a safety margin in order to avoid building damage. By means of vibration measurement and damage surveys in adjacent buildings, a relationship between the peak particle velocity and the possible risk for damage has been established. This has increased the need for vibration measurement and in turn facilitated the management of blasting work in order to prevent damage. Control of the particle velocity is important, as it has been shown to be directly proportional to the stress to which the building material is exposed. The relationship between particle velocity and stress in an idealized case, when a plane shock wave passes through an infinite elastic medium, can be expressed as

$$\sigma = \epsilon E \qquad (13.10)$$

$$\epsilon = \frac{v}{c} \qquad (13.11)$$

$$\sigma = \frac{vE}{c} \qquad (13.12)$$

where σ is the stress, ϵ is the dilation, E is the modulus of elasticity, v is the particle velocity, and c is the propagation velocity of the seismic wave.

13.4. Limiting Vibration Levels for Buildings and Installations

Table 13.1. Limit values for the vertical particle velocity v (in mm/s) for building damage [Langefors and Kihlström, 1963].

Ground material beneath buildings	Sand, gravel, clay (mm/s)	Moraine, slate, soft limestone (mm/s)	Granite, gneiss, hard limestone, quartzite sandstone, diabase (mm/s)	Results in normal residential area
	18	35	70	No noticeable cracking
	30	55	110	Fine cracks, and fall of plaster (threshold value)
	40	80	160	Cracking
	60	115	230	Serious cracking

Table 13.2. Recommended peak particle velocities in Germany according to DIN 4150 [1975].

Building class	Maximum resultant of the particle velocities v_r (mm/s)	Estimated maximum vertical particle velocity v_z (mm/s)
I Residential buildings, offices and others similarly built in the conventional way and being in normal condition	8	4.8–8
II Stable buildings in normal condition	30	18–30
III Other buildings and historical monuments	4	2.4–44

The formulae indicate that the propagation velocity of the waves is very important; however, for a given underground rock mass on which a given building is founded, the propagation velocity as well as modulus of elasticity in the building material are mainly constant in magnitude. Thus, for a given building, the stress level primarily depends on the particle velocity.

In order to recommend realistic limiting values for buildings, experience with blasting and vibration measurement is necessary. Restrictions (e.g., low particle velocity limiting values) can considerably increase the excavation costs. Therefore, before starting and during the early planning stage of a blasting job, it is important to first carry out visual inspection and risk analysis and then, based upon these results, to judge the sensitivity of the buildings and their foundations to vibrations. Many factors influence the permitted vibration values. Some of the most important are:

- Vibration resistance of the building materials
- General condition of the building
- Duration and character of the vibrations
- Presence of sensitive vibration equipment
- Foundation of the buildings
- Condition of the foundations
- Propagation characteristics of the wave in rock, earth, and building material, respectively
- Replacemest cost/highest likely repair cost

Table 13.1 shows a compilation of recommended limit values used in Swedish rock blasting when judging the risk for damage by ground vibrations in normal residential areas. Normal residential areas indicate houses with foundations and joists of concrete, outer walls of brick, and intermediate partitions of plastered, compact light concrete. Table 13.2 shows German recommended three component resultant peak particle velocity limits according to the DIN 4150 standard [1975].

For old buildings of inferior quality on rock foundations, it is common to reduce the permitted particle velocity from 70 mm/s to 50 mm/s and to 35 mm/s for light concrete buildings. Corresponding adjustments are also made for other types of foundations. Many values exceeding 110 mm/s have also been registered without causing damage to solidly constructed buildings. Heavy concrete constructions founded directly on rock can withstand single blasts up to 150 mm/s. The values indicated in Table 13.1 are empirical and based on more than 100,000 measuremests.

If the limit values indicated in Table 13.1 regarding "no noticeable cracking" are transferred into a tripartite diagram, they will fall on curve 2 in Figure 13.6. However, in the curve the limit value 70 mm/s for buildings founded on rock has been reduced to 50 mm/s. Curve 2 can be said to represent the recommended limit values for normal residential areas. For frequencies exceeding 40 Hz, the particle velocity is the criterion; but at lower frequencies, the displacemest represents the criterion.

The dominant frequency for vibrations transmitted through soft rock, moraine, sand, gravel, and clay is lower than that for vibrations transmitted through hard rock such as granite. This appears from Table 13.1 and curve 2 reflects this for low frequencies where displacement rather than vibration velocity is used as the criterion. Curve 1 in Figure 13.6 represents values at which buildings certainly get damaged, [Langefors and Kihlström, 1963].

In many parts of the world, there are no regulations defining limits to the vibration levels in buildings. The above recommendations can serve as a guideline to prevent or limit building damage to a practical level.

In connection with blasting near telecommunication offices, relay stations, or buildings containing other sensitive equipment such as computers, electron microscopes, scale weighing machines, grinding machines, turbines generators, and the like, consideration should also be given to the acceleration levels generated.

Extensive studies have been carried out by the US Bureau of Mines [Siskind et al., 1980] to review earlier research concerning structural damage from vibrations, to carefully monitor building response, and to make recommendations for what peak particle velocities the blaster will not exceed. The recommendations are given in Figure 13.7.

A maximum peak particle of 0.75 in/sec (19 mm/s) is recommended in the frequency range $f = 4$–12 Hz when gypsum board is used for the interior walls of the structure. A lower peak particle velocity of 0.5 in/sec (12.7 mm/s) is recommended when plaster cracking can occur. The recommended value of 2 in/sec (50.8 mm/s) for $f > 40$ Hz is based on the lowest vibration velocity level at which damage was observed, 2.2 in/s (56 mm/s).

13.5 Guidance Levels for Buildings According to Swedish Standard

For buildings founded on Scandinavian bedrock, the criteria for estimating ground vibrations and their building damage are well established. Evaluation of hundreds of thousands of measurements has given a reliable base which authorities can use to determine the necessary level of restrictions when permits are given for any blasting operation. Depending upon the geology, various limits have been established for different types of structures. The coupling to geology mirrors the considerations that were given to the influence of the frequency as the limits always were given as peak particle velocity only.

13.5. Guidance Levels for Buildings According to Swedish Standard

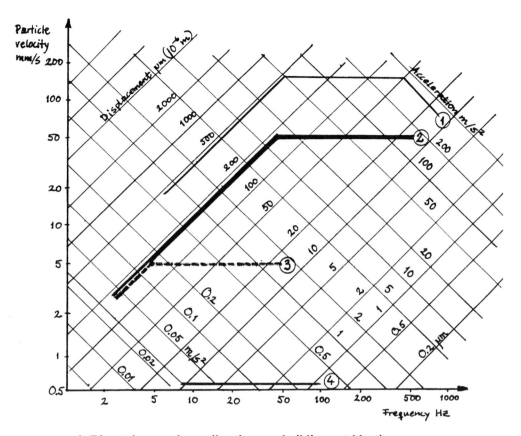

1) Direct damage from vibrations on buildings at blasting.
2) Recommended upper limit for blasting.
3) Recommended upper limit for piling, sheet piling, vibratory compactors, dynamic deep compaction, and traffic.
4) Disturbing vibrations. Human being. Remark: The curves 1,3 refer to vibrations registered in the foundation of the building. The above-standard values (2, 3) indicate recommended limit values for normal circumstances. The limit values always have to be complemented with expert judgments on the condition and the foundation of the building.

Figure 13.6. Criteria for damage and recommendations.

However, over the years, it was seen that the authorities successively lowered the limits and this made it very difficult for the blasters to keep the blasting operations cost effective. There was a trend — from the ongoing standardization work on vibration limits in Europe, from national standards from other countries in Europe with distinctively different geology and building standards, and from discussions concerning human annoyance — to apply even lower limits. The industry made an effort to discuss with and finally agree with the authorities that it would be beneficial for all parties involved if a Swedish standard could be established.

In Norway, similar work was going on in parallel, and the Norwegians, in an explicit way, had described parts of the earlier Langefors-Kihlström criteria used for estimation of ground vibration limits.

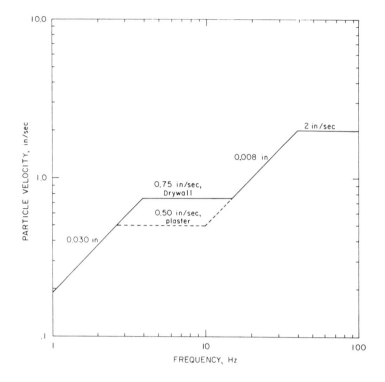

Figure 13.7. US Bureau of Mines recommendations for ground vibrations.

Finally in 1989, a Swedish Standard, SS 460 48 66, was accepted; it was revised in 1991. This standard is extremely helpful for calculation of guidance levels for blast-induced vibrations for buildings. The guidance levels do not consider human annoyance or risk for damage to vibration-sensitive equipment such as electron microscopes, computers (especially mass memory disks), relays, ... Guidance levels given should be used for establishing permitted vibration levels or threshold values. The standard is valid for all types of blasting operations, for example, tunneling, mining, leveling, etc.

The standard used only the vertical peak particle velocity but prescribes, for certain situations, that three-component measurements must be performed. Risk analysis, pre- and post-blast inspections, documentation, and instrument specification are mentioned. The vibrations shall, if possible, always be measured at the position of the foundation of the building where the vibration wave is transferred to the building.

13.5.1 Guidance Levels

These values are based on a broad, well-documented correlation between the vertical peak particle velocity component and induced damage to buildings founded on various types of geological ground. The guidance level v is given by

$$v = v_0 F_k F_d F_t$$

where v_0 denotes the uncorrected vertical peak particle velocity, F_k is a construction quality factor, F_d is the distance factor which considers the distance from the round to

13.5. Guidance Levels for Buildings According to Swedish Standard

Table 13.3. Uncorrected vertical peak particle velocity v_0.

Ground	v_0 (mm/s)
Loose morain, sand, gravel, clay	18
Firm morain, shale stone, soft limestone	35
Granite, gneiss, firm limestone, quartzite, sandstone, diabase	70

Table 13.4. Building factor F_b.

Class	Type of building or construction	F_b
1	Heavy constructions, such as bridges, harbor piers, and civil defense constructions	1.70
2	Industrial and office buildings	1.20
3	Standard living houses	1.00
4	Sensitive specially designed buildings with high arches, or constructions with large spans, for example, churches and museums	0.65
5	Historical buildings in damageable condition, and certain ruins	0.50

Table 13.5. Construction material factor F_m.

Class	Type of construction material	F_m
1	Reinforced concrete, steel, or wood	1.20
2	Not-reinforced concrete, brick, or clinker	1.00
3	Autoclave porous concrete	0.75
4	Mexi-brick (artificial limestone brick)	0.65

Table 13.6. Project time factor F_t for stationary works, use a sliding scale F_t for times up to 1 year, and $F_t = 0.75$ for project times over 5 years.

Type of blasting activity	F_t
Construction works such as tunnels, caverns, road cuts, and leveling	1.0
Stationary works such as quarries and mines	0.75–1.0

the measuring gauge, and F_t is a factor describing the project time for blasting work, for example, long-term mining activities or short-term construction activities.

13.5.1.1 Uncorrected Vertical Peak Particle Velocity

The values for the vertical peak particle velocity v_0 in Table 13.3 should be used. A more detailed value of v_0 can be achieved if the P-wave propagation velocity c_p is measured in the ground where the building is founded using

$$v_0 = \frac{C_p}{65}$$

where v_0 is in mm/s and C_p is in m/s.

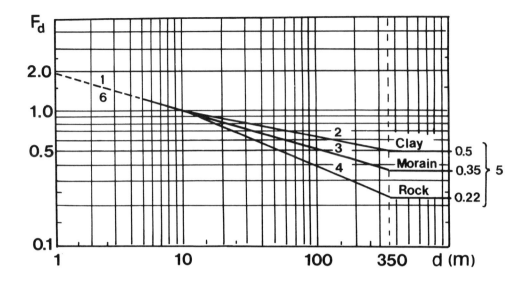

Figure 13.8. The distance factor F_d.

13.5.2 Construction Quality Factor

The *construction quality factor* F_k is the product of the *building factor* F_b and the *construction material factor* F_m; that is,

$$F_k = F_b F_m.$$

Buildings are described by five classes (Table 13.4). Classes 1 to 4 are valid for buildings of good standard. The construction material factors are shown in Table 13.5. When choosing the factor F_m, considerations should be given to the construction material with the lowest F_m value integrated in the building.

13.5.3 Distance Factor

The *distance factor* F_d can be determined using the diagram in Figure 13.8. It is a function of the shortest distance between the round and the building of concern.

The following six relations (with d, the distance in meters) are valid for the curves in Figure 13.8:

1. $F_d = 1.91\, d^{-0.28}$
2. $F_d = 1.56\, d^{-0.19}$
3. $F_d = 1.91\, d^{-0.29}$
4. $F_d = 2.57\, d^{-0.42}$
5. The distance factors $F_d = 0.22$, $F_d = 0.35$, and $F_d = 0.50$ for distances over 350 m are calculated to make the product $v_0 \cdot F_d = 18$ mm/s for rock, 15 mm/s for morain, and 9 mm/s for clay when using $v_0 = 70$, 35, and 18 mm/s, respectively.
6. For distances less than 10 m, special problems might occur. For example, expansion products can penetrate into existing joints, causing severe damage

to the structure due to large displacements. If problems are likely to occur, there is a need for more detailed monitoring to obtain the time history and the frequency domain for more than one component.

13.5.4 Project Time Factor

The *project time factor* F_t, which depends upon the duration of the blasting activity, is shown in Table 13.6.

13.5.5 Three Sample Calculations of the Guidance Level

Example 1. A Limestone Quarry Near Houses

A blast in a limestone quarry takes place 350 m from standard living houses. The houses are founded on gneiss and constructed of brick.
We calculate the guidance level v using

$$v = v_0 F_b F_m F_d F_t = 11.6 \text{ mm/s}$$

Given $v_0 = 70$, $F_b = 1.00$, $F_m = 1.00$, $F_d = 0.22$, and $F_t = 0.75$,

$$v = 11.6 \text{ mm/s}.$$

Example 2. Leveling Near a Brick Office Building

Leveling is to take place 6 m from a brick office building founded on solid rock. The guidance level is

$$v = 70 \cdot 1.2 \cdot 1.00 \cdot \left(19.1 \cdot (6)^{-0.28}\right) \cdot 1.0 = 97.2 \text{ mm/s}$$

Example 3. Construction Work Near a Brick Church

A blast from a construction work takes place 75 m from a brick church founded on loose morain.
The Swedish standard gives $v_0 = 18$ mm/s, $F_b = 0.65$, $F_d = 0.69$, $F_m = 1.00$, and $F_t = 1.0$. The guidance level is 8.1 mm/s.

13.5.6 Computers

The technical development in the electronic industry has been very fast and vibration-sensitive equipment such as computers and their accessories are rapidly becoming every man's property.

Threshold values for allowable vibrations given by manufacturers, suppliers, and users of computer equipment are defined in various ways. Some say the acceleration is most important, while some say the displacement is most important. Some use the RMS value and some the peak value. In many cases, it is not possible to determine if the values given are related to a certain shock profile or if they are related to continuous vibrations, etc. Seldom are recommendations given as to where to measure the incoming vibration [see Holmberg et al., 1983]. The fixed disk storage units are usually the most vibration sensitive equipment in a data processing system.

For several years, damage threshold values as low as 0.25 or 0.1 g have been used. These very limiting restrictions sometimes causes severe problem for the construction contractor, who then sometimes must use special equipment for pile driving, sometimes must change his entire blasting procedure, and sometimes must change techniques — from his ordinary technique to a slower and more expensive one. Often the equipment still needs to be shut down during the blasting event which, of course, can be very inconvenient for the real-time computer users. Vibration damping of the disk drives is often required, which introduces additional high costs.

By the use of isolators, it is possible to reduce the shock severity by absorbing the shock energy within the isolators and subsequently releasing the energy over a longer time. It is, of course, necessary to carefully choose the resonant frequency of the isolation system away from all resonances within the disk drive unit itself. Otherwise, the isolator may actually amplify the incoming shock pulse and the vibrations generated by the disk drive itself.

13.6 Planning for Blasting Work

Near most construction jobs where blasting is necessary, there are structures whose sensitivity will limit the maximum permissible vibrations. Client or contractor must then decide on the maximum charge that can be detonated without causing damage in the neighborhood. Where small-diameter holes are used, this maximum charge may be the charge having the same vibration effect as that from several holes detonated with detonators having the same nominal time delay. It is called the *maximum cooperating charge*. Because the cost increases considerably if the maximum cooperating charge has to be reduced, the economy of the blasting job is greatly influenced by that decision. Too large a cooperating charge may result in damages to neighboring buildings, damage claims, and even court disputes. Too conservative a decision on the size of the maximum cooperating charge will result in excessively increased costs and project time.

To optimize the blasting work, it is necessary to carry out risk analysis in order to determine, first, what size vibrations the environment will accept and second, how large a charge can be blasted at a certain distance without exceeding that vibration limit.

Before the blasting operations can begin, a risk analysis should be made involving a careful examination of the factors that can affect the blasting operations. The probability that a correct decision will be made increases as more information is made available. The decision data should be based on as many points as possible in the list below. The elements of risk analysis are illustrated in Figure 13.9.

13.6. Planning for Blasting Work

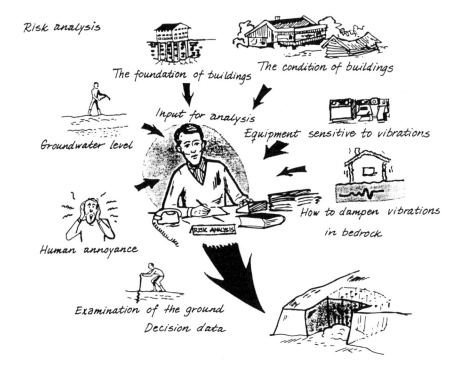

Figure 13.9. Risk analysis.

Checklist for risk analysis:

1. Has a geological examination been made regarding working site as a risk area?
2. Is there a potential risk for lowering the groundwater level?
3. What is the nature of foundations and underground parts of buildings in the area?
4. What is the type of construction and the condition of buildings within the risk zone?
5. Is there any equipment (such as computers, electron microscopes, laser equipment, relays, ...) which are sensitive to vibrations in the neighborhood?
6. Are there any underground objects (tunnels, cable trenches, telegraph cables, oil cisterns, district heating culverts, ...) that might be damaged by blasting?
7. What connections are there between vibration values, cooperating charges, and distances?
8. How are the inhabitants in the neighborhood influenced?
9. Has information about the blasting job been distributed to the neighbors?

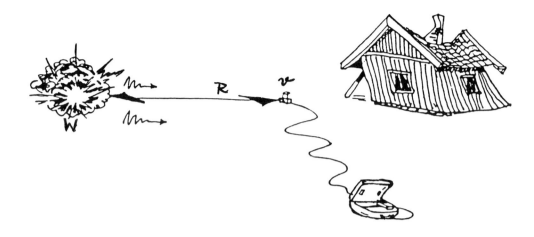

Figure 13.10. Test blasting.

13.7 Vibration Level, Distance, and Charge Weight

As rock is not an isotropic medium, it is often hard to predict the vibration level at a given distance. It is apparent that the different types of waves generated are dampened differently, depending on the strike and dip of occurring foliation or bedding. The schistosity, the distance between existing planes of weakness, and fluctuations in the groundwater level can affect the vibrations. Additionally, the blasting geometry, the performance of the explosives, the coupling between the explosive and the rock, and the distribution of ignition time intervals in the round also affect the vibrations.

In most cases, the maximum particle velocity (mm/s) is used to express what vibrations the actual object can withstand without being damaged when blasting is carried out. The investigations made show that a usable empirical relationship between particle velocity v, weight of charge W, and the distance R is

$$v = \frac{K}{\left[\dfrac{R}{\sqrt{W}}\right]^\alpha} \tag{13.13}$$

where the constants K and α vary with the foundation conditions, blasting geometry, and type of explosives.

To use the empirical formula (Equation 13.13) to reliably predict the vibration level for a determined distance, the constants K and α should have been determined by test blasting in the vicinity of the blasting place. In the test blast illustrated in Figure 13.10, small explosive charges were set off in advance to determine the transmission properties of the rock mass and thus provide a basis for determining the charge sizes in the main blast. In this way, any possible damage to neighboring structures could be prevented.

It is preferable that the test charges used in the test blast be placed on the line between the object that can be damaged and the blasting site. In connection with the test blasts, the actual vibration intensity is registered (e.g., the peak value of the particle velocity). In certain cases, it is insufficient to measure only the peak values of the acceleration, particle velocity, or displacement. In such cases, the time history of the vibration must be recorded and stored for later analysis.

13.7. Vibration Level, Distance, and Charge Weight

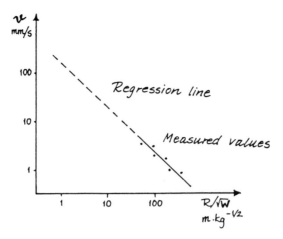

Figure 13.11. Adaptation of regression line in log-log diagrams.

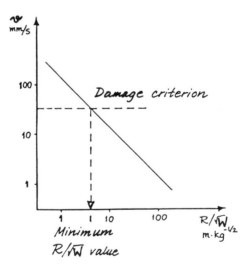

Figure 13.12. Determination of the lowest $R/W^{1/2}$ value to be used.

After recording the particle velocity, the measured peak values are plotted into a log-log diagram as in shown in Figure 13.11. A regression line is drawn through the number of discreet points.

As part of the risk analysis, an investigation should be made of the neighboring built-up area for equipment sensitive to vibrations, and a maximum vibration level for nearby buildings and other sensitive objects should be determined. This level (damage criterion) is then put into the diagram in Figure 13.11. The intersection between the damage criterion and the regression line gives the lowest $R/W^{1/2}$ value allowed to be used (Figure 13.12). For each distance, we have thus determined a unique weight of charge that must not be exceeded.

It must be remembered that the regression line is not an upper limit for measured particle velocity. Therefore, it is probable that the damage criterion will be exceeded if the intersection point between the damage criterion and the regression line is used to determine maximum weight of charge. However, the person who carries out the risk analysis should be aware of this and either add a safety factor to the damage criterion (and thus decrease the weight of the charge), or use another line with a higher confidence limit than the regression line. One should also be aware of the fact that charges of ANFO, dynamite, and an aluminized TNT-slurry might result in different particle velocities although the charge weights are the same.

To find the best regression line, $v = f(R, W)$ to the variables, we have to find the constants a, b, and c for the equation

$$v = a\, W^b R^c \tag{13.14}$$

$$\log v = \log a + b \log W + c \log R. \tag{13.15}$$

The constants are given when

$$\sum_{i=1}^{v} \left[\log v_i - \log a - b \log W_i - c \log R_i\right]^2 = \min. \tag{13.16}$$

Usually, it is assumed that constant b in Equation 13.15 is equal to $-\dfrac{c}{2}$, which gives us the square root scaling

$$v = a \left(\frac{R}{\sqrt{W}}\right)^c \tag{13.17}$$

or,

$$\log v = \log a + c \log \left(\frac{R}{\sqrt{W}}\right) \tag{13.18}$$

which has the form $y = \alpha + cx$; that is, a straight line. The constants can easily be determined with the help of a simple computer program for the least squares fitting.

13.8 Scatter of the Peak Particle Velocity

When peak particle velocities are plotted vs. scaled distance (e.g., $R/W^{1/2}$), it will be seen that there exists a fairly large scatter about the mean regression line achieved for a least squares fit of the logarithmic values. A number of factors are responsible for the scatter at the given scaled distance; these include geology, confinement conditions, type of explosive, rock properties, cooperating charges, delay times, different wave types, and errors in measurements and analysis.

Holmberg [1977] carried out a field experiment in order to investigate the stemming effect for vertical drilled holes with diameters 25, 38, and 76 mm and infinite burdens. The peak particle velocities were monitored, with geophones fastened to the surface rock with expansion bolts for all the single shots. Five test sites were used within a radius of 1 km. The same type of explosive was used for all shots. The weight of the explosive column (0.5–25 kg) and the stemming was varied. The hole depths were from 1.2 to 4.9 m. Two triaxial geophones were mounted on a line directed toward the blast hole.

Figure 13.13 shows the vertical peak particle velocities vs. the scaled distance at test site A. At this particular site, the scatter was low — probably indicating that the geology was consistent. In Figure 13.14, the measured values from the test sites

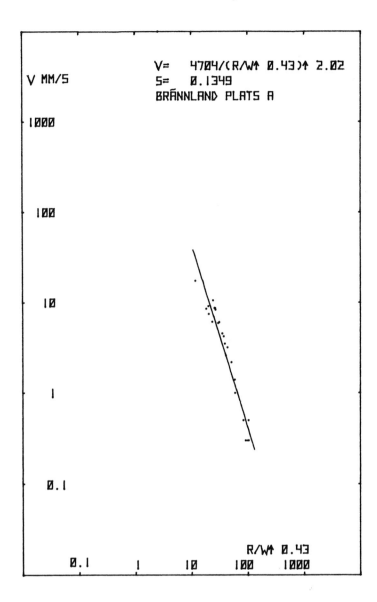

Figure 13.13. Best least squares fit showing attenuation relations change for test site A. Vertical peak particle velocities.

A through E are plotted and the scatter here is larger. A horizontal joint system was found at some of the sites, and they evidently affected the ground vibration waves and caused the scatter. It was not possible to notice any significant difference between the shots that cratered and those that did not. Figure 13.15 illustrates that the horizontal joint system forced the longitudinal velocity to be considerably higher (a factor of 1 to 8) than the vertical velocity component. When the charges were placed deeper than the joint system, the ratio became consistent at about 1 to 2.

The scatter has to be considered if a prediction of possible peak particle velocity is to be made for a given scaled distance. Instead of using the median regression line (50 %), it is safer to use a prediction line with a higher degree of confidence.

360 *Chapter 13. Ground Vibrations*

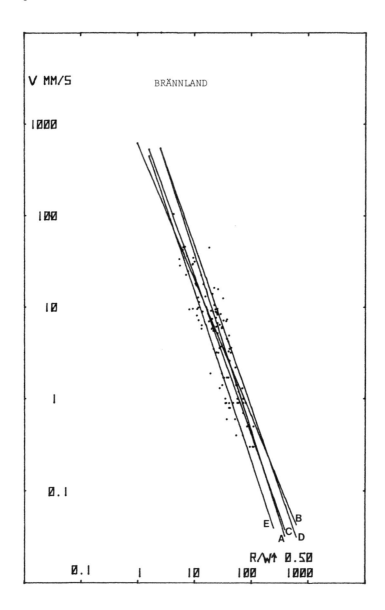

Figure 13.14. Attenuation relations change for sites A through E. Vertical peak particle velocities.

Figure 13.16 shows the attenuation relation change for 1363 data points given by the US Bureau of Mines [Nicholls *et al.*, 1971] and various confidence lines. The 90 % confidence line tells us that, in 90 % of the blasts, the particle velocity will not exceed the line for the various scaled distances. However, the figure should just be seen as an example for various confidence lines. The data used in the figure are not taken from blasts with the same blast geometry or at the same site. So, if we were to make a prediction for a certain work site, we should only use data collected from that particular site or from the close neighborhood.

Table 13.7. Reduction factors for the number of charges that may be initiated by detonators with the same interval number at an assumed frequency >60 Hz for different types of detonators.

Interval time	Interval No.	Reduction factor
25 ms	1— 10	1/2
25 ms	11— 20	1/3
500 ms	> 20	1/6

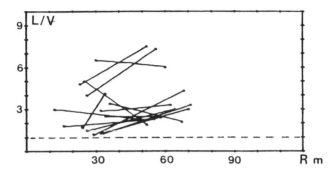

Figure 13.15. Ratio between peak values of longitudinal (L) and vertical velocities (V) at two distances from single test shots ($d = 25$ mm, hole depth $= 1.6$ m).

Figure 13.17 shows the scatter of regression lines (50% confidence lines) from different blasting geometrics and test sites. Finally, Figure 13.18 shows a diagram that gives all the relations between distance, charge weight, and peak particle velocity for various confidence lines. The diagram is based upon the US Bureau of Mines (USBM) data from Figure 13.16 and can be used as a rough guide for predicting peak particle velocities.

From the diagrams, one can see that if 100 kg is detonated at a distance of 300 m, the 99% confidence line will give a predicted peak particle velocity of 10 mm/s. This value would statistically be exceeded once for every 1000 rounds fired.

13.9 Cooperating Charges

In the preceding section, it was shown how the maximum detonating weight of charge can be estimated when the vibration level and the distance are known. The weight of the charge in question is the maximum total weight of charges that can be initiated at the same time. Using delay detonators, it is possible to blast rounds with considerably higher total charge weights per delay interval. The higher the interval number, the larger a total charge weight per interval can be used because the scatter of the delay time increases with the nominal delay time.

The *cooperating charge* is defined as the total charge per interval, multiplied by the reduction factor appropriate for the interval used.

In the USA, the USBM states that the delay interval should be equal to, or greater than, 8 ms in order that the two charges be considered separate charges. The size of the reduction factor is shown in Table 13.7 for a case where the dominating frequency band of the ground vibrations is not lower than 60 Hz. As the dominant frequency becomes lower, the reduction factor approaches the value of 1. Langefors and Kihlström [1973]

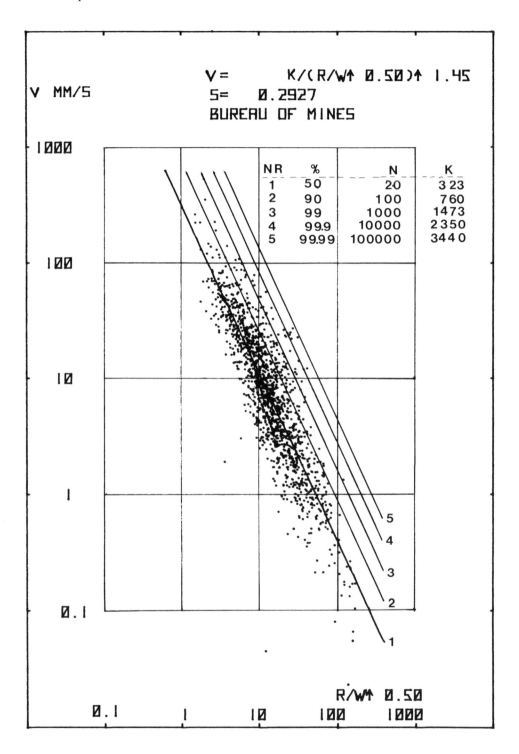

Figure 13.16. Confidence lines for US Bureau of Mines data.

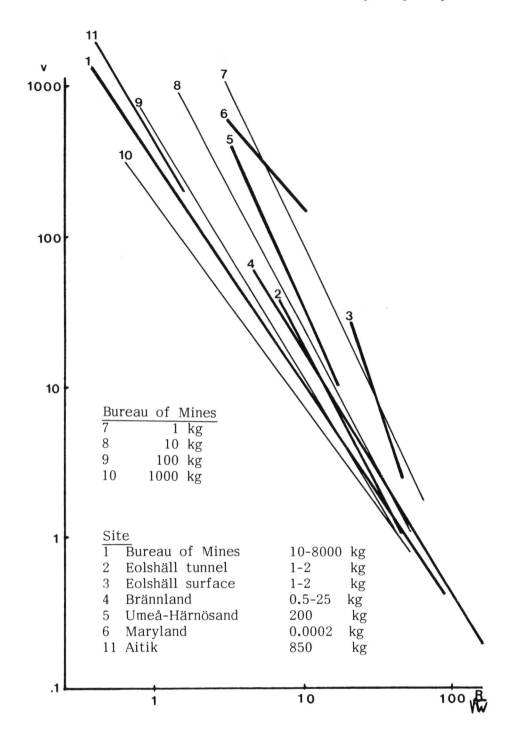

Figure 13.17. Regression lines for some various test sites, charge weights, and blast geometries.

364 Chapter 13. Ground Vibrations

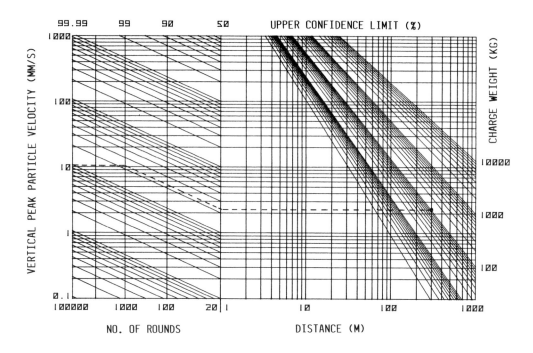

Figure 13.18. Diagram for finding the relations between distance, charge weight, confidence line, velocity, and number of rounds.

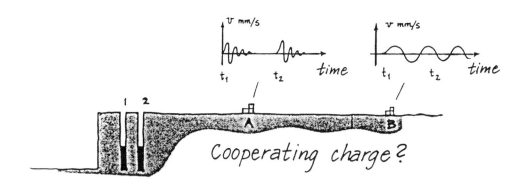

Figure 13.19. Cooperating charge is related to distance.

give reduction factors for detonators with varying scattering at different frequencies. Table 13.7 cannot be used if frequencies are below 60 Hz.

The expression cooperating charge is somewhat inappropriate as it is only applicable at certain distances. When blasting two separate charges with detonators having the same interval number (Figure 13.19), if the vibrations are observed at a short distance (A), the charges do not cooperate; while at a long distance (B), interference and reinforcement of the two vibrations may occur. Whether two charges, initiated one after the other, cooperate or not depends on the following factors:

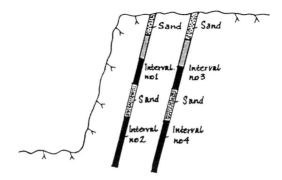

Figure 13.20. One method to decrease ground vibrations is to distribute the charge weight per hole into several parts initiated separately at different intervals. Note, however, that much larger vibrations than expected could occur if the sand stemming does not prevent transmission of detonation.

- Time interval between the initiations
- Velocity of propagation of the vibrations in the bedrock
- The decay time of the vibrations
- Distance from the observation point to the charge
- The geometry of the charge
- Velocity of detonation
- Confinement

13.10 Methods to Reduce the Vibration Level

By adapting the blasting method, drilling pattern, charging scheme, and ignition pattern, the size of the vibrations can be controlled. Among other things, the size of the vibrations depends on:

- Cooperating charge (see Figure 13.19)
- Confinement conditions
- The character of the rock
- The distance from the blasting site
- The geology, e.g., overlay soil types

Practical methods to reduce the ground vibrations by limiting the cooperating charging weight per interval are:

- Adapt the ignition pattern so that the charging level is spread over more intervals and the scattering in the delay elements of the detonators is utilized
- Reduce the number of holes and the hole diameter
- Use decked charges by dividing up the necessary charge level in a drillhole into more ignition intervals by means of sand plugs (see Figure 13.20)
- Use decoupled charges; charge diameter smaller than hole diameter
- Divide the bench into more benches Do not blast to the final depth at once.

At the moment of detonation, there should be as little confinement as possible. This can be obtained by:

- A carefully adapted ignition pattern, so that all the holes will break the burden in the easiest way
- Increased hole inclination (of the drillhole)
- Avoiding too large burdens and choke blasting

For blasting at a shorter distance than 100 m, the risk for interaction between the different intervals is small. The risk for cooperation between the intervals increases with large blast in quarries: for instance, where any structures sensitive to vibrations are situated a large distance away. The size of the vibrations is then influenced by:

- Charging level
- Interval times
- The resonance frequency of the ground (which depends on the depth and the character of the ground)
- The local geology

13.10.1 Costs When the Vibration Level Has to be Reduced

The cost for careful blasting near built-up areas increases very rapidly with decreasing permissible vibration level. The increase in costs primarily depends on the following factors:

- Drilling — smaller or greater number of drillholes.
- Charging — more detonators and higher cost of labor
- Blasting — more rounds and longer stand-up time

The costs of planning and control work will also increase in:

- Blasting
- Visual inspection
- Vibration measurement
- Blasting record
- Insurance administration

In addition to the costs mentioned, there are also a number of practical problems with varying degree of difficulty, the value of which is hard to estimate in terms of money.

13.10.2 Ground Excavation

Excavation must be carried out in many different materials and under widely different circumstances. For instance, there is a great difference between the vibration levels generated when excavating clay in the summer and in the winter when the surface is frozen.

Vibrations in excavation result from strikes with the bucket or when boulders are pried loose. Moving the excavator also causes vibrations, but they are generally much smaller and will therefore not be discussed here. The individual vibrations are usually of short duration and have the character of a shock, but they may be repeated over and over again during a period of hours or days. Even at small distances, the vibration levels are low and rarely exceed 3.5 mm/s; still, they can be very irritating.

As the vibration levels are highest when excavation takes place in loose soils, the dominating frequencies will be relatively low although the different wave types are rather

Table 13.8. The transmission ratio for various foundation depths.

Type of foundation	Transmission ratio $r = v_f/v_m$	
	(Mean)	(Standard deviation)
Foundation in clay on an edge-stiffened foot plate	0.62	0.15
Foundation in clay with cellar on the foot plate	0.40	0.18
Foundation in clay on piles or pedestals down to firm bottom	0.31	0.11

equally represented. A large part of the vibration consists of volume waves (P- and S-waves), which means that the vibrations are rapidly attenuated with increasing distance.

The circumstances are somewhat different when a trench blast fails, so that boulders have to be pried loose with an excavator. The vibrations then are of a higher frequency than those from common excavation but usually lower than from blasting. They are damped more slowly in the soil layers on top. As a result, higher vibrations have sometimes been registered from the excavation work than from the blasting itself.

Vibrations from excavation are generally no problem. Should it be necessary, however, to diminish the vibration levels, strikes and shocks with the bucket should be avoided as much as possible. In the winter when the ground frost is thin, it could be thawed out. Loosening of the soil should be considered before the actual excavation work commences.

13.11 General Methods to Reduce the Vibration Level
Transmission of Vibration from Ground to Building

In a project concerning traffic vibrations, an examination [Lande, 1981a] was carried out, the scope of which was to study the transmission of ground vibrations from the ground to the buildings. For the measurements, the vertical particle velocity in the ground surface v_m outside the building was registered and compared with the vertical particle velocity in the foundation of the building v_f. The buildings consisted of small terrace houses and apartment houses with up to 7 stories. The transmission number r is defined as

$$r = \frac{v_f}{v_m}. \tag{13.19}$$

The results reported by Lande show that the transmission number diminishes with greater foundation depth indicating that it is mainly the surface waves (Rayleigh waves) which dominate. The vertical amplitude of the Rayleigh wave has a maximum just below the ground surface and then the amplitude decreases with larger depth.

The results, compiled in Table 13.8, are based on measured values from 24 measurement sites. The table indicates that the vibrations (vertical) are damped when passing from the ground to the building and that the damping increases when the foundation depth increases. The study shows that the transmission number decreases when the frequency increases (i.e., the wavelength decreases).

Generally, for transmission of vibrations from the ground to the building, the particle velocity is lower in the foundation of the building than in the ground. Intensification of vibrations may occur higher in the building dependent upon the occurrence of resonance.

By making a slot between the vibration source and the affected building object, it is possible to further dampen the vibrations before they reach the building.

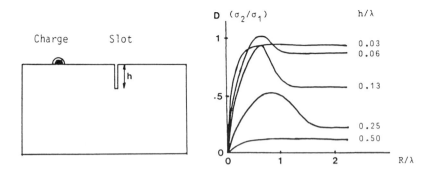

Figure 13.21. The damping factor as a function of the distance for different h/λ.

13.12 Model Studies with a Slot

Woods [1968] performed extensive half-scale experiments where vibration damping by means of a slot in fine, gravely sand was studied. Woods divided his experiments into two parts:

1. Active isolation, where the vibrations are reduced by a slot in the direct path close to the excitation source
2. Passive isolation, where a slot is placed at a larger distance from the source, close to the arrival site, thus helping to prevent the vibrations from reaching the area to be protected

For active isolation, Woods indicates:

    a. Using a depth of slot/wavelength ratio 0.6 creates a damping factor of 0.25
    b. A slot around the charge has an effect up to a distance equal to 10 times the wavelength
    c. At a slot which does not surround the excitation source, damping takes place in an area equal to the slot angle reduced by 45° of each end
    d. Intensification of the vibrations takes place in a direction toward the open part of the slot

For passive isolation, a protected area arises within a semicircle behind the slot. This area has, however, no distinct limits. For passive isolation, it is shown that:

    a. For greater distances from the charge, a larger slot is necessary
    b. Intensification of vertical vibrations takes place in front of and beside the slot
    c. Wall barriers are less effective than open slots

Reinhardt and Dally [1970] studied the effect in plates of photoelastic material. Figure 13.21 shows the arrangement of the charge to slot and resultant damping factors D obtained at different distances R for varying depth of slot h and wavelengths λ.

Figure 13.22. Slot drilling at the Naturhistoriska Museet in Gothenburg made it possible to reduce the vibration level in the building from blasting in the near vicinity.

13.13 Full-Scale Experiments with Slots in Rock

One method to reduce the vibrations from blasting is to drill a slot in the rock between the vibration source and the building. If the slot is drilled along the foundation and close to it, the house is protected not only against vibrations but also against back break and heaving of the rock below the foundation (see Figure 13.22). By carrying out slot drilling in front of a building or a plant which is sensitive to vibrations, as much as 80 % damping of the vibration level has been achieved, [Ekeroth, 1981].

Expansion of the Natural History Museum in Gothenburg, Sweden, required removal of about 5000 m^3 of rock. The main body of the rock was closer than 15 m to the existing building. The bench height was 6 m and blasting would take place as near as about 0.6 m from the nearest wall of the building. The construction of the building was judged to be capable of withstanding a particle velocity equal to 70 mm/s, which is the recommended value for normal buildings. On the other hand, such vibrations could not be allowed inside the building where fragile exhibits were exposed in glass cases. Here, the particle velocity was reduced to $v = 35$ mm/s. The vibrations were monitored at 12 measuring points.

The slot was drilled with 64 mm diameter overlapping holes, which meant that the rock adjacent to the building could be excavated in one bench by using decked charges

Table 13.9 Measured damping behind the slot.

Distance from slot(m)	Ratio (distance ÷ depth of slot)	Damping (%)
< 3	0.5	> 80
3–6	0.5–1.0	65–80
6–9	1.0–1.5	50–65
> 9	> 1.5	< 50

instead of three benches if the conventional method had been used. It was advantageous that the blasting would not take place closer to the slot than 0.7 to 1.0 m. The rock adjacent to the slot was easily excavated using a shovel excavator.

The slot was drilled to the depth of about 1.5 m below the theoretical base explosive charge influence and at a distance of 0.6 m from the dry wall. The length of the slot totaled 38 m and it was extended about 1 m beyond the corners of the building. Vibration measurements carried out behind the slot for blasts with charges at 3 to 35 m in front of the slot show that the damping is considerable. The results are shown in Table 13.9.

The precise extent of the slot's effectiveness in decreasing the vibrations at this particular site could not be determined. Subjectively, the slot was deemed effective by the experienced blasting professionals who analyzed the vibration records.

At distances larger than twice the depth of the slot, it is doubtful if the slot is cost effective. For short distances, vibration damping by slot drilling is very effective, but to be fully efficient, the slot must have the following properties:

- It should have parallel holes, with the slot left open
- It should be completely free from cuttings and water
- It should have its depth well below the theoretical basement for the explosive influence
- The extension of the slot beyond the protected object should be at least equal to the depth of the slot

Further savings for the client can be made using the slot by reducing or eliminating costs otherwise incurred for rock reinforcement and foundation strengthening of adjacent buildings. The rock surface exposed by the predrilled slot can be used as a mold when grouting.

13.13.1 Measures for Damping Vibrations which Reach Computers and Auxiliary Equipment

Construction work and, especially, blasting operations in built-up areas can be complicated by the proximity of installations sensitive to vibration such as computers and their auxiliary equipment. There are often several computer systems from different manufacturers close to each other and this complicates the risk analysis and subsequent damage prevention measures to be taken. The availability and continued operation of computers is very important as they are usually part of a real-time system, where the computers are needed to obtain information and to register transactions. (Some examples are banks, airline ticket sales, automated teller machines, and weather and environment data centers). The consequences of even a short halt can be devastating and very expensive. The manufacturers are very restrictive with regard to permitted vibrations and, therefore, considerable cost increases and delays result when blasting must be carried out in the neighborhood. It is often difficult for purchasers of blasting work and for contractors to foresee these problems before they are confronted with them.

In connection with the bidding, contracting, and subsequent performance of blasting operations in built-up areas, it is important to have a clear view of the prerequisites and possible restrictions. The builder/contractor needs a lot of information to be able to successfully plan and execute the project. To obtain the necessary overview and to accurately determine the technical and economical variables, it is important to invest in a so-called risk analysis at an early stage of the planning [Lande, 1981b].

Within the framework of risk analysis, sensitive vibration equipment is mapped in relation to the blasting site. At the same time, a study of the applicable restrictions is made and the permitted limiting values are noted for each of the computer installations. From the limiting values, the blasting work is planned. Dependent on the extent, the character, and the cost of the work, the quantity of explosives is optimized to blast in the most economic way.

The quantity of explosives per interval and the interval time determine the magnitude and the characteristics of the vibrations reaching the building foundations. From the foundation, the vibrations propagate throughout the building, interact with the dynamic properties of the structure, and finally are transmitted to the computer room. In cases where blasting vibrations have frequencies near the resonance frequency of the floor, magnification of the vibrations may occur. The recommended method for determining the vibration magnitude in floor surfaces is to make a prognosis based upon vibration measurements of the test blasts including:

- Recording and analysis of the dominant frequency range in the foundation of the building and on the floor surface in the computer room
- Calculation of the law of uniformity for the foundation and floor
- Calculation of the transmission function between the foundation and floor
- Calculation of duration and damping for the vibration period in the floor
- Calculation of the degree of damping for the computer installation

In cases where there is a risk that incoming vibrations from the floor could become too large in relation to the permitted values for computers, the computer can be isolated from floor vibrations by means of special dampers. Unfortunately, it is not possible to recommend placement of such dampers beforehand as there are different resonance frequencies for different floors, dependent on the dimension and properties of the floor.

Vibration damping of computers can result in an increase in the permitted limiting vibration values for the floor where the computer stands. Special springs are used for the vibration damping to ensure that the environmental demands — specified in the contract — are not exceeded. An increase in and subsequent control of the permitted vibration values make it economically possible to undertake blasting operations in built-up areas.

The method used to dampen vibrations reaching computers can also be applicable to other sensitive equipment (such as electron microscopes).

13.13.2 Vibration Damping for Buildings

There are techniques that can be applied to prevent large vibrations from damaging important buildings. The basic method is, in principle, similar to that used for damping computers.

At the foundation, the whole building is placed on a large number of "spring packages" selected to fit the weight of the building and to the likely frequencies and amplitudes of the expected incoming vibrations. The method is very expensive, and is mainly used for strategic buildings that must be protected, such as national defense information or coordination centers.

13.13.3 Practical Views on Measurement Techniques

Vibration measurements are made to monitor and control the effect of vibrations on buildings, installations in buildings, and the degree of disturbance to man in order to register how the effects differ with respect to the location of the measurement points, the vibration parameters, and the recording method. The underlying principle is to measure the vibrations where they are first transmitted to the object of concern. However, in the USA., the gauges usually need to be attached to the ground outside the foundation, as house owners seldom allow the mounting of the gauges at or inside the building.

13.13.4 Effect on Buildings

Incoming vibrations into and under a building cause widely varied effects. To calculate them, relatively comprehensive measurements must be made. From experience, we know that buildings often crack at vibration levels which are considerably below those that they should theoretically withstand. This very often has to do with the existence of previously induced stresses. The damage criteria are based on past experience, most of which is from measurements of the vertical particle velocity at the base of the building nearest to the vibration source. To effectively control the effect of vibrations on buildings, the measurements ought to be made in the same way. The transducer is then mounted onto a special attachment cube, fixed by means of an expansion-shell bolt in a supporting part of the framework. Recording can be made by instruments indicating peak values or analog values.

13.13.5 Effects on Building Installations

In principle, the vibrations should be measured in that part of the installation which can be affected by them. In practice, this is often impossible with, for instance, different types of relays, computers, balances, or other accurate measuring instruments. Therefore, the measurements are usually made in the foundation or at the site of the installation. The attachment method for the transducer varies from case to case. For low-frequency vibrations (< 30 Hz), double-sided adhesive tape could be used; while for higher frequencies, the transmitter must be glued or screwed on.

Which one of the parameters (displacement, particle velocity, or acceleration) should be used can only be determined by careful analysis of the type of installation and the equipment maker's specifications. It is rather common, though, to measure the acceleration of the vibration. In evaluating the recorded accelerations, one needs to be aware of the fact that high-frequency noise can give high acceleration levels, but owing to the negligible energy content, it is mostly a completely harmless vibration. Therefore, it is often suitable for acceleration measurements to filter out frequencies exceeding 100 to 200 Hz. For documenting the time of each blast and the overall amplitude of the associated vibrations, instruments which record peak values as well as the analog values can be used.

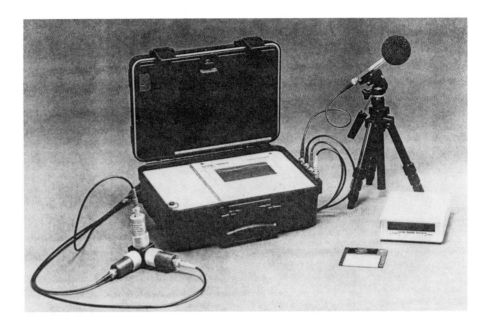

Figure 13.23. Vibration monitoring by unattended instruments.

13.13.6 Degree of Disturbance to Man

Investigations into man's sensitivity to vibrations have established that within the frequency range 8 to 80 Hz, the particle velocity of the vibration is the best measure of the degree of disturbance. As vibrations connected with traffic and construction activities lie mainly in this frequency range, it is obvious that the particle velocity should be measured. The velocity transducer should be placed where man comes into contact with the vibration, e.g., on floor surfaces in the building. They should also be placed as close to the center of the floor as possible, as this is where the vibrations are most intensive.

If the peak values of the vibrations are recorded by unattended instruments (Figure 13.23), the measuring point on the floor must be accompanied by a measuring point at the building foundation. This is necessary because normal use of the building often causes larger vibrations than those induced by blasting or traffic. Thus, it is important to be able to distinguish an incoming blast-generated vibration transmitted *from* the foundation of the building from those generated *within* the building. On hard floor surfaces, the recording equipment can be held firmly in place by double-adhesive tape; while on carpet, a heavy three-point support can often be used.

374   *Chapter 13. Ground Vibrations*

Figure 13.24. In-city blasting directly below family housing (Ullern, Norway).

Figure 13.24 shows tunnel blasting directly below an area with family housing. This is a typical in-city blasting operation where success depends on closely controlling the ground vibrations by the use of instruments such as shown in Figure 13.23.

Chapter 14

Air Blast Effects

14.1 Introduction

One undesirable side effect of blasting operations is the generation of air blasts. Although the air blast seldom causes structural damage, the sudden great noise may perturb neighbors and raise complaints.

A well-confined and covered charge, such as those used in construction blasting, is usually of no concern. Noise from large-hole open pit and coal mine blasts with unstemmed charges, on the other side, can strongly affect people and strengthen their fears that property damage may occur, even though the actual ground vibration and the air blast when measured may be very small disturbances.

Proper blast design, covering, and monitoring of air blast and ground vibration amplitudes together with objective information given to neighbors about the blasting operation will reduce complaints and make the overall operation more efficient.

The following terms are used in describing air blasts:

Air Blast is the common term for pressure waves in air emanating from explosions.

Noise is the audible part of the air blast; it has frequencies from 20 to 20,000 Hz.

Concussion is used for that (inaudible) part of the air blast having a frequency content below 20 Hz.

14.2 Characteristics of Pressure Waves in Air

When an explosive charge detonates in air, the rapidly expanding gaseous reaction products compress the surrounding air and move it outwards with a high velocity, initially equal to the detonation velocity. The resulting shock wave has a steep shock front, followed by a rapidly decreasing pressure. The surrounding air provides very little resistance to the expansion of the reaction products, which thus expand to pressures lower than the ambient atmospheric pressure. Propagation of air blast waves is preferably reported as a function of distance, normalized using the cube root of the charge mass. Figure 14.1 shows the shock front velocity and the particle velocity as functions of the reduced distance for a TNT charge in air. R denotes the distance to the charge and W the charge weight.

Let us assume that a spherical TNT charge is detonated in undisturbed air sufficiently high above the ground and away from any structures to prevent shock reflections from obstructions or from the ground from interfering with the blast wave arriving at

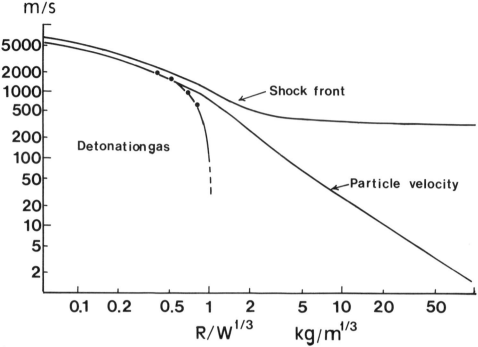

Figure 14.1. Shock front velocity, the associated particle velocity, and gas front velocity reduced distance from TNT charges in air [after Weibull, 1947, and Granström, 1956].

the gauge used to measure its pressure. The time history picked up by a pressure transducer positioned 5 m away from the point of detonation of a 1 kg TNT charge is shown in Figure 14.2.

As seen from the graph the pressure rises instantaneously from ambient pressure to its peak value $p_+ = 0.27$ bar. The pressure then decays gradually and, after time T_+, (the duration of the first phase) has returned to the ambient pressure. After time T_+, the pressure drops even further to a minimum pressure p_- of 0.06 bar (0.006 GPa), below the atmospheric pressure. The duration T_- of the negative phase (sometimes called the suction phase) is longer than the duration of the positive phase.

The front pressure p_+ is the maximum static pressure at distance $R = 5$ m. Due to the spherical propagation of the blast wave and due to losses to the surrounding air this front pressure will decrease with distance. In its passage over a surface parallel to the direction of propagation, the surface will be exposed to a static pressure equal to the front pressure of the blast wave.

If the wave is disturbed by a structure, the wave will be reflected. The reflected wave pressure depends on the initial shock front pressure and the angle of reflection. The maximum pressure will occur if the wave impacts a structure perpendicularly to for example a plane wall. This head-on reflection pressure will be twice the front pressure of the wave.

The blast effect on buildings and other structures depends on the amplitude and duration of the air pressure pulse as it arrives at the target.

The side-on positive momentum I_+ is defined as

$$I_+ = \int_0^{T_+} p(t)dt \qquad (14.1)$$

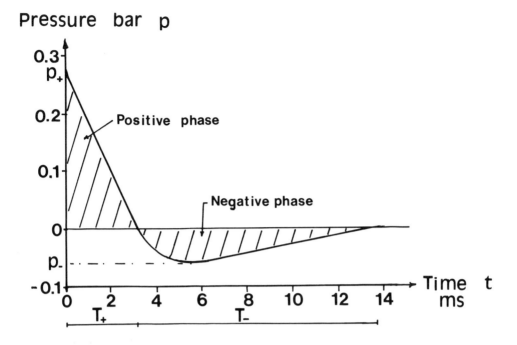

Figure 14.2. Shock wave over-pressure versus time at distance 5 m from 1 kg TNT in air. $p = 0$ is one atmosphere pressure of ambient air [after Granström, 1956].

and the side-on negative momentum I_- is defined as

$$I_- = \int_{T_+}^{T_+ + T_-} p(t) dt \qquad (14.2)$$

14.3 Blast Scaling

Cube-root scaling can be employed when comparing blast wave characteristics from charges of different sizes W detonated at different distances R from a point of observation. The cube root scaling law, formulated by Hopkinson [1915], states that self-similar blast waves are produced at identical scaled distances when two different-sized explosive charges of similar geometry and of the same explosive are detonated in the same atmosphere.

Any distance R from a explosive charge W then can be transformed to a characteristic scaled distance r

$$r = \frac{R}{W^{1/3}}. \qquad (14.3)$$

By representing the specific wave values for pressure, momentum, and duration as a function of the scaled distance r, it is possible to compare these values with each other.

Eriksson [1987] compared several published sources regarding wave characteristics such as front pressure, duration, and impulse in the undisturbed air blast from a spherical TNT charge. According to Eriksson, the agreement is good for pressure measurements (see Figure 14.3) but, for the duration, the discrepancy between values given by different authors can be greater than 1 order of magnitude.

As can be seen from Figure 14.3, the relation between pressure and scaled distance is not linear in the log-log diagram. For small pressures, it is common practice to estimate the front pressure for an unconfined charge as

$$p_+ = 0.7 \frac{W^{1/3}}{R}. \tag{14.4}$$

The formula above should be used as an estimate for values of $\frac{R}{W^{1/3}} > 50$.

Low air blast pressures are often given in the logarithmic decibel scale, or in millibars. The following is a convenient translation table to other pressure units:

1 atmosphere (atm)	= 1.01325 bar	= 1,013.25 millibar
1 atm	= 1.013 kg f/cm	= 101,325 N/m² (Pa)
1 atm	= 14.696 psi	
1 kg f/cm²	= 14.223 psi	
1 kg f/m²	= 9.81 N/m² (Pa)	

where Pa means Pascals and kg f/m² means kilogram-force per square meter.

The logarithmic decibel scale is convenient to use for acoustic measurements where the disturbance affects humans. The audible part of the air blast shows a wide spectrum of different amplitudes and frequencies and the decibel scale is widely accepted for describing the sound intensity as it is experienced by the human ear:

$$db = 20 \log \left(\frac{p}{p_o}\right) \tag{14.5}$$

where p_o is the reference pressure. $p_o = 2 \times 10^{-5}$ Pa $= 2 \times 10^{-10}$ bar $= 2.9 \times 10^{-9}$ psi.

Figure 14.4 shows the relation between db and pressures expressed in psi and bars. The solid line represents Equation 14.6 where the front pressure can be estimated from the scaled distances:

$$p = 0.7 \frac{W^{1/3}}{R} \tag{14.6}$$

where p is in bars, W in kg, and R in m, or

$$p' = 25.57 \frac{(W')^{1/3}}{R'} \tag{14.7}$$

where p' is in psi, W' is in lbs., and R' is in ft.

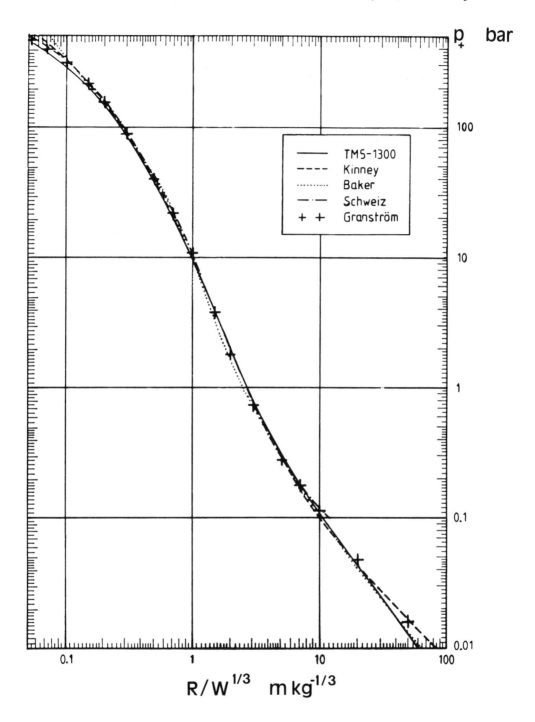

Figure 14.3. Front pressure from a 1 kg spherical TNT charge [from Eriksson, 1987].

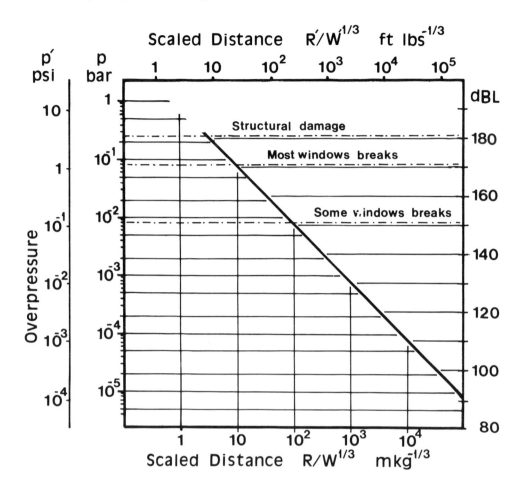

Figure 14.4. Nomogram for determining air blast overpressures (expressed in psi, bar, and db) as a function of the scaled distance (expressed in m/kg$^{1/3}$ and ft/lb$^{1/3}$). The solid diagonal line is $p = 0.7 \dfrac{W^{1/3}}{R}$.

14.4 Air Blast Induced Damage

The blast effect on buildings and other structures depends on the amplitude and duration of the air pressure pulse as it arrives at the target. The effect also depends on whether the exposed area of the building structure is facing the blast or if it is parallel to the direction toward the explosion.

Structures of different rigidity react differently to the air blast depending on the duration (frequency content) of the blast wave. For example, at intermediate distances from a large blast, a structure such as a glass window may withstand the peak pressure and the positive momentum of the wave, but may fracture as a result of the negative momentum. The walls of a blast containment chamber usually have to be designed to withstand the positive momentum (or the equilibrium static overpressure), rather than the shock front pressure (positive overpressure). The positive overpressure is of less importance because of its short duration; there is not enough time for the deformations under the positive overpressure load to reach breaking strain. Later, as the structure

14.4. Air Blast Induced Damage

continues to move with the momentum imparted to it, the maximum strain occurs, and the risk for damage reaches its maximum.

The blast effect also depends on whether the exposed area of the building structure is facing the blast or if it is parallel to the direction toward the explosion. The damage potential of the blast wave at the target at different distances must therefore be described by several different characteristics of the wave, namely:

- The head-on reflected wave peak overpressure
- The head-on reflected wave positive momentum
- The head-on reflected wave negative momentum
- The side-on peak overpressure
- The side-on positive momentum
- The side-on negative momentum

Thus, depending on its distance from the blast and the rigidity of its structural parts, a building generally responds to an air blast at different — and generally higher — frequencies than the main harmonic frequency of the air blast. At long distances, the air blast has a low main harmonic frequency, in the range 0.1 to 10 Hertz, which is mostly outside of the audible region of frequencies. The typical building response at sound pressure levels below 130 db is a clearly audible rattle, mainly from windows and doors and from objects standing on shelves. With increasing amplitude, window breaks begin to occur at about 152 db. Most windows in a area would break at an amplitude of 172 db, and structure damage would occur at 182 db or over. Figure 14.4 shows the damage potential in pressure (bar) or sound overpressure (db) as a function of the scaled distance.

The US Bureau of Mines recommends the following safe levels for air blast:

- 135 db when using a 0.1 Hz high pass linear measurement method
- 133 db when using a 2.0 Hz linear peak response
- 129 db when using a 5 or 6 Hz linear peak response
- 105 db when using the db-C weighting scale and when events do not exceed a 2 second duration

The authors [Siskind *et al.*, 1980] of the US Bureau of Mines report state

> "The four air blast levels and measurement methods above are equivalent in terms of structure response and any one can be used as a safe-level criterion. Of the four methods, only the 0.1 Hz high-pass linear method accurately measures the total air blast energy present; however, the other three were found to adequately quantify the structure response and also repression techniques that are readily available to industry. Where a single air blast measuring system must be used, the 2-Hz linear peak response is the best overall compromise. The human response and annoyance problem from air blast is probably caused primarily by wall rattling and the secondary noises. Although these will not entirely be precluded by the keeping below the four air blast levels given above, they are low enough to preclude damage to residential structures and any possible human injury over the long term."

Often, the distance between the blasting operation and the nearest residential building is so long that air blast monitoring is not necessary. Siskind *et al.* [1980] recommend that, in the absence of monitoring, the following cube root scaled distances should be maintained as a minimum distance between residential buildings and different types of blasting operations:

- Coal highwall 70 m kg$^{-1/3}$ $=$ 180 ft lb$^{-1/3}$
- Coal parting 200 m kg$^{-1/3}$ $=$ 500 ft lb$^{-1/3}$
- Quarries and mines 100 m kg$^{-1/3}$ $=$ 250 ft lb$^{-1/3}$
- Construction and excavation 200 m kg$^{-1/3}$ $=$ 500 ft lb$^{-1/3}$
- Unconfined blasting 320 m kg$^{-1/3}$ $=$ 800 ft lb$^{-1/3}$

14.5 Reduction of Air Blasts

If, for example, the explosive charge is confined in rock, a large reduction of overpressure and momentum is achieved. We can often judge directly by listening if the charge is stemmed or if it is detonated with a small burden. Wiss and Linehan [1978] and Siskind *et al.* [1980] divided the causes of an airblast into several mechanisms. The most important are:

- The air pressure pulse
- The rock pressure pulse
- The gas release pulse
- The stemming release pulse

The *air pressure pulse* usually has a large amplitude and it is caused by direct rock displacement at the rock face and in the cratering zone. A deeper charge or a better-confined charge produces a lower pressure pulse. The frequency is considerably lower than the frequencies from the rock pressure pulse, the gas release pressure pulse, or the stemming release pulse. It arrives at the measuring guage after the rock pressure pulse.

The seismically-induced *rock pressure pulse* is usually of such low amplitude that it can be neglected. It arrives first of all pulses at the monitoring equipment.

The *gas release pulse* is born when the confined expanding detonation gaseous products vent through the fractured rock. The pressure pulse arrives after the air pressure pulse. Together with the stemming release pulse, the gas release pulse is the pulse which disturbs the people in the neighborhood.

The *stemming release pulse* is caused by gas escaping from the blown-out stemming. It is characterized by a high-frequency wave superimposed on the air pressure pulse.

Detonating cord is still, to some extent, used to interconnect downlines for initiating detonation in the explosive in drillholes. However, since the sharp air shock wave from the detonating cord is a major cause of disturbance and is annoying to the neighbors, the practice is rapidly being discontinued except in remote locations.

Detonating cord that runs through the stemming into the explosive and the primer should be avoided if there are complaints from the neighborhood. When the detonating cord detonates through the stemming, a channel is created which will increase the stemming release pulse amplitude.

As each detonating hole will produce its own fingerprint in the air pressure time history diagram, it is possible to arrange the delays in such a way that annoyance can be minimized. For example, a long round with 17 ms delay between the holes in the row can be connected and sequentially fired in such a way that the initiation sequence is moving away from the sensitive object, thereby minimizing the risk for building up more

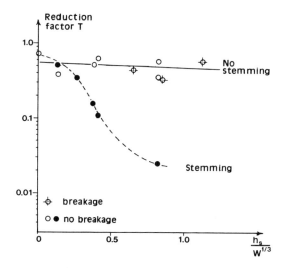

Figure 14.5. Reduction factor T as a function of reduced unloaded hole depth.

severe blast waves propagating in a direction toward the sensitive object. Alternatively, the delay times can be increased to reduce interaction.

R. Holmberg and G. Persson [1978] investigated the reduction factor for stemmed and unstemmed single shots. A reference charge of 28 g TNT was initiated some 10 to 20 ms before the explosive charge in the borehole was initiated. This 28 g TNT reference charge was shot in order to correct for any influence of wind and temperature. Three holes had such a burden that breakage occurred. The other shots were totally confined so that only stemming ejection and cratering could take place. Hole diameters were 38 and 76 mm. The Dynamex charges varied in weight from 1.88 to 25.2 kg, and the unloaded hole lengths varied from 0 to 1.38 m for $\phi 38$ mm and from 0 to 2 m for $\phi 76$ mm. Pressure measurements were made at a distance of about 20 m from the 28 g TNT charge and at a distance of 20 to 85 m from the boreholes.

The results of the tests are described in terms of a reduction factor T which is the ratio between the front pressure of the blast wave from the explosive charge in the borehole and the estimated front pressure of the blast wave from an unconfined explosive charge with the same weight:

$$T = \frac{p_c}{p_u}. \qquad (14.10)$$

The estimated front pressure p_u was calculated from the formula

$$p_u = k \cdot 0.7 \frac{W^{1/3}}{R} \qquad (14.11)$$

where k is a constant related to the type of explosive; for Dynamex, it was measured to be 0.75.

Figure 14.5 shows the results. The reduction factor T is plotted as a function of the reduced unloaded hole depth $h_s/W^{1/3}$. The stemming material had a grain size in the range 0 to 18 mm. A charge in a drillhole without stemming gave a front pressure which was 40 to 50 % of the front pressure from an unconfined charge with the same weight.

For a Dynamex charge in a ϕ76 mm hole with a 1 m unloaded hole length, the front pressure was reduced to 48%; for the same diameter hole with a 1 m stemming length, the pressure was reduced to 15% of the front pressure for an unconfined charge of the same weight. If a stemming of 2 meters was used, the reduction was even greater.

14.6 Focusing Effects

A blast wave emitted from a bench blast will propagate through the air for long distances. Close to the blast, the shock velocity is supersonic, but it quickly decreases to the speed of sound. If the air were a homogeneous medium without temperature or humidity gradients and without wind, the velocity would be constant and the shock front would spread outward in a spherical symmetric pattern. The over-pressure would decay as the inverse of the distance squared.

Predictions of air blasts from blasting operations must always take into account the wind speed and the temperature in the surrounding air.

The velocity of sound in air at standard temperature and pressure (STP) is 340 m/s (331 m/s at 0°C). The density of air at STP is 1.29 kg/m^3. In air or other gases, the velocity of sound increases proportionally with the square root of the absolute temperature; the velocity increase is approximately 2% for each 10°C temperature increase.

Usually, the temperature decreases with altitude — an average of the gradient is 0.6°C per 100 m. This is, however, just an average and large discrepancies may occur. Sometimes, the temperature is constant with altitude (isothermal) and, sometimes, it will even increase with altitude (temperature inversion).

Wind and temperature inversions are the most frequent causes of unexpected blast overpressures long distances away from a blasting site. Sometimes overpressures can occur which are as much as an order of magnitude larger than under normal weather conditions.

A sailor, an airplane pilot, and a golfer are aware that windspeed (usually) increases with altitude. (This is the reason why tall masts are common on fast sailing boats.) This increase in windspeed with altitude can adversely increase the air blast pressure at a point situated downwind from the blast site.

Baker [1973] reported the successful use of multiplication factors for estimating the air blast pressure under various sound velocity gradient conditions relative to the case of a homogeneous atmosphere (see Figure 14.6).

The nearest airport is usually able to give information on the local atmospheric conditions, although short-duration and short-range variations in atmospheric conditions make these unreliable. The best information is that obtained immediately before the blast from a small shot or from a careful logging of the atmosphere at the site. This is seldom a practical possibility for the blaster as it is tedious and expensive work to log these atmospheric conditions just before every shot. The quarry of the open pit mine blaster who carries out blasting at the same site over a long period of time can learn empirically through air blast measurements how the weather influences the magnitude of the air blast pressures. Then, with good planning, a round may be postponed a few hours or to the next day when the atmospheric conditions may be more favorable.

A typical sound velocity gradient and the corresponding blast wave ray paths are shown in Figures 14.7 and 14.8.

Temperature inversions are most likely to occur in the early morning or late afternoon. It might be a good idea to avoid blasting at these times.

CATEGORY	DESCRIPTION		MULTIPLICATION FACTOR
1	SINGLE NEGATIVE GRADIENT	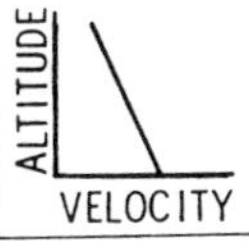	0
2	SINGLE POSITIVE GRADIENT	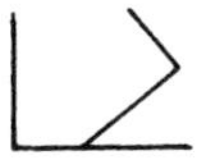	5
3	ZERO GRADIENT NEAR SURFACE WITH POSITIVE GRADIENT ABOVE	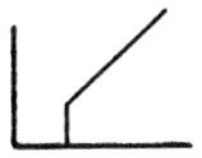	10
4	WEAK POSITIVE GRADIENT NEAR SURFACE WITH STRONG POSITIVE GRADIENT ABOVE		25
5	NEGATIVE GRADIENT NEAR SURFACE WITH STRONG POSITIVE GRADIENT ABOVE	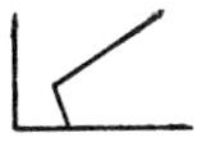	100

Figure 14.6. Various types of velocity gradients and increases in intensity at a focus for each type [from Baker, 1973].

14.7 Inform the Neighbors

Many precautions can and should be taken to avoid air blasts; but when all is said and done, it is important to remember that good relations with the neighbors is probably the most effective insurance against complaints. A neighbor who knows what the blasting operation is all about and is well informed about its benefits to the area and society as a whole is far less likely to complain about the noise from an unusually loud blast. It is a very inexpensive courtesy that pays back handsomely in many ways to INFORM all neighbors within earshot in advance, and at regular intervals thereafter, of the purpose of the blasting operation and the way it will affect the neighbors. Point out the positive effects as well as the negative. Often, the arrival of a blasting operation means new jobs and additional business that can generate tax dollars to pay for schools and other benefits in the area. An open day, with hot dogs and balloons for the kids, allowing the neighbors to see for themselves what the blasting site looks like, is often a very effective medicine against complaints about air blast. If, in spite of all precautions damage does occur, BE GENEROUS. The immediate repair of a window or replacement of a broken picture frame is a negligible cost in most large-scale blasting operations. The later repair of a festering complaints situation that was not taken care of immediately can be extremely costly and time consuming. Also, whenever possible, INVOLVE neighbors in the operation. This may not be possible when blasting in a city, but in a remote area with just a few neighbors, a little added business or a part-time job goes a long way toward creating a positive atmosphere and open lines of communication — these are the best guards against the fear and worry of the unknown which is the cause of almost all complaints involving air blast damage.

386   Chapter 14. Air Blast Effects

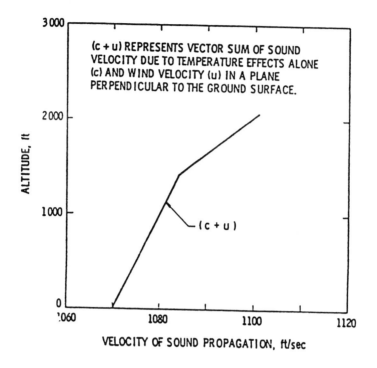

Figure 14.7. Typical sound velocity gradient due to temperature and wind velocity effects.

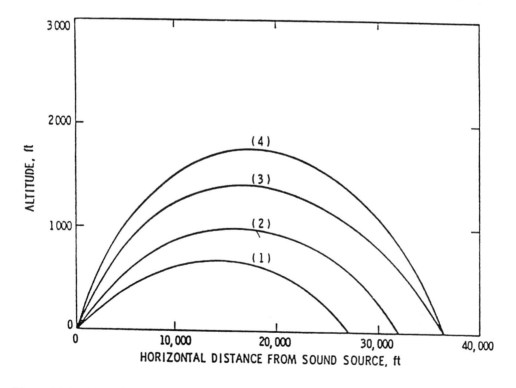

Figure 14.8. Paths of sound waves in the atmosphere for the gradient shown in Figure 14.7.

Chapter 15

Toxic Fumes

This chapter deals with the fumes created by detonating explosives and blasting agents which may contain reaction products that pose a health or environmental hazard. The bulk of the reaction products from a detonating oxygen-balanced explosive are harmless. Small amounts of toxic reaction products will, however, be formed as a result of deviations from oxygen balance, incomplete reaction, or secondary reactions with the atmospheric air. We will briefly discuss these effects here, and also techniques for measurement and control of toxic fumes.

In an ideal detonation, an oxygen-balanced explosive with the atomic constituents carbon, hydrogen, nitrogen, and oxygen (a $CHNO$ explosive) would form only the gaseous reaction products carbon dioxide (CO_2), water vapor (H_2O), and nitrogen (N_2). In real detonations, due to incomplete reaction of the explosive and subsequent reactions with the surrounding air, other reaction products will always be present, and some of these are toxic if the concentration becomes high enough. The primary toxic fumes produced are carbon monoxide (CO), nitrous oxide (NO), and nitric oxide (NO_2). The total content of the latter two is often jointly called the NO_x content. The maximum allowable content of CO in air (in Sweden) is 10 parts per million (ppm), that of NO_x is 2 ppm. When the deviation from oxygen balance is large, other toxic products may also be formed. Some explosives or explosive ingredients are toxic in themselves: for example, nitroglycerin (NG) or nitroglycol (EGDN). In incomplete reaction of dynamite, detectable amounts of NG and/or EGDN are often found in the reaction products — breathing air contaminated with the reaction products from dynamite may lead to a headache of the same kind as that produced by breathing air containing NG vapor in a magazine or by skin contact with dynamite. In incomplete reaction, other molecule fragments of the ingredient explosives, oxidizers, or fuels may also be formed, some of which may also be toxic.

The amount of nonideal detonation products formed depends on a number of factors: the explosive composition and its homogeneity, the effect of water on the explosive after being loaded into a wet drillhole (water resistance), the velocity of detonation, the charge diameter, the loading density, the type of initiation, the type of cartridge wrapper, and especially the confinement of the explosive. Before and during the detonation, additional reactions can also occur between the explosive and the surrounding rock; for example, when the rock contains sulphide or other reactive components.

Table 15.1. Fumes classification of explosives in the USA.

A. Permissible explosives rated according to the US Bureau of Mines classification (Noxious gases measured: CO, NO_2. NO, and H_2S).

Fume Class	Noxious Gases (cu. ft./lb.)	Noxious Gases (l/kg)
A	< 1.25	< 78
B	1.25–2.50	78–156

B. Non-permissible explosives rated according to the Institute of Makers of Explosives classification. (Noxious gases measured: CO, and H_2S).

Fume Class	Noxious Gases (cu. ft./lb.)	Noxious Gases (l/kg)
1	< 0.36	< 22.5
2	0.36–0.75	22.5–46.8
3	0.75–1.52	46.8–94.9

15.1 Fume Classification of Explosives

Several laboratory methods have been developed in order to classify explosives for underground blasting. Among these are the Bichel Gauge, the Crawshaw-Jones Apparatus, and the Ardeer Tank method. In these methods, a small sample of the explosive is detonated in a closed tank, and the reaction product composition is then analyzed for its content of toxic fumes. The concentrations measured with these laboratory-scale methods are generally quite different from the real life concentrations measured in full-scale field experiments. Most of them work with a very small quantity of the test explosive to be detonated, which influences the detonation performance. Many blasting agents with a large critical diameter cannot be tested because they do not detonate at all in the small sample size of these tests.

In the USA, there are two different fume classifications for explosives. One concerns permissible explosives for use in underground coal mines and one concerns non-permissibles. Both classifications are based on the Bichel Gauge Test, which is a heavy closed bomb with an inner volume of about 15 l where one cartridge of the tested explosive is fired. The gases are collected and analyzed. Table 15.1 shows the US allowable limit values. The US Bureau of Mines measures CO, NO_2, NO, and H_2S. The Institute of Makers of Explosives (IME) values are only for CO and H_2S.

Elitz and Zimmerman [1978] report on a West German test method for simulation of explosive used in rock blasting. The test is carried out in a rock chamber with a volume of 30 to 100 m^3. A total charge of about 7 kg explosive to be tested are fired in 7 to 10 shotholes drilled in the rock wall of the chamber. Although the confinement is not always constant as the burden and spacing might vary somewhat, the method gives more reliable results than the laboratory-scale experiments. If the round generates less than 32 l/kg CO and less than 4 l/kg NO_x, the explosive is considered acceptable for underground use.

Table 15.2. NITRODYNE calculated fumes from ANFO with various oxygen balances.

Explosive mixture	Oxygen balance	Calculated gaseous detonation products (mol/kg)						Heat of reaction at constant pressure
ANFO	(%)	H_2O	CO_2	N_2	CO	H_2	O_2	CH_4	(MJ/kg)
92/8	-9.41	26.46	2.42	11.49	3.19	2.58	—	0.03	3.34
94/6	-2.06	27.59	3.42	11.74	0.81	0.48	—	—	3.69
94.56/5.44	0	27.78	3.83	11.81	—	—	—	—	3.79
95/5	1.61	27.55	3.52	11.87	—	—	0.50	—	3.72
96/4	5.29	27.04	2.82	11.99	—	—	1.65	—	3.18

15.2 Computer Calculations of Reaction Products

A number of thermodynamic computer codes are available for calculation of the equilibrium reaction product composition and explosive performance. Codes such as the BKW code or the TIGER code can be used for modeling detonation performance. Other codes, such as the NITRODYNE code, can be used to model the constant volume explosion performance.

These codes are not useful for making predictions of the actual fumes content of the reaction products from detonation of a commercial explosive in a drillhole. They compute the reaction product composition under the theoretical assumption that all reactions go to completion, and that all reaction products are in thermodynamic equilibrium with each other. They do not consider the reaction kinetics, i.e., that reactions take time, so that some reaction products do not have time to be formed, and others are retained because they are slow to react. Nor do they take into account the reaction between the escaping hot detonation products and the surrounding air. Therefore, they can merely be used for predicting trends, such as the likely direction of change in toxic fumes content when the explosive composition is changed, or when the type of cartridge wrapper material is changed. The equilibrium thermodynamic calculations of reaction product compositions generally give very low contents of toxic fumes — much lower than the real fumes contents of even the best commercial explosives — because they only consider complete reaction.

Table 15.2 shows some calculated results for ANFO when the oxygen balance is shifted from a deficit to an excess of oxygen. The program used was NITRODYNE which does not take into account the formation of NO.

15.3 On-Site Measurements

Gunnar Persson [1983] at the Swedish Detonic Research Foundation carried out extensive measurements of toxic fumes and developed a test method which has found extensive use in Sweden. A test chamber in the Stripa Mine in Sweden (see Figure 15.1) was established in 1976. The rock chamber had a volume of 240 m^3. A second chamber was established at LKAB in Malmberget, and a third is now available at Nitro Nobel AB's test site in Västra Sund. Blasting experiments were carried out at these sites using unconfined charges, charges confined in steel tube, single charges in drillholes in rock with and without breakage, and in full-scale tunnel rounds.

Immediately after a blast, the chamber was sealed off and a fan was started to thoroughly mix the fumes with the air inside the chamber. As the volume of the

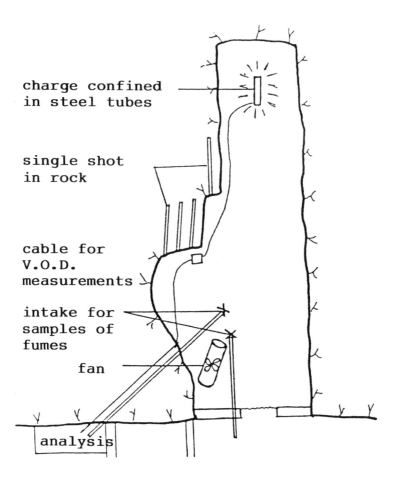

Figure 15.1. Test chamber in Stripa Mine.

chamber was well known, analysis of the fumes could be made at various points in time after the blast (Table 15.3). The fumes escaping from a tunnel round expand after the detonation. 1 kg (2.2 lbs) of explosives will result in about 0.9 m^3 (32 cu. ft.) of gaseous detonation products at STP. This is enough to pollute almost 1000 m^3 of air, based on the toxic fume limit values established by Swedish authorities. Since the temperature in the gas is initially above the ambient temperature, the fume cloud will first flow close to the roof of the tunnel until it has been diluted with air and cooled off. The forced ventilation system installed in the tunnel is normally shut down before the blast, and started soon after. It blows fresh air into the area of the blast so that the fume cloud will be further diluted with air and forced to flow slowly out through the tunnel. Figure 15.2 shows a fume cloud moving with the speed of 0.5 m/s. At higher flow velocities, the front is steeper and the distance between front and tail is shorter.

The oxygen content in the explosive, the pressure-time profile in the detonation, and the flame temperature of the detonation products all affect the formation of nitric oxide (NO) from nitrogen (N_2). The NO evolved will be further oxidized to nitrogen dioxide (NO_2) when it comes into contact with the oxygen in the mine atmosphere by the reaction

$$2NO + O_2 \longrightarrow 2NO_2.$$

Table 15.3. Average emission of fumes from tested commercial explosives [from G. Persson, 1983].

Type of gas	% of total volume emitted gas	Comment
H_2O, steam	35–65	
N_2, nitrogen	20–30	
CO_2, carbon dioxide	10–40	Stable gas
H_2, hydrogen	0.5–10	Explosive with an oxygen excess
O_2, oxygen	0.1–3.0	
CH_4, methane	<0.5	
CO, carbon monoxide	0.5–10	Stable gas
NO_x, nitric oxides	0.1–2.0	NO is oxidized to NO_2 which reacts with H_2O and forms HNO_3 and HNO_2
NH_3, ammonia	0.0–0.1	
Other gaseous organic species	<0.1	
Lead dust	40 mg/cap	From the caps

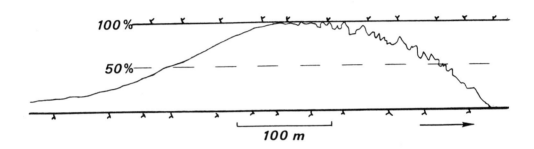

Figure 15.2. Shape of a fume cloud 300 m away from a blasted tunnel round.

Figure 15.3 shows the oxidation rate of NO as a function of the initial concentration of NO at a temperature of 9°C. The line shows at what time 50% of the NO has formed NO_2. High NO_2 concentrations combined with a high relative humidity can result in the formation of small amounts of nitric acid (HNO_3). The reaction proceeds according to the formula

$$2NO_2 + H_2O \longrightarrow HNO_2 + HNO_3.$$

Ammonia (NH_3) will form in oxygen-deficient explosives, although it usually disappears after 10 to 15 minutes — probably due to its reaction with the relatively acid water vapor produced in the detonation. NH_3 can also be formed during charging of drillholes, if ammonium nitrate comes into contact with calcium carbonate in cement or shotcrete used for support work. This is immediately detectable by the smell from this pungent gas. The reaction formula is

$$NH_4NO_3 \longrightarrow NH_3 + HNO_3.$$

Table 15.4. Threshold values for toxic fumes.

Fume	Sweden (8 hr working day)	USA (Threshold values, ppm)
CO	25	50
NO_x	20	
NO_2	2	5
NO		25
NH_3		50
CO_2		5000

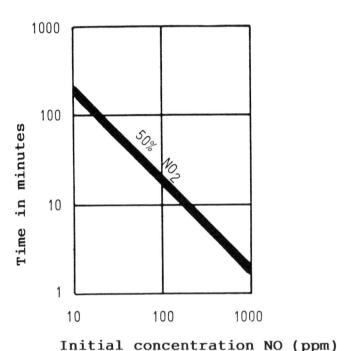

Figure 15.3. Oxidation rate of NO as a function of its initial concentration [from G. Persson, 1983].

Typically, only the carbon monoxide (CO) and the nitrogen oxides (NO and NO_2) reach such high concentrations that it becomes dangerous to be exposed to the fumes after a blast. Carbon monoxide (CO) is a colorless, odorless gas formed by the incomplete combustion of organic material. Its density is slightly lower (1.25 kg/m^3) than that of atmospheric air (1.293 kg/m^3), but it is extremely unlikely that separation occurs in the mine or in the tunnel. The gas is highly toxic if inhaled. Carbon dioxide's affinity for blood hemoglobin is considerably higher than that of oxygen. Inhalation can cause suffocation if the CO concentration is high enough.

Nitric oxide (NO) is also colorless and highly toxic if inhaled. NO is a strong irritant to skin and mucous membranes. Through oxidation, it forms nitrogen dioxide (NO_2). NO_2 is a reddish-brown gas, and inhalation is fatal at high concentrations. If the fume cloud has NO_2 concentrations exceeding 30 ppm, the color is orange/red when one looks through the cloud against a white light source. Table 15.4 reviews threshold values of exposure concentrations established in Sweden by the Labor Safety Board [Arbetarskyddsstyrelsen, 1974] and in the USA [ACGIM, 1970].

Investigations made by G. Persson showed that 4 to 8 % of the fumes from tunnel rounds is trapped in the muck pile. By intensive watering of the muck pile for 1 hour, it is possible to evacuate 25 to 50 % of the trapped gas.

Fumes coming from nitroglycerin explosives contain small amounts of nitroglycerin and/or nitroglycol which both have the effect of dilating blood vessels and thereby cause headaches and lower the blood pressure. Skin contact with the explosive is perhaps the most important form of exposure; but inhalation of the vapors during charging or inhalation of the vapors of unreacted or partially reacted explosive after the blast will also cause headaches.

The long-term effects of exposure to nitroglycerin have been studied extensively [Hogstedt and Axelson, 1977; Hogstedt and Andersson, 1979; and Hogstedt 1980] following reports that personnel regularly exposed to nitroglycerin for a long time during working hours might suffer from symptoms of abstinence when no longer exposed. Several cases were reported from Germany by Symanski [1952]. They described cases of sudden death in dynamite workers 24 to 72 hrs after the last exposure to dynamite, such as during a holiday or leave period. The studies by Hogstedt suggested that dynamite factory workers who had been exposed to the nitrate esters NG and EGDN for periods exceeding 20 years during the 1950 and 1960 time periods might have experienced an increased risk of dying from diseases of the cardiovascular system. It is to be noted that both the German and Swedish dynamite workers had been exposed for very long time periods in plants where the air concentration of nitrate esters had likely been considerably higher than 1 mg/m^3. Also, in the past, no protective clothing or gloves to reduce skin contact were worn in these factories. It is unlikely, however, that even long time exposure to the low levels of nitrate ester vapors now allowed in modern dynamite plants or the even lower levels found in modern well-ventilated underground blasting sites would cause the type of health hazards indicated in these studies.

15.4 Influence of Confinement

Formation of the toxic fumes NO_x is heavily influenced by the confinement of the charge. Figure 15.4 shows the results from various blasting geometries. The highest NO_x concentrations are found when firing an unconfined charge in air. The tendency to form CO is also highest for the unconfined charge, but CO formation is less affected than the formation of NO_x by the degree of confinement. It is obvious that confinement must be considered for any type of fume characterization of explosives. During the test explosion, the explosive and its reaction products should be exposed to, as closely as possible, the same pressure time history as occurs in the drillhole in the actual detonation of the explosive in a production blast. Figure 15.5 shows two methods proposed by G. Persson for fume classification of explosives. The more promising method is the gas pendulum in which the explosive is placed in a cavity at the interface between two steel weights, suspended in wires so that they initially touch. As the explosive is detonated, the weights are pushed by the reaction product pressure and swing freely away from each other, releasing the reaction product gas to the atmosphere. The gas pendulum has a specially prepared casing to hold the explosive, fabricated of a porous sand material. In addition to accelerating the two weights, the explosive fired in the pendulum also performs work by crushing the inner tubing.

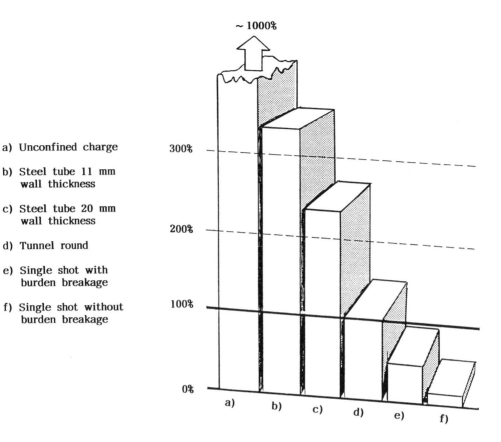

Figure 15.4. Influence of confinement upon the formation of NO_x from a dynamite. 100 % is equal to the NO_x formed at blasting a 10 m² tunnel round with a specific charge of 3 kg/m³ [from G. Persson, 1983].

It should be noted that the influence of the confinement is to increase the time during which the pressure in the reaction volume is high enough to allow the reactions to proceed rapidly. As the reaction rates in composite explosives with granular ingredients is governed by the rate of grain burning within the reacting explosive, the grain size of ingredients is of great importance to the explosive's fume characteristics. An explosive containing coarse-grained ingredients takes longer to burn at a given pressure, and therefore generally produces greater amounts of toxic fumes than one containing finer-grained ingredients under a given confinement, even if both explosives are perfectly oxygen balanced.

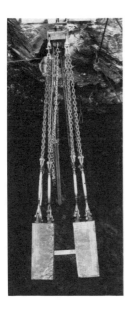 

Figure 15.5. Methods suggested for fume classification of explosives: (a) gas pendulum, (b) gas mortar [from G. Persson, 1983].

15.5 Recommendations

It is possible to considerably reduce the amount of toxic fumes formed in tunneling if the following recommendations and procedures are adhered to during the planning and charging of the blast:

1. Use an oxygen-balanced explosive with good fume characteristics.
2. Use alignment devices when drilling so most holes are slanted slightly upward. This prevents water from accumulating in the drillhole, contaminating the explosive and affecting its detonation performance.
3. Do not use detonating cords in ANFO-loaded small diameter holes as it might not initiate the ANFO to complete reaction, resulting in large amounts of toxic fumes. The cord itself is strongly oxygen deficient, and by itself generates about 3 l CO per meter length of cord.
4. Leave an unloaded (or preferably stemmed) hole length equal to 10 times the hole diameter at the mouth of the hole in tunneling and drifting. Filling explosives all the way to the collar does not contribute significantly to the breakage process, but it produces a lot of toxic fumes.

Chapter 16

Metal Acceleration, Fragment Throw, Metal Jets, and Penetration

It is part of a well-rounded education in rock blasting and explosives engineering to know the basics of how metal parts and fragments (and for that matter parts and fragments made of other materials as well) are accelerated by an explosive charge; It is important to know, given their initial velocity, how far a metal fragment will fly in air; and how deep will it penetrate into a target. These matters, apart from being part of a general education, also often come up in considering how long safety distances should be and in designing safety barriers. On another plane they are also, of course, the fundamentals for the design of warheads and armor protection for military purposes.

16.1 The Gurney Equations for Metal Plate Acceleration

This section summarizes the elegant application of classical mechanics to the problem of how to approximately calculate the acceleration of a metal part in contact with an explosive charge, which was first given by Ronald W. Gurney in a report of restricted circulation published during the Second World War [Gurney, 1943]. A detailed account of Gurney's equations and a complete collection of literature references to subsequent work in this area is given in an article by Kennedy [1972] and in a more recent article by Jones, Kennedy, and Bertholf [1980]. Gurney considered first the simple case of an explosive slab having metal plates of equal thickness on both sides. He then treated other, asymmetrical cases of metal covered slabs of explosive, and the cylindrical and spherical cases of a metal tube or a metal sphere filled with explosive.

As shown by Kennedy [1972], the Gurney approximations are in very good agreement with computer modeling results. They are also amply verified by experiment and provide a useful tool in engineering calculations in a whole range of explosive/metal interaction design problems.

16.1.1 Explosive Slab between Metal Plates of Equal Thickness

Consider, as in Figure 16.1, an infinite slab of explosive with initial thickness $2x_o$, initial density of ρ_o and mass per unit area $2C$, with both surfaces are covered with sheets of metal, each of mass M per unit area. We are interested in finding a simple way of approximately calculating the plate velocity V to which the metal is accelerated following the detonation of the charge. We will make the following assumptions regarding the acoustic time τ_a (which is defined as the time for a sound wave to traverse the explosive's reaction products between plates), the time for expansion $\tau_{expansion}$, and the gas velocity u:

1. Assume that τ_a is much smaller than $\tau_{expansion}$.
2. Assume that the expansion of the explosive's reaction products is isentropic.
3. Assume that the gas pressure is constant across the gas between the two plates, i.e., that the density of the gas between the plates is only a function of time and that the gas velocity u therefore varies linearly from zero on the line of symmetry to equal the metal plate velocity at each plate surface.
4. Assume that the internal energy of the metal after it has been accelerated is much smaller than its kinetic energy.

In other words, in the first approximation, Gurney neglected the angle between the plates (this was calculated separately as shown below) and regarded only the dynamic one-dimensional motion of the plates and the reaction product gases between them. He also neglected the thermal energy deposited in the plates as a result of the shock compression during the initial passage of the detonation front along the plate surfaces. A further assumption is that the expansion of the gas is slow enough for pressure equilibration throughout the space between the plates to occur at all times. This assumption is particularly important for the elegance of Gurney's treatment since it results in a gas velocity that varies linearly with the position between the plates. These simplifying assumptions were checked and found reasonable over a wide range of conditions. A good, detailed discussion of these and other assumptions in the Gurney approach can be found in Kennedy [1972].

It is obvious that no material will cross the plane of symmetry which, therefore, can be replaced by a rigid barrier. As the gas expands and accelerates the metal plates, let us consider the stage where the metal plates are separated by a distance $2L$. At that point, the constant density of the gas between the plates is ρ, so that

$$\rho = \rho_o \frac{x_o}{L} = \frac{C}{L} \tag{16.1}$$

$$u = V \frac{x}{L}. \tag{16.2}$$

We introduce the effective specific internal energy E_g of the explosive, the Gurney energy (i.e., that part of the explosion energy which is transformed into kinetic energy of the metal and gas):

$$CE_g = \frac{1}{2}MV^2 + \int_0^L \frac{1}{2}\frac{C}{L}V^2\frac{x^2}{L^2}dx = \frac{1}{2}MV^2 + \frac{1}{6}CV^2 \tag{16.3}$$

$$V^2 = \frac{2E_g}{\frac{M}{C} + \frac{1}{3}} \tag{16.4}$$

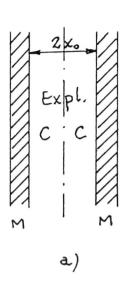 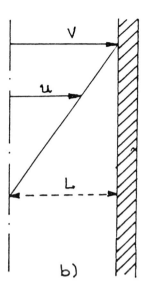

Figure 16.1. Symmetric acceleration of two metal plates by a slab of explosive (a) at the initial condition and (b) during acceleration. The plane of symmetry has been replaced by a rigid barrier.

$$\frac{V}{\sqrt{2E_g}} = \left[\frac{M}{C} + \frac{1}{3}\right]^{-\frac{1}{2}}. \tag{16.5}$$

16.1.2 Explosive Slab between Metal Plates of Unequal Thickness and Explosive Slab with a Single Metal Plate

The case described in the previous section is often called the *flat, symmetric sandwich* case. The case of an explosive slab between metal plates of unequal thickness is called the *flat, asymmetric sandwich* case. A special subcase of the flat asymmetric sandwich is the *open-faced sandwich*, in which one of the metal plates is of infinitely small thickness. These cases can be solved in a similar manner to that given above.

16.1.3 Explosive-Filled Metal Tube and Metal Sphere

For the cylindrical case of a metal tube filled with explosive (Figure 16.2), we can make similar assumptions as for the symmetric slab case, i.e., that there is pressure equilibration within the tube at all times, and that therefore the density at any given time during the acceleration is constant within the tube and the gas velocity varies linearly with radial position, that the expansion of the gas is isentropic, without shocks, and that the shock heating of the metal is negligible. The variable R describes the position of the inner surface of the metal tube as a function of time, and replaces L used in the parallel plate case.

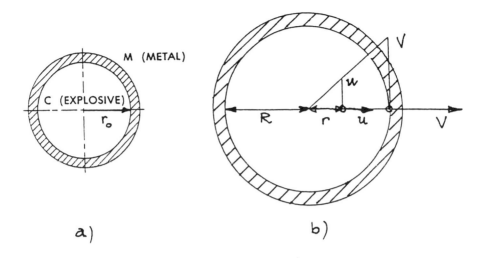

Figure 16.2. Symmetric acceleration of a metal tube filled by a detonating explosive. a) The initial condition, b) during acceleration.

$$\rho = \rho_o \frac{\pi r_o^2}{\pi R^2} = \frac{C}{\pi R^2}. \tag{16.6}$$

$$CE_g = \frac{1}{2} MV^2 + \int_0^R \frac{1}{2} \frac{r^2 V^2}{R^2} \frac{C}{\pi R^2} \cdot 2\pi r\, dr$$

$$= \frac{1}{2} MV^2 + \frac{1}{4} CV^2 \tag{16.7}$$

$$V^2 = \frac{2E_g}{\dfrac{M}{C} + \dfrac{1}{2}} \tag{16.8}$$

$$\frac{V}{\sqrt{2E_g}} = \left[\frac{M}{C} + \frac{1}{2}\right]^{-\frac{1}{2}} \tag{16.9}$$

The case of an explosive-filled spherical shell has a similar solution:

$$\frac{V}{\sqrt{2E_g}} = \left[\frac{M}{C} + \frac{3}{5}\right]^{-\frac{1}{2}} \tag{16.10}$$

Since the main part of the energy transfer to the metal plates occurs in the pressure region above about 1 kbar, the Gurney energy E_g, which represents the energy actually transferred to the plates, is generally found to be between 60 and 70% of the total explosion energy. The remainder of the energy is expended in expansion of the gas to its very large final volume and is not effective in accelerating the plates.

For commercial blasting explosives, E_g is only 10 to 15% of the total explosion energy, because their CJ-pressures are low and the pressure falls below the cutoff pressure at an early stage.

16.1.4 Summary of Gurney Equations and Resulting Velocities

The results of the Gurney calculations are presented below. The solutions for the dimensionless Gurney velocity $\dfrac{V}{\sqrt{2E_g}}$ as functions of the metal loading factor M/C are given for the five simple geometries — symmetric sandwich, asymmetric sandwich, open-faced sandwich, cylinder, and sphere — in Figure 16.3. The Gurney energies and specific impulses for some different explosives are shown in Table 16.1. Figure 16.4 gives the dimensionless Gurney velocity as a function of the metal loading factor for the four cases: symmetric sandwich, cylinder, sphere, and open-faced sandwich.

16.1.5 Specific Impulse Delivered to a Large Metal Object

The equation for the metal plate velocity in the open sandwich case can be used to derive the specific impulse I_{sp}, i.e., the impulse per unit explosive charge mass delivered to a large object in contact with the explosive slab, an often occurring case, for example in demolition work. For example, in the demolition of a bridge, explosive charges 25 mm thick, 50 mm wide and 100 mm long may be attached to a heavy structural member. The important measure of the ultimate effect of the charge is the momentum transferred to the structure during the short time the gas pressure acts on the structure. The actual motion of the structure during this time is negligible, but the heavy structure has acquired the momentum which is equal to a small velocity multiplied by the large mass, and may continue moving until it fractures. The specific impulse for this case can be written

$$I_{sp} = \frac{M}{C} V. \tag{16.11}$$

Using V for the open-faced sandwich from Figure 16.3, for $M/C \gg 1$, we find

$$V = \sqrt{2E_g} \frac{C}{M} \sqrt{\frac{3}{4}}. \tag{16.12}$$

$$I_{sp} = \sqrt{1.5 E_g}. \tag{16.13}$$

Thus, the specific impulse is a function only of the explosive's Gurney energy.

16.1.6 Direction of Motion of the Accelerated Metal Plate.

An interesting observation is that, while the metal is thrown out to form an angle ϕ with the original plane of the plate, the direction of motion of each particle in the plate is in a direction which forms an angle $\phi/2$ with the normal to the moving plate. This can be understood if we assume that the metal plate does not change thickness or length as a result of the acceleration. This assumption was verified by several experiments by Lundberg and Gyldén [1962] in which a slab of explosive accelerated a metal plate with several fine grooves cut at right angles to the detonation propagation direction. The plate was photographed using a flash X-ray at a stage when the region of the plate with the grooves was in flight. The original position of the grooves was marked by the fine jet of metal emerging from each groove. From measurements of the still picture and the picture of the plate in motion taken from the same position, they determined that the distance between the groove positions on the plate in flight was equal to that on

Flat, symmetric sandwich:

$$\frac{V}{\sqrt{2E_g}} = \left[\frac{M}{C} + \frac{1}{3}\right]^{-\frac{1}{2}}$$

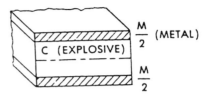

Asymmetric sandwich:

$$\frac{V_m}{\sqrt{2E_g}} = \left[\frac{1+A^3}{3(1+A)}\frac{N}{C}A^2 + \frac{M}{C}\right]^{-\frac{1}{2}}$$

where $\quad A = \dfrac{1 + 2\dfrac{M}{C}}{1 + 2\dfrac{N}{C}}$

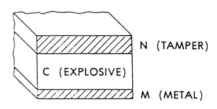

Open-faced sandwich:

$$\frac{V}{\sqrt{2E_g}} = \left[\frac{\left(1+2\dfrac{M}{C}\right)^3 + 1}{b\left(1+\dfrac{M}{C}\right)} + \frac{M}{C}\right]^{-\frac{1}{2}}$$

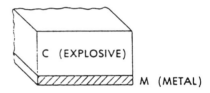

Cylinder:

$$\frac{V}{\sqrt{2E_g}} = \left[\frac{M}{C} + \frac{1}{2}\right]^{-\frac{1}{2}}$$

Sphere:

$$\frac{V}{\sqrt{2E_g}} = \left[\frac{M}{C} + \frac{3}{5}\right]^{-\frac{1}{2}}$$

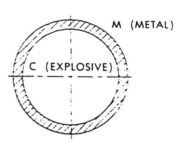

Figure 16.3. Metal acceleration by explosives. The Gurney solutions.

Table 16.1. Gurney energies and specific impulses of explosives [After Kennedy, 1972].

Explosive	Density (g/sm^3)	$\sqrt{2E_g}$ (km/sec)	I_{sp} (km/sec)	E_g (MJ/kg)	ΔH_d (MJ/kg)	$E_g/\Delta H_d$
RDX	1.77	2.93	254	4.28	6.28	0.68
Comp. C-3	1.60	2.83	245	3.99		0.64
TNT	1.63	2.68	232	3.58	—	—
		2.44	211	2.79	4.53	0.61
Tritonal (80/20)[1]	1.72	2.32	201	2.68	7.41	0.36
Comp. B	1.72	2.71	235	3.64	5.02	0.72
		2.77	240	3.81		0.76
		2.70	234	3.64		0.72
		2.68	232	3.60		0.72
HMX	1.89	2.97	257	4.44	6.20	0.72
PBX-9404	1.84	2.90	251	4.23	5.73	0.74
Tetryl	1.62	2.50	217	3.14	4.86	0.65
TACOT	1.61	2.12	184	2.26	4.10	0.55
Nitromethane	1.14	2.41	209	2.89	5.15	0.56
PETN	1.76	2.93	254	4.31	6.24	0.69
EL506D (DuPont Sheet Explosive)	1.46	2.28	197	2.60	—	—
		2.50	217	3.14	—	—
EL506L (DuPont Sheet Explosive)	1.56	2.20	191	2.43	—	—
Trimonite No. 1[1]	1.1	1.04	90	0.54	5.27	0.10

[1] Detonates nonideally.

Note: The preferred value of $\sqrt{2E_g}$ for each explosive is listed first. Equation 15.13 was used to calculate values of I_{sp} from values of $\sqrt{2E_g}$, or vice versa. Tritonal 80/20 is TNT/Al = 80/20.

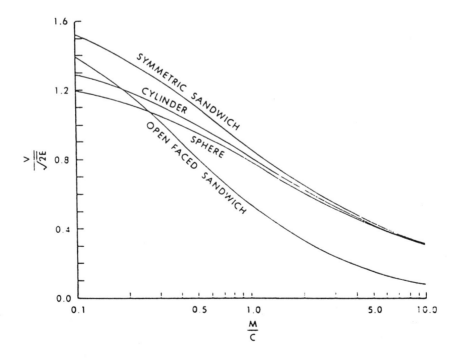

Figure 16.4. Dimensionless velocity of metal as a function of loading factor, M/C.

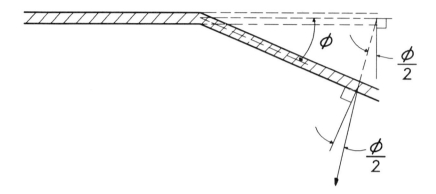

Figure 16.5. The direction of motion of the plate forms an angle $\phi/2$ with the normal to the moving plate.

the plate at rest before the shot. Also, within the accuracy of these measurements, the thickness of the plate was found to be the same in flight as at rest before the shot.

In other words, the direction of motion of the plate is along the base of an equisided triangle with the plate's original and moving surfaces as equal sides. The direction of motion forms an angle $\phi/2$ both with the plane of the original plate and with the plane of the plate in flight (See Figure 16.5).

The detonation velocity D, the plate velocity V, and the plate angle ϕ are related through the equation:

$$V = 2D \sin\left(\frac{\phi}{2}\right). \tag{16.14}$$

16.2 Shaped Charge Jet Formation

Having mastered the technique for calculating the velocity of a metal layer in contact with an explosive charge, we now can apply a similar technique to calculate approximately the velocity of the straight cone-shaped metal liner in a shaped charge, the velocity with which it converges on the axis, and the subsequent formation of a metal slug and fast moving metal jet. The simple theory presented here follows the original theory of conical-shaped charges proposed by Birkhoff et al. [1948]. It assumes, as can be amply justified, that the pressures encountered by the metal first in the explosive acceleration and then in the jet formation process are so much larger than the material's strength that the material, although still in the solid state, flows like an inviscid, incompressible fluid. An excellent, recent review of later, more detailed theories and the computational treatment of shaped charges can be found in the book by Walters and Zukas [1989].

A simple, cylindrical shaped charge is initiated to detonate from the center of the flat end surface of the charge. It has a straight conical metal liner. Knowing the metal liner thickness and density, and the half top angle of the cone, and using the Gurney equations, we can in principle calculate the velocity of each portion of the metal liner, and the direction of that velocity velocity vector. We may assume the thickness of explosive in the Gurney equations to be the thickness at right angles to the liner. (Using a similar approach, we can use the Gurney equations to calculate the velocity of each portion of the liner; even if the metal thickness varies, the liner is not a straight cone, and/or the explosive charge is not a straight cylinder.)

Obviously, one major component of the velocity of each liner element is radially toward the axis (another is in the forward, axial direction), and because of the rotational symmetry, the liner elements on a given radius of the liner will all arrive at the axis at the same time with the same velocity. All liner elements will successively flow toward the axis and collide along the axis. The point of collision will move forward relative to the liner's original position as liner elements from successively outer portions of the liner arrive at the axis.

Let us simplify the problem by following the flow in a coordinate system that moves with the collision point. Figure 16.6 shows the simple picture of the resulting flow. The material velocity toward the collision point is u, the angle between the flow direction and the normal to the charge axis is φ. In the moving coordinate system, the point of collision is a stagnation point which separates the liner in such a way that the side of the liner that was against the explosive now flows rearward relative to the stagnation point, while the side of the liner that formed the inside of the liner flows in the forward direction. The inside of the cone becomes the outside of the forward moving jet, the outside of the cone, that was against the explosive, becomes the outside of the rearward moving material, called the slug, because relative to the jet it is slow-moving. In the laboratory coordinate system, of course, the entire flow is in the forward direction.

For the proper functioning of the jet formation process, it is a necessary condition that the flow of material everywhere is subsonic. Thus, we can apply the well-known Bernoulli equation to write the energy of the flowing material

$$E + pv + \frac{1}{2}u^2 = \text{constant} \tag{16.15}$$

where E is the specific internal energy of the metal, p its pressure, and v its specific volume. In a similar way as when deriving the Gurney equations, we may safely assume the internal energy and the product of pressure and specific volume of the metal to be small compared to its kinetic energy $\frac{1}{2}u^2$. Therefore, the velocity of the slug and the velocity of the jet in the moving coordinate system are equal to the inflow velocity u of the cone material. Let us now assume that a fraction δ of the liner mass forms the jet, while the fraction $(1-\delta)$ forms the slug. We can now write the equation for conservation of momentum in the axial forward direction

$$-u \sin \varphi = \delta u - (1-\delta)u \tag{16.16}$$

from which we find

$$\sin \varphi = 1 - 2\delta. \tag{16.17}$$

Table 16.2. Relation between the influx angle φ and the jet mass fraction δ.

Influx angle φ	Fraction of liner mass in jet δ
0	1/2
30	1/4
90	0

Table 16.2 shows the mass fraction δ of the liner which forms the jet as a function of the inflow angle φ. (Note that the inflow angle is different from the conical liner's apex angle ϕ.) Normal shaped charges have values of φ around 30°. The tip velocity of the jet from a copper liner can reach velocities over 10 km/s. Since the explosive thickness (and therefore the liner velocity) decreases toward the periphery of the charge, the inflow velocity which is equal to the jet velocity becomes lower and lower; therefore, the jet will stretch and become longer and longer as it flies toward its target. For maximum penetration, a *standoff distance* has to be arranged between the charge and the target at the moment of firing, so that the jet can stretch to its most effective length. Note that the metal of the liner is in the solid state (the material strength is small compared to the inertial forces in the jet formation); therefore, the material can only stretch up to a certain limit, set by the ductility of the jet material. Thereafter, if the velocity gradient is sufficiently great, the jet breaks up into fragments, which may tumble in flight, reducing the penetration because the fragment hitting broadside may not pass through the hole cut by previous jet elements. Figure 16.7 shows a regular shaped charge jet in two late stages of formation. The last stage includes fragmentation of the jet after it has been stretched to its maximum continuous length. This jet was produced by a shaped charge with a straight conical copper liner. The intense shear of the material at the stagnation point can lead to local melting at the axis, and some regular shaped charge jets are indeed found to be hollow, thick-walled tubes rather than massive rods.

16.2.1 Explosively Formed Fragments

If the liner is made hemispherical or shaped like a forward concave disc, the stretch of the jet can be reduced at the expense of the velocity, and the resulting short *explosively formed fragment* or projectile can be formed into a shape similar to a long rod, with a length-to-diameter ratio of 10–30. Often, the projectile becomes hollow. Explosively formed projectiles can be made in different shapes. Figure 16.8 shows a high-speed photograph of a short explosively formed projectile in flight. Such projectiles can attain velocities of 3000 m/s or more. The photograph in Figure 16.8 was obtained using a streak camera with the slit at right angles to the path of the projectile. The velocity of the projectile and the slit in the film plane were synchronized in such a way that a true image of the projectile was formed on the film.

16.3. Target Penetration   407

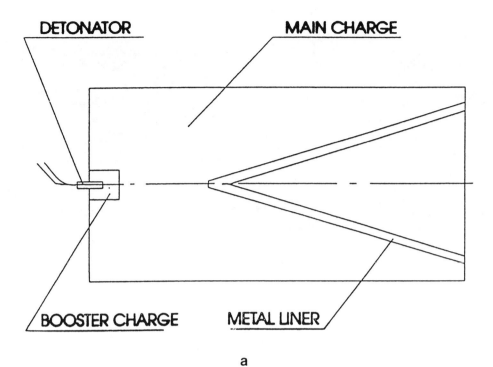

a

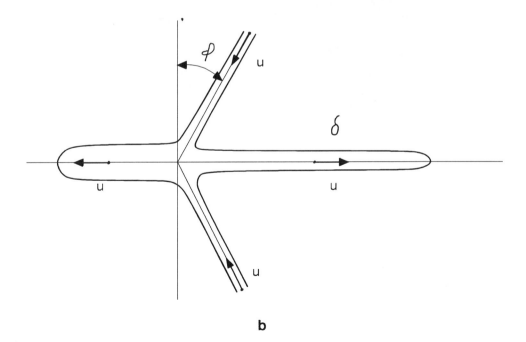

b

Figure 16.6 (a) An experimental shaped charge of simple design with low-density liner. (b) The jet formation process in a coordinate system that follows the collision point.

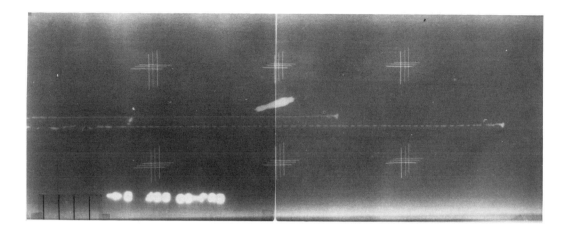

Figure 16.7. A shaped charge jet at two stages of extension (Photo courtesy of Gencorp Aerojet).

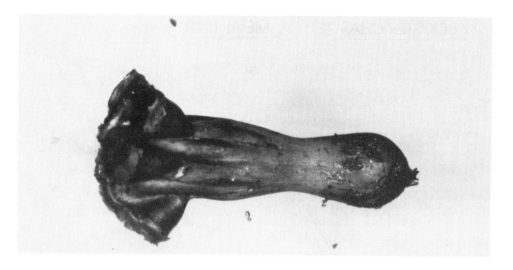

Figure 16.8. High-speed camera photograph of an explosively formed projectile in flight. Syncroballistic photo, effective exposure time is 0.4 μsec (Photo courtesy of Gencorp Aerojet).

16.3 Target Penetration

Let us now consider the high-velocity impact of a metal rod penetrating into a metal target. This is in a way similar to the jet formation process in reverse. In the same way as we did previously for the shaped charge jet formation process, we can use the Bernoulli equation to calculate the penetration of a rod-shaped metal projectile into a solid metal target. For a sufficiently high projectile velocity u, just as in the case of the jet formation described above, material strength effects are negligible. The collision of the rod with the material forms a crater, and the rod material and the displaced target material flow radially out from a stagnation point on the axis. In the first approximation, if the two materials have the same density, the length of rod consumed in producing a hole equals

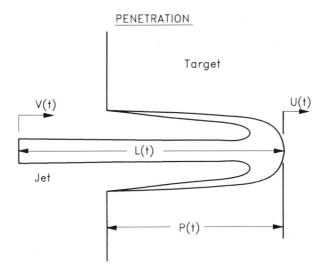

Figure 16.9. The collision process in rod penetration seen in a coordinate system which follows the stagnation point.

the depth of the hole. This becomes apparent if we consider the collision at the bottom of the hole in a coordinate system that moves with the stagnation point (Figure 16.9).

If the velocity of the collision point, the *penetration velocity* is U and the rod velocity is V, and if the target material and the rod have different densities, ρ_r and ρ_t, respectively, we can write the pressure p at the collision point equal to the stagnation pressure $\rho U^2/2$ for each of the two materials, the rod moving with velocity $V - u$ and the target material moving with velocity U

$$p = \frac{1}{2}\rho_r(V - U)^2 = \frac{1}{2}\rho_t U^2. \tag{16.18}$$

Thus, a projectile of length l_p will penetrate to a depth P, *the penetration*, into the target, where

$$P = l_p \frac{U}{V - U} = l_p \left(\frac{\rho_r}{\rho_t}\right)^{\frac{1}{2}}. \tag{16.19}$$

This treatment is approximately valid provided the velocity of the projectile is great enough so that the stagnation pressure p at the collision point is greater than the strengths of the two materials involved.

The expression in Equation 16.19 can be used, for example, to determine approximately how thick a steel shield is needed to stop aluminum fragments from a detonator. Assume that the largest fragment has a length in the direction of motion of 2 mm and that its velocity is 1000 m/s. The density of aluminum is 2.70 g/cm^3; that of steel is 7.8 g/cm^3. We may safely assume that the velocity is high enough for the stagnation pressure to be above the strength of steel. From Equation 16.19, the penetration is 1.2 mm. With a good safety margin, a 3 mm thick steel plate should thus be sufficient to stop the fragments from a detonator, provided the thickness of the fragments is less than 2 mm. Some commercial blasting caps have a depression at the base so that the end of the detonator becomes a nearly-perfect shaped charge which forms a long, coherent jet of the cap metal a fraction of a millimeter thick. That jet, of course, may penetrate a thicker shield. For such a case, an experimental determination of the required steel shield thickness is very cheap and far more accurate than a back-of-the-envelope calculation such as outlined above, even if several detonators must be used to ascertain reproducibility of the results.

16.4 Safety Distance for Metal Fragments

Metal fragments of very high velocity, such as those thrown by a thick-walled steel tube filled with a detonating high explosive can fly for a long distance. How long depends more on the shape of the fragment, its density, and its thickness than on the very high initial velocity. This is because the air resistance, which increases with the square of the velocity, reduces the high velocity very quickly. The main distance of throw is then covered by the fragment at a moderate velocity where the air resistance has less radical influence. The following treatment provided by Davis [1980] provides an elegant, simple, and therefore very useful rule of thumb for estimating the safe distance a person has to stay from a fragment generating explosive charge in the open.

First, let us see how long the distance of throw would be if there were no air resistance. If a fragment were thrown at an initial velocity V in a direction approximately 45° above the horizontal, the maximum length of throw L under the influence of gravity alone would be given by the classical expression

$$L = \frac{V^2}{g} \qquad (16.20)$$

where g is the gravitational acceleration, 9.81 m/s². The maximum distance of throw without air resistance would thus be approximately 1/10 of the square of the velocity, in metric units. For example, given an initial velocity of 3130 m/s, the distance of throw would be 10^6 m, or 1000 km. This is more than 10 times the range of a hypervelocity gun with that muzzle velocity. Obviously, the air resistance is all important for understanding the actual distances of throw at these very high velocities.

The rate of reduction in velocity is of course a function of the velocity itself, as well as the shape and thickness of the fragment. (An irregularly shaped metal fragment, of course, will slow down much more rapidly than a projectile carefully designed for minimum air resistance.) To find out more about the effect of air resistance, let us write the equation of motion for the other extreme case where air drag is the only retarding force:

$$M\left(\frac{dV}{dt}\right) + \frac{1}{2}c_d A \rho_a V^2 = 0 \qquad (16.21)$$

where M is the mass of the fragment, V its velocity, t the time, c_d the air drag coefficient, A the area of the fragment at right angles to the direction of motion, and ρ_a the density of the air. The equation of motion equates the product of the fragment mass and its acceleration (deceleration) with the (retarding) force on the fragment due to air drag.

The mass of the fragment can be written approximately as

$$M = \rho_m A h \qquad (16.22)$$

where ρ_m is the density of the fragment metal and h is the thickness of the fragment in the direction of motion. We find by rearranging the two equations that

Table 16.3. Safe distances from fragment generating explosive charges.

Material	Thickness h (mm)	Density ρ_m (kg/m^3)	Air density ρ_a (kg/m^3)	Characteristic distance L (m)	Safety distance $8L$ (m)
Dural	3.2	2780	1.3	6.8	54.3
Brass	12.7	8500	1.3	83.0	664
Steel	25.4	7830	1.3	153	1224
Lead	25.4	11350	1.3	222	1774
Tungsten	25.4	19300	1.3	377	3017

$$\frac{V^2}{\frac{dV}{dt}} = \frac{2h\left[\frac{\rho_m}{\rho_a}\right]}{C_d}. \qquad (16.23)$$

The expression on the left has the dimension of distance and is in fact a characteristic length L which is proportional to the length of throw. The expression for L from Equation 16.23 can be further simplified if we assume the air drag coefficient to be of the order of 2:

$$L = h\frac{\rho_m}{\rho_a}. \qquad (16.24)$$

From experimental observations of a great many firings of fragment-generating charges, Davis concluded that a distance equal to $5L$ is just about the maximum range of the fragments. Including a safety factor, Davis recommends that the safety distance for people be set at $8L$.

For explosively accelerated fragments, the initial velocity V_o has little influence on the distance of throw. This is because the air drag force, proportional to the square of the velocity, is very large during the first part of the fragment travel. The main part of the travel then is covered by the fragment traveling at a moderate velocity.

Table 16.3 shows the safe distances for fragments of different metals and different thickness.

For the most common fragment, 1 in thick steel, the safe distance is about 2/3 of a mile. Interestingly, this is about the same as the distance of throw of rock fragments from a 10 in diameter drillhole. (Compare Chapter 12, Figure 12.10.)

The data presented in Table 16.3 have to be taken for what they are, namely a guide for understanding how different factors such as metal density and the size of the fragment influence the length of throw. One must be particularly careful in defining the thickness of the fragment in the direction of motion. In real life, fragments may tumble as they travel through the air and they may experience lift in certain aspect angles, allowing their trajectory to become different from that indicated by the above simple theory. A human life is worth more than any gain that the strict adherence to a theoretical formula may bring. Therefore, a range safety officer has to find out for himself by his own experience or by consulting the most knowledgable people available, the actual safe distance for each given experiment, and take appropriate action to make sure no accident will occur. For in the final reckoning, there is no reason at all (except for the fact that everyone wants to watch the shot!) for having people in the open exposed to the risk, however small, of being hit by a metal fragment from an explosives experiment. Explosive charges with external metal parts should not be fired at all until everybody within range (perhaps as much as several kilometers) is in a strong concrete shelter.

Chapter 17

Explosive Art, Explosive Metal Forming, Welding, Powder Compaction, and Reaction Sintering

There is a strong relationship between art, science, and engineering. In these creative human activities, more real progress is made through the use of intuitive, rather than logical mental processes. In all three, the knowledge base from which new developments can grow is so wide, so complex, and has so many variables that conscious logical thought is too limited as a tool for creation and must be augmented by unconscious mental processes. In the unconscious, previously unknown combinations and relationships can be discovered within a myriad of facts, painstakingly collected through hard work over periods of years — sometimes a lifetime — in the brain of gifted individuals. By the process of intuition, often perceived as a flash of sudden insight, such individuals can see a pattern where no one had seen it before, or they can create a work of fine art, understand or explain a new mechanism, invent a new process, create a new theory, or propose a new model. Often, the gift those individuals have is the perseverance to keep working with the same problem until all the facts are firmly established in their minds. This often takes a very long time, and most people lose patience and go on to other things long before they have accumulated all the facts.

In the world of explosives, the close ties between art, science, and engineering are particularly clearly visible in the dynamic working of metals and other materials by the dynamic forces created by explosion.

In this chapter, we will introduce the reader to some results of the inventive process. We will show that explosives can be used creatively not only to produce useful and practical road cuts and underground caverns in rock, but also to play or to please the eye with works of artistic beauty, or to make new materials, the real usefulness of which may still remain to be discovered. We will show some examples of explosive art, such as an ancient, explosively-formed embossed silver plate, and we will introduce and show works by two artists who have used explosives in their art. Then, we will introduce the fundamentals of several techniques of working materials in which the force of explosives is used as a power source. These include explosive forming of metals; explosive welding (the process by which first multilayered coins in the USA were made); explosive powder compaction (a technique for making new metal alloys and other materials with unusual properties); and the emerging technique of explosive reaction sintering of materials (a technique where shock-induced chemical reactions are used to create new solid materials, intermetallic alloys, ceramics, semiconductors, and even low critical-temperature superconductors).

414   Chapter 17. Explosive Art and Working of Materials

Figure 17.1. Silver plate explosively formed using forms produced in the 1880s at Vinterviken, Stockholm (Gösta Larsson, Vinterviken, 1962).

17.1 Explosive Art

A major part of this book has been devoted to explain how the force of explosives can be harnessed to do constructive, peaceful work, in mining, minerals excavation, and building construction. In this section we will go one step further and illustrate ways that explosive forces can be used to create not only useful but artistic effects — even works of fine art. Figure 17.1 shows a silver plate, made by the process of explosive forming (for details of the process, see Section 17.2 below). The plate has the approximate dimensions 450 by 350 mm and is made from 1 mm thick, solid silver. The forms and the first plates made in the forms were produced at Vinterviken in Stockholm during the 1880s, as a demonstration of the force of the new dynamite explosives. The plate in the picture is a later production, made in 1962 by Gösta Larsson at Vinterviken using the old forms, which have now been deposited at the Technical Museum in Stockholm.

In the following, we will describe two different techniques, one used by the Swedish artist Verner Molin, the other by New Mexico artist Evelyn Rosenberg. Each in his or her own way has achieved the ability of self-expression, and create truly unique pieces of fine art by the unusual medium of explosive working of metals.

Figure 17.2. Explosive engraving "Alfred Nobel" by Verner Molin, 1967.

17.1.1 Verner Molin

The discovery of the first explosive art form we will discuss here was probably incidental: a leaf, caught between a detonating explosive and a metal plate, was found to leave an imprint on the metal in which the finest detail of the soft leaf appeared to be engraved in the strong metal. It was first mentioned by Munroe in 1888. The actual engraving is due to jet formation. Wherever there is a sudden change in thickness of the leaf, such as at both sides of a leaf vein, a jet is formed cutting a hair-fine groove in the metal.

The late Swedish artist and mystic Verner Molin [1968] became fascinated by the phenomenon and made himself a master of what he called *explosive engraving*. Figure 17.2 shows an example of Molin's art, fittingly in the likeness of Alfred Nobel. It is a print, made by the same ink transfer process that is used in the reproduction of traditional copper engravings. In traditional engraving, a pattern is formed in the copper plate by acid etching in the scratches made by the engraving needle in the thin layer of wax covering the copper surface. Printers' ink is then deposited in the grooves etched out in the copper surface, and the ink is finally transferred to the paper in the printing press. Molin, working with experimentalists of the Detonics Laboratory of Nitro Nobel AB at Vinterviken, Stockholm (the site of Alfred Nobel's early development and production of nitroglycerin for use as a rock blasting explosive), replaced the slow hand scratching of lines, one by one, with the almost instantaneous explosive engraving into the surface of the soft copper sheet of the pattern he had previously painted on it. He covered the painted surface of the copper with a thin sheet of plexiglas. The plastic explosive (similar to Composition C-3) was placed as a thin layer on top of the plexiglas. The secret of the process is that any obstacle, however soft, in the way of the impact of the explosive or the plexiglas on the metal surface will produce the above-mentioned

Figure 17.3. Detail from the mural "Evolutionary Landscape" made by the process of *Detonography* [Evelyn Rosenberg, 1986].

jetting reproducing the pattern with the resulting jet cuts in the metal. The fibers or grit particles which Molin mixed into his paints provided the obstacles, in the same way as the veins of the leaf, create the fine cuts in the metal. These fine cuts absorb the printer's ink making it possible to transfer the image to the paper in the printing process.

17.1.2 Evelyn Rosenberg

Not content with the fine relief provided by the explosive engraving technique, Evelyn Rosenberg [see Wolkomir, 1987], initially assisted by fellow artist Alice Seely, developed a powerful and expressive deep relief artform she calls *detonography*. She makes a clay original from which a negative Plaster of Paris form is made. A thin sheet of metal (such as mild steel, copper, brass, stainless steel, or combinations of these), together with additional resistances such as straws of grass or grit, is suspended above the Plaster of Paris form. A thin layer of explosive, such as DuPont's sheet explosive, is separated from the metal by a thin layer of protective rubber. The sheet metal is accelerated to impact into the form. Before the form is destroyed by the impact, the metal takes on its shape, and any resistances in its way are engraved into the metal surface. Rosenberg has developed techniques by which she can simultaneously explosively weld different metal foils to the accelerated sheet metal. Figure 17.3 shows an example of Rosenberg's work, a mural placed above the entrance to the Natural Science Museum in Albuquerque, New Mexico. Rosenberg has become a very successful artist — now almost exclusively working with the explosive art she created — and is represented in many public buildings throughout the country.

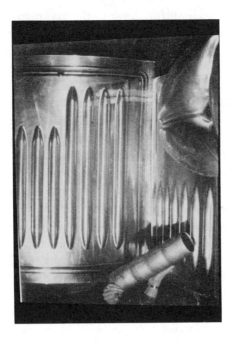

Figure 17.4. Explosively formed rocket nose cone.

17.2 Explosive Forming

The technique used by Rosenberg is essentially a combination of the explosive engraving technique with the technique we normally call *explosive forming*. For engineering applications, of course, there is a need for a more precise reproduction of a predetermined shape; therefore, rigid forms are used. The space between the sheet metal blank and the form is evacuated to remove the resistance the compressed air otherwise would make, and water is used as a medium to transfer momentum from the explosive to the metal, which is pressed into the form rapidly to take the form of the form. Most explosive forming operations are carried out under water in a tank or a pond, using heavy steel forms.

An advantage of the explosive forming process over conventional deep drawing methods is that bulbous, complicated shapes can be produced, using external forms that can be taken apart. Figure 17.4 shows such a shape.

17.3 Explosive Welding

The impact of a fast-moving metal plate hitting a stationary metal plate simultaneously over the entire surface creates a high pressure at the interface between the two plates and a shock wave is transmitted through each plate. Afterward, the two plates separate at the interface.

If the impact is at an angle, the conditions for forming a jet as described in Chapter 15 may occur, dependent upon the impact angle. If the velocity of the collision point is well below the velocity of sound in each metal, the flow of material at the collision point is isentropic, and no shock wave is formed. The jet, consisting of material from both plates, removes the oxide surface layer of the two materials, and as a result, the two metals are cold-welded together, with no melting. Unless the impact is symmetric, the jet material will ultimately collide again with one of the metal surfaces, inserting the broken-up oxide layer into the metal.

Cowan, Douglas, and Holtzman [1964, 1966] at DuPont invented and patented the process of parallel plate explosive welding. Later improvements are due to Cowan, Bergmann, and Holtzman [1968, 1970]. The two plates to be welded together are placed parallel to each other, separated by a stand-off distance. An explosive with a detonation velocity equal to the intended collision point velocity is placed as a layer on top of the upper, thinner plate. Upon detonation, the upper plate is deflected toward the lower plate which it hits at an angle ϕ determined by the detonation velocity and the explosive-to-metal mass loading ratio of the upper plate. Variables for controlling the process are the thickness of the explosive in relation to the upper metal plate, the detonation velocity, and the stand-off distance. As described in Chapter 15, the velocity of the accelerated plate is directed at angle $\phi/2$ to the originally horizontal plate surface; thus, the plate is given a velocity component in the forward (detonation) direction. Thus, as the upper plate impacts the lower plate, there is a sliding motion at the interface which may create waves at the interface by the same mechanism (interface instability) which creates waves at the surface of water when the wind blows over it. The sliding and wave motion also creates local deformation heating, which may melt pockets of material at the interface. Figure 17.5 shows an example of a wavy interface with small pockets of once molten material at the wave "crests". However, with careful selection of variables, a cold weld without waves can be created.

This is sometimes a necessary goal. When welding metals which form hard, brittle intermetallics (such as, for example, iron and titanium), extensive melting can occur causing inferior welds of low strength.

The forward velocity component of the accelerated plate creates an accumulating momentum in the upper plate, which propagates with sound velocity in the upper plate ahead of the detonation front and can cause tensile cracking of the plates, with cracks running parallel to the detonation front, starting at the edge of the plate package farthest away from the point of initiation. To minimize this effect, a momentum trap in the form of a strip of metal, is loosely spot-welded along three edge sides of the upper plate, in contact with the upper plate. Upon arrival of the tensile momentum, the momentum trap metal strips separate from the upper plate, preventing the cracking. To further minimize this effect, initiation of detonation is often done at the center of the long side of the upper plate.

When welding large, thick metal plates together (for example, slabs of steel of several square meter surface, 200 mm thick, clad with 30 mm thick stainless steel, having a 125 mm thick layer of explosive covering the upper plate), the air blast wave and sound of the explosion travels a long way. A safety distance of the order of 10 km is necessary around a facility where naked charges weighing 500 kg or more are fired to prevent building damage and window breakage as well as human irritation. Considerable reduction in the required safety distance can be obtained by damping the sound by the simple method of placing a layer of water on top of the explosive (invented by I. Persson). The water not only reduces the air blast from a given mass of explosive, but it also acts as a tamper to reduce the mass of explosive required for accelerating a plate of given thickness.

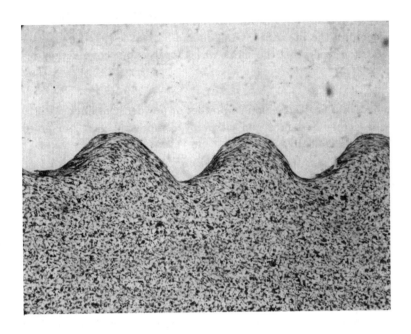

Figure 17.5. Optical micrograph showing the wavy interface produced by explosive welding. (Photo courtesy of Explosives Fabricators, Inc.).

17.4 Explosive Powder Compaction and Reaction Sintering

Processing of materials by shock compression utilizes the extremely rapid and intense deposition of energy in powders by shock waves. In the rapid deformation of powder grains to fill out the pores between grains, plastic deformation heating under the high pressure of the high shock wave can melt and fuse the grains together into a solid pore-free material. Because of the short duration of the heat and pressure, dissimilar materials can be fused together into a polycrystalline material without the growth of large crystals and separation of phases which would accompany any effort to hot press or melt them together using conventional metallurgical techniques. Ceramics so hard and brittle that engineers find them impossible to machine can be molded, with explosives, into the desired shapes. A variety of processing approaches are being developed that utilize this concept. These include shock-induced solid-state polymorphic phase transformations, shock compaction, shock modification, shock enhanced sintering, and shock-induced reaction synthesis. The behavior of porous materials under shock compression is significantly different from that of solid-density materials in that large amounts of extra energy are required to plastically deform and crush the particles in the process of void collapse. The relationship between pressure and specific volume is described by the Hugoniot curves for the mixtures of materials used [Thadhani, 1993].

A variety of geometries and fixtures for dynamic consolidation of powders have been documented in the literature [Graham and Sawaoka, 1987]) and patented. Three major methods for generating the shock waves for dynamic compaction of powders are [Thadhani, 1988]:

1. Impact of a projectile against powder or powder container, with the projectile being accelerated in a gun either by compressed gas, gun powder, or even electromagnetically

2. Detonation of explosive in either direct contact with the powder (or container) or explosively accelerated plate impact

3. Semi-automated commercial machines for accelerating a piston at low impact velocities

A comprehensive review and bibliography of the work on shock-compression chemistry and shock synthesis of materials was provided by Thadhani [1993]. Graham [1993] documented shock-compression science in book form. His is the most definitive work on the subject at the present time. The technology of shock-compressed matter is truly multidisciplinary and involves the fields of physics, electrical engineering, solid mechanics, metallurgy, geophysics, materials science, and explosive engineering. There is great potential for using explosives to create new materials. Also, from the study of the behavior, melting, and resolidification of metals and other inert materials when shock compressed, we can draw useful conclusions on the likely early stages in the shock compression and shock initiation of porous explosives.

17.4.1 Explosive Powder Compaction

In shock compaction, the consolidation of powders occurs when shock energy is preferentially deposited at particle surfaces resulting in local plastic deformation, heating, and local melting of a small fraction of the material at the particle surfaces. Subsequently, particle fracture (in the case of hard and brittle materials) and cleansing and exposure of fresh surfaces occurs, resulting in partial melting and welding or even solid-state interparticle bonding [Thadhani, 1993].

In dynamic compaction the consolidation of the powders is accomplished by the passage of a strong shock wave through the powders; the welding between them is a result of a dynamic process similar to explosive welding. The rapidity of the process allows material at the powder surfaces to melt, while the interior remains at a relatively low temperature [Meyers and Murr, 1981].

In the process of dynamic compaction, an explosive or a composite of different explosives is packed into a cylindrical container surrounding a metal pipe containing a mix of metal powders. Upon detonation, the explosive causes the sample material to implode, compacting the powder grains into a solid. The explosive acts like a giant high speed mechanical press, forcing the powder particles together under tremendous pressure; the powder surfaces melt, then solidify quickly, bonding them together into solids that are difficult to make otherwise. For some classes of materials such as alloys, ceramics, and polycrystalline diamond, explosive shock compaction or synthesis is the best, or only, way to form parts.

17.4.2 Shock-Induced Chemical Reaction Synthesis and Sintering

Shock-induced reaction synthesis or sintering is a process in which chemical reactions in powder mixtures — started by the local heating that accompanies shock compression — leads to chemical reactions resulting in the synthesis of compounds which are consolidated at the same time by the shock wave. Under shock compression, interesting chemical reactions can occur. For instance, normally incompatible metals such as titanium and aluminum react under explosive shock to form titanium aluminide, an extremely strong and heat-resistant intermetallic alloy that's being considered for use in the skins of future hypersonic airplanes such as the National Aerospace Plane, or X-30. Other shock-reaction sintered materials include nickel-based superalloys for jet engines and aluminum-lithium alloys – forming a light, stiff airframe material. The list of materials that have been made experimentally by shock-induced chemical reaction sintering is long: super-tough polycrystalline diamond and cubic boron nitride cutting tools, gallium arsenide semiconductors, ultra-hard boron carbide and titanium diboride ceramic plates for tank armor, and high-temperature superconductors [Ashley, 1989].

A useful, commercially viable product is DuPont's explosively synthesised polycrystalline diamond powder. Graphite is the raw material, and the end product is even more valuable than natural diamond powder since there is no other way of making poly-crystalline diamond. Polycrystalline diamond is far superior to monocrystalline diamond for the fine polishing of brittle, hard-to-work materials such as ruby, sapphires, and the ceramics and ferrites used in electronics. DuPont makes its diamond in a system of concentric steel pipes. The inner pipe is filled with a mixture of graphite and metal powders. This pipe then sits inside a driver tube that has a larger diameter and thick wall. Finally, explosive is packed around the driver tube. Upon detonation of the explosive, the collapsing driver tube generates pressures of 10 to 30 GPa within the product tube. The size of the charge setup is large and, at present, a few shots provide the entire US market with its requirements of polycrystalline diamond powder. Because of the large size of the charge, the shock wave and pressure pulse lasts long enough (many microseconds) for crystals of diamond to grow to a size of about 10 nm [Beard, 1988].

Another interesting new technique for producing fine powder diamond is to utilize the regular detonation of carbon-rich explosives such as TNT and TNT-based explosives. TNT has a great excess of carbon in its molecular composition, and shows an irregular behavior of the detonation velocity at high intital charge density. At a density of 1.55 g/cm^3, the linear relation between detonation velocity and inititial charge density suddenly changes slope from 3163 to 1700 m/s/g/cm^3, as reported by Urizar, James, and Smith [1961]. It was concluded at an early stage that this change in slope could be the result of a change in the heat of formation of carbon by +6 kcal/mole from the low density side of 1.55 g/cm^3 to the high density side. One possible mechanism for such a change would be if some of the excess carbon condensed to a more strongly bound structure than regular soot which is similar to graphite. We now know that free carbon, formed in the detonation reaction, is transformed to extremely fine grained diamond, of grainsize as small as 40 Å, under the conditions of relatively high temperature and pressures nearly 20 GPa, where the formation of diamond is favored. Yields of diamond as high as 25 to 30 % of the excess carbon (which translates into 8-9 % of the total explosives weight) have been reported by Lyamkin *et al.* [1988] and Greiner *et al.* [1988].

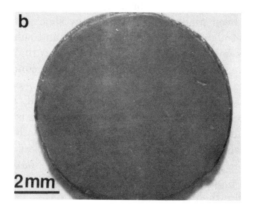

Figure 17.6. Diamond disc made by shock compacting diamond powder: (a) the diamond powder before compaction (micrograph); (b) the 12.5 mm diameter, 1 mm thick compact.

Careful structure-determinations, confirming that the crystal structure of the white powder extracted from the soot indeed was that of diamond were carried out by Greiner and Hermes [1989]. Extensive earlier work was reported by Titov et al. [1989].

Staver et al. [1984] reported that the addition of extra carbon to an explosive composition increase the yield of diamond. In Russia, the technique has been developed to the point of useful industrial production of ultrafine diamond powder, and work is in progress to find ways of sintering this powder into larger grains.

Shock compression processing is emerging as a novel technique for the fabrication of esoteric materials. Although shock compression processing imparts unique characteristics to materials and economic considerations suggest that the commercialization of the technology may indeed be feasible, actual industrialization on a large scale has been slow to develop [Thadhani, 1988].

Presently at New Mexico Tech, diamond powders are being shock-consolidated to nearly crystal density (84 to 95% of theoretical maximum density), 12 mm round compacts. One possible application of great current interest is for high thermal conductivity substrates for high power (or power density) semiconductor devices, such as laser diodes. Efforts are being made by Joshi et al. [1993] to obtain millimeter-thick compacts with high thermal conductivity. Figure 17.6 shows the diamond powder raw material and the resulting 12.5 mm diameter, 1 mm thick compact.

Chapter 18

Safety Precautions, Rules, and Regulations

The manufacturing, transportation, storage, and use of explosives in mining and building construction are among the safest industrial operations, if one considers the number of working hours lost due to accidents and also if one considers the number of fatal accidents.

However, accidents involving explosives are set apart from other industrial activities by the suddenness with which an accident may happen and the devastating damage caused by the explosion. Also, unless precautions had been taken to protect personnel in case of an accident, explosion accidents are often fatal or cause serious injury. Anyone who has seen or experienced the effects of explosions or inspected the scene of an explosion accident will forever remember to treat explosives, explosive devices, and energetic materials with great care.

There are several, very different ways in which the safety of an industrial operation involving explosives can be enhanced. Most of these fall within one of the following groups: personal precautions, inofficial recommendations, local safe operating procedures, and government regulations.

Personal precautions are extremely important because they are what the person directly involved with the explosive on site can actually do to maintain a safe operation and to avoid accidents. The personal precautions taken by an individual in an industrial setting are strongly dependent on the management (employer) and its ability to create a system for educating and training employees. They also depend for encouragement on the general atmosphere of confidence and sharing of responsibility between employer and employees that has been established in a workplace.

Inofficial recommendations include, for example, precautions recommended by an insurance company (often as a condition for providing insurance), and the threat by an employer to terminate an employee who does not follow local safety rules and regulations. Such inofficial recommendations are often very effective and can be adjusted to new conditions much more rapidly than can government regulations.

Safe operating procedures (SOPs), sometimes referred to as standard operating procedures, are, in the authors' opinion, the most important means of maintaining a safe operation. They specifically address, the situation at a given workplace during the performance of a specific procedure. They do, however, require an atmosphere at the workplace that allows, and requires, every employee to be a participant in the safety assurance effort.

Government regulations provide a framework for enforcing safe behavior, a legal system for prevention of unsafe procedures, and prosecution after the fact in those cases where an accident may have been caused by a violation of specific regulations.

In this chapter, we will provide some additional information on these four ways of enhancing safety.

18.1 Personal Precautions

The first item under the heading of personal precautions is to realize that you, personally, are the key link in the chain of safety. A safe operation requires that each and everyone involved in the operation does his or her part in maintaining safety. All those involved are linked together to form a chain, the safety chain, which will be no stronger than its weakest link. The goal is to maintain its strength. If one link in the safety chain breaks, the whole chain of safety breaks down. Whatever you do, and whoever you are, you are the link that must not break. If you are a manager, make sure that each and everyone of those working for you are empowered and required to work safely. If you are not a manager, know that you are empowered and required to work safely. This includes everyone being required, not just having the right, to stop the operation, whatever that operation is, if he of she does not feel safe. It includes the unconditional requirement — that everyone *without exception* — to follow safety rules, regulations, and safe operating procedures. It includes the requirement to never use the killing words: "It is not my responsibility" or "It has worked before, even if it is against the stupid regulations". It also includes the hard action of removing anyone from his or her job who does not comply with all the requirements of safety.

A useful rule in personal precautions is: "Think First". In all dealings with explosive materials, this rule is a golden rule. After an accident has happened, it is too late to think — and if you were on site, you may not even have a second chance. The time to think carefully about how to perform a given operation is before you start it. "Think first" is the same as careful and thoughtful planning, consulting with others, checking out the rules and regulations, and reading the safe operating procedures for the operation in question, so that nothing is left to chance.

A tidy and well-maintained workplace is also a safe workplace. It is safe for several reasons. First, it speaks of the work being well organized. Second, it is easy to work safely in a tidy workplace. It is less likely that a spanner will find its way into a mixer, causing a major accident, if there is no spanner lying around where it should not be in the first place, during normal operation. Third, a tidy workplace makes for calm and relaxed working conditions.

18.2 Inoffical Recommendations

A very powerful role in all safety work is played by inofficial recommendations. These can take many different forms. A simple rule that it is a condition for employment that you leave cigarettes, matches, and lighters at the gate of the explosives plant is an example of an inoffical recommendation. The demand by an insurance company that it is a condition for their policy to be valid at a specific workplace that everyone wears a helmet is another. An insurance company may stipulate that it will not insure a contractor for damages he may inflict on a third party unless he follows a set of inofficial recommendations, perhaps issued jointly by several insurance companies.

Such recommendations are powerful because, in fact, there is no way around them. In the workplace, either you leave the matches behind or you have no job. As a contractor, you either follow the inofficial recommendations or you have no insurance. Neither the employee nor the contractor can afford not to follow these recommendations.

18.3 Safe Operating Procedures (SOPs)

A Safe Operating Procedure (SOP) is a written document setting out the procedure that must be followed for carrying out a specific operation. A well-organized workplace has a set of SOPs which covers all aspects of its operation, and no new operation is allowed to be started without first having an SOP written and agreed on beforehand. An important by-product of the system of having SOPs for every operation is that it forces those responsible for starting a new operation to think first — to go through the proposed new operation step by step to make sure there is no hidden danger. There should also be a set mechanism for agreement and modification of an SOP, by a Safety Committee, and by those administratively and operationally responsible for the operation.

18.4 Government Regulations

Government regulations are in part different in different countries; and in the USA. they are also in part different in different states. Increasingly, however, government regulations regarding explosive materials in many countries now take as their starting point the recommendations provided by the United Nations. For example, the recommendations on storage and transportation of dangerous substances issued by the United Nations Committee of experts on the Transportation of Dangerous Goods are now written into the government regulations for transport and storage set out in the US Code of Federal Regulations.

The main documentation for the current federal regulations in the US is the Code of Federal Regulations. The key regulations regarding explosives are 30 CFR 56 and 57 (Federal Metal and Nonmetallic Mine Safety and Health Regulations) and 49 CFR Parts 100 through 177, Part 383, and Parts 390 through 399. These and other CFRs are the documents which form the basis for different federal agencies to enforce regulations and for the legal system to prosecute. They are very important documents to know about, to understand the meaning of, and to follow. More details of what they contain are given in the listing of "REFERENCES FOR SAFE STORAGE, HANDLING, AND TRANSPORTATION OF AMMUNITION AND EXPLOSIVES" below.

In the USA, several different federal authorities are empowered to supervise and enforce the federal regulations.

Matters pertaining to the manufacturing and storage of explosives are entrusted to the Bureau of Alcohol, Tobacco, and Firearms (ATF). A permit to manufacture explosives has to be obtained from the ATF before any manufacturing of explosives may take place.

Matters pertaining to surface mining are supervised and enforced by the Office of Surface Mining (OSM).

Matters pertaining to the safety and health of employees generally, and also in the energetic materials industry are entrusted to the Office of Safety and Health Administration (OSHA). In Sweden, the equivalent of the OSHA is the Arbetarskyddsstyrelsen (Worker Safety Administration).

In the USA, it is a condition for receiving funding for a contract with the Department of Defense or the Department of Energy that the contractor follow the stipulations in the Department of Defense Contractors' Safety Manual.

18.4.1 General References for Safe Storage, Handling, and Transportation of Explosives

1. 26 CODE OF FEDERAL REGULATIONS COMMERCE IN EXPLOSIVES — Part 181 — Regulations to regulate interstate or foreign commerce in explosives.
2. 27 CODE OF FEDERAL REGULATION Part 55, Public Law 91-452, Chapter 40, Importation, Manufacture, Distribution, and Storage of Explosive Materials — Regulations designed to protect interstate and foreign commerce against interference and interruption by reducing the hazards to personnel and property arising from misuse and unsafe or insecure storage of explosive materials.
3. 29 CODE OF FEDERAL REGULATIONS — Parts 1910 and 1926 Occupational Safety and Health Standards (Continued) — Regulations for the safety and health standards for the protection of labor. This includes protection from hazardous materials and dangers associated with work involving hazardous materials.
4. 30 CODE OF FEDERAL REGULATIONS (CFR-30) — Parts 56 and 57 set forth mandatory safety and health standards for surface and underground metal or nonmetal mines, including open pit mines, which are subject to the Federal Mine Safety and Health Act of 1977. The purpose is the protection of life, the promotion of health and safety, and the prevention of accidents.
5. 49 CODE OF FEDERAL REGULATIONS (CFR-49) — Parts 100 through 177 Transportation — Federal rules and regulations for the transportation of hazardous goods which include explosives. Included are rules to identify, classify, package, test, mark, and ship hazardous goods by vessel, and public highway.
Part 383 Commercial Drivers License Standards — Defines the requirements for a Commercial Drivers License which is required for personnel involved in the transportation of hazardous goods which includes Class 1 explosives
Part 390 through 399 Federal Motor Carrier Safety Regulations, General — General safety, vehicle, driver, training, and reporting requirements for transportation of explosives.
6. UNITED NATIONS PUBLICATIONS ON TRANSPORT OF DANGEROUS GOODS ST/SG/AC.10/1 — Developed by the Committee of experts on the transportation of Dangerous Goods to provide governments and international organizations with information to classify, package, test, mark, label or placard, and ship hazardous goods.
ST/SG/AC.1O/11 — Developed by the Committee of experts on the transportation of Dangerous Goods to provide international test methods and procedures to be the most useful to properly classify explosives or hazardous articles for transportation.
7. INTERNATIONAL CIVIL AVIATION ORGANIZATION Technical Instructions For Transportation Of Dangerous Goods By Air — General approach to regulating the transport of dangerous goods by air. The standard UN International methods have been modified to cover the particular requirements for air transport.
8. NFPA 495 MANUFACTURE, TRANSPORTATION, STORAGE, AND USE OF EXPLOSIVES — The National Fire Prevention Association (NFPA) code is intended to provide reasonable safety in the manufacturing, storage, transportation, and use of explosive materials.
9. BUREAU OF ALCOHOL, TOBACCO, AND FIREARMS, DEPARTMENT OF THE TREASURY ATF 5400.7 (The ATF is the regulatory agency responsible for the enforcement of the regulations in 27 CFR regarding importation, manufacture, distribution, and storage of explosive materials.)

10. INSTITUTE OF MAKERS OP EXPLOSIVES (IME) — Safety Library Publications developed by the commercial industry concerned with safety in the manufacture, transportation, storage, handling, and use of explosives. The IME provides advisory services to legislative committees and regulatory bodies engaged in matters related to explosive safety.

11. DEPARTMENT OF TRANSPORTATION EMERGENCY RESPONSE GUIDE, P 5800.5 — Guidebook developed by the Department of Transportation for the use by First Responders, such as fire fighters, police, and emergency services personnel. Procedures are listed for initial action to be taken to protect an individual and the general public when responding or handling an incident involving hazardous materials.

18.4.2 Military Standards, Publications, and Regulations for Safety in the Manufacture, Transportation, Storage, Handling, and Use of Explosives

1. DEPARTMENT OF DEFENSE 6055.9 — STD, AMMUNITION AND EXPLOSIVES SAFETY STANDARDS — DOD established uniform safety standards applicable to ammunition and explosives, to associated personnel and property, and to unrelated personnel and property exposed to the potential damaging effect of an accident involving ammunition and explosives during their development, manufacturing, testing, transportation, handling, storage, maintenance, demilitarization, and disposal.

2. US ARMY MATERIAL COMMAND (AMC) SAFETY MANUAL AMC 385-100 — Safety Manual — General safety guidelines for ammunition and explosives. Utilizes DOD 6055.9-STD for a primary source.

3. DEPARTMENT OF DEFENSE 4145.26,M, DOD CONTRACTORS SAFETY MANUAL FOR AMMUNITION AND EXPLOSIVES — This manual provides reasonable and standardized ammunition and explosive safety principles, methods, practices, requirements, and information for contractual work or services performed in connection with contracts involving ammunition and explosives. This manual applies to work or services for DOD. Utilities DOD 6055.9-STD for a primary source.

4. US DEPARTMENT OF ENERGY DOE/EV/06194 — DOE Explosive Safety Manual — This manual describes the DOE explosive safety requirements applicable to operations involving the development, testing, handling, and processing of explosives or assemblies containing explosives.

5. DEPARTMENT OF THE NAVY, NAVSEA OP 5 VOLUME 1, AND EXPLOSIVE ASHORE — Standardized safety regulations for the production, renovation, care, handling, storage, preparation for shipment, and disposal of ammunition and explosives.

6. DEPARTMENT OF THE AIR FORCE — Explosives Safety Manual AFR 127-100 — General guidelines for the storage, handling, transportation and disposal of ammunition and explosives.

7. DEPARTMENT OF DEFENSE — Test Protocol for Insensitive Munitions 2105A — Descriptions of test procedures for determining if a Munition item or package satisfies the requirements for insensitive munitions, i.e., munitions with increased safety against mass detonation.

Chapter 19

Safety in Production of Explosives

As an example of the practical application of the general precautions, rules, and regulations described in the previous chapter, we will discuss and describe below in some considerable detail the safety philosophy and safety tests and criteria used in one commercial explosives company. This is done in the hope that some of the good practices already in use in one commercial enterprise may spread to others, and also in the hope that by discussing these practices, members of the energetic materials community may develop even better and safer tests and practices.

The world is never stationary, and progress comes faster, the more freely new developments can become available to others. In many areas of explosives development, proprietary interests limit the availability of new developments to others. This is a natural part of the product development process in a free society based on competition in the marketplace. Fortunately, in the field of explosives safety, it has long been established that new developments and information on accidents and their causes should be shared freely and promptly with other competing companies because this may save lives. The authors are grateful to the Nitro Nobel AB and the DYNO international group of explosives companies for permission to include this section in the book.

19.1 Introduction

In the past, new explosive formulations were developed by a trial and error procedure in which the testing of new production methods and the design of new process equipment were an integrated part. Today, engineering design and computer modeling coupled with field experience and accumulated know-how largely replace trial and error in the development of new explosives compositions and production methods. This greatly reduces uncertainty about performance and eliminates accident risks long before a new explosive is ready to be given to the customer for field trials.

A long-established safety consciousness in the explosives industry has also played a significant role in avoiding mistakes that could have caused injury to the people involved in the production process and damage to production facilities. In this chapter, we will outline some of the tests used to avoid accidents.

19.2 Hazards Prediction

Tests must be carried out by the manufacturer of explosives and accessories to assess the sensitivity of new explosives products to impact, friction, shock, and thermal stimuli in order to safeguard against inadvertent explosions.

Standardized tests not only measure the sensitivity of a new explosive to a selection of specific stimuli which model real life situations, but are also used to provide data upon which can be based a meaningful dialogue on the overall safety in the manufacture and use of explosives. Often, such dialogue can reveal the need for additional experiments to throw light upon a new explosive formulation's response to the environment in a proposed new production facility. The additional experiments sometimes lead to the development of a new standardized sensitivity test method. The UN Recommendation on the Transport of Dangerous Goods — Tests and Criteria is a test battery that has grown out of an international cooperative effort, the Round Robin Series, in which tests with different methods were carried out on the same selection of standard explosive substances in different laboratories around the world. Another useful test battery is the Nordtest methods extensively used in the Scandinavian countries. By precisely specifying both the test equipment and the method for carrying out the test, these standardized tests ensure that the results of tests made in different laboratories give closely similar results for materials of similar sensitivity.

Accident prevention depends on the ability to predict the possible sources of explosion and the means to avoid them. The traditional means of hazard prediction was based on experience, which is a purely empirical approach. Such an approach cannot be extended to new manufacturing processes and machinery with any great degree of confidence. This fact has been revealed through many tragic accidents throughout the explosives industry in the past when new types of explosives have been put into production too hastily and with insufficient safety testing.

Waterbased explosives, for example, were long considered to be a quantum step toward greater safety in comparison with the nitroglycerin-based explosives. They did not respond at all in the standard friction, impact, and open burning tests. Being liquid or semi-liquid, they could be produced in large quantity using new types of modern manufacturing processes, similar to those used in the food industry, employing mechanical pumps to convey the explosives into high-speed cartridging machines or directly from a pump truck into the drillhole. Even if made cap sensitive by the addition of microballoons or chemical gassing, these explosives did not respond to the stimuli simulated by the small-scale traditional hazard tests.

But the new manufacturing processes expose explosives to new stimuli which were not present in the older processing techniques. One such new stimulus is heating to burn under confinement. The waterbased explosives, which do not sustain burning under atmospheric pressure if tested in a limited size package, burn readily at a pressure of a few atmospheres, and therefore can easily cause explosion if heated in the confined space of a pump or steel pipe. As we now know, any assumption, explicit or implicit, that the traditional tests could safeguard against accidents caused by exposing the new explosives to such entirely new kinds of stimuli as those present in pumping and extruding was wrong. As a result, several unforeseen accidents occurred with the cap-sensitive waterbased explosives, which appeared so extremely insensitive when tested using the traditional test battery.

(As discussed in more detail in Chapter 3, it is possible that a similar misjudgment regarding the safety level of a new untried explosive was made when nitroglycerin was first introduced as a safer rock blasting explosive. Compared to the extremely friction-sensitive black powder then used in blasting, the oily liquid nitroglycerin — which did not respond even to being hit with a hammer — may have appeared as a very safe explosive, indeed. The failure to realize that the new explosive brought with it new risks unique to its specific characteristics, such as the thermal instability of an insufficiently washed nitroglycerin containing traces of spent acid, may have been one possible cause of the early accidents with Alfred Nobel's new product.)

In the manufacturing, transportation, and handling of explosives, unforeseen incidents may occur as a result of physical stimuli which typically include unintended impact, friction, or heat. These causes, individually or jointly, may result either in direct detonation or in a combustion which builds up pressure to transit from deflagration to detonation, resulting in possible injury to personnel and severe damage to facilities and production capabilities. Each new production method may expose the explosive to its own collection of stimuli, different from those experienced in the past.

It is therefore important that the "battery" of sensitivity tests includes test procedures that simulate what may happen when something goes wrong in any piece of equipment and in any handling situation.

Individual manufacturers, in addition to internationally standardized tests, often have developed their own testing procedures for investigating and designing with the choice of materials that can be used in their specific explosives manufacturing equipment.

Many manufacturers also have their own testing procedures for investigating and selecting the materials used in their explosives manufacturing equipment. Studies must often be performed on any newly designed equipment intended for use with a specific explosive, as material properties and designs in, for example, a pump may affect different types of explosive in different ways. Materials which may be comparatively safe when used with one kind of explosive may be less safe when used with another.

When designing a new production line, careful studies must also be carried out for intermediate products throughout the entire production line to ensure that a process component does not influence or alter the properties of an intermediate product in a critical manner. For example, the unsensitized emulsion, often called *matrix*, is a very insensitive energetic material if at room temperature and free from air bubbles. But if overheated, or if air is drawn into it through the seals of an ill-designed pump or in a blender, the emulsion matrix can be transformed into a very sensitive material indeed.

Concerns over what might happen if a breakage occurs in each component of a machine must be taken seriously and tests should be carried out to simulate what could happen with the specific matrix or explosive to be used. It is also well to remember that no piece of equipment is proof against misuse, and that individuals involved in operating the equipment do not always act in the way first anticipated.

19.3 Waterbased Explosives

Large-scale manufacture of waterbased explosives developed during the 1960s. The concept at that time was to use high strength, high density aluminized TNT-sensitized watergels. Following the invention of water-in-oil emulsions and emulsion/ANFO mixtures in the mid-1960s, the heavy TNT slurries were replaced by the new, lower density water-in-oil emulsions. The emulsion technology has proven to be very successful and the explosives users are extremely happy with the rock breakage performance achieved by these highly efficient explosives and blasting agents.

In the process employed by Nitro Nobel AB in Sweden, the unsensitized emulsion matrix is manufactured in the plant, where it may be mixed with AN prills and pumped through a detonation trap into an overhead storage bin from which it is fed into the bulk transportation or loading truck by gravity. Before loading into the drillholes, the gassing agent is introduced and evenly distributed into the matrix to produce a sensitized blasting agent of explosive.

Experience with these explosives over a long time span has shown that not one accidental detonation has occurred in the Nitro Nobel plants in Sweden during processing, transportation, or delivery. This has been made possible through a combination of these products' unique safety characteristics, the skill and safety awareness of the operators developed through systematic training, and the careful selection and engineering of the manufacturing and delivery equipment used. Nonetheless, one must always remember that all explosives are powerful and that the potential for catastrophic accident is ever present.

From the laboratory work in the research and development department where new formulations are tested, through the pilot plant test, and into the large scale commercial processing of each new explosive, a continuous safety analysis is performed to ensure a high level of safety.

From the transportation point of view, it is beneficial if the matrix can be classified as a nonexplosive, and this is generally possible with the new water-in-oil matrices. A combination of several factors — the most important ones being a relatively high water content, a high density, and a lack of any air bubbles or other sensitizers — provide the criteria for classifying the practically nondetonable matrix as a nonexplosive.

19.4 Mechanical Sensitivity Testing

As an example, we will briefly describe the tests presently used within Nitro Nobel AB for determining the sensitivity of the emulsion matrices and final emulsion explosives products. These tests are carried out as part of the process of obtaining approval to manufacture, store, ship, and use these explosives by the authorities, such as the Inspectorate of Explosives and Flammable Substances. Tests are also carried out to obtain measured values of the sensitivity of experimental new products to judge their merit with respect to different manufacturing, transportation, and blasting operations and situations.

19.4.1 Shock and Impact Tests

A feared event in processing and transportation of explosives is the potential detonation or explosive reaction due to accidental impact. All explosives must be tested in order to determine their sensitivity to shock and impact.

Air Gap Test

The air gap detonation transmission test (Figure 19.1) is carried out to predict the propensity of the explosive to undergo sympathetic detonation. This is a property of an explosive of major concern in processing and transportation, and also from a functional point of view.

The test determines the ability of an explosive to transmit detonation through air from one charge to another some distance away. The test is not a compulsory one for government approval.

19.4. Mechanical Sensitivity Testing

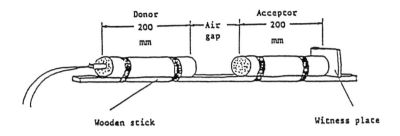

Figure 19.1. Set-up

Booster Strength	Booster size
Cap No.2	ϕ7.1 mm
Cap No.4	ϕ7.1 mm
Cap No.6	ϕ7.1 mm
Cap No.8	ϕ7.1 mm
2 g Comp. C-4[1]	ϕ12.4×12.4 mm
4 g Comp. C-4[1]	ϕ15.6×15.6 mm
8 g Comp. C-4[1]	ϕ19.6×19.6 mm
16 g Comp. C-4[1]	ϕ24.7×24.7 mm
32 g Comp. C-4[1]	ϕ31.1×31.1 mm
64 g Comp. C-4[1]	ϕ39.2×39.2 mm
128 g Comp. C-4[1]	ϕ49.4 × 49.4 mm
256 g Comp. C-4[1]	ϕ62.3 × 62.3 mm
512 g Comp. C-4[1]	ϕ78.5 × 78.5 mm
1024 g Comp. C-4[1]	ϕ98.8×98.8 mm

[1] The booster density is 1350 kg/m^3.

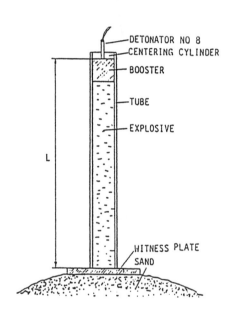

Figure 19.2. Test arrangement using a witness plate.

Minimum Booster Test

This test (Figure 19.2), combined with a measurement of the detonation velocity in a standard charge size, is often a part of the production control. It also indicates the sensitivity to minimum shock stimuli. The test determines the minimum charge needed for initiation to stable detonation.

434 Chapter 19. Safety in Production of Explosives

(a) Ballistic pendulum
(b) Sample
(c) Support block
(d) Steel support and sample level adjust
(e) Iron stand
(f) Bullet catcher

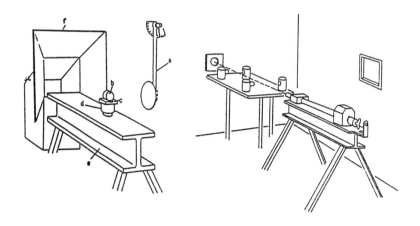

Figure 19.3. Gun and sample set-up.

Cap Sensitivity Test

The cap sensitivity test is designed to determine the sensitiveness or sensitivity of a substance to the shock from a standard detonator (blasting cap). It is used as one of the criteria for classifying an energetic material as a "very insensitive" explosive substance in the "UN HAZARD Division 1.5".

BAM Fallhammer Test

The BAM impact test yields quantitative results in the form of the impact energy needed for initiation. It is used to determine if the explosive is too hazardous for transport according to the UN classification.

Projectile Impact Test

This test (Figure 19.3) determines the sensitivity of explosives to projectile impact, specifically by a brass bullet having a diameter of 15 mm.

(a) Detonator wires
(b) Detonator
(c) Screw cap of malleable cast iron
(d) Booster (or relay) of RDX/Wax (95/5)
(e) Substance under test
(f) Steel tube to DIN 2448 specification, material ST.00
(g) Welded steel base

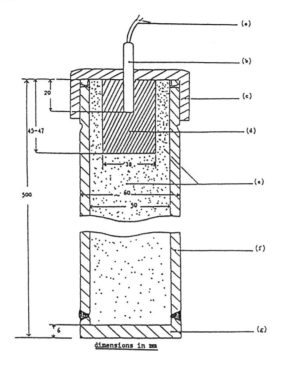

Figure 19.4. Steel Tube Test set-up.

BAM 50/60 Steel Tube Test

The sensitiveness of solid and liquid substances to detonation shock under confinement is measured in this test. The test is needed for the UN classification (Figure 19.4). It determines the sensitiveness of solid and liquid substances to detonation initiation by shock under confinement.

19.5 Heat and Friction Tests

Heating can affect every explosive and cause thermal decomposition which may lead to detonation. At room temperature, waterbased explosives do not burn stably under atmospheric pressure; therefore, the significant tests must provide some degree of confinement to allow for a pressure build-up.

A worst case in processing may be when a pump or a blender with moving parts is working the explosive against a dead end so that all the mechanical energy is transferred into heat. The temperature in the explosive within the mechanical equipment may rise locally very quickly, and a thermal decomposition can rapidly cause a pressure build-up which then can start a deflagration-to-detonation transition.

Even without confinement, a fire in a manufacturing plant for emulsion explosives can become extremely critical when very large quantities of explosive are set on fire, or when heat is transferred from the fire to explosives confined in bins. Several cases have been reported where quantities of waste emulsions, intended to be destroyed by burning at a special burning site, have instead detonated on the burning site. It is possible that the intense heat transfer from the flames above a very large surface of burning emulsion explosive with added fuel can lead to a deflagration to detonation transition even at atmospheric pressure. Contamination with more sensitive explosives or detonators is also a possible cause for such phenomena.

Koenen Test

This test is compulsory for the UN classification. The sensitivity (sensitiveness) to the effect of intense heat under partial and defined confinement is measured in this test.

Princess Incendiary Spark Test

This test is used to assess the case of ignition of an explosive substance by incendiary sparks produced by a length of safety fuse. If a substance ignites in the test, it is assumed that it is not "very insensitive". The test will be compulsory in the future for UN classification.

External Fire Test for Hazard Division 1.5

This is a test on a possible UN Division 1.5 substance, in the packaging in which it will be transported to determine whether the substance explodes when involved in a fire. The volume of the package or packages shall not be less than 0.15 m^3, but the substance weight need not exceed 200 g.

Deflagration to Detonation Transition (DDT) Test

This test (Figure 19.5), according to the new (1993) UN classification, will be compulsory in the future for approval for Hazard Division 1.5. The test is used to determine the tendency for the substance to undergo transition from deflagration to detonation. The test substance is filled into a closed tube and ignition is by a hot wire.

Woods Test

This test is used to determine the minimum temperature needed for a reaction to start in the explosive. A small amount of explosive is put into the $+75°C$ melt of Woods metal. The temperature is increased until reaction occurs.

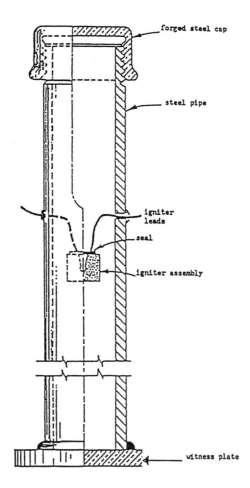

Figure 19.5. DDT Test set-up.

19.6 Detonation Stability Testing

The detonation stability is always measured during the R&D work as a production control and also to control the ageing of a product.

Velocity of Detonation (VOD) Test

The detonation velocity should be measured (Figure 19.6) in the explosive after the detonation has reached a steady-state detonation.

Critical Diameter Test

This test determines the detonability of an explosive by means of determining the minimum diameter at which a stable detonation can propagate.

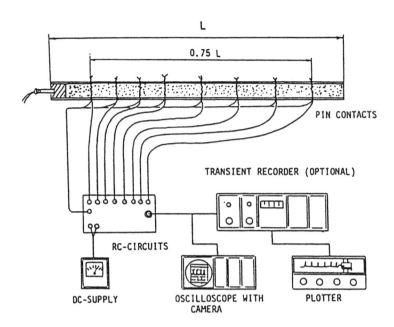

Figure 19.6. Example of set-up for VOD. measurement.

19.7 Testing of Unsensitized Emulsion Matrix for Transportation

The testing to be performed for an approval for transporting an unsensitized emulsion matrix as a nonexplosive product can be found in the Manual of Tests for the Hazard Classification of Explosives by the UN Committee of Experts.

The question *"Is it an explosive substance?"* is answered on the basis of national and international definitions of an explosive substance and the results of two types of tests used to asses possible explosive effects.

Type 1(a) Tests

These are shock tests with a defined booster and confinement to determine the ability of the substance to propagate a detonation. Alternative tests are:

Test 1(a) (*i*)	BAM 50/60	Steel Tube Test
Test 1(a) (*ii*)	TNO 50/70	Steel Tube Test
Test 1(c) (*iii*)		Gap Tests for Solids and Liquids
Test 1(d) (*iv*)		Gap Tests for Solids and Liquids

Type 1(b) Tests

These are combustion or thermal tests to determine the thermal response of the substance. Alternative tests are:

Test 1(b) (*i*)   Koenen Test
Test 1(b) (*ii*)   Internal Ignition Test
Test 1(b) (*iii*)   SCB Test

The emulsion matrix is classified as too insensitive for inclusion in Class 1 if a Type 1(a) test and a Type 1(b) test do not indicate any reaction.

It must be emphasized that the emulsion matrix must be tested at a temperature above the temperature at which it is fed from the storage bin into the transportation vehicle. The density should not be less than what is specified for the matrix.

At Nitro Nobel AB, the emulsion matrix is tested according to the BAM 50/60 Steel Tube Test and the Koenen Test. The tests are described in the UN recommendations.

19.8 Conclusion

This chapter has outlined some of the hazard tests carried out in order to get approval from authorities for the manufacturing, transportation, and handling of explosives. Apart from this, these tests give useful information for the research, product development, and production departments to make professional judgements of the safety of present and new products and processes.

Chapter 20

United Nations Recommendations on the Transport of Dangerous Goods

Tests and Criteria
Second Edition
United Nations, New York, 1990

20.1 General Introduction

The United Nations **Recommendations on the Transport of Dangerous Goods** has been adopted by most countries, including the USA and Sweden, as a standard for classifying and testing not only explosives but a wide range of "dangerous" substances and articles that may have explosive or flammable properties depending on the circumstances. The recommendations are intended to serve as a guideline for *"the competent authority"* in each country to make decisions regarding the classification of dangerous substances with respect to transportation hazards.

The Recommendations, developed by the United Nations Committee of Experts on the Transportation of Dangerous Goods, are in two parts. They are bound separately in paperback volumes with orange covers and are therefore often referred to as the *"Orange Books"*. **Volume 1, ST/SG/AC.10/1/Rev.7 Recommendations**, provides information on how to classify, package, test, mark, label or placard, and ship hazardous goods. **Volume 2, ST/SG/ AC.10/11/Rev.1 Tests and Criteria**, provides internationally accepted test methods and procedures for properly classifying explosives or hazardous articles for transportation. The following excerpts from the Tests and Criteria volume are included to make the reader familiar with the procedure for classification of a dangerous substance and to illustrate some of the tests. The two volumes are living documents and change regularly as new editions are published. Therefore, the reader, when the need arises to consult the volumes, **must always obtain the current edition of the Orange Books**.

Please Note: Never use the following excerpts for actual classification of an explosive, but only to understand the general idea of classification.

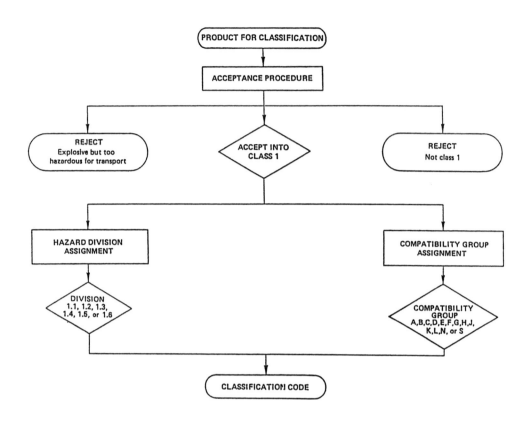

Figure 20.1. Scheme of procedure for classifying a substance or article.

20.2 Classifying a Dangerous Substance or Article

For determining whether an unknown substance or article which is presumed dangerous can be transported, and if so how, the United Nations protocol relies on the procedure for classifying a substance or article as set out in the scheme of Figure 20.1 and the detailed schemes of Figures 20.2 and 20.3.

Dangerous goods are divided into 10 classes, Classes 1–10 (see Table 20.1), with respect to the different general types of dangers encountered with different materials. Once a material has been classified, it is included in the long list of dangerous goods (Table 20.2), an alphabetical list of common articles and substances providing details of classification, labeling, and rules regarding its transportation.

Class 1 contains all explosives, and is subdivided into six divisions, Divisions 1.1 to 1.6 (see Table 20.3) with respect to the materials' behavior in a fire, such as its propensity for mass detonation or mass explosion or for creating a mass fire hazard. Table 20.4 gives examples of explosive materials and articles within the six groups. Note that the term "explosives" here comprises all substances and articles that can be the cause of an explosion; namely, explosives, propellants, and pyrotechnic materials

20.2. Classifying a Dangerous Substance or Article

Table 20.1. Dangerous goods

Class 1	Explosives
Class 2	Gases compressed, liquified, dissolved under pressure, or deeply refrigerated
Class 3	Flammable liquids
Class 4	Flammable solids/substances liable to spontaneous combustion/substances which (in contact with water) emit flammable gases
Class 5	Oxidizing substances; organic peroxides
Class 6	Poisonous (toxic) and infectious substances
Class 7	Radioactive materials
Class 8	Corrosives Group I, II, and III
Class 9	Miscellaneous
Class 10	Substances and articles with multiple hazards

and articles manufactured with the intent to produce an explosive or pyrotechnic effect, as well as substances that may be explosive although they were not manufactured with that intent.

In addition to the divisions, materials belonging to Class 1: Explosives are also divided into 13 compatibility divisions, Divisions A–N (see Table 20.5) with respect to the level of risk for explosion following flame initiation.

After having determined that a substance is neither too hazardous for transportation, nor falls outside the criteria for classification as an explosive, it is accepted into Class 1: Explosives (see Figure 20.1). It is then assigned a hazards division and a compatibility group, and finally given a classification code.

The acceptance procedure, for deciding whether a product can be accepted for classifying as a possible explosive and the procedure for assignment of hazard division and compatibility group, is performed with the help of two connected flowcharts, Figures 20.2 and 20.3. Before reaching each decision point, one or more tests may be prescribed, the results of which will provide the basis for the decision. There are 7 test series in all, one for each decision point. Within each test series, there are a number of alternative tests, from which the competent authority may select one or more for testing any one substance or article, depending on the availability of test facilities, the tradition in the country, and other considerations. The tests and the associated criteria are described in detail in the Orange Book, Volume 2, Tests and Criteria. In the following are provided only the names of the tests included in the tests Series 1, 2, and 3. The reader is encouraged to study the Orange Books for details on these tests and the criteria according to which the test results are judged in making classification judgements. In general, the results of the tests determine whether the material is an explosive substance (Test Series 1); if it is too insensitive to be included in Class 1 (Test Series 2); or if it is too hazardous for transportation at all (Test Series 3). Further tests are prescribed for determining the division within Class 1 to which the material belongs.

Table 20.2. UN Dangerous Goods list. Sample page from the alphabetical listing of common articles and substances.

Name	UN No.	Class or division	Sub-sidiary risk	Labels	State varia-tions	Special provi-sions	UN packing group	Passenger Aircraft Packing instruc-tions	Passenger Aircraft Max. net quantity per package	Cargo Aircraft Packing instruc-tions	Cargo Aircraft Max. net quantity per package
1	2	3	4	5	6	7	8	9	10	11	12
Ammonium nitrite	forbidden										
Ammonium perchlorate	0402	1.1D				A22		forbidden		forbidden	
Ammonium perchlorate	1442	5.1		Oxidizer	JP 18	A22	II	509	5 kg	512	25 kg
Ammonium permanganate	forbidden										
Ammonium persulphate	1444	5.1		Oxidizer			III	516	25 kg	518	100 kg
Ammonium picrate, dry or wetted with less than 10% water by mass	0004	1.1D						forbidden		forbidden	
Ammonium picrate, wetted with less than 10% water by mass	1310	4.1		Solid flammable	BE 3	A40	I	416	0.5 kg	416	0.5 kg
Ammonium polysulphide solution	2818	8	6.1	Corrosive & Poison			II	808	1 liter	812	30 liter
Ammonium polyvanadate	2861	6.1		Poison			II	613	25 kg	615	100 kg
Ammonium sulphide solution	2683	8	3 6.1	Corrosive & Liquid flammable & Poison			II	808	1 liter	812	30 liter

Table 20.3. Hazard Divisions. Class 1: Explosives.

1.1	Mass detonating explosive hazard
1.2	Non-mass detonating fragment-producing explosive hazard
1.3	Mass fire, minor blast, or fragment hazard
1.4	Moderate fire, no blast or fragment hazard
1.5	Explosive substance, mass explosion, or ammunition article, unit risk
1.6	Extremely insensitive substances

Table 20.4. Hazard Divisions Examples.

1.1	Dynamite, cast boosters, cap sensitive emulsions, water gels and slurries, Class A detonators
1.2	Ammunition with projectile hazard
1.3	Non-detonating propellants and explosives
1.4	Class C detonators, safety fuses, other Class C explosives
1.5	ANFO, non-cap sensitive emulsions, water gels, slurries, packaged blasting agents
1.6	Currently no commercial explosives

Table 20.5. Compatibility Groups. Class 1: Explosives.

Group A	Primary explosive substance
Group B	Primary explosive substance without at least two safety features
Group C	Propellant explosives
Group D	Secondary detonating explosives without initiator or propelling charge, or article with two or more safety features
Group E	Secondary detonating charge without initiator, but with propelling charge
Group F	Secondary detonating charge with initiator and propelling charge
Group G	Pyrotechnics
Group H	Explosive and white phosphorous
Group J	Explosive and flammable liquid
Group K	Explosive and toxic chemical agent
Group L	Explosive with special risk (hypergolic, water activated, phosphides)
Group S	Package protects against all hazards
Group N	Extremely insensitive substances

20.3 Test Series 1, 2, and 3

The tests in Series 1 are intended to provide experimental evidence to help decide whether the substance in question is an explosive. The question *"Is it an explosive substance?"* (Box 4, Figure 20.1) has to be answered on the basis of national and international definitions of an explosive substance and the results of two types of tests; namely, Test Series 1 type (a) and Test Series 1 type (b) in order to assess possible explosive effects.

The tests in Series 2 are intended to help decide whether the substance in question is too insensitive for acceptance into Class 1. The question *"Is the substance too insensitive for acceptance into Class 1?"* (Box 6, Figure 20.1) has to be answered on the basis of the results of these tests.

The tests in Series 3 are intended to indicate whether the substance is too hazardous for transport. The question *"Is the substance too hazardous for transport (in the form in which it was tested)?"* (Box 10, Figure 20.2) has to be answered on the basis of these tests.

446   Chapter 20. UN Recommendations on the Transport of Dangerous Goods

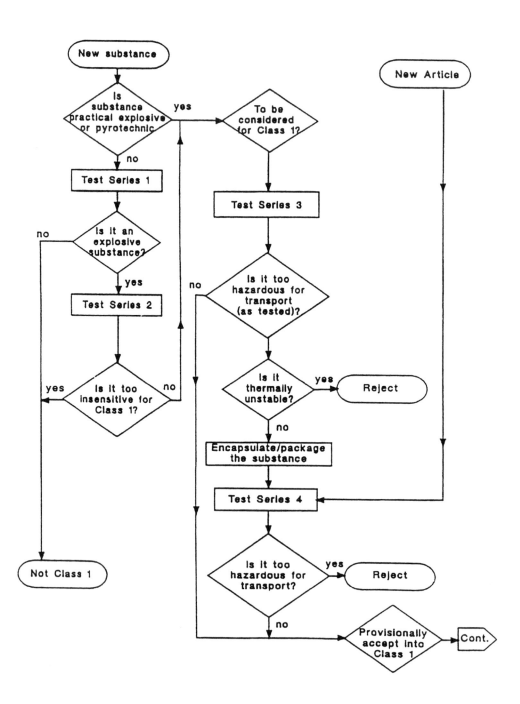

Figure 20.2. Acceptance procedure.

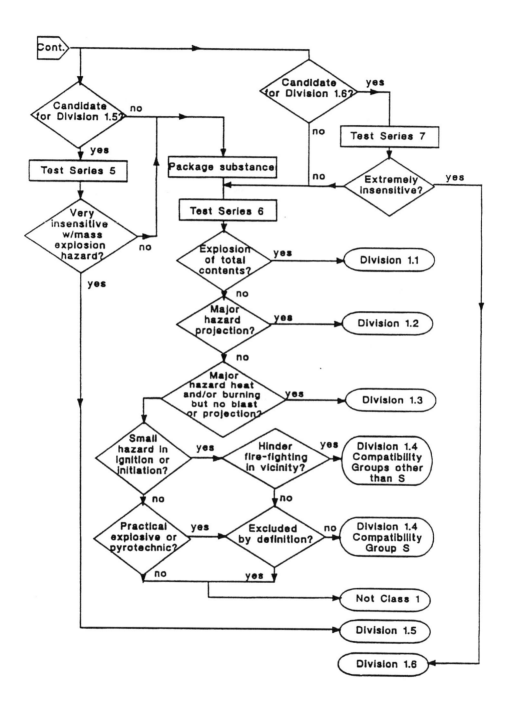

Figure 20.3. Procedure for assignment of Hazard Division.

20.3.1 Test Series 1

Test Series 1 Type (a): These are shock tests with a defined booster and confinement to determine the ability of the substance to propagate a detonation.

    Test 1 (a) (*i*)   BAM 50/60 Steel Tube Test
    Test 1 (a) (*ii*)   TNO 50/70 Steel Tube Test
    Test 1 (a) (*iii*)   Gap Test for Solids
    Test 1 (a) (*iv*)   Gap Test for Liquids

Test Series 1 Type (b): These are combustion or thermal tests to determine the thermal response of the substance.

    Test 1 (b) (*i*)   Koenen Test
    Test 1 (b) (*ii*)   Internal Ignition Test
    Test 1 (b) (*iii*)   SCB Test

The question in Box 4 of Figure 20.2 is answered "yes" if a "+" is obtained in either test type 1 (a) or 1 (b).

20.3.2 Tests Series 2

Test Series 2 Type (a): These are shock tests with a defined booster and confinement to determine the ability of the substance to propagate a detonation.

    Test 2 (a) (*i*)   BAM 50/60 Steel Tube Test
    Test 2 (a) (*ii*)   TNO 50/70 Steel Tube Test
    Test 2 (a) (*iii*)   Gap Test for Solids
    Test 2 (a) (*iv*)   Gap Test for Liquids

Test Series 2 Type (b): These are enclosed burning tests with a defined heating rate or igniter inside a confinement to determine the ability of the substance to burn violently or explosively under confinement.

    Test 2 (b) (*i*)   Koenen Test
    Test 2 (b) (*ii*)   Internal Ignition Test
    Test 2 (b) (*iii*)   Time/Pressure Test
    Test 2 (b) (*iv*)   SCB Test

20.3.3 Tests Series 3

Test Series 3 Type (a): These are fallhammer tests designed to determine the substance's sensitivity to ignition by low velocity impact or pinching between metal surfaces.

    Test 3 (a) (*i*)   Bureau of Explosives Machine
    Test 3 (a) (*ii*)   BAM Fallhammer
    Test 3 (a) (*iii*)   Rotter Test
    Test 3 (a) (*iv*)   30 kg Fallhammer Test
    Test 3 (a) (*v*)   Modified Type 12 Impact Tool

Test Series 3 Type (b): These are friction tests to determine the substance's sensitivity to friction or rubbing.

    Test 3 (b) (*i*)   BAM Friction Apparatus
    Test 3 (b) (*ii*)   Rotary Friction Test
    Test 3 (b) (*iii*)   ABL Friction Test

Test Series 3 Type (c): This is a test for determining the substance's thermal stability at a relatively low temperature, 75° C.

    Test 3 (c) Thermal Stability Tests at 75° C

Test Series 3 Type (d): These are burning tests to determine the substance's intensity of burning.

    Test 3 (d) (*i*)   Small-Scale Burning Test
    Test 3 (d) (*ii*) Small Scale Burning Test

The question in Box 10 is answered "yes" if a "+" is obtained in either test type 3 (a), 3 (b), 3 (c), or 3 (d).

20.4 Test Series 4

The tests in Series 4 are intended to provide experimental evidence to help decide whether the article or package in question is too hazardous for transport. The question *"Is the article, packaged article or packaged substance too hazardous for transport?"* (Box 16, Figure 20.1) has to be answered on the basis of these tests. If the answer is no, then the article or package is provisionally accepted into Class 1.

20.5 Test Series 5, 6, and 7

The tests in Series 5, 6, and 7 are intended to provide experimental evidence to help assign the article or substance into the hazards divisions within Class 1 Explosives; namely Divisions 1.1 to 1.6.

A substance is first exposed to the question *"Is the substance a candidate for Division 1.5?"* (Box 21, Figure 20.3). If it is, then it is exposed to the tests Series 5. If not, the substance is first packaged in the box or container in which it is intended to be transported. It is then exposed to the test Series 6.

An article is first exposed to the question *"Is the article a candidate for Division 1.6?"* (Box 24, Figure 20.3). If yes, then the article is explosed to the tests Series 7; then the question is asked: *"Is it an extremely insensitive article?"* (Box 40, Figure 20.3). If it is, then the article belongs to division 1.6.

The question *"Is it a very insensitive explosive substance with a mass explosion hazard?"* (Box 21, Figure 20.2) has to be answered on the basis of the results of the tests Series 5. If the answer is yes, then the substance belongs to Division 1.5. If the answer is no, then the substance is packaged in the box or container in which it is intended to be shipped, and the packages are then subjected to the tests in Series 6.

The tests in Series 6 are the crucial ones for deciding which of the four Divisions 1.1 through 1.4 the material belongs to. Several questions are decided on the basis of these test results.

The question *"Is the result (of the Test Series 6) an explosion of the total contents?"* (Box 26, Figure 20.3) decides if the substance or articles in the package belongs to Division 1.1.

If not, the result of the question *"Is the major hazard (encountered in the test Series 6) that from dangerous projections?"* (Box 28, Figure 20.3) decides if the substance or articles in the package belongs to Division 1.2.

If not, the result of the question *"Is the major hazard radiant heat and/or violent burning but with no dangerous blast or projection hazard?"* (Box 30, Figure 20.3) decides whether the substance or articles in the package belongs to Division 1.3. Four further questions decide whether the correct division is 1.4, and if so, to which compatibility group the material belongs. There is also a possibility that the material will be rejected at this point as not belonging to Class 1 after all.

20.6 Decision-Making Agencies and Approved Laboratories

A company may acquire the capability to carry out one or more of the tests recommended and described by the United Nations Recommendations on the Transport of Dangerous Goods. Such capability is valuable for the internal judgements and decisions that need to be made on a continuing basis in the development of new products or when changing operational procedures. However, it is to be noted that the final judgement and decision on classification of a dangerous substance or article in most countries is the responsibility of one or more government agency or office. For example, in the USA, decisions regarding classification of a material for transportation by road or rail is the responsibility of the Department of Transportation. In Sweden, such decisions are made by the Inspector of Explosives and Flammable Substances. In both countries, the tests used to provide a basis for the decisions are carried out according to the United Nations recommended Tests and Criteria, although in special cases, additional test may be required. In the USA., only a very few laboratories are approved for carrying out the tests; The Bureau of Mines of the US Department of the Interior is one such laboratory. In Sweden, a manufacturer is often required to carry out the tests, under the supervision of the government Inspectorate.

Exercises

Exercises are numbered with the chapter number to which they apply followed by a consecutive number within each chapter. Solutions to problems marked with a * are provided in the Solutions chapter.

Exercise 1.1 * Lundborg described the shear strength τ of rock materials as a function of the normal pressure σ_n by the equation

$$\tau = \tau_0 + \frac{\mu \sigma_n}{1 + \dfrac{\mu \sigma_n}{\tau_i - \tau_0}}$$

where τ_i is the limit value of the shear strength at very large σ_n, τ_0 is the shear strength at zero normal pressure, and μ is the coefficient of friction between crack surfaces. A particular type of sandstone has $\tau_i = 900$ MPa, $\tau_0 = 20$ MPa, $\mu = 0.7$. What is the shear strength of this sandstone at normal pressure 100 MPa?

Exercise 1.2 * In Lundborg's statistical theory of the strength of brittle materials, the strength is described by two parameters: μ which has the form of a friction coefficient and M which represents the degree of statistical scatter of the strength over the space angle. The theory provides a particularly simple expression, in μ and M, for the ratio between the uniaxial strength σ_{10} and the biaxial strength σ_{20}

$$\left(\frac{\sigma_{20}}{\sigma_{10}}\right)^M = \sqrt{1 + \mu^2} + \mu$$

For a material with $\mu = 0.9$ and $M = 2$, what is that ratio?

Exercise 1.3
  a. Use Lundborg's equation for the confined shear strength of brittle materials

$$\tau = \tau_0 + \frac{\mu \sigma_n}{1 + \dfrac{\mu \sigma_n}{\tau_i - \tau_0}}$$

  to calculate the shear strength of flintstone (using $\mu = 2.5$, $\tau_0 = 0.15$ GPa, and $\tau_i = 2.14$ GPa) at normal stress 0 and 1 GPa (compressive).
  b. What is a typical value of the shear strength of steel expressed in GPa?
  c. Explain the difference between the shear strength of steel and that of flintstone at 1 GPa (compressive) normal stress.

Exercise 1.4 List typical values of the following strength values for granite and steel in units of MPa and psi:

Strength property	Granite		Steel	
	MPa	psi	MPa	psi
Tensile strength				
Unconfined compressive strength				
Shear strength at infinite confinement				
Hugoniot elastic limit				
Elastic modulus				

a. What are the ratios between the Hugoniot elastic limit and the elastic modulus for granite and steel?
b. Why are these ratios so different?

Exercise 1.5 The stress at which a crack of given geometry in a given material will propagate is determined by the critical value of the stress intensity factor, K_{IC}, which is a material property, equal to 50 MN/m$^{3/2}$ for steel and 3 MN/m$^{3/2}$ for a particular type of granite. For a crack stressed in uniaxial tension perpendicular to the crack direction under a load at which the crack is just beginning to extend, the tensile stresses in the area ahead of the crack are given as a function of the distance x from the crack tip by the approximate relation

$$\sigma(x) = \frac{K_I C}{\sqrt{2\pi x}}.$$

Sketch, in one diagram with approximately correct equal scale, the stress distribution ahead of the crack for steel and the granite.

Exercise 1.6 * The theoretical maximum shear strength of a perfect crystalline material is often given as a fraction of the elastic modulus E, and can be estimated to be somewhere in the region $E/10$ to $E/30$, depending on the form of the interatomic forces. The Hugoniot Elastic Limit (HEL) is defined as the maximum shock amplitude for which the material can deform elastically, without any plastic deformation. The HEL for a given kind of steel is 800 MPa, its E is 210,000 MPa, while the HEL for a type of granite is 3000 MPa and its E is 60000 MPa. Discuss why the ratio HEL/E for the two materials differs so much.

Exercise 2.1 *
  a. Explain why the cost K for excavating a unit volume of rock (by drilling, full face boring, or by drilling and blasting) is very approximately inversely proportional to the diameter d of the hole produced.
  b. Using the Kallin expression for the volume excavation cost

$$K = \frac{k_1}{d}$$

with the value $k_1 = 80$ \$/m^3 for drilling in granite, list the approximate cost per meter hole depth for three drilled holes of diameter 50 mm; 100 mm; and 2.5 m.

Exercise 2.2 Provide a firing times plan for an open pit blast of 50 drillholes of diameter 300 mm, arranged in five rows, with a 7 m burden and a 9 m spacing, and using pyrotechnic delay detonators, for

    a. electric detonators
    b. NONEL detonators (without use of surface delays)
    c. NONEL detonators (with use of surface delay connectors)

Exercise 2.3 Sketch in a log-log diagram the approximate cost for excavating one unit volume (liter or cubic meter) of rock as a function of the diameter of the cavity diameter, extending from a drillhole of a few millimeters diameter to an open pit mine of 1 km diameter.

    a. Give a plausible explanation for the cost - diameter relationship in terms of the energy required for extending a drilled cavity a unit distance.
    b. How would, in your opinion, a 10-fold increase in the price of oil affect the cost-scale in the cost vs. diameter diagram?

Exercise 2.4
    a. Make a simple derivation of a relationship between the unit hole volume cost when drilling or boring holes in rock and the diameter of the hole, based on the assumption that in most rock drilling, the perimeter of the hole bottom is where most of the drilling energy and tool bit wear occurs.
    b. If the cost of drilling a 4 1/2 in diameter hole in rock at a quarry is $1.50 per foot hole length, what hole length cost would you predict for an 8 in diameter hole in rock under similar conditions, using the relationship derived under part (a).

Exercise 3.1 * Explosive or energetic materials can be divided, depending on the use for which they are manufactured, into three main groups, namely (high) explosives, propellants, and pyrotechnics. High explosives can be divided into three subgroups: namely primary explosives, secondary explosives, and tertiary explosives. Propellants can be divided into liquid and solid propellants; solid propellants can be divided into single base propellants, double base propellants, and composite propellants.

    a. Give an example of an energetic material from each of the three high explosive subgroups, and in the case of secondary explosives, explain the difference between a single explosive and a composite explosive.
    b. How would you classify black powder, which consists of hard, compact pellets of a fine-grained mixture of the ingredients sulphur, charcoal, and potassium nitrate?

Exercise 3.2 * What is a permissible explosive, and which US authority has the power to decide whether an explosive is permissible or not?

Exercise 3.3 * Describe and draw schematics of two different ways of grouping explosive materials:
  a. with respect to ease of initiation of explosion by fire
  b. with respect to end use of the material

Exercise 3.4 * Explain in a few words the meaning of the following terms:
  a. Energetic material
  b. Explosive
  c. High explosive
  d. Class A Explosive
  e. Blasting agent
  f. Oxidizer

Exercise 3.5 * What property of chemical reactions is fundamental to all decomposition processes involving explosives, such as slow thermal degradation, burning (in air or in an inert atmosphere), burning- (or deflagration)-to-detonation transition (DDT), shock initiation, and detonation?

Exercise 3.6 Plot an approximate curve showing the half-life (or life to 5% decomposition) of ammonium nitrate as a function of temperature, indicating the approximate range of reaction times (lifetime) for the following decomposition processes:
  a. Long-term storage at room temperature.
  b. The Henkin-McGill time-to-explosion test.
  c. The drop-weight impact test.
  d. Shock initiation.
  e. Detonation.

Exercise 3.7 * Describe the mechanisms for initiation of chemical reaction in an explosive containing gas-filled spherical voids (gas-bubbles or glass microballoons)
  a. By impact compression, such as in a drop-weight impact test (compare with the Brower compression test)
  b. In shock initiation or detonation

Exercise 3.8 Describe, with the help of a simple sketch, the main features of surface burning (without atmospheric oxygen) of a propellant or explosive. You should give approximate values of the temperature, and indicate temperature gradients, in the solid and in the gaseous material.

Exercise 3.9 What is a permissible explosive and where are such explosives used and why?

Exercise 3.10 Describe the process of producing a pumpable emulsion explosive (in the U.S. classified as a blasting agent) in a fixed plant.

Exercise 3.11 Describe and sketch three common types of trucks used to deliver bulk blasting agents into large-diameter drillholes.

Exercises

Exercise 3.12 * The rate law for the reaction of an explosive is given by

$$\frac{d\lambda}{dt} = k(1-\lambda)e^{-T^\dagger/T}$$

where λ is the extent of reaction. $k = 10^{20}\text{s}^{-1}$ is a constant, $T^\dagger = 25000$ K is the activation temperature (another constant), and T is the temperature.

a. What is the ratio of the reaction rate at 600 K to the reaction rate at 590 K?
b. At what temperature does the rate become 10^6s^{-1}, a rate fast enough for detonation?

Exercise 3.13 Provide a schematic picture of the process of heating, evaporation, decomposition and chemical reaction associated with a surface burning of an energetic material. Explain how an increase in the pressure of the surrounding gas would affect the linear rate of burning at right angle to the surface.

Exercise 4.1 * Derive for a plane shock and a plane detonation the conservation equations for mass, momentum, and energy, and also the simple approximate relations between internal energy increase and particle velocity.

Exercise 4.2 ANFO detonating in large drillholes has an explosion energy of 4 MJ/kg, a detonation velocity of 5.3 km/s, and a detonation pressure of 6.3 GPa at an initial density of 850 kg/m³. Assuming a value of the average specific heat $C_v = 1.7$ kJ/kg°C, what is the approximate temperature of the reaction products of ANFO at the detonation (Chapman-Jouguet) state?

Exercise 4.3 Using the table for computing explosion energy and gas volume of oxygen-balanced explosives, Table 4.5, calculate the oxygen-balanced composition and the explosion energy of two ammonium nitrate-based explosive compositions, $Al + AN$ and $FO + AN$.

Exercise 4.4 Like most homogeneous secondary explosives, the liquid explosive nitromethane, when free of bubbles or other inhomogeneities, is relatively insensitive to shock initiation.

a. Use the shock Hugoniot for unreacted nitromethane in the form

$$U = C + Su$$

with the parameter values $C = 2$ km/s; $S = 1.38$; together with the shock conservation equations, to calculate the shock pressure and the increase in internal energy when nitromethane of initial density 1.14 g/cm³, initially at rest, is shock compressed to a particle velocity of 1 km/s.

b. Assuming a specific heat $c_v = 1.25$ kJ/kg°C, what is the approximate shock temperature increase of the nitromethane shocked to particle velocity 1 km/s?

Exercise 4.5 The heat of formation ΔH_f° of water (in gaseous form) is equal to -241.8 kJ/mole. Assuming that liquid hydrogen and solid oxygen could be safely mixed and pumped into drillholes in the form of a detonable slurry of density 0.45 g/cm^3, would this be a better rock blasting explosive than ANFO at density 0.82 g/cm^3? (You may assume, for simplicity, that the explosion energy per unit volume of explosive can be used to compare the strength of the two explosives). For the purpose of calculating the explosion energy of ANFO, which is an oxygen-balanced mixture of ammonium nitrate and fuel oil (diesel oil, approximately similar to the atomic composition of dodecane), the data of Table 4.5 may be used.

Exercise 4.6 What is the origin of the toxic fumes CO and NO_x in rock blasting using an oxygen-balanced explosive?

Exercise 4.7 * Calculate, with the help of tabulated values of the heats of formation of ingredients and reaction products, the explosion energy of ANFO. In this example, ANFO is assumed to be an oxygen-balanced (to CO_2) mix of ammonium nitrate and dodecane. (Dodecane is an alkane with 12 carbon atoms, similar in chemical composition to fuel oil.)

Exercise 4.8 * Explosives for use in underground blasting are designed to be nearly oxygen-balanced to CO_2 in order to minimize the reaction products' content of the main toxic reaction products CO and NO_x.

(If the ingredient composition is nearly but not quite oxygen-balanced and has some excess oxygen, more NO and NO_2 (together called NO_x) than CO is found in the reaction products; if the composition is slightly deficient in oxygen, more CO than NO is found. Considering that the allowable limit concentration of NO in workplace air is five times less than that of CO, the oxygen-balance is usually adjusted to be a few percent on the negative side, so that more CO than NO_x is formed.)

Disregarding these small deviations from oxygen-balanced composition, what is the CO_2 oxygen-balanced composition of ammonium nitrate mixed with

  a. TNT
  b. Nitromethane
  c. Carbon, for example, lamp black
  d. Aluminum

Exercise 4.9 * Using the diagram of explosive expansion work as a function of volume expansion ration, explain the differences in performance between the three explosives — emulsion, ANFO, and nitromethane — in blasting in hard rock, soft porous rock, and ditching in sand.

  a. Consider and define the effective expansion ratio for blasting in each of these geomaterials; then rank the three explosives with respect to useful expansion work per unit drillhole volume. Use the following values of density: emulsion, 1.25 g/cm^3; ANFO, 0.82 g/cm^3; and nitromethane, 1.14 g/cm^3.
  b. Assuming that the cost of the finished drillhole is \$0.15 per liter hole volume, and the price of the explosives delivered free in holc is \$0.50/kg for the emulsion, \$0.25/kg for the ANFO, and \$3.60/kg for the nitromethane, which of the three is most cost-effective for blasting in soft porous rock? (Take their different loading densities and effective expansion work into account.)

Exercise 4.10 * For the five explosives: (1) cartridged dynamite, (2) cartridged emulsion, (3) ANFO, (4) bulk emulsion, and (5) a 50/50 emulsion/ANFO mix, list approximately the following five characteristics:
  a. Weight percent composition
  b. Explosion energy (in MJ/kg and cal/g)
  c. Density (in kg/liter and g/cm^3)
  d. Infinite diameter detonation velocity (in km/s and ft/sec)
  e. Price (in \$/kg and \$/lb to a large consumer)

Exercise 4.11 * Discuss the possible causes and likely composition of some excessive amounts of toxic fumes that might be detected in underground blasting using

  a. An oxygen-balanced dynamite
  b. ANFO
  c. A pumpable emulsion explosive

Exercise 4.12 * The measured detonation velocity of ANFO made from AN prills mixed with diesel fuel oil was found to decrease toward the periphery of the container. The explosive was filled into a very large (25 m diameter) hemispherical container using a pneumatic loading technique, in which the prills, carried by a high-velocity air stream, were blown radially into the container through a series of evenly spaced holes in the container wall. Discuss the possible cause of this change in the detonation velocity at what should be near infinite diameter (ideal detonation) charge size.

Exercise 4.13 The liquid explosive nitromethane can withstand a shock pressure of 8.5 GPa for nearly a microsecond without appreciable chemical decomposition. The relation between shock velocity D and particle velocity u for nitromethane of initial density 1.128 g/cm^3 is

$$D = 2.00 + 1.38u$$

Calculate the temperature of the liquid, assuming no chemical reaction takes place, at the shock pressure 8.5 GPa, if the average specific heat is $c_v = 1250$ J/kg/K. (Check point: $u = 1.72$ km/s.)

Exercise 4.14 The measured detonation velocity of nitromethane of initial density 1.128 g/cm^3 is 6.0 km/s. The calculated $y = -(\partial \ln P/\partial \ln v)_s$ for nitromethane at the detonation state is 2.17.

  a. Calculate the pressure, density, specific volume, particle velocity, and increase in internal energy at the Chapman-Jouguet state.
  b. Assuming that the average specific heat of the nitromethane reaction products is 2250 J/kg°C, and the explosion energy is 6.2 MJ/kg, calculate the reaction product temperature at the Chapman-Jouguet state.

For this problem (and the previous problem), it is useful to know that

$$P_1 u = \rho_0 D \left(E_1 - E_0 + \frac{u^2}{2} \right)$$

but

$$P_1 = \rho_0 D u$$

so

$$\frac{P_1 u}{\rho_0 D} = u^2.$$

Thus

$$u^2 = E_1 - E_0 + \frac{u^2}{2}$$

or

$$E_1 - E_0 = \frac{u^2}{2}.$$

Units: $m^2/s^2 = J/kg$ and $(km/s)^2 = MJ/kg$.

Exercise 4.15 For Westerley granite of initial density 2630 kg/m^3, there is a linear relationship between shock wave velocity D and particle velocity u.

$$D = C + S u$$

where $C = 2.1$ km/s and $S = 1.63$. The relationship is valid in the pressure region from 5 to 50 GPa. Calculate the shock wave velocity D, pressure P, density ρ, and internal energy increase $E_1 - E_0$ for a shock wave in Westerley granite having a particle velocity of 1 km/s.

Exercise 4.16
   a. Make, using the data derived in Problems 4.14 and 4.15, an approximate plot in the pressure vs. specific volume plane of the shock Hugoniot for unreacted nitromethane, and the detonation Rayleigh line and the detonation Hugoniot for nitromethane. From the diagram thus constructed, what is the Chapman-Jouguet pressure of nitromethane?
   b. Calculate, using the data given in Problem 4.14, the pressure and specific volume for nitromethane at the von Neumann point. Check the position of the point calculated in part (a) by seeking the intersection of the unreacted Hugoniot and the detonation Rayleigh line, which is the von Neumann point.
   c. What is the physical significance of the von Neumann point?

Exercise 4.17 * A plane copper projectile strikes a plane copper target. The velocity of the projectile before impact was 800 m/s. Data for shocks in copper have shown that the shock velocity and particle velocity of the Hugoniot curve are related by $U = c + su$, with $c = 3940$ m/s and $s = 1.489$. The density of copper is 8930 kg/m^3.
[On a Hugoniot curve, $p = p_0 + \rho_0 U u$ and $\frac{\rho}{\rho_0} = \frac{U}{U - u}$.]

   a. What is the velocity of the interface after the collision?
   b. What is the pressure in the shocked copper target after the collision?
   c. What is the density of the shocked copper?
   d. If the target is 4.5 mm thick, what is the transit time for the shock wave through the copper?

Exercise 4.18 * *The Shock Tube Problem.* The complete theoretical background for this problem is not altogether covered in Chapter 4 (or elsewhere in this book). It is, however, a very instructive problem for advanced study of the properties of shocks and expansion waves. It was originally proposed by W.C. Davis and heightened noticeably the awareness and excitement of students in the class we taught together in 1992.

(PAP's remark)

A shock tube, with the high-pressure reservoir on the left, is filled with air at room temperature, 293 K. The high pressure is 5 MPa (about 50 atm), and the low pressure is 50 kPa (about 0.5 atm). Assume adiabatic $\gamma = 1.4 =$ a constant, and that the gas behaves as an ideal polytropic gas. The density of the high-pressure air is 60 kg/m^3, and that of the low-pressure air is 0.6 kg/m^3.

a. Make a qualitative sketch of the x-t plane, with the diaphragm separating the two reservoirs at $x = 0$ and $t = 0$ corresponding to the time the diaphragm is ruptured. Show the head (lead characteristic) of the rarefaction region, the end (tail characteristic) of the rarefaction region, the interface (contact discontinuity) between shocked gas and rarefied gas, and the shock wave in the low-pressure gas. Note that the state of the material is uniform between the contact discontinuity and the shock.

b. Make a qualitative sketch of the pressure, velocity, and the density of the gas as a function of distance at some time after the diaphragm is ruptured. Mark the interface between shocked gas and rarefied gas in each diagram.

c. For an ideal gas, $p = \rho R T$, $c^2 = \frac{\gamma p}{\rho}$, and $c_v = \frac{R}{\gamma - 1}$. In the isentropic expansion region, $p/p_0 = \left(\frac{\rho}{\rho_0}\right)^\gamma$. Show that

$$\frac{c}{c_0} = \left(\frac{p}{p_o}\right)^{(\gamma-1)/2\gamma} = \left(\frac{\rho}{\rho_o}\right)^{(\gamma-1)/2} = \left(\frac{T}{T_0}\right)^{1/2}.$$

(Note that the subscript 0 denotes the values in some chosen reference state. In our problem, the reference state will be the state of the high-pressure gas before the diaphragm is ruptured.)

d. In the isentropic expansion (the rarefaction), the velocity and sound speed were related by

$$u = \frac{2}{\gamma - 1}(c_0 - c).$$

The characteristic curves were found to be

$$\frac{dx}{dt} = u - c.$$

Write the expression for pressure vs particle velocity in the expansion.

e. The state of the shocked gas, originally at low pressure, is described by the Hugoniot jump conditions. Show that the relationship between shock and particle velocity is given by

$$U = \frac{(\gamma + 1)u}{4} + \sqrt{\left(\frac{(\gamma + 1)u}{4}\right)^2 + \frac{\gamma p_0}{\rho_0}}.$$

f. Use the relationship found in (e) with the conservation of momentum condition to get the Hugoniot curve in the p-u plane. Notice that the subscript 0 in the jump conditions denotes the initial state of the low-pressure gas. Do not confuse it with the subscript 0 for the expansion, where it denotes the initial state of the high-pressure gas. Find the match at the interface between the Hugoniot curve for the shocked low-pressure gas and the expansion curve for the high-pressure gas found in (d).

g. Find the values of the state variables in (c) in the constant state region after the expansion.

h. Solve for the trajectory of the interface and the tail characteristic of the expansion.

i. Find the shock velocity and the values of the state variables in the shocked low-pressure gas. Use the Hugoniot jump conditions, not the isentropic relationships of (c).

Exercise 4.19 *

a. Determine the CO_2-oxygen-balanced mixture of nitromethane (CH_3NO_2) and ammonium nitrate (NH_4NO_2), and calculate the difference between the sum of the heats of formation of the ingredients and that of the reaction products.

b. What is the relationship between the heat of formation difference calculated in part (a) and the "explosion energy" when the composition reacts at constant volume?

Exercise 4.20 The detonation velocity of crushed prills ANFO at density 1 g/cm^3 in a steel tube of inner diameter 52 mm is 4100 m/s, while that calculated for ANFO of the same density using the BKW computer code using the BKW equation of state with parameters adjusted to fit RDX is 5750 m/s. Explain the difference, and draw a schematic P,v diagram to support your reasoning.

Exercise 4.21 *

a. Derive the conservation equations for a one-dimensional detonation in their simplest form using the detonation energy Q.

b. For nitromethane at density 1.13 g/cm^3 in a charge of large diameter, the detonation velocity was measured to be 6,29 km/s. The experimental detonation pressure of nitromethane is given by one author as 14.1 GPa. Derive a simple expression that allows you to determine the slope γ of the isentrope of the detonation reaction products at the Chapman-Jouguet point and use the expression to calculate the value of γ for nitromethane at that point.

Exercise 4.22 * The parameters C and S in the linear relationship between shock velocity and particle velocity in plane (one-dimensional) shock waves in Westerley granite of initial density 2,630 kg/m^3 are $C = 2.1$ km/s and $S = 1.63$. The linear relationship with these parameters is valid in the shock pressure region from 5 to 50 GPa.

a. Calculate the shock wave velocity D, the pressure P, the density ρ, and the increase in internal energy $E - E_0$ for a plane shock wave in Westerley granite having a particle velocity of 2 km/s.

b. Assuming an average value for the specific heat of Westerley granite of 0.73 kJ/kg°C, what is the shock temperature?

Exercise 4.23 * ANFO, the most commonly used explosive for rock blasting, for computational purposes can be approximated by the mixture of 94.6 wt % ammonium nitrate with 5.4 wt % dodecane.

  a. Calculate the explosion energy of this mix, which can be considered oxygen-balanced.
  b. In a large-diameter charge of ANFO at an initial density of 850 kg/m^3, detonation velocity was found to be 5.3 km/s, and the detonation pressure 6.3 GPa. Assuming a value of the average specific heat $c_V = 1.7$ kJ/kg°C, what is the approximate temperature of the reaction products of ANFO at the detonation (Chapman-Jouguet) state?

Exercise 4.24 * For Westerley granite of initial density 2630 kg/m3, there is a linear relationship between shock wave velocity D and particle velocity u in a plane (one-dimensional) shock,

$$D = C + Su$$

where $C = 2.1$ km/s and $S = 1.63$. The relationship is valid in the shock pressure region from 5 to 50 GPa. Calculate the shock wave velocity D, the pressure P, the density ρ, and the internal energy increase $E - E_0$ for a shock wave in Westerley granite having a particle velocity of 1.5 km/s. Also, assuming the average specific heat of the granite is 0.73 kJ/kg°C, what is the shock temperature?

Exercise 4.25 * ANFO of initial density $\rho_0 = 850$ kg/m^3 detonating in large-diameter drillholes has an explosion energy $Q = 4$ MJ/kg, a detonation velocity of 5.3 km/s, and a detonation pressure $P_{cj} = 6.3$ GPa. Assuming that the average specific heat of the reaction products of ANFO is $c_V = 1.7$ kJ/kg°C, what is the approximate temperature of the reaction products of ANFO at the detonation (Chapman-Jouguet) state?

Exercise 4.26 * Using thermodynamic properties listed in various tables, calculate the oxygen-balanced composition and the explosion energy for two ammonium nitrate-based energetic materials, namely

  a. Aluminum + ammonium nitrate
  b. Fuel oil + ammonium nitrate.

For fuel oil, you may use data for dodecane.

Exercise 4.27 * Why is aluminum + ammonium nitrate not used much as a rock blasting agent?

Exercise 5.1 * The time scatter in the delay time for a specific, normal type of pyrotechnic delay detonator is of the order of 5% of the delay time. The delay time scatter of electronic clock delay detonators can be made less than 10 μs. Discuss the possible technical advantages that could conceivably be derived from using electronic delay detonators in the two different types of rock blasting operations; namely,

  a. Large-diameter drillhole, open pit blasting
  b. Tunneling, using 51 mm diameter drillholes

Exercise 5.2 For the following four detonators (a-d), make simple schematic drawings showing the main elements of their interior design and explain, in a few words, their functioning:

a. A standard electric detonator with pyrotechnic delay
b. A NONEL detonator with pyrotechnic delay
c. An electronic delay detonator
d. An exploding bridgewire detonator
e. What is a No. 20 detonator and what could be its delay time?

Exercise 6.1 * For concentrated charges having a burden of rock B, the charge weight W_0 required for breaking the burden of rock was expressed by Langefors and Kihlström as

$$W_0 = a_2 B^2 + a_3 B^3 + a_4 B^4$$

where $a_2 = 0.07$ kg/m^2, $a_3 = 0.4$ kg/m^3, and $a_4 = 0.004$ kg/m^4. (a) Give a physical explanation of the three terms. (b) Calculate the charge weights required to break loose a burden of 0.1, 1, and 10 m. (c) If an explosive of bulk density 1,250 kg/m^3 were poured into a drillhole with diameter 10 inches and a burden of 10 m, how long should the charge be?

Exercise 6.2 * For an extended charge in bench blasting, the breaking effect at the "*toe*" (the bottom of the bench) decreases with increasing charge length. What is the ratio between the weights of the concentrated charge W_0 and the elongated charge W of length B that have the same breaking capacity at the toe of the bench?

Exercise 6.3 In an open pit copper mine, blasting is done using drillholes of 12 in (305 mm) diameter and 50 ft (15 m) bench height. The subdrilling is 0.3 times the burden, the unloaded hole length is also 0.3 times the burden, and the ratio of hole spacing to burden is 1.25. Dividing the total explosives consumption during the year with the total solid rock volume removed, the resident drill and blast engineer arrives at an average specific charge $q = 0.45$ kg/m^3 ANFO equivalent.

a. If ANFO of density 0.82 g/cm^3 is the only explosive used, what burden and spacing were used?
b. What is the linear charge concentration for a 12 in diameter drillhole filled with ANFO at a loading density of 0.82 g/cm^3?
c. Using the equation for the linear charge concentration l_b for a full bottom charge:

$$l_b = 1.11 \left(0.07 B + 0.35 B^2 + 0.004 B^3 \right),$$

what is the required l_b for the burden calculated under part (a)?
d. Discuss the answers you have derived for parts (a-c).

Exercise 6.4 What is the purpose of (a) stemming the unloaded part of the drillhole, and (b) what is the preferred material to use for stemming?

Exercise 6.5 * The cost of the drilling and blasting operation in open pit fragmented rock also influences another major part of the mining cost, namely the cost of loading and transporting the fragmented rock to the crusher station. Poorly fragmented rock takes longer to load, and costs more in terms of equipment wear and tear in the loading, transportation, and crushing operations than well-fragmented rock.

a. List a set of criteria, equations, or diagrams that can be used to support the cost optimization decisions for a new open pit copper mine with an expected annual production of 10 million metric tons of ore.
b. Using these criteria, equations, or diagrams, select the following parameters for the operation:
   i. Drillhole diameter (within the range from 75 to 300 mm, for which drill rigs are available)
   ii. Explosive type and loading method (discuss the selection for wet and dry holes, selecting from the previously mentioned types of explosives: cartridged dynamite, cartridged emulsion, ANFO, bulk emulsion, and a 50/50 emulsion/ANFO mix)
   iii. Hole burden and spacing (assume a bench height of 15 m), and specific charge (overall consumption of explosive per cubic meter of solid rock).

Exercise 8.1 * The average boulder size L produced in a blasting round is defined as the size of the sieve opening through which 90 to 95 % of the fragments from the round will pass. For a given rock mass, L is a function of the specific charge q and the burden B. For the type of rock mass (hard granitic rock) used to derive the relationship between L, q, and B shown in Figure 8.13, at a burden of 8 m, a) what are the L values for specific charges $q = 0.4$, 0.6 and 0.8 kg/m^3? b) Using Larsson's empirical expression for L as a function of the average fragment size k_{50}, what are the average fragment sizes for these three specific charge values?

Exercise 8.2 * The damage to the remaining rock mass near an extended charge in a drillhole can be estimated by the integration of the peak vibration particle velocity over the charge length. Using the data in Figure 9.2, and assuming incipient damage will occur at a peak vibration particle velocity of 700 mm/s, at what distance from the charge does damage begin (a) for a 3 m long charge of linear charge density 10 kg/m; (b) for a 15 m long charge of linear charge density 75 kg/m?

Exercise 12.1 * When blasting in hard competent rock throw of flyrock, R_{max} is given by the expression

$$R_{max} = 260 \, d^{2/3}$$

and the corresponding diameter ϕ of the longest flying boulder is given by

$$\phi = 0.1 \, d^{2/3}$$

where R_{max} and ϕ are in meters, and the charge diameter d is in inches. What is the maximum throw length and the corresponding longest flying boulder size for an 8 in diameter charge in a drillhole?

Exercise 12.2 What is the maximum expected length of throw of flyrock from a rock blasting operation using 10 in diameter drillholes?

Exercise 13.1 A building exposed to ground vibrations from blasting will experience a dynamic strain that is proportional to the ratio between the vibration peak particle velocity and the vibration wave velocity in the underground supporting the building. Thus, if critical elements of the building can withstand only a limited strain without damage, then this strain limit defines a limit value for the ground vibration peak particle velocity for each type of underground wave velocity. If 70 mm/s is the limiting value for the peak particle velocity to avoid damage to a building founded on hard rock with a Rayleigh wave velocity of 3000 m/s, what would be the limiting value of particle velocity for the same building founded on wet clay with a Rayleigh wave velocity of 750 m/s?

Exercise 13.2 When predicting ground vibration damage caused by nearby blasting with a single concentrated charge, the following equation may be used to estimate the local vibration particle velocity v:

$$v = K \left(\frac{W^\alpha}{R^\beta} \right)$$

where K, α, and β may have the values 0.7, 0.7, and 1.4 m/s, respectively (in MKS units).

Damage occurs when v exceeds a critical value, which depends on the wave velocity of the rock mass and the strength of the structure.

When blasting in granitic rock masses, the critical value of v for incipient building damage may be 70 mm/s for a building founded on granite. The critical value of v for incipient damage to the granitic rock mass itself may be 700 mm/s.

   a. How far away from a concentrated 1 kg charge in a drillhole in granite does building damage begin to occur under these circumstances?
   b. How far away from a concentrated 1 kg charge in a drillhole does damage to the remaining rock begin to occur?
   c. Using Figure 9.3, how far away from a 3 m long extended charge in a drillhole in rock, with a linear charge concentration of 0.5 kg/m, will rock damage begin to occur, if the critical value of v is 700 mm/s?

Exercise 16.1 * A square slab of TNT 300 mm on a side with a thickness of 25 mm is lying on a massive steel plate. It is covered by a steel plate also 300 mm square that is 5 mm thick. The TNT has a density of 1.62 g/cm^3 and an explosion energy of 1000 cal/g. The steel has a density of 7.83 g/cm^3. Make the Gurney assumptions (1) the energy at late time is all kinetic energy of the products gases and the metal plate, (2) the gas density is constant so the particle velocity is a linear function of distance, and (3) the massive base plate does not move.

   a. Show that the Gurney formula for the plate velocity is

$$v = \frac{\sqrt{2E}}{\sqrt{\frac{M}{C} + \frac{1}{3}}}$$

where M is the mass (per unit area) of the plate, C is the mass (per unit area) of the explosive, and E is the useful specific energy of the explosive.

b. Suppose the Gurney energy E is 2/3 of the chemical energy of the TNT. (Use 4.184 J/cal to make the conversion.) Calculate the plate velocity.

c. Now assume the TNT is initiated along one edge of the square. The plate will be bent upward at an angle as the detonation runs across the explosive. Remember that the motion of a metal particle is in a direction at half the deflection angle. The detonation velocity of the TNT is 6.9 km/s. What is the deflection angle? Hint: In a given time, the detonation front and the point at which the plate bends move a distance given by the time multiplied by the detonation velocity, while the particles of the plate move up at an angle at their Gurney velocity calculated in part (b).

Exercise 18.1 *

a. In what publication series can we find the main U.S. Government regulations for the transportation of hazardous materials such as explosives and blasting agents,

b. Which agencies are responsible for enforcing and controlling the compliance of manufacturers and users of commercial explosives with these regulations?

c. Describe the US Government required tests and procedures for determining whether, for the purpose of road transportation, an energetic material is a Class A explosive or a blasting agent.

d. What is the relationship between the UN recommendations for the transport of dangerous goods and the US Government regulations as set out in the publications mentioned in part (a)?

Exercise 19.1 * Discuss (a) the key safety considerations and (b) the explosion risk generators to be considered in designing, building, and operating an explosives plant.

Solutions to Selected Exercises

Solution 1.1

$$\tau = \tau_0 + \frac{\mu \sigma_n}{1 + \left(\frac{\mu \sigma_n}{\tau_i - \tau_0}\right)} = 20 + \frac{0.7 \times 100}{1 + \frac{0.7 \times 100}{900 - 20}}$$

$$= 20 + \frac{70}{1.0795} = 20 + 64.8 = \boxed{84.8 \text{ MPa}}$$

Solution 1.2 Using $\mu = 0.9$ and $M = 2$, the ratio is

$$\left(\frac{\sigma_{20}}{\sigma_{10}}\right) = \left(\sqrt{1+u^2} + \mu\right)^{1/M} = \left(\sqrt{1+0.81} + 0.9\right)^{1/M} = 2.245^{1/2} = \boxed{1.498}$$

Solution 1.6

	Steel	Granite
$\frac{\text{HEL}}{E}$:	$\frac{800}{210,000} = \frac{1}{263}$	$\frac{3,000}{60,000} = \frac{1}{20}$
Maximum strength:	$\frac{\sigma}{E} = \frac{1}{10}$ to $\frac{1}{20}$	$\frac{\sigma}{E} = \frac{1}{7}$ to $\frac{1}{20}$

Steel deforms by the motion of dislocations, which begin to move at stress levels far lower than those required for motion of entire crystal planes past each other. Granite, being a brittle material, does not deform plastically, except locally, and increasing confinement has the effect of closing existing cracks to increase internal friction. Ultimately, at very high pressures, granite will deform plastically (or preferentially) by motion of crystal planes past each other, rather than by dislocation motion. The HEL is the stress situation with maximum confinement, i.e., with no strain in the two directions in the plane parallel to the shock front. Therefore, the ratio of HEL to E for granite $\left(\frac{1}{20}\right)$ approaches the maximum strength in the range of $\frac{\sigma}{E}\left(\frac{1}{20} \div \frac{1}{20}\right)$, while steel's $\frac{\text{HEL}}{E}$ is much smaller than the maximum strength $\frac{\sigma}{E}\left(\frac{1}{10} \div \frac{1}{20}\right)$.

Solution 2.1

a. Hole volume cost is mainly energy and tool wear. Both are primarily expended at periphery of hole bottom. Thus, approximately, cost per unit length of a hole of diameter d is

$$K = 1 \cdot \pi d \approx d.$$

Volume per unit hole length is

$$K = 1 \cdot \pi \frac{d^2}{4} \approx d^2.$$

Cost per volume

$$k = \frac{K}{V} = \frac{d}{d^2} = \frac{1}{d}.$$

b.

$$K = \$1.50 \text{ per ft. hole length at } d = 4.5 \text{ in}$$

$$K = \frac{\$1.5}{0.3} = \$5 \text{ per m hole length}$$

$$V = \text{Vol/m hole length} = \frac{1}{2} \times 4.5^2 = \frac{20.25}{2} = 10.125 \text{ liter/m}$$

$$k = \frac{K}{V} = \frac{\$5}{10.125} = \$0.49 \text{ per liter at } d = 4.5 \text{ in}$$

$$V_8 = \frac{4.5}{8} \times \$0.49 = \$0.29 \text{ per liter at } d = 8 \text{ in}$$

$$k_{8\text{in}} = \$0.276 \times 32 = \$8.832 \text{ per m} = 0.3 \times \$8.832 = \$2.65 \text{ per ft}.$$

Solution 3.1

a. Primary explosive: lead azide, PbN_6.
   Secondary explosive: TNT, trinitrotoluene, $C_4H_5O_6N_3$.
   Tertiary explosive: AP, ammonium perchlorate, NH_4ClO_4.

b. Black powder is an energetic material that burns rapidly even at atmospheric pressure, by surface burning of the grains, producing both gaseous and condensed reaction products. Its burning rate increases approximately linearly with increasing pressure. Thus, it may properly be classed as a propellant. (Of course, black powder was used extensively both as a gun propellant and as a rock blasting agent for many centuries before Alfred Nobel's inventions of the nitroglycerin-based dynamites and smokeless gun propellants).

Sometimes, as in a black powder fuse (safety fuse), it is produced to propagate initiation by flame from one place to another, or to produce light or sound effects. Thus, it is also a pyrotechnic material.

When used in rock blasting, it produces an explosion by rapid burning under high pressure in the drillhole. Thus, it is certainly an explosive material, but unless it could detonate, it would not be considered a high explosive.

To the authors' knowledge, it has *not* been reported anywhere that detonation in black powder has ever been observed. The reaction products of black powder contain a large proportion solids, which would reduce its ability to detonate. Possibly, a sufficiently large-diameter charge of medium-density black powder of small grain size exposed to a large booster might detonate. (Whatever the reaction mechanism of black powder, whether detonation or rapid grain burning, a fire in an old-fashioned production plant for black powder invariably led to a devastating explosion, the overall effects of which were not much different from a plant-explosion involving a detonating explosive).

Solution 3.2 A permissible explosive is one in which the flame temperature is low enough for the explosive to pass a series of tests (different for different countries depending on their particular underground coal mining conditions), called gallery tests, designed to ensure that the flame emitted from a drillhole in blasting does not ignite an explosive methane/air or coal dust/air mixture. The final decision to allow an explosive for underground use in a specified mining conditions class (also different in different countries) is made by the state-appointed director of mine safety (or equivalent) (in the U.S. the Director of Safety in Mines, Department of the Interior, Bureau of Mines).

Solution 3.3
  a. Primary, secondary, tertiary explosive materials.
   - *Primary explosives* will detonate by flame heating, or by a spark, even if in a size only a fraction of a millmeter. Even a single grain will detonate. The primary explosives are consequently very hazardous to handle.
   - *Secondary explosives* do not detonate by flame heating or by spark unless confined or exposed to shock compression. The quantity needed for transition from burning to explosion or detonation is in the range of kilograms to tons.
   - *Tertiary explosives* are extremely difficult to explode by fire alone, although a very large fire involving as much as several thousands of tons of these materials can indeed lead to explosion and detonation, with devastating effects because of the large quantities involved.

  b. High explosives, propellants, pyrotechnics.

Solution 3.4
  a. An energetic material is one that can decompose and thereby become very hot. This is a very general definition; an exploding wire can then also be regarded as an energetic material. Often, the term energetic material is used more narrowly to indicate an explosive material, or as a common name for explosives, propellants, and pyrotechnics.
  b. An explosive is a material that, upon decomposition or chemical reaction, attains a high temperature *and* a high pressure with gaseous reaction products.
  c. High explosives are those explosives which are intended for use as explosives to do explosive work.
  d. Class A Explosives (in the USA) are explosives that pass certain DOT tests, including that of being sensitive to a No. 8 detonator.
  e. A blasting agent is an explosive material that passes certain DOT tests, including that of *not* being cap sensitive.
  f. Oxidizer defined by DOT tests as not posing an explosion danger upon fire. Note though that some oxidizers, such as AN and AP, can explode and detonate in large quantities.

Solution 3.5 The property of having a rate which very rapidly (exponentially) increases with temperature.

Solution 3.7
a. Upon impact compression of a gas bubble, with the pressure kept on for several milliseconds, the adiabatically heated gas (several thousand degrees) heats the surface of the explosive and starts a surface burning, which continues until all the surrounding material is consumed, or until the pressure (and thereby the temperature) is released, causing the flame to go out.
b. In shock compression of a cavity, the material on the side of the cavity first met by the shock wave is accelerated into the cavity and impacts the opposite wall of the cavity. There, because of the focusing (shaped charge effect) of kinetic energy, a collision pressure shock wave is generated which has a higher pressure (and temperature) than the incoming shock wave and which creates a region within the explosive with temperature sufficiently high for rapid reaction, to enchance the shock wave pressure (hot spot).

Solution 3.12
a. $T_1 = 600$ K, $T_2 = 590$ K

$$\text{Ratio} = \frac{e^{-T^\dagger/T_1}}{e^{-T^\dagger/T_2}} = e^{-T^\dagger \left(\frac{1}{T_1} - \frac{1}{T_2}\right)}$$

$$= e^{25000 \left(\frac{590-600}{600 \times 590}\right)} = e^{\frac{250,000}{590354,000}} = e^{0.70621} = \boxed{2.026}.$$

The ratio is 2.03, i.e., the reaction rate doubles for a $10°$ temperature increase.

b.
$$\frac{d\lambda}{dt} = \kappa (1 - \lambda) e^{-T^\dagger/T} = 10^6.$$
$$k = 10^{20} \text{ s}^{-1}.$$

The largest rate at $\lambda = 0$:

$$\frac{d\lambda}{dt} = 10^6 = 10^{20} \times 1 \times e^{-25,000/T}.$$
$$e^{-25,000/T} = 10^{-14}.$$

Now if $e^x = 10^{-14}$, then $\ln e^x = \ln 10^{-14}$, so $x = \ln 10^{-14}$. Hence, $-\frac{25000}{T} = \ln 10^{-14}$. Thus,

$$T = \frac{25,000}{14 \times \ln 10} = \boxed{776 \text{ K}}$$

Solution 4.1

Mass:
$$\boxed{\rho_0 D = \rho(D - u)} \tag{1}$$

Momentum:
$$\boxed{P - P_0 = \rho_0 D u} \tag{2}$$

Energy (double boxed for detonation):
$$P u \boxed{\boxed{+ P_0 D Q}} = \rho_D (E - E_0) + \rho_0 D \frac{u^2}{2} \tag{3}$$

But according to (2)
$$\rho_0 D = \frac{P - P_0}{u} \tag{4}$$

Then, from (3)
$$P u \boxed{\boxed{+ P_0 D Q}} = \frac{P - P_0}{u}(E - E_0) + \frac{P - P_0}{\rho u} \frac{u^2}{2} \tag{5}$$

$$\frac{P u^2}{P - P_0} = E - E_0 + \frac{u^2}{2} \boxed{\boxed{+ Q}} \tag{6}$$

$$E - E_0 = \frac{u^2}{2}\left(\frac{2P}{P - P_0} - 1\right) + Q = \frac{u^2}{2}\left(\frac{P + P_0}{P - P_0}\right) + Q \tag{7}$$

$$\boxed{E - E_0 = \frac{P + P_0}{P - P_0} \frac{u^2}{2} \boxed{+ Q}} \tag{8}$$

For $P \gg P_0$:
$$\boxed{E - E_0 = \frac{u^2}{2} \boxed{+ Q}} \tag{9}$$

Solution 4.7
Assume x is the number of moles of AN in oxygen balanced ANFO:

$$x\,NH_4NO_3 + C_{12}H_{26} = x\,N_2 + (13+2x)\,H_2O + 12\,CO_2$$

H_2: $\qquad\qquad\qquad 2x+13$
O: $\qquad\qquad\qquad 3x$
N_2: $\qquad\qquad\qquad x$

Oxygen balance requires that:

$$3x = (13+2x) + 24$$

Thus the number of moles of AN and reaction product H_2O are:

moles AN: $\qquad\qquad x = 13 + 24 = 37$
moles H_2O: $\qquad\quad 13 + 2x = 13 + 2 \times 37 = 87$

The mass ratio of AN to $C_{12}H_{26}$ is:

mass of AN: $\qquad\qquad 37 \times 80.04 = 2961.48$
mass of $C_{12}H_{26}$: $\qquad 1 \times 170.34 = \underline{170.34}$
Total mass: $\qquad\qquad\qquad\qquad\qquad 3131.82$

Fraction $C_{12}H_{26}$: $\qquad 170.34/3131.82 = 0.05439$
Fraction AN: $\qquad\qquad 1 - 0.05439 = 0.9456$
Mass percent: $\qquad\qquad C_{12}H_{26} = 5.44\%$

Explosion energy:

$$-37 \times 365.7 - 1 \times 351.0 + 87 \times 241.8 + 12 \times 393.7$$

$$= -13530.9 - 351.0 + 21036.6 + 4724.4$$
$$= 11879.1 \text{ kJ per mole } C_{12}H_{26}$$

Per kg explosive, there is 54.4 g $C_{12}H_{26}$, or

$$54.4/170.34 = 0.3194 \text{ moles } C_{12}H_{26}$$

Therefore,
Expl. energy/kg expl: $0.3194 \times 11879.1 = 3794.2$ kJ/kg = $\boxed{907.7 \text{ kcal/kg}}$

Solution 4.8

a. TNT/AN:
$$x \text{ TNT} + (1-x) \text{ AN} = 0.$$
$$-0.7398\, x + 0.20000(1-x) = 0.$$
$$0.9398\, x = 0.2$$
$$x = \frac{0.2}{0.9398} = \boxed{21.3\% \text{ TNT}}$$

b. NM/AN:
$$-0.393\, x + 0.2(1-x) = 0$$
$$0.593\, x = 0.2$$
$$x = \boxed{33.7\% \text{ NM}}$$

c. C/AN:
$$-2.977 \times x + 0.2(1-x) = 0$$
$$2.867\, x = 0.2$$
$$x = \frac{0.2}{2.867} = \boxed{6.98\% \text{ C}}$$

d. Al/AN:
$$-0.8889 \times x + 0.2(1-x) = 0$$
$$1.0889\, x = 0.2$$
$$x = \boxed{18.37\% \text{ Al}}$$

Solution 4.9

a. Effective expansion ratios:

Hard rock	3–5
Soft porous rock	5–15
Sand	400

Expansion work in MJ/kg and MJ/l at $\dfrac{v}{v_0} = 4, 10, 400$.

Explosive	ρ_0	4		10		400		Ranking
	(kg/l)	(MJ/kg)	(MJ/l)	(MJ/kg)	(MJ/l)	(MJ/kg)	(MJ/l)	
Emulsion	1.25	2.6	3.25	2.8	3.5	2.8	3.5	2
ANFO	0.82	2.6	2.13	3.2	2.6	3.9	3.2	3
NM	1.14	3.7	4.22	4.4	5.0	5.3	6.0	1

b. Ranking in price per liter drillhole.

$$\text{Emulsion:} \quad \frac{3.5}{0.15 + 0.5 \times 1.25} = 4.51 \quad \text{Effective MJ/\$}$$

$$\text{ANFO:} \quad \frac{2.6}{0.15 + 0.25 \times 0.82} = 7.32 \quad \text{Effective MJ/\$}$$

$$\text{NM:} \quad \frac{5.0}{0.15 + 3.60 \times 1.14} = 0.847 \quad \text{Effective MJ/\$}$$

Solution 4.10

Dynamite:

AN/SN/NG/TNT/woodmeal/NC/chalk 40/17/30/10/2/1.5/0.5

$Q = 4.5$ MJ/kg $= 1077$ cal/g

$\rho_0 = 1.45$ g/cm^3 $= 1.45$ kg/l

$D_\infty = 6.3$ km/s $= 21,000$ ft/s

$P = \$2$/kg $= \$0.91$/lb

Emulsion:

AN/SN/H_2O/Wax/Oil/Emulsifier 67/13/12/3/3/2

$Q = 2.8$ MJ/kg $= 680$ cal/g

$\rho_0 = 1.1$ g/cm^3 $= 1.1$ kg/l

$D_\infty = 5.8$ km/s $= 19,300$ ft/s

$P = \$1.50$/kg $= \$0.68$/lb

ANFO:

AN/FO 94.4/5.6

$Q = 4.0$ MJ/kg $= 957$ cal/g

$\rho_0 = 0.82$ g/cm^3 $= 0.82$ kg/l

$D_\infty = 5.7$ km/s $= 19,000$ ft/s

$P = \$0.30$/kg $= \$0.13$/lb

Bulk emulsion:

AN/H_2O/Oil/Emulsifier 76/16/6/2

$Q = 2.8$ MJ/kg $= 670$ cal/g

$\rho_0 = 1.25$ g/cm^3 $= 1.25$ kg/l

$D_\infty = 6.3$ km/s $= 21,000$ ft/s

$P = \$0.60$/kg $= \$0.26$/lb

"Heavy ANFO" 50/50 ANFO/Bulk emulsion:

AN/H_2O/FO/Emulsifier 85.2/8/5/1

$Q = 3.4$ MJ/kg $= 815$ cal/g

$\rho_0 = 1.26$ g/cm^3 $= 1.26$ kg/l

$D_\infty = 6.8$ km/s $= 22,600$ ft/s

$P = \$0.45$/kg $= \$0.20$/lb

Solution 4.11

a. CO, H_2, NO_x, and even unreacted NG or EGDN may result from such causes as dead-pressing due to detonation in adjacent holes, or a discontinuous string of cartridges. When detonation is nonideal, as in such cases, excessive amounts of the just-mentioned toxic fumes may result.

b. Wet ANFO may give excessive amounts of NO_x. The smaller the drillhole, the greater is this effect. Off-specification composition:

$$\text{Too high FO\,\% will give } CO$$
$$\text{Too low FO\,\% will give } NO_x$$

c. Pumpable emulsion can be desensitized by shock from nearby holes. Then, incomplete combustion and highly nonideal detonation may result, producing both CO and NO_x.

Solution 4.12

The air-carried prills might possibly be sorted by the different air resistance, so that large prills preferentially fall near the container wall; whereas a mix (with higher packing density) of large and small prills would fall at the center. This would result in a lower detonation velocity toward the periphery.

Solution 4.17

a. Interface velocity is $\frac{1}{2}v_0 = \boxed{400 \text{ m/s}}$

b. $p = \rho_0 U u = 8930 \times 4535.6 \times 400 = \boxed{16.201 \text{ GPa}}$

c. $p = 8930 \dfrac{4535.6}{4535.6 - 400} = \boxed{9794 \text{ kg/m}^2}$

d. $\dfrac{0.0045}{4535.6} = \boxed{0.992 \text{ }\mu\text{s}}$

Solution 4.18 a. and b.

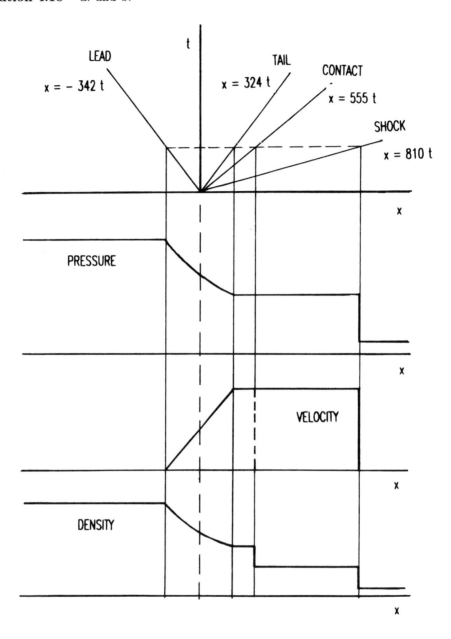

Solution 4.18. a. and b. The Shock Tube Problem. At time $t = 0$, the region of the tube ($x < 0$) is filled with a high-pressure gas; the region ($x > 0$) is filled with a low-pressure gas. At time $t = 0$, the membrane separating the two regions is removed and, as a result, the high-pressure gas expands and drives a shock wave into the low-pressure gas. An expansion wave, bounded by a lead characteristic (which propagates to the left into the high-pressure gas) and a tail characteristic (which propagates in the expanded high-pressure gas), lowers the pressure in the high-pressure region. There is a contact surface between the two gases, across which the pressure and velocity are constant but the density of the expanded high-pressure gas is higher at the constant pressure than the shock-compressed, initially low-pressure gas. Note that the flow of the expanded high-pressure gas is supersonic so that the tail characteristic actually moves to the right in a fixed coordinate system.

Solution 4.18 c.)

$$\frac{c}{c_0} = \left[\left(\frac{p}{p_0}\right)\left(\frac{\rho_0}{\rho}\right)\right]^{\frac{1}{2}} \qquad \frac{p}{p_0} = \left(\frac{\rho}{\rho_0}\right)^\gamma \qquad \frac{\rho}{\rho_0} = \left(\frac{p}{p_0}\right)^{\frac{1}{\gamma}}$$

$$\frac{c}{c_0} = \left[\frac{p}{p_0}\cdot\left(\frac{p}{p_0}\right)^{-\frac{1}{\gamma}}\right]^{\frac{1}{2}} = \left[\left(\frac{p}{p_0}\right)^{\frac{\gamma-1}{\gamma}}\right]^{\frac{1}{2}} = \left(\frac{p}{p_0}\right)^{\frac{\gamma-1}{2\gamma}}$$

$$\frac{c}{c_0} = \left[\left(\frac{\rho}{\rho_0}\right)^\gamma\left(\frac{\rho_0}{\rho}\right)^{-1}\right]^{\frac{1}{2}} = \left(\frac{\rho}{\rho_0}\right)^{\frac{\gamma-1}{2}}$$

$$\frac{c}{c_0} = \left[\frac{T}{T_0}\right]^{\frac{1}{2}}$$

Note that c_0 is the same in the high and low pressure regions.

d.

$$c = c_0 - \frac{\gamma-1}{2}u$$

$$\frac{c}{c_0} = 1 - \frac{\gamma-1}{2}\cdot\frac{u}{c_0}$$

$$\frac{p}{p_0} = \left(\frac{c}{c_0}\right)^{\frac{2\gamma}{\gamma-1}} = \left[1 - \frac{\gamma-1}{2}\cdot\frac{u}{c_0}\right]^{\frac{2\gamma}{\gamma-1}}$$

e. Conservation of mass and momentum:

$$\frac{v}{v_0} = 1 - \frac{u}{U}$$

and

$$p - p_0 = \rho_0 U u.$$

Define

$$c_0^2 = \frac{\gamma p_0}{\rho_0}$$

then

$$\frac{p}{p_0} = 1 + \frac{\gamma U u}{c_0^2}.$$

Conservation of energy:

$$pu = \rho_0 U\left[\left(E + \frac{1}{2}u^2\right) - E_0\right].$$

Solutions to Selected Exercises

Equation of state:

$$pv = RT \qquad E = c_v T \qquad c_v = \frac{R}{\gamma - 1} = \text{constant}$$

$$E = \frac{pv}{\gamma - 1}$$

$$pu = \rho_0 U \left\{ \frac{pv}{\gamma - 1} + \frac{1}{2} u^2 - \frac{p_0 v_0}{\gamma - 1} \right\}$$

Divide by $p_0 c_0$, and multiply by the terms within the brace by ρ_0

$$\frac{p}{p_0} \cdot \frac{u}{c_0} = \frac{U}{c_0} \left\{ \frac{\left(\frac{p}{p_0}\right) \cdot \left(\frac{v}{v_0}\right)}{\gamma - 1} + \frac{\gamma u^2}{2c_0^2} - \frac{1}{\gamma - 1} \right\}$$

Substitute for $\dfrac{p}{p_0}$ and $\dfrac{v}{v_0}$, and multiply by $(\gamma - 1)$:

$$(\gamma - 1)\frac{u}{c_0}\left(1 + \frac{\gamma U u}{c_0^2}\right) = \frac{U}{c_0}\left(1 + \frac{\gamma U u}{c_0^2}\right)\left(1 + \frac{u}{U}\right) + \frac{\gamma(\gamma - 1)u^2 U}{2c_0^3} - \frac{U}{c_0}$$

$$(\gamma - 1)\frac{u}{c_0} + \frac{\gamma(\gamma - 1)Uu^2}{c_0^3} = \frac{U}{c_0} - \frac{u}{c_0} + \frac{\gamma U^2 u}{c_0^3} - \frac{\gamma U u^2}{c_0^3} + \frac{\gamma(\gamma - 1)Uu^2}{2c_0^3} - \frac{U}{c_0}$$

Collecting terms:

$$\gamma \frac{u}{c_0} + \frac{\gamma(\gamma - 1)Uu^2}{2c_0^3} = \frac{\gamma U^2 u}{c_0^3} - \frac{Uu}{c_0^3}$$

Divide by $\gamma \dfrac{u}{c_0}$

$$1 + \frac{(\gamma - 1)Uu}{2c_0^2} = \frac{U^2}{c_0^2} - \frac{Uu}{c_0^2}$$

$$1 + \left(\frac{\gamma - 1}{2} + 1\right)\frac{Uu}{c_0^2} = \frac{U^2}{c_0^2}$$

$$\left(\frac{U}{c_0}\right)^2 - \left(\frac{\gamma + 1}{2} \cdot \frac{u}{c_0}\right)\left(\frac{U}{c_0}\right) - 1 = 0$$

$$\frac{1}{2}\left(\frac{U}{c_0}\right)^2 - \left(\frac{\gamma + 1}{4} \cdot \frac{u}{c_0}\right)\frac{U}{c_0} - \frac{1}{2} = 0$$

$$\frac{U}{c_0} = \left(\frac{\gamma + 1}{4} \cdot \frac{u}{c_0}\right) \pm \left[\left(\frac{\gamma + 1}{4} \cdot \frac{u}{c_0}\right)^2 + 1\right]^{\frac{1}{2}}$$

$$U = \frac{\gamma + 1}{4} u + \left[\left(\frac{\gamma + 1}{4} \cdot u\right)^2 + c_0^2\right]^{\frac{1}{2}}$$

f. Shock Hugoniot curve:

$$p = p_L + \rho_L u \left[\frac{\gamma+1}{4} u + \sqrt{\left(\frac{\gamma+1}{4} u\right)^2 + c_0^2} \right]$$

where p_L and ρ_L are the initial state of the low pressure gas.
Expansion curve:

$$p = p_H \left[1 - \frac{\gamma-1}{2} \cdot \frac{u}{c_0} \right]^{\frac{2\gamma}{\gamma-1}}$$

Note that c_0 is the same in both reservoirs because the temperature is the same. Solve iteratively for p and u:

$$p_L = 5 \times 10^4 \text{ Pa} \qquad p_H = 5 \times 10^6 \text{ Pa}$$

$$\rho_L = 0.6 \text{ kg/m}^3 \qquad \rho_H = 60 \text{ kg/m}^3$$

$$\gamma = 1.4 \qquad \gamma = 1.4$$

$$c_0 = 342 \text{ m/s} \qquad c_0 = 342 \text{ m/s}$$

$$p = 5 \times 10^4 + 0.6 \times u \left[0.6u + \sqrt{0.36u^2 + (342)^2} \right]$$

$$p = 5 \times 10^6 \left[1 - 0.2 \frac{u}{342} \right]^7$$

u	p_1	p_2
1000	824375	10637
500	276480	444144
600	358359	242805
550	315733	330524
560	328987	311087
555	319843	320679

Solution: $u = 555$ m/s $\qquad p = 320$ kPa

$$U = 0.6u + \sqrt{0.36u^2 + (342)^2}$$

$$U = \boxed{810 \text{ m/s}}$$

Check: $p = p_L + 0.6 \times 555 \times 810 = 320$ kPa.

g. $u = 555$ m/s.   Expansion region:

$$c = c_0 - \frac{\gamma - 1}{2} u$$
$$c_0 = 342 \text{ m/s}$$
$$\frac{\gamma - 1}{2} = 0.2$$
$$c = 342 - 0.2 \cdot 555 = \boxed{231 \text{ m/s}}$$

$$\frac{p}{p_0} = \left(\frac{c}{c_0}\right)^{\frac{2\gamma}{\gamma - 1}} = \left(\frac{231}{342}\right)^7 = 0.06414, \quad p = 5 \cdot 10^6 \cdot 0.06414 = \boxed{320 \text{ kPa}}$$

$$\frac{\rho}{\rho_0} = \left(\frac{c}{c_0}\right)^{\frac{2}{\gamma - 1}} = \left(\frac{231}{342}\right)^5 = 0.14058, \quad \rho = 60 \cdot 0.14058 = \boxed{8.43 \text{ kg/m}^3}$$

$$\frac{T}{T_0} = \left(\frac{c}{c_0}\right)^2 = \left(\frac{231}{342}\right)^2 = 0.45622, \quad T = 293 \times 0.45622 = \boxed{134 \text{ K}}$$

h. Trajectory of interface:
$$\frac{dx}{dt} = 555 \text{ m/s}.$$
$$x = \boxed{555\, t}$$

Tail characteristic:
$$\frac{dx}{dt} = 555 - 231 = 324 \text{ m/s}.$$
$$\boxed{x = 324\, t}$$

i. Shocked gas:
$$u = \boxed{555 \text{ m/s}}$$
$$U = \boxed{810 \text{ m/s}}$$
$$p = \boxed{320 \text{ kPa}}$$

$$\frac{\rho}{\rho_0} = \frac{U}{U - u} = \frac{810}{810 - 555} = 3.176, \quad \rho = 3.176 \times 0.6 = \boxed{1.91 \text{ kg/m}^3}$$

$$\frac{T}{T_0} = \frac{\left(\frac{p}{\rho}\right)}{\left(\frac{p_0}{\rho_0}\right)} = \frac{320 \text{ kPa}}{50 \text{ kPa}} \times \frac{0.6 \text{ kg/m}^3}{1.91 \text{ kg/m}^3} = 2.01, \quad T = 293 \times 2.01 = \boxed{589 \text{ K}}$$

Solution 4.19 Assume x is the weight fraction of NM in AN.
a. Oxygen balance:
$$x(-0.393) + (1-x)\,0.200 = 0$$
$$x = \frac{0.200}{0.593} = 0.337$$

$$\boxed{\begin{array}{l} 33.7\%\ \text{NM} \\ 66.3\%\ \text{AN} \end{array}}$$

$$Q = 0.337 \times 10.589 + 0.663 \times 1.583 = 3.568 + 4.050 = \boxed{4.618\ \text{MJ/kg}}$$

b. Difference: $\Delta H = Q + \Delta n RT$ where Δn = change in number of moles.

Solution 4.21

a. Mass: $\rho_0 D = \rho(D-u)$

Momentum: $P = \rho_0 Du$

Energy: $E - E_0 + \tfrac{1}{2} u^2 = \dfrac{Pu}{\rho_0 D} + Q$

b.
$$P = \frac{\rho_0 D^2}{\gamma + 1}$$
$$\gamma + 1 = \frac{\rho_0 D^2}{P} = \frac{1.13 \times 6.29^2}{14.1} = 3.17$$
$$\gamma = \boxed{2.17}$$

Solution 4.22

a. $C = 2.1$ km/s, $S = 1.63$, and $\rho_0 = 2630$ kg/m^3.

$$\begin{cases} D = C + Su \\[4pt] \rho_0 D = \rho(D-u) \\[4pt] P = \rho_0 Du \\[4pt] E - E_0 = \dfrac{u^2}{2} \end{cases} \qquad \frac{\rho_0}{\rho} = 1 - \frac{u}{D}$$

$u = 2.0$ km/s

$D = 2.1 + 1.63 \times 2 = 2.1 + 3.26 = \boxed{5.36\ \text{km/s}}$

$$\frac{\rho_0}{\rho} = 1 - \frac{2}{5.36} = 0.6269$$

$$\frac{\rho}{\rho_0} = \frac{1}{0.6269} = 1.595$$

$$\rho = 1.595 \times 2630 = \boxed{4195 \text{ kg/m}^3}$$

$$P = 2630 \times 5.36 \times 2 = \boxed{28.19 \text{ GPa}}$$

$$E - E_0 = \frac{1}{2}u^2 = \boxed{2 \text{ MJ/kg}}$$

b.

$$T - T_0 = \frac{E - E_0}{C_v} = \frac{2 \times 10^3}{0.73} = 2740 \text{ K}$$

Assume $T_0 = 293$ K $(T_0 = 293$ K)

$$T = 2740 + 293 = \boxed{3033 \text{ K.}}$$

$$t = 3033 - 273 = \boxed{2760°\text{C.}}$$

Solution 4.23

a. $Q = 0.946 \times 1.583 + 0.54 \times 44.226 = 1.498 + 2.388 = \boxed{3.886 \text{ Mj/kg.}}$

b. $Pu + \rho_0 DQ = \rho_0 D(E - E_0) + \rho_0 D\frac{u^2}{2}$.

$$\frac{Pu}{\rho_0 D} = E - E_0 + \frac{u^2}{2} - Q.$$

$$P = \rho_0 D u$$

$$\frac{Pu}{\rho D} = u^2$$

$$E - E_0 = \frac{u^2}{2} + Q$$

$$P = \rho_0 D u$$

Given $P = 6.3$ GPa, $D = 5.3$ km/s, and $\rho_0 = 850$ kg/m^3.

$$u = \frac{6.3 \times 10^3}{850 \times 5.3} = 1.398 \text{ km/s}$$

$$E - E_0 = \frac{u^2}{2} + Q = \frac{(1.398)^2}{2} + Q = 0.977 + 3.886 = 4.863 \text{ Mj/kg}$$

$$C_v = 1.7 \text{ kJ/kg}$$

$$T - T_0 = \frac{4.863 \times 10^3}{1.7} = 2860°$$

Assume $T_0 - 293$ K,

$$T = 2860 + 293 = \boxed{3154 \text{ K}}$$

$$t = 3154 - 273 = \boxed{2881°C}$$

Solution 4.24

$$D = C + su$$

$$\rho_0 = 2,630 \text{ kg/m}^3 \qquad u = 1.5 \text{ km/s}$$

$$s = 1.63$$

Shock wave velocity:

$$D = 2.1 + 1.63 \times 1.5 = \boxed{4.545 \text{ km/s}}$$

Pressure:

$$P = 2630 \times 4.545 \times 10^3 \times 1.5 \times 10^3 = 17.93 \times 10^8 \text{ Pa} = \boxed{17.93 \text{ Gpa}}$$

Density:

$$\rho = \frac{2630}{1 - \frac{1.5}{4.545}} = \frac{2630}{0.670} = \boxed{3925 \text{ kg/m}^3}$$

Energy:

$$E - E_0 = \frac{1}{2} u^2 = \frac{1}{2} 1.5^2 \times 10^6 = \boxed{1.125 \text{ MJ/kg}}$$

Shock temperature:

$$T - T_0 = \frac{1125}{0.73} = 1541°K$$

$$T_0 = 293°K$$

$$T = \boxed{1834 \text{ K}} \quad \text{or} \quad t = \boxed{1561°C}$$

Solution 4.25

ANFO at $\rho_0 = 850$ kg/m^3; $D = 5.3$ km/s; $Q = 4$ MJ/kg; $E - E_0 = \frac{1}{2}u^2 + Q$; $P_C = 6.3$ GPa:

$$6.3 \times 10^9 = 850 \times 5.3 \times 10^3 \times u$$

$$u = 1.4 \times 10^3 \text{ m/s} = 1.4 \text{ km/s}$$

$$E - E_0 = \frac{1}{2} \times 1.4^2 + 4 = 4.98 \text{ km/s}$$

$$T - T_0 = \frac{4.98 \times 10^3}{1.7} = \boxed{2930 \text{ K}}$$

$$t = \boxed{2657 °C}$$

Solution 4.26

a. $Al + AN$

Oxygen balance to CO_2, Al_2O_3:

Assume the weight fraction of aluminum in the mix is x.
g O_2 per g of the mix $xAl + (1-x)AN$

Oxygen balance requires:

$$-0.8889x + 0.2(1-x) = 0$$

$$-10.0889x = -0.02$$

$$x = \frac{0.2}{1.0889} = 0.184$$

Oxygen-balanced composition in wt%:

Al/AN $\boxed{18.4/81.6 \text{ wt \%}}$

Explosion energy:

$$0.184 \times 30.880 + 0.816 \times 1.583 = 5.68 + 1.29 = \boxed{6.97 \text{ Mj/kg}}$$

b. FO/AN

Oxygen balance:

$$-3.48 \cdot x + 0.2(1-x) = 0$$

$$x = \frac{0.2}{3.68} = 0.054$$

Oxygen balanced composition in weight %:

FO/AN $\boxed{5.4/94.6}$

Explosion energy:

$$0.054 \times 44 \times 226 + 0.946 \times 1.583 = 2.39 + 1.53 = \boxed{3.89 \text{ Mj/kg}}$$

Solution 4.27

Al/AN gives a lot of thermal energy and, therefore, has a high flame temperature, but a lot of solid reaction products. Thus, the major part of the energy comes out upon expansion to very large volume expansion ratios $(V/V_0 > 10000)$ — much larger than those $(10 < V/V_0 < 20)$ where the effective rock fragmentation work is completed and gas starts leaking out to the atmosphere rather than doing useful fragmentation work on the rock. Most of the extra energy obtained by replacing FO with Al is thus wasted. (See Figure 4.14.)

Solution 5.1

a. No great advantages are gained with high-precision timing because of the large time for acceleration of rock. Large scatter (<25 msec) may even be an advantage in minimizing vibration due to cooperating charges from several holes.

b. Possibly, great improvement in contour blasting where 10 μsec precision might give better results. Contour holes are set off at a late time (0.5–1.5 μs delays) and thus, with 5% scatter which is normal for pyrotechnic delay detonators, the holes in the contour row scatter by 25–75 ms.

Solution 6.1

a. Given $W_0 = a_2 B^2 + a_3 B^3 + a_4 B^4$ where $a_2 = 0.7, a_3 = 0.4, a_4 = 0.004$.

The term $a_2 B^2$ is representative for material strength effects such as strain and fracture.

The term $a_3 B^3$ is representative of material bulk effects such as density and bulk deformation energy, and also for kinetic energy.

The term $a_4 B^4$ is representative of gravity effects such as energy for lifting the rock center of gravity due to swelling or for dragging rock along the ground (friction force is proportional to the weight of rock (mass × gravity); friction energy is mass × distance, thus has dimension m^4).

b.

$B(m)$	$a_2 B^2$	$a_3 B^3$	$a_4 B^4$	W_0 (kg)
0.1	0.0007	0.0004	0.004×10^{-4}	0.0011
1.0	0.07	0.4	0.004	0.474
10	7	400	40	447

c.

$$\phi = 10 \text{ in}$$

$$v \approx \frac{10^2}{2} = 50 \text{ l/m}$$

$$Q = 447 \text{ kg}$$

$$\ell = 50 \times 1.25 = 62.5 \text{ kg/m}$$

$$\text{Charge length} = \frac{447}{62.5} = \boxed{7.15 \text{ m}}$$

Solution 6.2

$$W = \frac{W_0}{0.6}$$

Ratio: $\quad \dfrac{W_0}{W} = \boxed{0.6}$

Solution 6.5

a. Drilling cost as a function of hole diameter (see Figure 2.25).

The equation relating charge weight Q to burden B is

$$Q = a_2 B^2 + a_3 B^3 + a_4 B^4 = 0.07\,B^2 + 0.4\,B^3 + 0.004\,B^4.$$

Assume B is not very important, then

$$q = \frac{Q}{B^3} = \frac{0.07}{B} + 0.04 + 0.004\,B$$

We assume the cost of explosive in \$/kg to be independent of d

$$q = \frac{Q}{B^3} = 0.04 \text{ kg/m}^3$$

b. i. Select d as large as possible to have lowest drilling cost, $\boxed{d = 300 \text{ mm}}$.

ii. Cost of explosive per m^3 rock in the first approximation is 0.4 times the cost per kg/explosive.

$$Q = \frac{Q}{B^3} \approx 0.4 \times \text{ kg/m}^3.$$

Cost per m^3 rock $= 0.4\times$ cost per kg.

Hole volume (expressed in l/m) required per m^3 rock for ANFO is different from that of the other explosives:

$$\frac{1}{0.82} = 1.22 \quad \text{vs.} \quad \frac{1}{1.25} = 0.8.$$

Thus drillhole volume cost for ANFO is 1.525 times higher than for the others.

Explosive	Cost of explosives Assumed price (\$/kg)	Explosive cost/m^3 rock
Dynamite	2.00	0.8
Emulsion (EM)	1.50	0.6
ANFO	0.30	$0.12 \times 1.525 = 0.18$
Bulk Emulsion	0.60	0.24
50/50 EM/AN (Emulex)	0.45	0.18

> For dry holes, select ANFO, auger loaded.
> For wet holes, select 50/50 emulsion/ANFO, pump loaded.

iii. Emulex:
For full length bottom charge, the burden $B = 40 \times d = 40 \times 0.3 = 12$ m.
Full bottom charge length $= 1.3B = 1.3 \times 12 = 15.6$ m.
But bench height is only 15 m. Therefore, reduce the bottom charge length to 12m.
Charge weight per loaded meter hole length ($d = 300$ mm ≈ 12 in) is

$$l_b = \frac{12^2}{2} \times 1.250 = 90 \text{ kg/m}.$$

Total charge per hole $= 12 \times 90 = 1080$ kg/hole.
Assume the specific charge $q = 0.4$ kg/m^3. The rock volume per hole then is

$$\frac{1080}{0.4} = 2700 \text{ m}^3/\text{hole}.$$

The surface area per hole is

$$\frac{2700}{15} = 180 \text{ m}^2/\text{hole}.$$

The burden B is now

$$B = \sqrt{\frac{180}{1.25}} = \boxed{12 \text{ m.}}$$

The Hole spacing is

$$S = 1.25 B = \boxed{15 \text{ m.}}$$

Test the result by comparing with the method shown in figure 6.13.

Solution 8.1

$$k_{50} = \frac{L}{2.6}.$$

q (kg/m^3)	L (m)	k_{50}
0.4	1.45	0.56
0.6	0.9	0.35
0.8	0.63	0.25

Solution 8.2   a. $\boxed{2.4 \text{ m}}$ b. $\boxed{26 \text{ m.}}$

Solution 12.1 Note: d is in inches, R and ϕ are in meters.

$$R_{max} = 260\, d^{2/3}$$

$$d = 8 \text{ in}$$

$$d^{2/3} = \sqrt[3]{8^2} = \sqrt[3]{64} = 4$$

$$R_{max} = 260 \times 4 = \boxed{1040 \text{ m}}$$

$$\phi = 0.1 \times 4 = \boxed{0.4 \text{ m}}$$

Solution 16.1

a.

$$CE = \frac{1}{2} M v^2 + \frac{1}{2} \int_0^{x_1} C \frac{v^2}{x_1^3} x^2\, dx$$

$$CE = \frac{1}{2} M v^2 + \frac{1}{2} \left(\frac{C v^2}{3} \right)$$

$$v^2 = \frac{2E}{\dfrac{M}{C} + \dfrac{1}{3}}$$

$$v = \frac{\sqrt{2E}}{\sqrt{\dfrac{M}{C} + \dfrac{1}{3}}}$$

b.

$$E = 4184 \times 10^3 \times \frac{2}{3} = 2,789,300$$

$$M = 1 \times 1 \times 0.005 \times 7830 = 39.15 \text{ kg/m}^2$$

$$C = 1 \times 1 \times 0.025 \times 1620 = 40.5 \text{ kg/m}^2$$

$$\frac{M}{C} + \frac{1}{3} = 0.9667 + 0.33333 = 2.071.5 \text{ m/s}$$

$$v = \left[\frac{2 \times 2789.3}{1.300} \right]^{\frac{1}{2}} = \boxed{2.02 \text{ m/s}}$$

c.

$$\sin \frac{\alpha}{2} = \frac{\frac{1}{2} \times 2071.5}{6900}$$

$$\frac{\alpha}{2} = 8.63323°$$

$$\alpha = \boxed{17.3°}$$

Solution 18.1

a. Code of Federal Regulations (CFR).
b. BATF (Bureau of Alcohol, Tobacco, and Firearms); DOT; OSHA; MSHA; USCG; OSM; EPA; and NFPA.
c.

Cap sensitive	Detonates with No. 8 detonator unconfined, or then Class A explosive if less than 4 in drop weight test initiates in more than 50% of tests
Not cap sensitive	Initiated by a No. 8 detonator in a $3\frac{3}{8}$ in diameter, $6\frac{3}{8}$ in long container offering no confinement with lead cylinder 4 in long by 2 in diameter, if no change in length of the lead cylinder occurs
DTA	Less than 100°C (212°F), first exothermic
Thermal stability	50 g, 48 hours, 75°C (167°F), no fumes, discoloration, or other change
Electrostatic	Less than 0.006 J
Impact	More than 10 meters
Fire	Bonfire, no explosion

d. The US Government regulations are partly, but not yet (in 1993) totally, in harmony with the UN recommendations on the Transport of Danger Goods. The CFR contains tests that are different from (but provide essentially the same information as) the tests described in the UN recommended tests and criteria.

Solution 19.1

a. Key safety considerations:
       1. Remote control
       2. Minimize amount of material in process (continuous processing where possible)
       3. Separate different process steps into different buildings

b. Explosion risk generators:
       1. Intermediate product buffers (tanks, etc.)
       2. Reduce building cost
       3. Reduce equipment cost.

References

Abel, F.A., 1866, "Researches on Gun-cotton: on the Manufacture and Composition of Gun-cotton", *Philosophical Transactions* 156, p. 268-308.

ACGIH, 1970, "Threshold Limit Value of Airborne Contaminants for 1970", American Conference of Governmental Industrial Hygienists.

Akai, K. and Mori, H., 1967, "Study of Failure Mechanism of Sandstone under Combined Compressive Stress," *Proc. Soc. Civil Eng.*, No. 147. (In Japanese.)

Almgren, G. and Benedik, R., 1968, "How Boliden's Crater Cut Slashes Raise Cost," *World Mining* 4, Feb. 1968, p.38-42.

Al'tshuler, L.V., Kormer, S.B., Bakanova, A.A., and Trunin, R. F., 1960, "Equation of State of Aluminum, Copper and Lead in the High Pressure Region," *Soviet Phys. JETP*, 11, p. 573.

Al'tshuler, L.V., 1965, "Use of Shock Waves in High Pressure Physics," *Soviet Phys. Usp.*, 8, p. 52.

Ambraseys, N.N., and Hendron, A.J., Jr., 1968, "Dynamic Behaviour of Rock Masses," *Rock Mechanics in Engineering Practice*, Eds., Staff, K.G., and Zienkiewcz, O.C., John Wiley & and Sons, pp. 203-236.

Anderson, D.A., Winzer, S.R., and Ritter, A.D., 1982, "Blast Design for Optimizing Fragmentation while Controlling Frequency of Ground Vibration," *Proceedings of the Eighth Conference on Explosives and Blasting Techniques*, Ed., Konya, C.J., Society of Explosives Engineers, Montville, OH, pp. 69-89.

Andrews, A.B., 1980, "Design Criteria for Blasting," *Proceedings of the Seventh Conference on Explosives and Blasting Techniques*, Ed., Konya, C.J., Society of Explosives Engineers, Montville, OH, pp. 173-192.

Arbetarskyddsstyrelsen, 1974, "Directions for rock works (Berganvisningar)," *Arbetarskyddsstyrelsens anvisningar* nr 67, (In Swedish).

Arnberg, P. W., 1983, *Human Annoyance Caused by Vibrations in a Building*, National Swedish Road and Traffic Research Institute, Report NR 339, Linköping, Sweden. (In Swedish.)

Ashley, S., 1989, "Dynamite Metals," *Popular Science*, 3, p. 104.

Atlas Powder Company, 1987, *Explosives and Rock Blasting*, Dallas, TX.

Bachmann, W.E., and Sheenan, J.C., 1949, *J. Am. Chem. Soc.*, 71, p. 1812.

Baker, W.E., 1973, *Explosions in Air*, University of Texas Press, Austin, TX.

Barker, D.B., Fourney, W.L., and Holloway, D.C., 1979, "Photoelastic Investigation of Flaw Initiated Cracks and Their Contribution to the Mechanism of Fragmentation," *The Twentieth US Symposium on Rock Mechanics*, Austin, TX, June 4-6.

Bauer, A., Harries, G., Lang, R., Prezios, L., and Selleck, D.J., 1965, "How IOC puts crater research to work", *Eng. Min. J.*, 166 (9), p. 117-121.

Bauer, P.A., Dabora, E.K., and Manson, N., 1990, *Chronology of Early Research on Detonation Wave*, American Institure of Aeronautics and Astronautics, Inc.

Bdzil, J.B., 1981, "Steady-State Two-Dimensional Detonation," *J. Fluid Mech.*, 108, pp. 195-226.

Bdzil, J.B., Fickett, W., and Stewart, D.S., 1989, "Detonation Shock Dynamics: A New Approach to Modeling Multidimensional Detonation Waves," *Preprints of the Ninth Symposium (International) on Detonation*, Portland, OR, Aug. 28-Sept. 1.

Beard, J.A., 1988, "Explosive Mixtures," *New Scientist*, **11**, p. 44.

Beattie, T.A., and Grant, J.R., 1988, "The Effect of Inter-Hole Timing on Heave," *Explo '88 Workshop*, Melbourne, Australia.

Belidor, B.F., 1725, *Noveau cours mathématique à l'usage d'Artillerie et du Genie*, Paris, France, p. 505. (In French.)

Béranger, R., 1971, Notes from a lecture: "Damping of airblasts from buried explosive charges," given at the Institut fur Chemie der Treib- und Explosivstoffe, Jahrestagung 1970, Karlsruhe, Bundesrepublik Deutschland. (Swedish translation by Algot Persson, Swedish Detonic Research Foundation, Report DS 1971:39, Stockholm, Sweden.)

Bergengren, E., 1960, *Alfred Nobel*, Thomas Nelson and Sons Ltd., (translated by Alan Blair).

Bergmann, O.R., Wu, F.E., and Edl, J.W., 1974, "Model Rock Blasting Measures Effect of Delays and Hole Patterns on Rock Fragmentation," *Eng. Min. J.*, **174**, p. 124-127.

Berning, W.W., 1948. *Investigation of the Propagation of Blast Waves over Relatively Large Distances and the Damaging Possibilities of Such Propagations*, US Army Ballistic Research Laboratory, Report No. 657, Aberdeen Proving Ground, MD.

Berthelot, M., and Vieille, P., 1882, "Sur la vitesse de propagation des phenomenes explosifs dans les gaz", *C. R. Accad. Sci., Paris* **94**, pp. 101-108, seance du 16 Janvier, 1882; **95** pp. 151-157, seance de 24 Juillet, 1882.

Bhandari, S. 1974, "Blasting in Non-homogeneous Rocks", *Australian Mining* **66**(5), pp. 43-48.

Biasutti, G.S., 1985, *History of Accidents in the Explosives Industry*, published by the author.

Birkhoff, G., MacDougall, D., Pugh, E., and Taylor, G., 1948, "Explosives with Lined Cavities", *J. Appl. Phys.*, **19**(6), p. 563.

Bjarnholt, G., and Holmberg, R., 1976, "Explosive Expansion Works in Underwater Detonations," *Proceedings of the Sixth Symposium on Detonation*, ACR-221, Office of Naval Research, Department of the Navy, Arlington, VA, pp. 540-550.

Bjarnholt, G., 1977, *Explosive Performance in Rock Blasting*, Swedish Detonic Research Foundation, Report DS 1977:8, Stockholm, Sweden. (In Swedish.)

Bjarnholt, G., Holmberg, R., and Ouchterlony, F., 1982, *A System for Contour Blasting with Directed Crack Initiation*, Swedish Detonic Research Foundation, Report DS 1982:3, Stockholm, Sweden. (In Swedish.)

Bjarnholt, G., Holloway, D.C., Wilson, W.H., and Mäki, K., 1988, *Smooth Wall Blasting Using Notched Boreholes — A Field Study*, Swedish Detonic Research Foundation, Report DS 1988:3, Stockholm, Sweden.

Blindheim, O.T. 1973, "New experience from full-face boring," Conference in Rock Blasting Techniques, Oslo, Norway. (In Norwegian.)

Blindheim, O.T., 1979, "Drillability Predictions in Hard Rock Tunnelling," The Second International Symposium on Tunnelling, 1979., Institute of Mining and Metallurgy, London.

Bluhm, H.F., 1969, *Ammonium Nitrate Emulsion Blasting Agent and Method of Preparing Same*, U.S. Patent 3,447,978, June 6, 1969, filed August 3, 1967.

Bollinger, G. A., 1980, *Blast Vibration Analysis*, Southern Illinois Press, Carbondale, IL, pp. 137.

Brännfors, S., 1977, "Rock tunnelling in Sweden — from conception to a completed product," *Proceedings of the First US-Swedish Underground Workshop: Mechanical Boring or Drill- and Blasting Tunnelling*, Stockholm, December 5–10, 1976, Document D3:1977, Statens Rad for Byggnads-forskning, Stockholm.

Brinkley, S.R., and Gordon, W.E., 1972, *Engineering Design Handbook: Principles of Explosives Behaviour*, AMC Pamphlet No. 706-180, Headquarters, US Army Material Command, Washington, D.C.

Chapman, D.L., 1899, "On the Rate of Explosion in Gases," *Phil. Mag.*, **47**, p. 90.

Chiappetta, R.F., and Borg, D.G., 1983, "Rock Fragmentation by Explosives," *Preprints of the First International Symposium on Rock Fragmentation by Blasting*, Eds., Holmberg, R., and Rustan, A., Lulea University of Technology, Lulea, Sweden, pp. 301-331.

Clark, D., Larsson, B., and Lande, G., 1983, "Vibration: Its Effect and Measurement Techniques at or near Dwellings," *Proceedings of the Ninth Conference on Explosives and Blasting Technique*, Ed., Konya, C.J., Society of Explosives Engineers, Montville, OH, pp. 27-62.

Clay, R.B., 1978, *Water-In-Oil Blasting Composition*, U.S. Patent 4,111,727, August 05, 1978, filed September 19, 1977.

Cole, R.H., 1948, *Underwater Explosions*, Princeton University Press, Princeton, NJ.

Contestabile, E., and Craig, T.R., 1985, "Use of Shape Charges in Mining Operations," *CIM Bulletin*, **78**, December 1985, p. 41-46.

Cook, M.A., 1958, *The Science of High Explosives*, Reinhold Publishing Company, New York, NY.

Cook, M.A., 1974, *The Science of Industrial Explosives*, Graphic Service and Supply, Inc., USA.

Cooper, T., 1981, *An Implementation of the Cundall Block Program on the HP 2100 with Extensions to the Physical Model*, Swedish Detonic Research Foundation, Report DS 1981:7, Stockholm, Sweden.

Courant, R., and Friedrichs, K.O., 1950, *Supersonic Flow and Shock Waves*, Interscience, New York, NY.

Cowan, G.R., Douglas, J., and Holtzman, A.H., 1964, *Explosion Bonding of Metals*, Basic Process US Patent 3,137,937 by E. I. DuPont De Nemours & Co., expired June 19, 1981.

Cowan, G.R., Douglas, J., and Holtzman, A.H., 1966, *Explosion Bonding of Metals*, Basic Product US Patent 3,233,312 by E. I. DuPont De Nemours & Co., expired Feb. 18, 1983.

Cowan, G.R., Bergmann, O.R., and Holtzman, A.H., 1968, *Low Velocity Explosion Bonding/Process*, US Patent 3,397,444 by E. I. DuPont De Nemours & Co., expired Aug., 1985.

Cowan, G.R., Bergmann, O.R., and Holtzman, A.H., 1970, "Low velocity explosion bonding/product," Basic Product US Patent 3,493,353 by E. I. DuPont De Nemours & Co., expired Feb 19, 1987.

Cowperthwaite, M., and Zwisler, W.H., 1973, *TIGER Computer Program Documentation*, Stanford Research Institute, Publication No. Z106, Menro Park, CA.

CRC Press, *Handbook of Chemistry and Physics*, 64th Edition.

Crocker, C.S., 1979, "Vertical Crater Retreat Mining at the Centennial Mine of Hudson Bay Mining and Smelting Co., Limited," *CIM Bulletin*, Jan. 1979, p. 90-94.

Cundall, P.A., 1971, "A computer model for simulating progressive, large-scale movements in blocky rock systems", *Symposium of ISRM*, Nancy, France.

Dally, J.W., and Fourney, W.L., 1977, "The Influence of Flaws on Fragmentation," *Proceedings of the Sixth International Colloquium on Gasdynamic of Explosives and Reactive Systems*, Stockholm, Sweden, Aug. 22-26.

Davis, W.C., 1980, personal communication.

D'Andrea, D.V., Fischer, R.L., and Hendrickson, A.D., 1970, *Crater Scaling in Granite for Small Charges*, US Bureau of Mines, Report of Investigation 7409, 28 pages.

DIN4150, 1975, Deutsche Normen, Vornorm, Erschütterungen im Bauwesen, Teil 3.

Dixon, H., 1893, "The Rate of Explosion in Gases", *Philosophical Transactions A*, **184**, p. 97-188.

Döring, W., 1943, "On Detonation Processes in Gases," *Ann. Phys.*, **43**, p. 421-436.

Dowding, C.H., 1985, *Blast Vibration Monitoring and Control*, Prentice-Hall, Inc. NJ.

Dremin, A. N., 1991, American Physical Soc. Topical Meeting on Shock Waves, Albuquerque, NM.

DuPont, 1983, *Blaster's Handbook*, 175th Anniversary Edition, E.I. DuPont De Nemours & Co., Wilmington, DE.

Edwards, A.T., and Northwood, T.D., 1960, "Experimental Studies of the Effects on Blasting on Structures," *The Engineer* **210**, p. 538-546.

Egly, R.S. and Neckar, A.E., 1964, *Ammonium Nitrate-Containing Emulsion Sensitizers for Blasting Agents*, U.S. Patent 3,161,551, December 15, 1964, filed April 7, 1961.

Ekeroth, R., 1981, *Damping of Blast Vibrations by Slot Drilling*, The Swedish Rock Blasting Committee 1981, Stockholm, Sweden. (In Swedish.)

Elitz, E., and Zimmerman, R., 1978, "Die in der Bundes Republik Deutschland Benutzte Metode zur Bestimmung der Toxische Bestandteile der Sprengschwaden-Extest 76," *Propellants and Explosives* **3**, p. 17. (In German.)

Eriksson, B., and Ladegaard-Pedersen, A., 1971, *Flyrock in Blasting I, Interviews*, Swedish Detonic Research Foundation, Report DS 1971:32, Stockholm, Sweden. (In Swedish.)

Eriksson, S., 1987, *The Air Blast from a TNT charge. A Comparison between some sources*, Fort F, Report A4:87, Eskilstuna, Sweden.

Falkdal, J.E., 1985, "Test of Open Stoping in the Kristineberg Mine," *Proceedings of the International Symposium on Large Scale Underground Mining*, Ed., Almgren, G., CENTEK Publishers, Lulea, Sweden, pp. 119-126.

Federenko, P. I., and Kovtun, I.N., 1977, "Influence of the Shape and Size of the Compensation Spaces on the Crushing Quality in Ore Breaking by Boreholes," *Fiziko Tekhnicheskie Problemy Razrabotki Poleznykh Iskopaemykh*, No. 1, Jan-Feb., p. 38-42.

Fedoroff, B.T., and Sheffield, O.E., 1962, *Encyclopedia of of Explosives and related items*, PATR2700, Volume 2, Table 1, pp. B266, Picatinny Arsenal, Dover, NJ.

Fickett, W., and Davis, W.C., 1979, *Detonation*, University of California Press, Berkeley, CA.

Field, J.E., and Ladegaard-Pedersen, A., 1969, *Controlled Fracture Growth in Rock Blasting*, Swedish Detonic Research Foundation, Report DS 1969:8, Stockholm, Sweden.

Field, J.E., and Ladegaard-Pedersen, A., 1969, *The Importance of the Reflected Stress Wave in Rock Blasting*, Swedish Detonic Research Foundation, Report DL 7, Stockholm, Sweden.

Field, J.E., and Ladegaard-Pedersen, A., 1972, "Fragmentation Processes in Rock Blasting," *Dachema-Monographier*, 69.

Fraenkel, K.H., 1958, *Handbook: Rock Blasting Technique*, AB Masse W Tullberg, Esselte AB, Stockholm. (In Swedish.)

From, T., 1976, *Model Scale Blast Experiments in Jointed Material*, Swedish Detonic Research Foundation, Report DS 1976:11, Stockholm. (In Swedish.)

Gehrig, N.F., 1965, *Aqueous Emulsified Ammonium Nitrate Blasting Agents Containing Nitric Acid*, U.S. Patent 3,164,503, January 5, 1965, filed May 13, 1963.

Gibbs, T.R., and Popolato, A., 1980, LASL Explosive Property Data, University of California Press, Berkeley, CA, pp. 281-282.

Goodier, A., 1982, "Mining Narrow Veins by Vertical Crater Retreat at the Radiore No. 2 Mine," *CIM Bulletin* **75**, June, 1982, p. 65.

Graham, R.A., 1993, *Solids Under High-Pressure Shock Compression: Mechanics, Physics, and Chemistry*, Springer-Verlag, New York, NY., 221 pages.

Graham, R.A., and Sawaoka, A.B., 1987, "Explosive Processing of Ceramics," *Explosive Processing Methods*, Eds: R.A. Graham and A.B. Sawaoka, Trans. Tech. Publications, p. 31.

Granström, S.A., 1956. "Loading Characteristics of Air Blasts from Detonating Charges," *Handlingar nr 100*, The Royal Institute of Technology, Stockholm, Sweden.

Grant, J.R., 1990, "Initiation Systems — What does the Future Hold?" *Third International Symposium on Rock Fragmentation by Blasting — Fragblast '90*, Brisbane, Australia, August 26-31.

Greiner, N.R., and Hermes, R., 1989, "Chemistry of Detonation Soot: Diamonds, Graphite, and Volatites", *Ninth Symp. (Int.) on Detonation*, Aug. 28-Sept 1, Portland, OR.

Greiner, N.R., Phillips, D.S., Johnson, J.D., and Volk, F., 1984, "Diamonds in Detonation Soot", *Nature*, **333**, pp. 440-442.

Gurney, R.W., 1943, *The Initial Velocities of Fragment of Bombs, Shells, and Grenades*, US Army Ballistic Research Lab., Report No. 405, Aberdeen Proving Ground, MD.

Gustafsson, R., 1973, *Swedish Blasting Technique*, SPI, Gothenburg, Sweden.

Gustafsson, R., 1976, "Smooth Blasting," *Proceedings, International Symposium Tunnelling 76*, London, March 1-5, 1976, Institute of Mining and Metallurgy, London, UK.

Gustafsson, R., 1981, *Blasting Technique*, Dynamit Nobel Wein GmbH, Vienna, Austria.

Hansson, I., 1985, "New Blasting Technique for Mass Blast of Pillars," *Proceedings of the International Symposium on Large Scale Underground Mining*, Ed., Almgren, G., Lulea University of Technology, Lulea, Sweden, pp. 41-45.

Harries, G., and Beattie, T., 1988, "The Underwater Testing of Explosives and Blasting," *The AusIMM Explosives in Mining Workshop*, Melbourne, Victoria, Australia, pp. 23-25.

Hatheway, A.W., and Kirsch, G.A., 1982, "Engineering Properties of Rock," *Handbook of Physical Properties of Rock,* Vol. II, Ed., Carmichael, R.S., CRC Press, Inc., Boca Raton, FL, pp. 307-323.

Hinzen, K.-G., Lüdeling, R., Heinemeyer, F., Röh P., and Steiner, U., 1987, "A New Approach to Predict and Reduce Blast Vibration by Modeling of Seismograms and Using a New Electronic Initiation System," *Proceedings of the Thirteenth Conference on Explosives and Blasting Technique, Miami, FL.,* Ed., Boddorff, R.D., Society of Explosives Engineers, Montville, OH, pp. 144-161.

Hogstedt, C.H., and Axelson, O. 1977, "Nitroglycerin-notroglycol explosure and the mortality in cardi-cerebrovascular diseases among dynamite workers", *J. of Occupational Medicine,* **19**, pp. 675-678.

Hogstedt, C.H. and Andersson, K., 1979, "A cohort study on mortality among dynamite workers", *J. of Occupational Medicine,* **21**, pp. 553-556.

Hogstedt, C.H., 1980, *Dynamite — occupational exposure and health effects*, Linköping Univ. Medical Dissertation No. 84, Linköping, Sweden

Holdo, J., 1980, personal communications (Atlas Copco AB).

Holloway, D.C., Bjarnholt, G., and Wilson, W.H., 1987, "A Field Study of Fracture Control Techniques for Smooth Wall Blasting: Part 2," *Proceedings of the Second Symposium on Rock Fragmentation by Blasting*, Eds., W.L. Fourney and R.D. Dick, SEM, Bethel, CT, pp. 646-656.

Holmberg, R., 1975, "Computer Calculations of Drilling Patterns for Surface and Underground Blastings," *Design Methods in Rock Mechanics*, Eds., C. Fairhurst and S. Crouch, *The Sixteenth Symposium on Rock Mechanics*, University of Minnesota, Minneapolis, MN.

Holmberg, R., 1975, *Fragmentation as a Function of Charge Concentration and Hole Pattern*, The Swedish Rock Blasting Committee, Stockholm, Sweden, pp. 271-291. (In Swedish.)

Holmberg, R., 1977, *Computer Program for Calculation of Explosion Energies*, Swedish Detonic Research Foundation, Report DS 1977:16, Stockholm, Sweden.

Holmberg, R., 1977, *Some Single Shot Ground Vibration Measurements*, Swedish Detonic Research Foundation, Report DS 1977:9, Stockholm, Sweden. (In Swedish.)

Holmberg, R., 1978, *Flyrock and Noise in Blasting*, Swedish Detonic Research Foundation, Report DS 1978:15, Stockholm, Sweden (In Swedish).

Holmberg, R., 1978, "Measurements and Limitations of Rock Damage in Remaining Rock," *Rock Mechanics Meeting*, Stockholm, Sweden.

Holmberg, R., 1981, "Optimum Blasting in Bedded and Jointed Oil Shale Formations," *A fragmentation review and a proposal for fragmentation test in oil shale*, PL.92216 for US Dept. of Energy, Laramie Energy Technology Center, USA.

Holmberg, R., 1983, *Hard Rock Excavation at the CMS/OCRD Test Site Using Swedish Blast Design Techniques*, Report BMI/OCRD-4(3), NTIS, US Dept. of Commerce, Springfield, VA.

Holmberg, R., 1984, *Vibrations Generated by Traffic and Building Construction Activities*, Swedish Council for Building Research (BFR), Report D15:1984, Stockholm, Sweden.

Holmberg, R., Ekman, G., and Sandström, H., 1983, *Comments on Present Criteria for Vibration Sensitive Electronic Equipment*, Swedish Detonic Research Foundation, Report DS 1983:3, Stockholm, Sweden.

Holmberg, R., and Hustrulid, W., 1981, "Swedish Cautious Blast Excavations at the CSM/ONWI Test Site in Colorado," *Proceedings of the Seventh Conference of Explosives and Blasting Technique*, Phoenix, AZ.

Holmberg, R., Lundborg, N., and Rundqvist, G., 1981, *Ground Vibrations and Damage Criteria*, Swedish Council for Building Research (BFR), Report R85:1981, Stockholm, Sweden. (In Swedish.)

Holmberg, R., and Mäki, K., 1982, "Case Examples of Blasting Damage and its influence on Slope Stability," *Int. Conf. on Stability in Surface Mining*, Ed. Brawner, C.O., SME, New York, pp. 773-793.

Holmberg, R., Naarttijärvi, T., and Nilsson, L., 1982, "LKAB Introduces Large Hole Diameter Stoping at the Fabian Mine," *Mining Engineering*, 34, p. 72-78.

Holmberg, R., and Persson, G., 1976, *The Effect of Stemming on the Distance of Throw of Flyrock in Connection with Hole Diameters of 25, 36 and 76 mm*, Swedish Detonic Research Foundation, Report DS 1976:1, Stockholm, Sweden. (In Swedish.)

Holmberg, R., and Persson, G., 1978, *Flyrock and air shock waves in blasting*, Swedish Detonic Research Foundation, Report DS 1978:15, Stockholm, Sweden. (In Swedish.)

Holmberg, R., and Persson, P.-A., 1978, "The Swedish Approach to Contour Blasting," *Proceedings of the Fourth Conference on Explosives and Blasting Technique,* Society of Explosives Engineers, New Orleans, LA, February 1-3.

Holmberg, R., and Persson, P.-A., 1979, "Design of Tunnel Perimeter Blasthole Patterns to Prevent Rock Damage," *Proceedings, Tunnelling '79,,* London, March 12-16, Ed., Jones, M.J., Institution of Mining and Metallurgy, London, UK.

Holmberg, R., Rustan, A., Naarttijärvi, T., and Mäki, K., 1980, *Driving a Raise with VCR in the LKAB-mine in Malmberget,* Swedish Detonic Research Foundation, Report DS 1980:12, Stockholm, Sweden. (In Swedish.)

Hopkinson, B., 1915, *British Ordnance Board Minutes,* 13565.

Hopler, R.B., 1980, "The Hercudet Nonelectric Delay Blasting Cap System. New Developments in Field Usage," *Proceedings of the Sixth Conference on Explosives and Blasting Techniques,* Ed., Konya, C.J., Society of Explosives Engineers, Montville, OH, pp. 430-455.

Hoskins, E.R., 1969, "The failure of thick-walled hollow cylinders of isotropic rock," *Int. J. Rock Mech. Min. Sci.,* **6**, p. 99.

IME, 1988, *Safety Guide for the Prevention of Radio Frequency Radiation Hazards in the Use of Commercial Electric Detonators,* Institute of Makers of Explosives, Dec. 1988, Washington, USA.

IVA, 1979, *Soil and Rock Dynamics,* The Royal Swedish Academy of Engineering Science, Report 225, Stockholm, Sweden. (In Swedish.)

Jarlenfors, J., and Holmberg, R., 1980, *Blasting against a compressible rock mass. Model scale experiments in PMMA,* Swedish Detonic Research Foundation, Report DS 1980:5, Stockholm, Sweden. (In Swedish.)

Jerberyd, L., and Persson, A., 1972. *Measurements of Maximum Air Shock Wave Pressure from Various Explosives,* Swedish Detonic Research Foundation, Report DS 1972:17, Stockholm, Sweden. (In Swedish.)

Johansson, C.H., and Persson, P.-A., 1970, *Detonics of High Explosives,* Academic Press, London, UK.

Johansson, C.H, Selberg, H.L., Persson, A., and Sjölin, T., (1958). "Channel Effect in Detonation Tubes with an Open Space between the Charge and the Tube Wall," Proceedings, 31st International Congress for Industrial Chemistry, Liège, vol.II, p. 258.

Jouguet, E., 1905, "On the Propagation of Chemical Reaction in Gases," *Journal de Math. Pures et Appliquées,* **60**, p. 347-425.

Jones, G.E., Kennedy, J.E., and Bertholf, L.D., 1980, "Ballistic Calculations of R.W. Gurney", Amer. J. Phys., **48**(4), pp. 264-269.

Joshi, V. S., Grebe, H. A., Thadhani, N. N., and Iqbal, Z., 1993, "Effect of Packing Density on Shock-Consolidation of diamond Powders", *Proceeding of Joint AIRPT/APS Conference,* June 28-July 2, Colorado Springs, CO.

Karnelo, E., and Mäki, K., 1988, *Drifting and Tunneling by Rotary Cutting Technique,* Swedish Detonic Research Foundation, Report DS 1988:5, Stockholm, Sweden. (In Swedish.)

Kennedy, D.L., and Jones, D.A., 1993, "Modeling Shock Initiation and Detonation in the Nonideal Explosive PBXW-115," *Tenth International Symposium on Detonation,* July 16, 1993, Boston, MA.

Kennedy, J.E., 1972, "Explosive Output for Driving Metal," *Proceedings of the Twelfth Annual Symposium, Behavior and Utilization of Explosives in Engineering Design,* Albuquerque, NM, 2-3 March, 1972, ASME and UNM.

Kirby, I.J., and Leiper, G.A., 1985, "A Small Divergent Detonation Theory for Intermolecular Explosives," *Proceedings of the Eighth Symposium (International) on Detonation,* Albuquerque, NM, July 15-19, pp. 176-186.

Kolsky, H., 1953, *Stress Waves in Solids,* Oxford University Press, London, UK, pp. 213.

König, R., 1991, "Firing with Electronic Detonators in Quarries," *Nobel Heft,* **1.** (In German).

Krauland, N., 1985, "Test with Open Stoping at the Kristineberg Mine — Rock Mechanic Aspects," *Proceedings of the International Symposium on Large Scale Underground Mining,* Ed., Almgren, G., CENTEK Publishers, Lulea, Sweden, pp. 127-138.

Ladegaard-Pedersen, A., and Persson, A., 1973, *Flyrock in Blasting II, Experimental Investigation,* Swedish Detonic Research Foundation, Report DS 1973:13, Stockholm, Sweden. (In Swedish.)

Ladegaard-Pedersen, A., and Persson, A., 1968, "Photographic Studies of the Movement of the Blasted Material in Small Scale Blasting", Swedish Detonic Research Foundation, Report DL-1968:34, Stockholm, Sweden. (In Swedish.)

Ladegaard-Pedersen, A., and Holmberg, R., 1973, *The Dependence of Charge Geometry on Flyrock, Caused by Crater Effects in Bench Blasting,* Swedish Detonic Research Foundation, Report DS 1973:38, Stockholm, Sweden. (In Swedish.)

Lande, G., 1981a, *Vibrations in Soil and Frozen Soil-Measurements in Orebro,* Institute of Technology, Uppsala University, UPTEC 81 34 R, TRAVI-4, Sweden. (In Swedish.)

Lande, G., 1981b, *Computers — A Problem When Blasting in Built-Up Areas,* The Swedish Rock Blasting Committee 1981, Stockholm, Sweden. (In Swedish.)

Lang, L.C., 1962, "A Blasting Theory and its Application," Iron Ore Company of Canada, PR(R) 10/62, Canada.

Lang, L.C., 1976, "The Application of Spherical Charge Technology in Stope and Pillar Mining," *Eng. Min. J.* **177,** May 1976, p. 98.

Lang, L.C., 1978, "Cratering Theory Evolves into New Underground Mining Technique," *Rock Breaking — Equipment and Techniques,* The Australasian Institute of Mining and Metallurgy, Australia, pp. 115-124.

Lang, L.C., 1981a, "Driving Underground Raises with VCR," *SEE News,* **6**(3), Sept. 1981, USA.

Lang, L.C., 1981b, "VCR Used Successfully from Surface in Underground Stoping in Australia," *SEE News,* **6**(4) Dec. 1981, USA.

Lang, L.C., 1983, *A Brief Review of Livingston's Cratering Theory,* Swedish Detonic Research Foundation, Report DS 1983:1, Stockholm, Sweden.

Lang, L.C., Holmberg, R., and Niklasson, B., 1982, *A Proposal for the Design of a VCR Stope at the Luossavaara Research Mine,* Swedish Detonic Research Foundation, Report DS 1982:20, Stockholm, Sweden.

Lang, L.C., Roach, R.J., and Osoko, M.N., 1977, "Vertical Crater Retreat, An Important Mining Method," *Can. Mining J.,* **98,** Sept. 1977, p. 69.

Langefors, U., and Kihlström, B., 1963, *The Modern Technique of Rock Blasting,* John Wiley & Sons, Inc., New York, and Almqvist & Wiksell, Stockholm.

Larsson, B., 1974, *Blasting of Low and High Benches, Fragmentation from Production Blasting,* The Swedish Rock Blasting Committee, Stockholm, Sweden, pp. 247-271. (In Swedish.)

Larsson, B., Holmberg, R., and Westberg, J., 1988, "Super Accurate Detonators — A Rockblaster's Dream," *Proceedings of the Fourteenth Conference on Explosives and Blasting Technique,* Anaheim, CA, pp. 44-58.

Lavrentiev, V., 1974, personal communication.

Lee, J., 1990, *Detonation Shock Dynamics of Composite Energetic Materials,* Ph.D. Thesis, New Mexico Institute of Mining and Technology, Socorro, NM.

Lee, J., Bdzil, J.B., and Persson, P.-A., 1993, in preparation.

Leiper, G.A., and Cooper, J., 1990, "Reaction Rates and the Charge Diameter Effect in Heterogeneous Explosives," *Proceedings of the Ninth Symposium (International) on Detonation*, Portland, OR, pp. 197-208.
Livingston, C. W., 1962, *A Blasting Theory and Its Applications*, notes from a course given in Georgetown, CO.
Lundberg, B., 1974, "Penetration of Rock by Conical Indenters," *Int. J. Rock Mech. Min. Sci. & Geomech. Abst.* **11**, p. 209-214.
Lundberg, L. and Gyldén, N., 1962, private communication.
Lundborg, N., 1967, "The Strength/Size Relation of Granite," *Int. J. Rock Mech. Min. Sci.* **4**, p. 269.
Lundborg, N., 1970, "Air Blasts, Supersonic Booms, and Noise," Swedish Detonic Research Foundation Report, DS 1970:11, Stockholm, Sweden. (In Swedish).
Lundborg, N., 1972, "A Statistical Theory of the Polyaxial Compressive Strength of Materials," *Int. J. Rock Mech. Min. Sci.* **9**, p. 617.
Lundborg, N., 1974, "A Statistical Theory of the Polyaxial Strength of Materials," in *Advances in Rock Mechanics, Proceedings of the Third Congress of the International Society for Rock Mechanics*, Denver CO, Sept. 1-7, 1974. Vol I, Part B, pp. 180.
Lundborg, N., 1974, *The Hazards of Flyrock in Rock Blasting*, Swedish Detonic Research Foundation, Report DS 1974:12, Stockholm, Sweden.
Lundborg, N., 1981, *Risk for Flyrock When Blasting*, The Swedish Council for Building Research, BFR-Report R29:1981, Stockholm, Sweden. (In Swedish.)
Lundborg, N., Holmberg, R., and Persson, P.-A., 1978, *Relation between Vibration Level, Distance and Charge Weight*, The Swedish Council for Building Research, BFR Report R11:1978, Stockholm, Sweden. (In Swedish.)
Lundborg, N., Persson, A., Ladegaard-Pedersen, A., and Holmberg, R., 1975, "Keeping the Lid on Flyrock in Open Pit Blasting," *Eng. Min. J.*, **176**, May 1975, p. 95-100.
Lyamkin, A.I., Petrov, E.A., Ershov, A.P., Sakovich, G.V., Staver, A.M., and Titov, V.M., 1988, "Diamonds Manufactured from Explosives", *Dokl. Akad. Nauk SSSR*, **302**(3), p. 611.
Mader, C.L., 1962, *STRETCH BKW, A Code for Computing the Detonation Properties of Explosives*, Los Alamos Scientific Laboratory report, LADC-5691, Los Alamos, NM.
Mader, C.L., 1963, "Detonation Properties of Condensed Explosives Computer Using the BKW Equation of State," Los Alamos Scientific Laboratory report LA-2900, Los Alamos, NM.
Mader, C.L., 1979, *Numerical Modeling of Detonations*, University of California Press, Berkeley, CA.
Mäki, K., 1982, *Characterization of Rock Structures at Crater Blasting Experiments in the Luossavaara Mine*, Swedish Detonic Research Foundation, Report DS 1982:17, Stockholm, Sweden.
Mallard, E. and Le Chatelier, H., 1881, "Sur la vitesse de propagation des phenomenes explosifs dans les gaz", *C. R. Accad. Sci. Paris*, **93**, p. 145-148, seance de 18 Juillet, 1882.
McAfee, J.M., Asay, B.W., and Bdzil, J.B., 1993, "Deflagration-to-Detonation in Granular HMX : Ignition, Kinetics, and Shock Formation", *Proc. 10th International Symp. on Detonation*, July 12-16, Boston, MA.
Medearis, K., 1977, "Development of Rational Damage Criteria for Low-Rise Structures Subjected to Blasting Vibrations," *Proceedings of the Eighteenth US Symposium on Rock Mechanics*, pp. 1-16.
Meyer, R., 1977, *Explosives*, Verlag Chemie, Weinheim, Germany.

Meyers, M.A., and Murr, L.E., Eds., 1981, *Shock Waves and High-Strain-Rate Phenomena in Metals*, Plenum Press, New York, pp. 828.

Miller, R.E., 1979, "Vertical crater retreat mining method as applied to L 519 slot 14B and 15D Mount Isa Mine," *Australian Mineral Foundation Workshop 120/79*, Adelaide, June 25-29.

Mitchell, S.T., 1980, "Vertical Crater Retreat Stoping Proves Successful at Homestake," *Min. Eng.*, Nov., pp. 1581-1586.

Molin, V., 1968, "Explosive Engraving," *Leonardo*.

Monahan, C.J., 1979, "The Crater Blasting Method Applied to Pillar Recovery at Falconbridge Nickel Mines Limited," *CIM Underground Operators Conference*, Feb. 19-21, Timmins, Ontario, Canada.

Munroe, C.E. 1888, "Modern Explosives," *Scribner's Magazine*, **3**, Jan-June, pp. 563-576.

Naarttijärvi, T., Rustan, A., Öqvist, J., and Ludvig, B., 1980a, *Laboratory Test in Controlled Blasting*, Lulea University of Technology, Report TULEA 1980:22, Lulea, Sweden. (In Swedish.)

Naarttijärvi, T., Rustan, A., Öqvist, J., and Ludvig, B., 1980b, *Field Experiments with Cautious Blasting*, Lulea University of Technology, Report TULEA 1980:26, Lulea, Sweden. (In Swedish.)

Naarttijärvi, T., Rustan, A., Öqvist, J., and Ludvig, B., 1980c, *Controlling Crack Initiation with Shaped Charges*, Lulea University of Technology, Report TULEA 1980:75T, Lulea, Sweden. (In Swedish.)

Nakano, M., and Ueada, T., 1983, "New Firing Method for Underwater Blasting," *Proceedings of the Ninth Conference on Explosives and Blasting Technique*, Ed., Konya, C.J., Society of Explosives Engineers, Montville, OH, pp. 146-167.

Nicholls, H. R., Johnson, C. F., and Duvall, W. I., 1971, *Blasting Vibrations and Their Effects on Structures*, US Bureau of Mines, Bulletin 656.

Niklasson, B., 1979, "Mining with Large Hole Diameters in Sullivan Mine, Cominco Ltd., Kimberley, BC, Canada," *Bergsskolan i Filipstad*, Sweden. (In Swedish.)

Niklasson, B., 1984, *Vertical Crater Retreat Mining at the Luossavaara Research Mine*, Swedish Detonic Research Foundation, Report DS 1984:5, Stockholm, Sweden.

Niklasson, B., Holmberg, R., Olsson, K., and Schörling, S., 1986, "Longer Rounds to Improve Tunneling and Development Work," *Proceedings of Tunnelling '88*, Inst. Min. Met. London, pp. 213-221.

Niklasson, B., and Keisu, M., 1991, *New Techniques for Drifting and Tunnel Blasting — The Sofia Project*, Swedish Detonic Research Foundation, Report DS 1991:10, Stockholm, Sweden. (In Swedish.)

Niklasson, B., and Keisu, M., 1992, "New Methods for Contour Blasting Using Electronic Detonators and Water-Notched Boreholes, Including Longer Drift Rounds and Cuts without Large Cut Holes," *Proceedings of High Tech Seminar on Blasting Technology, Instrumentation, and Explosives Applications*, Nashville, TN.

Norell, B., 1985, *The Effect of Delay Time on Fragmentation. Small Scale Experiment*, Swedish Detonic Research Foundation, Report DS 1985:1, Stockholm, Sweden. (In Swedish.)

Norén, C.H., 1956, "Blasting Experiments in Granite Rock," *Quarterly Colorado School of Mines*, **51**(3), p. 213-225.

Olson, D.B., 1992, Private communication.

Olson, J.J., Willard, R.J., Fogelson, D.E., and Hjelmstad, K.E., 1973, *Rock Damage from Small Charge Blasting in Granite*, US Bureau of Mines, Report of Investigation 7751, 44 p.

Olsson, M., 1987, *Blasting against a Compressible Bed of Fragmented Ore*, Swedish Detonic Research Foundation, Report 1987:5, Stockholm, Sweden. (In Swedish.)

Olsson, M., and Bergqvist, I., 1993, "Crack Lengths from Explosives," *BeFo Rock Mechanics Meeting in Stockholm*, Sweden, March 18, pp. 225-232. (In Swedish.)

Oppenheim, A.K. (Antoni), 1961, "Development and Structure of Plane Detonation Waves", Reprint from the Fourth AGARD Colloquium: *Combustion and Propulsion*, Milan April 4-6, 1960, Pergammonn Press.

Ouchterlony, F., 1972, *Analysis of the Stress State Around Some Expansion Loaded Radial Crack Systems in an Infinite Plane Medium*, Swedish Detonic Research Foundation, Report DS 1972:11, Stockholm, Sweden.

Ouchterlony, F., 1983, "Fracture Toughness Testng of Rock," *Rock Mechanics, CISM Courses and Lectures*, Ed., Rossmanith, H.P.) No. 275, Springer, Vienna, Austria, pp. 69-150.

Pershin, S.V., Tsaplin, D.N., Dremin, A.N., and Antipenko, A.G., 1991, "Possibility of diamond formation during detonation of picric acid," *Fiz. Goreniya Vzryva*, **27**(4), pp.117-121. (In Russian)

Persson, A., 1970, "High-Speed Photography of Scale-Model Rock Blasting," *Proceedings of the Ninth International Congress on High Speed Photography*, Denver, CO.

Persson, A., 1975, *ANFO-Explosives. Initiation, Detonation, and Toxic Fumes*, Swedish Detonic Research Foundation, Report TM1, Stockholm, Sweden.

Persson, G., 1983, *Methods to Reduce the Emission of Noxious Fumes when Blasting Underground, ASF-project 79/3270*, Swedish Detonic Research Foundation, Report DS 1983:18, Stockholm, Sweden. (In Swedish.)

Persson, G., and Höglund, P., 1991, *Method for Detonation Performance Control (Tracer)*, Swedish Detonic Research Foundation, Report DS 1991:8G, Stockholm, Sweden. (In Swedish.)

Persson, P.-A., 1967, *Fuse for Transmission or Generation of Detonation*, Swedish Patent 333321, July 20, 1968, Priority from July 20, 1967.

Persson, P.-A., 1968, *Fuse*, U.S. Patent 3,590,739, July 16, 1968, Priority from July 20, 1967, Issued July 6, 1971.

Persson, P.-A., 1973, *The Influence of Blasting on the Remaining Rock*, Swedish Detonic Research Foundation, Report DS 1973:15, Stockholm, Sweden. (In Swedish.)

Persson, P.-A., 1975, "Bench Drilling — an Important First Step in the Rock Fragmentation Process," *Proceedings Bench Drilling Days*, June 9-13, 1975, Atlas Copco, Stockholm, Sweden.

Persson, P.-A., Holmberg, R., and Persson, G., 1977, *Careful Blasting of Slopes in Open Pit Mines*, Swedish Detonic Research Foundation, Report DS 1977:4, Stockholm, Sweden. (In Swedish.)

Persson, P.-A., and Johansson, C.H., 1974, "Fragmentation Systems," *Proceedings of the Third Congress of the International Society for Rock Mechanics*, Denver, Colorado, September 1-7.

Persson, P.-A., Ladegaard-Pedersen, A., and Kihlström, B., 1969, "The Influence of Bore Hole Diameter on the Rock Blasting Capacity of an Extended Explosive Charge," *Int. J. Rock Mech. Min. Sci.*, **6**, p. 277.

Persson, P.-A., and Ladegaard-Pedersen, A., 1970, *Large Scale Experiments with the Wide Space Blasting Method*, Swedish Detonic Research Foundation, Report DS 1970:14, Stockholm, Sweden.

Persson, P.-A., Lundborg, N., and Johansson, C.H., 1970, "The Basic Mechanisms in Rock Blasting," *Proceedings of the Second Congress of the International Society for Rock Mechanics*, Beograd.

Persson, P.-A., and Mäki, K., 1973, Swedish Detonic Research Foundation, Report DS 1973:1, Stockholm, Sweden.

Persson, P.-A., and Persson, I., 1964, "Determination of Equation of State Data for Uranium at High Pressures by Shock Wave Experiments," FOA 2 Report A 2299-222. (In Swedish.)

Persson, P.-A., and Persson, I., 1965, "Equation of State Data for Palladium, Lead, Lithium Hydride, Plexiglass, and Glass at High Pressures by Shock Wave Experiments," FOA 2 Report A 2395-222. (In Swedish.)

Persson, P.-A., Schmidt, R.L., Eds, 1977 "Mechanical Boring or Drill & Blast Tunneling (MB or DB)," *Proceedings of the First US–Swedish Underground Workshop*, Stockholm, Sweden, December 5-10, 1976. Document D3:1977–Statens Rad For Byggnadsforskning (Swedish Council for Building Research), Stockholm, Sweden.

Poncelet, E.F., 1961, "Theoretical Aspects of Rock Behavior Under Stress," *Proceedings of the Fourth Symposium of Rock Mechanics*, Bull. Min. Ind. Exp. Station.

Project Report 1-76, 1976, *Fullface Boring of Tunnels*, Institutt for Anleggsdrift, Geologisk Institutt, NTH, 94 p.

Reiher, H., and Meiser, F.J., 1931, "Human Annoyance of Vibrations," *Forschung auf dem Gebeite des Ingenieurwesens*, 2(11). (In German).

Reinhardt, H. W., and Dally, J. W., 1970, "Some Characteristics of Rayleigh Wave Interaction with Surface Flaws," *Materials Evaluation*, Oct. 1970.

Rice, M.H., McQueen, R.G., and Walsh, J.M., 1958, "Compression of Solids by Strong Shock Waves," *Solid State Physics,* Volume 6, Eds., Seitz, F., and Turnbull, D., Academic Press, New York, NY, pp. 1-63.

Richart, F.E., Hall, J.R., and Woods, R.D., 1970, *Vibrations of Soil and Foundations*, Prentice-Hall, Inc., Englewood Cliffs, NJ.

Riemann, B., 1860, "Uber die Forpflanzung ebener Luftwellen von endlicher Schwingungwseite", (On the propagation of plane air waves of finite amplitude", *Sitz, Berichte der Wissenschaftlicher*, Ges. der Univ. Göttingen, (Proceedings of the Scientific Society, Univ. of Göttingen), Germany.

Rinehart, J.S., 1975, *Stress Transients in Solids*, Hyperdynamics, Santa Fe, NM.

Robbins, R.J., 1976, "Development Trends in Tunnel Boring Machines for Hard Rock Applications: Mechanical Boring or Drill and Blast Tunnelling," *First US–Swedish Underground Workshop*, Stockholm, December 5-10. Document D3:1977 — Statens rad for byggnadsforskning (Swedish Council for Building Research), Stockholm, 1977, pp. 27-39.

Rossmanith, H.P., and Knasmillner, R., 1983, "Spallation, Break-up, and Separation of Layers by oblique Stress-Wave Incidence," *Proceedings of the First International Symposium on Rock Fragmentation by Blasting*, Eds., Holmberg, R. and Rustan. A., Lulea Univeity of Technology, Lulea, Sweden, pp. 149-168.

Roth, J., 1979, *A Model for the Determination of Flyrock Range as a Function of Shot Conditions*, Management Science Associates, CA.

Rowlandson, P., 1979, "Applications of DTH Drilling at Pamour Porcupine Mines," *CIM Underground Operators Conference*, Feb. 19-21, Timmins, Ontario, Canada.

Sakamoto, M., Yamamoto, M., Aikov, K., Fukui, H., and Ichikawa, K., 1989, "A Study on High Accuracy Delay Detonator", *Proceedings of the Fifteenth Conference on Explosives and Blasting Techniques*, Society of Explosives Engineers, New Orleans, LA, pp 185-200.

Seinov, N.P., and Chevkin, A.F., 1968, "Effect of Fissure on the Fragmentation of a Medium by Blasting," *Fiziko-Tekhnicheskie Problemy Razrobotki Poleznykh Iskopaemykh*, No. 3, May-June, p. 57-64.

Selleck, D.J., 1962, "Basic Research Applied to the Blasting of Cherty Metallic Iron Formation," *International Symposium on Mining Research,* Ed., Clark, G.B., Volume 1, Pergamon Press, pp. 227-248.

Selmer-Olsen, R., and Blindheim, O.T., 1970, "On the Drillability of Rock by Percussive Drilling," *Proc. 2nd Congress Int. Soc. Rock Mechanics,* Belgrade.

Shuster, A. 1893, Reference cited by H. Dixon [1893].

Siskind, D.E., and Fumanti, R.R., 1974, *Blast-Produced Fractures in Lithonia Granite,* US Bureau of Mines, Report of Investigations 7901, Pittsburgh, PA.

Siskind, D.E., Stackora, V.J., Stagg, M.S., and Kopp, J.W., 1980, *Structure Response and Data produced by Air Blast From Surface Mining,* US Bureau of Mines, Report of Investigations 8485, Pittsburgh, PA.

Siskind, D.E., Stagg, M.S., Kopp, J.W., and Dowding, C.H., 1980, *Structure Response and Damage Produced by Ground Vibration from Surface Mine Blasting,* US Bureau of Mines, Report of Investigation 8507.

Siskind, D.E., Steckley, R.C., and Olson, J.J., 1973, *Fracturing in the Zone Around a Blast Hole,* US Bureau Mines, Report of Investigation 7753.

Siskind, D.E., Stagg, M.S., Kopp, J.W., and Dowding, C.M., 1980, *Structure Response and Damage Produced by Ground Vibration from Surface Mine Blasting,* US Bureau of Mines, Report of Investigation 8507.

Smith, L. C., 1967, "On Brisance and a Plate-Denting Test for the Estimation of Detonation Pressure", *Explosivstoffe,* **5**, p. 106, **6**, p. 130.

Stagg, M. S., Siskind, D. E., Stevens, M. G., and Dowding, C. H., 1984, *Effects of Repeated Blasting on a Wood-Frame House,* US Bureau of Mines, Report of Investigation 8896.

Staver, A.M., Gubareva, N.V., and Lyamkin, A.I. 1984, "Synthesis of Ultrafine Diamond Powders by Explosion", *Fiz. Goreniaivzryva,* **20**(5), pp. 100-104. (In Russian.)

Svanholm, B.O., Persson, P.-A., and Larsson, B., 1977, "Smooth Blasting for Reliable Underground Openings," *Rockstore 77,* Session 1, Sweden.

Svärd, J., 1992, "Possibilities with accurate delay times: Results of some field tests using electronic detonators." *Proceedings of the Fourth High Tech Seminar on Blasting Technology, Instrumentation, and Explosives Applications,* Nashville, TN.

Swanholm, B.O., Persson, P.-A., and Larsson, B., 1977, "Smooth Blasting for Reliable Underground Openings," *Rockstore 77: Storage in Excavated Rock Caverns, Proceedings of the First International Symposium,* Stockholm, Ed., Bergmann, M., Pergamon, 1977, **3**, pp. 573.

Swedish Standard SS 460 48 66, 1991, *Vibration and Shock-Guidance Levels for Blasting-Induced Vibration in Building,* SIS, Box 3295, 10366, Stockholm, Sweden.

Taylor, J., 1952, *Detonation in Condensed Explosives,* University Press, Oxford, UK.

Thadhani, N. N., 1988, "Shock Compression Processing of Powders," *Adv. Mater. and Manuf. Proc.,* 4(4), p. 493.

Thadhani, N.N., 1993, "Shock-induced chemical reactions and synthesis of materials," *Progress in Materials Science,* **37**(2), p. 130.

Thompson, D.E., 1979, *Field Evaluation of Fracture Control in Tunnel Blasting,* US Department of Transportation, Report No. UMTA-MA-06-0100-79-14, Washington, D.C. 20590.

Titov, V.M., Anisichkin, V.E., and Mal'kov, I. Yu., 1989, "Synthesis of Ultrafine Diamonds in Detonation Waves", *Ninth Symp. (Int.) on Detonation,* Aug. 28-Sept 1, Portland, OR.

Urizar, M.J., James, E., Jr., and Smith, L.C., 1961, "Detonation Velocity of Pressed TNT", *Phys. Fluids,* **4**, p. 262.

Vestre, J., 1987, "Underwater Testing of Explosives Energy Release under Critical Conditions," *Nordic Conference on Explosives Performance*, Swedish Detonic Research Foundation, May 21-22, pp. 35-62.

Volchenko, N.G., 1977, "Influence of Charge Arrangement Geometry and Short-Delay Blasting on the Crushing Indices in Compression Blasting," *Fiziko Tekhnicheskie Problemy Razrabotki Poleznykh Iskopaemykh*, No. 5, Sept-Oct., p. 57-63.

Von Neumann, J., 1942, *Theory of Detonation Waves*, Office of Scientific Research and Development, Report No. 549.

Voreb'ev, I.T., 1972, "Features of the Development and Propagation of the Rayleigh Surface Wave in the Dzhezkazgan Deposit," *Fiziko Tekhnicheskie Problemy Razrabotki Poleznykh Iskopaemykh*, 8(6), pp. 36-43; *Soviet Min. Sci.*, 8, 1972, pp. 634-639.

Wade, C.G., 1973a, *Water-In-Oil Emulsion Explosive Containing Entrapped Gas*, U.S. Patent 3,715,247, February 06, 1973, filed September 3, 1970.

Wade, C.G., 1973b, *Water-In-Oil Type Explosive Compositions Having Strontium-Ion Detonation Catalysts*, U.S. Patent 3,765,964, October 16, 1973.

Wade, C.G., 1978, *Water-In-Oil Emulsion Explosive Composition*, U.S. Patent 4,110,134, August 29, 1978, filed November 3, 1977.

Walters, W.P., and Zukas, J.A., 1989, *Fundamentals of Shaped Charges*, John Wiley & Sons, New York, NY.

Weibull, W., 1939, "A Statistical Theory of the Strength of Materials," *Proc. Roy. Swedish Acad. Eng. Sci.*, 151.

Weibull, W., 1947. "Detonation of Spherical Charges in Air: Travel Time, Front Velocity, and Wave Length of Emitted Wave," *Tidskrift for Kustartilleriet*, 5(44). (In Swedish.)

Wijk, G., 1982, "The Stamp Test for Rock Drillability Classification," Swedish Detonic Research Foundation, Report DS 1982:1, Stockholm, Sweden.

Winzer, S.R., Furth, W., and Ritter, A.D., 1979, "Initiator Firing Times and Their Relationship to Blasting Performance," *Proceedings of the Twentieth US Symposium on Rock Mechanics*, Austin, TX.

Winzer, S.R., Anderson, D.A., and Ritter, A.D., 1983, "Rock Fragmentation by Explosives," *Preprints of the First International Symposium on Rock Fragmentation by Blasting*, Eds., Holmberg, R. and Rustan, A., Lulea University of Technology, Lulea, Sweden, pp. 225-249.

Wiss, J.R., and Linehan, P., 1978, *Control of Vibration and Blast Noise from Surface Coal Mining*, US Bureau of Mines, Open File Reports 103(4)-79.

Wiss, J.F., and Parmelee, R.A., 1974, "Perception of Transient Vibrations," *J. Structural Div.*, ASCE, 100, No. ST4, Proc. Paper 10495, pp. 773-787.

Wolkomir, R., 1987, "She's an artist whose explosives make a lasting impression," *Smithsonian*, 18(9), Dec., pp. 166-173. Note: The artist is Evelyn Rosenberg.

Wood, W.W., and Kirkwood, J.G., 1954, "Diameter Effect in Condensed Explosives. The Relation between Velocity and Radius of Curvature of the Detonation Wave", *J. Chem. Phys.*, 22, p. 1920.

Woods, R. D., 1968, "Screening of Surface Waves in Soils," *J. Soil Mech. Fond. Div.*, 94, p. 951.

Worsey, P.N., and Lawson, J.T., 1983, "The Development Concept of the Integrated Electronic Detonator," *Preprints of the First International Symposium on Rock Fragmentation by Blasting*, Eds., Holmberg, R. and Rustan, A., Lulea University of Technology, Lulea, Sweden, pp. 251-258.

Wright, G., 1949, *J. Am. Chem. Soc.*

Zeldovich, Y.B., 1940, "On the Theory of the Propagation of Detonation in Gaseous Systems," *Zh. Eksp. Teor. Fiz.*, 10, pp. 542-568, USSR (English Translation, NACA TM 1261, 1950).

Zinn, J., and Rogers, R., 1962, "Thermal Initiation of Explosives", J. Phys. Chem., 66, p. 2646.

Units

It is important for students and engineers, even those in the USA, to learn to reap the advantages of the decimal metric system, which is the system of units adopted and used by the rest of the world. The metric system, or International System of Units (SI), is therefore used as the standard throughout this book. In the SI system, lengths are given in meters (m), masses in kilograms (kg), times in seconds (s), and electrical currents in Amperes (A). The standard derived unit for force is the Newton (N) (9.81 N = 1 kgm/s^2), that for pressure is the pascal (Pa) (1 Pa = 1 N/m^2), and that for energy is the Joule (J) (1 J = 1 Nm). For convenience, multiples or fractions of 1000 of the standard units are used, for example 1 km for 1000 m, 1 mm for 1/1000 m, 1 μs for 10^{-6} s, and 1 kPa, 1 MPa, and 1 GPa for 10^9, 10^9, and 10^9 Pa, respectively. Equations throughout are written so as not to be unit dependent, except in the rare case where an equation can provide a convenient rule of thumb, easy to remember, if expressed in mixed units.

(An example of such a rule of thumb is the expression in Equation 2.20, which gives the approximate volume in liters of 1 meter length of a drillhole as its diameter in inches, squared and divided by 2. A drillhole with a 4 in (4 inch) diameter thus has a volume of 8 l/m (8 liters per meter). The authors felt that providing the readers with such a convenient shortcut method, accurate at the percent level, was worth a small sacrifice of the otherwise strict adherence to metric units.)

Two concessions to practicality have been made:

1. Many practicing rock blasters around the world still exclusively use the English system of units, counting drillhole diameters in inches (in.), weights in pounds (lbs) or short tons (2000 lb), and volumes of rock in cubic yards (cu. yd.). For their benefit, and also for the benefit of students who need to be conversant with both systems, the English units are set out in brackets after the standard units in most of the applied parts of the book.

2. The explosives community inherited a very large and valuable treasure of experimental and computed data from the nuclear weapons research programs at Los Alamos, Sandia, and Lawrence Livermore National Laboratories. The researchers of these laboratories used (and to some extent still in 1992 use) a hybrid metric system which gives lengths in cm, times in s, and pressures in kbar or Mbar (1 bar = 10^6 dyne/cm^2 = 10^5 Pa). In this hybrid system, specific internal energy (energy per unit mass of a substance) is sometimes given in the compound units Mbar cm^3/g (1Mbar cm^3/g = 100 MJ/kg). Rather than redrawing diagrams taken from this era, with the risk of introducing errors, we have chosen to give such diagrams and some tables in the original hybrid metric units. (As a matter of historical curiosity, the reader might be interested to know of a short-lived but remarkable unit for a very short time, also introduced at Los Alamos, but no longer in use. It is the *shake*, named after the very short time of one shake of a lamb's tail; 1 shake = 10^{-8} s).

Conversion Factors

Physical Property	Symbol	Unit System			Multiplication Factor
		US/British	cgs	SI (m/kg/s)[a]	(US/British or cgs to SI)
Angle			deg	rad	1.745×10^{-2}
C-J pressure	P_{CJ}		bar	Pa	1.00×10^{5}
Density	ρ		g/cm^3	kg/m^3	1000
Detonation velocity	D		mm/μsec	km/s	1
Heat of detonation	ΔH_{det}		cal/g	J/kg	4.187×10^{3}
Heat of formation	ΔH_f		cal/g	J/kg	4.187×10^{3}
			kcal/mol	kJ/mol	4.187
Initial elastic modulus	E_0	psi		Pa	6.895×10^{3}
Internal Energy	E		cal/g	J/kg	4.187×10^{3}
Length		in (inch)		m	25.4×10^{-3}
Mass	M, C	lb		kg	0.4536
Pressure	P	psi		Pa	6.895×10^{3}
			atm	Pa	1.013×10^{5}
			bar	Pa	1.00×10^{5}
Sliding velocity	ν	in/min		m/s	4.233×10^{-4}
		ft/sec		m/s	3.048×10^{-1}
Specific heat	C_P		cal/g$^\circ$C	J/(kg K)	4.187×10^{3}
Specific Volume	v	cu. yd./lb		m^3/kg	1.686
Temperature[b]	T	$^\circ$F		K	$\dfrac{(T_F - 32)}{1.8} + 273.16$
			$^\circ$C	K	$t + 273.16$
Thermal conductivity	λ		cal/cm sec$^\circ$C	W/m-K	4.187×10^{2}
Thermal expansion	CTE	in/in$^\circ$F		m/m-K	$\dfrac{1}{1.8}$
			cm/cm$^\circ$C	m/m-K	1
Vapor pressure	v.p.		mm Hg, Torr	Pa	1.33×10^{2}
Volume	V		cm^3	m^3	1.0×10^{6}
			cu. yd.	m^3	0.76455
Weight		lb		kgf[c]	0.4536

[a] In this column, the abbreviations used are those of the International System of Units (SI). In the SI system, temperature is in K (degrees Kelvin) rather than the obsolete $^\circ$K.

[b] In this book, the symbols for temperature are T for K and t for $^\circ$C.

[c] The old metric unit for force, kgf = 9.81 N, does not belong in the SI system of units, and its continued use should not be encouraged.

List of Symbols

A	Displacement amplitude.
a	Crack length.
α_1	Hole deviation at the collar, collaring deviation.
α_2	Angular hole deviation at collar, initial direction error.
α_3	Additional hole deviation caused by "bending" (curving) of the hole.
B	Burden — the thickness of rock mass in front of a hole, distance between two rows of drillholes.
C	Explosive charge mass per unit surface area.
C	Constant, intercept in linear u_s vs. u_p relation.
c	Sound velocity.
CJ	Chapman-Jouguet. Mostly used as an index to indicate that the indexed variable is at its ideal detonation, Chapman-Jouguet state.
D	Detonation (wave) velocity.
d	Hole diameter.
d	Depth of burial.
D_{CJ}	Chapman-Jouguet detonation velocity.
d_o	Optimum depth of burial.
ϵ	Strain, dilatation.
$\sqrt{2E}$	Gurney energy, the explosive energy effective in accelerating metal.

E	Elastic modulus, Young's modulus.
E or E_i	Internal energy is defined by the First Law of Thermodynamics which states that heat Q added to a body together with the external work W done on the body equal the sum of the increase in its internal energy E_i (or E), kinetic energy E_k, and potential energy E_k: $Q + W = E_i + E_p + E_k$. (Also incorrectly, but frequently, used for specific internal energy, i.e., internal energy per unit mass of material.)
e	Specific internal energy, i.e., internal energy per unit mass of material. Specific internal energy is sometimes, (incorrectly and inconsistently in current literature) written E — even in *this* book.
E_s	Strain Energy Factor (in Livingston's cratering theory).
f	Vibration frequency.
F	Force.
F_d	Distance factor which considers the distance from the round to the measuring gauge.
F_k	Construction Quality Factor.
F_m	Construction Material Factor.
F_t	Project Time Factor.
ϕ	Diameter of empty hole in parallel hole cut, hole diameter. (Standard notation in Sweden and India for hole diameter, but not so common in the USA.)
ϕ	Boulder diameter.
γ	Lookout angle: In drilling for a tunnel blast, the angle between the tunnel's axial direction and the direction of the contour holes which cannot be drilled parallel to the tunnel axis because of the size of the drilling machine.
γ	Slope of the isentrope $\gamma = -\left(\dfrac{\partial \ln p}{\partial \ln v}\right)_s$
$-\Delta H_d$	Heat of detonation.
$-\Delta H_f^\circ$	Heat of formation.
H	Hole length.
h_s	Stemming length of hole, filled with sand or crushed rock, drill cuttings.

List of Symbols

I_{sp}	Specific impulse, the product of mass and velocity per unit mass of explosive.
i	Impulse density, integral of pressure over time.
K	Bench height.
ℓ	Linear charge concentration, charge mass per unit charge length.
ℓ_b	Linear charge concentration in the bottom part of a hole, linear bottom charge concentration.
ℓ_c	Linear charge concentration in column charge: $l_c = 0.4\, l_b$.
M	Metal mass per unit surface area.
ω	Angular frequency.
Ω	Solid angle.
P	Pressure.
P_e	Constant volume explosion pressure.
p_c	Estimated front pressure.
p_u	Front pressure.
π	Ratio of circumference to diameter of a circle, 3.14159.
ϕ	Phase angle.
Q	Charge mass.
Q_d	Detonation energy, i.e., the thermal energy released by chemical reaction in an ideal (CJ) detonation.
Q_e	Explosion energy, i.e., the thermal energy released by chemical reaction at constant volume.
R	Mass reaction rate.
R	Radius.
r	Radius.
ρ	Density, i.e., mass per unit volume.
S	Material constant, slope in linear u_s versus u_p relation.
S	Spacing, distance between two consecutive drillholes in a row.
σ	Standard deviation. If the mean value $\bar{x} = 0$, then the standard deviation is equivalent to the R.M.S.

List of Symbols

t	Temperature in °C.
t	Time.
T	Absolute temperature, in Kelvin (K). The symbol °K is no longer in common use.
T	Reduction factor.
U	Shock (wave front) velocity.
u	Velocity.
u_s	Shock (wave front) velocity.
u_p	Particle velocity (mostly used to denote the material velocity in shock waves, detonation waves).
V	(Metal plate) velocity.
$\dfrac{V}{\sqrt{2E_g}}$	Dimensionless Gurney velocity.
V	Volume, crater volume.
V_g	Gas volume (often incorrectly used for specific gas volume, i.e., gas volume per unit mass of gas).
v	Specific volume, i.e., volume per unit mass of a material.
v	Guidance level.
v	Vibration (material or particle) velocity.
v_0	Uncorrected vertical peak particle velocity.
W	Charge weight (mostly used for charge mass).
w	Vibration intensity.
$\hat{x}$	Peak value, i.e., the maximum absolute value of the vibration during a time interval. Additionally, the sign, indicating whether the peak values are positive or negative needs to be taken into account.
$|\bar{x}|$	Undirected mean value.
x_{eff}	R.M.S.-value of a signal.
$\bar{x}$	Mean value. The mean value for vibrations is often defined to be 0 ($\bar{x} = 0$).

Glossary

Advance
   In tunnel blasting, the effective increase of the tunnel length resulting from one round.

Air Blast
   The common term for pressure waves in air emanating from explosions. See also, *concussion* and *noise*.

Angular deviation
   The angle between the intended and the actual direction of the drillhole at the collar.

Bench blasting
   Open blasting with vertical or near vertical holes drilled from a horizontal surface, with a vertical or near vertical free surface in the forward direction.

Bit penetration
   In percussive drilling, penetration depth per impact.

Blast
   (1) To blast or fire a charge. (2) The detonation of one charge in the open or in a drillhole, (3) the effect transmitted through the air of a detonating charge, (4) sometimes, a round of several charges fired simultaneously or in one interval sequence.

Blast geometry
   Geometry of rock volume and drillholes.

Blast wave
   Air blast wave.

Booster charge
   See Primer charge.

Borehole
   Often used instead of drillhole, which is the preferred term for a hole drilled in rock, mostly for the purpose of being loaded with explosive.

Boring
   Mostly used for the process of full-face boring of tunnels or large (<0.5 m) diameter holes in rock.

Bottom charge
   That part of the charge effective in removing the burden at the bottom part of the drillhole, often the charge extending from the drillhole bottom a length equal to 1.3 times the burden.

Brisance
   The ability of an explosive to shatter and fragment steel, concrete, and other hard structures (obsolete).

Brower compression test (apparatus)
   Test for quantitative measurements of the sensitivity of explosives and other energetic materials to initiation by the adiabatic heating of a gas, for example the situation in bubble compression. The apparatus, developed by Kay Brower, consists of a short vertical hardened steel thick-walled cylinder, closed at its lower end with a sapphire window, and having a close-fitting 12.7 mm diameter piston with O-rings at its upper end, sealing off a volume of the cylinder bore. The piston is restricted in its upward motion so that it cannot leave the cylinder. A drop weight, or a fast moving projectile, is allowed to impact the exposed top end of the piston, pushing the piston down into the cylinder, compressing the gas (normally argon) which is enclosed in the cylinder together with a very small volume of the explosive to be tested for adiabatic gas compression sensitivity. The piston bounces on the compressed gas, which limits the time the explosive is exposed to the high gas temperature. The amount of chemical energy released by surface burning (or consumed by endothermic decomposition) during the short time the explosive's surface is exposed to the high temperature can be measured by the drop weight's rebound, and a sample of the reaction products remaining in the cylinder afterwards can be drawn through a valve for chemical analysis. Optical studies of the radiation can be made through the sapphire window.

Burden
   The thickness of rock between the drillhole and the free surface parallel to the hole, i.e., thickness of rock to be removed.

Button bits
   Percussive drill bits having tungsten carbide button inserts.

Channel effect
   The interaction between the detonating explosive charge in a drillhole and the shock wave in the air trapped in the channel between the charge and the drillhole wall, leading to compression, and sometimes dead-pressing, detonation failure of the explosive charge.

Charge weight
   Weight of the explosive charge in a drillhole.

Collar
   The mouth or open end of a drillhole.

Collaring deviation
   Distance between the intended and actual hole position at the collar.

Column charge
    Charge with reduced linear charge density in that part of the hole which is more than 1 burden length from the end of a tunnel or above the intended bottom of a bench.

Concussion
    That (inaudible) part of the air blast having a frequency content below 20 Hz.

Confined compressive strength
    Compressive strength of rock where lateral expansion is limited (by a confining wall or the surrounding rock itself).

Confined shear strength
    Shear strength of rock where lateral expansion is limited (by a confining wall or the surrounding rock itself).

Contour holes
    The drillholes along the perimeter (forming the contour) of the blast.

Covering
    Material such as rubber tire mats which is used to cover a blast to prevent or limit the throw of flyrock.

Cratering
    Blasting with a drillhole at right angle to the free surface, thus forming a crater.

Decoupled charges
    Charges that do not fill the drillhole, sometimes also used for charges with a low packing density.

Detonating cord
    Usually finely powdered PETN contained in a thread-wound soft textile cover overextruded with plastic. Strength 1–40 g PETN/m, most common 10 g/m.

Disc cutter
    In full face boring, discs, usually made of hardened, tough steel, sometimes with button bits around periphery. The discs are pressed against the rock face while being rolled along the rock face, breaking off hand-sized flakes of rock.

Drifting
    Europe usage, called tunnel excavation in the USA.

Drill bit
    The cutting part of a drill rod, most often having a tungsten carbide insert in the form of a vedge or multiple buttons which, when pressed against the rock by the percussive or rotating action of the drilling machine cuts loose fragments of the rock at the bottom of the drillhole.

Drillhole
    Hole drilled in rock for the purpose of being loaded with an explosive charge for rock blasting.

Drilling
> Mostly used for the process of drilling holes of diameter up to 0.5 m in rock for the purpose of blasting.

Drilling deviation
> Deviation of the drillhole from its intended position, direction, and path (see also collaring deviation, angular deviation, hole deviation).

Dynamite explosive
> Explosive based on nitroglycerin and/or nitroglycol. Derived from *"Dynamit"*, Alfred Nobel's trade name for his first plastic explosive, the guhr dynamite.

Electric detonator
> Detonator with a fine metal bridge wire, which is heated when an electric current is passed through it. The heated wire ignites burning in a layer of pyrotechnic or explosive, which in turn sets off the detonator's main charge.

Electronic detonator
> Detonator in which the time delay and other logic functions are provided by the microelectronic circuitry of a microchip, which feeds the main ignition current, usually from an internal capacitor.

Emulsion explosive
> Explosive in which a hot, concentrated solution of main oxidizer (for example, ammonium nitrate) in water is emulsified into the fuel components which contain an emulsifier, forming 1 to 10 μ sized droplets of the oxidizer/water solution which are separated from each other by thin films of the fuel/emulsifier.

Explosive weight
> See charge weight.

Fan cut
> In tunneling, opening at the center of the rock face formed by angling drillholes to form one or more fans.

Fixation factor (f)
> In bench blasting, describes the the ease with which the burden can be removed. The fixation factor f is a function of the hole inclination n by $f = 3/(3+n)$, where $n = 0$ is a vertical hole drilled from a horizontal free surface.

Flyrock
> Rock fragments thrown unpredictably from a blasting site by the force of the explosion.

Four Pillars (Major Elements) of the Science of Rock Blasting
> The rock mass. The mechanical processes of drilling and boring. The explosive material. The detonator and booster system.

Four-section cut
> In tunneling using a large empty central hole, opening in which four drillholes are drilled at equal distances from the center hole, then four more holes at equal, longer distances from the center hole, etc.

Fragmentation
 Degree to which the rock is fragmented by blasting.

Free surface
 A surface sufficiently close to the charge for the charge to break loose the rock in between.

Grade strength
 A measure of the strength of a straight sodium nitrate dynamite explosive originally based on its nitroglycerin content (obsolete usage).

Hole deviation
 Same as drilling deviation.

Hydrocode
 Hydrodynamic computing code. In this book, a computer program (or code) to calculate the motion of a material or the reaction products of detonation. (Originally, the term hydrocode referred to calculations of the motion of a material whose strength is negligible compared to the prevailing pressures and dynamic stresses. Present-day hydrocodes can deal effectively with materials in the pressure and stress range where the strength effects are not negligible.)

Ideal detonation
 Also called Chapman-Jouguet detonation. A detonation in which the chemical reaction goes to immediate completion forming reaction products that are in chemical equilibrium. Real detonations, even in charges of very large size approach, but never reach the ideal detonation state.

Insensitive munitions
 Munitions that fulfill the requirements of the U.S. Department of Defense test protocol 2105A, a test battery specifying the stimuli an insensitive munition effect should withstand without causing mass detonation.

Langefors-Kihlström weight strength
 Measure of explosive strength of Dynamite explosives derived from their calculated explosion energy and gas volume, based on comparative tests in bench blasting.

Lifter holes
 The holes in the bottom section of a tunnel round, where the explosive charges have to lift the rock they break loose.

Lifters
 See Lifter holes.

Linear charge concentration or Linear loading density
 Charge mass per unit length of loaded drillhole.

Loading density
 See linear charge concentration.

Lookout angle γ
In tunnel blasting the necessary angle between the contour holes and the axial direction of the tunnel, necessary because of the radial size of the drilling machine.

Maximum throw
The longest distance of throw of flyrock.

Multiple-row round
In bench blasting, a round with more than one row of holes parallel to the free surface.

Noise
By definition, the audible part of the air blast; it has frequencies from 20 Hz to 20,000 Hz.

Nonideal detonation
See ideal detonation.

One Man – One Machine
The (old) method of blasting where several hand-held drilling machines were used, operated by one man per machine.

P-wave
The (direct, radial) compressive wave.

Packing density
Loaded charge mass per unit drillhole volume.

Parallel hole cut
In tunneling, a cut with parallel drillholes, usually with a central, empty hole providing the free surfaces required for initial break-out.

Permissible explosive
Explosives (specially developed to have a low flame temperature that does not ignite certain specified methane/air or coal dust/air mixtures) that are permitted for use underground in a methane or coal dust atmosphere.

Plough cut
See plow cut. Plough is the usual British spelling.

Plow cut
In tunneling a pattern of drillholes at the center of the tunnel face where several rows of holes vertically above each other are drilled at an angle from the outside in, forming several V-shaped cuts, which are fired one after the other in sequence, creating an opening with free surfaces for the rest of the blast.

Practical burden
Burden measured in the plane of the top bench surface (differs, when inclined holes are used, from the real or *projected* burden which is measured perpendicularly between parallel holes or rows of holes).

Primer
>Explosive charge (initiated by detonating cord or a detonator) large enough and of sufficient detonation pressure to initiate the main charge (for example the blasting agent loaded into a drillhole).

Pumpable blasting agent
>Blasting agent capable of being safely pumped by a mechanical pump (for example into a drillhole).

R-wave
>The Rayleigh wave, a surface wave caused by the relaxation of the elastic stresses set up by the interaction of the shear wave with the ground free surface.

Rayleigh wave
>See R-wave.

Rock constant
>A measure of the mass of a standard explosive needed for fragmenting a unit burden of rock in a standard geometrical charge configuration. Constant in the fundamental equation giving the charge mass as a function of the burden.

Rock mass damage
>Damage to the remaining, not fragmented rock mass after the fragmented rock has been removed.

Round
>The blasting of a rock mass with multiple drillholes loaded with explosive charges, all set off in a predetermined time interval pattern (or multiple charges fired simultaneously).

S-wave
>Shear wave.

Safety fuse
>Black powder fuse with tarred textile cover.

Shock wave
>Compression wave propagating pressure well above the strength of the material and therefore having a very steep, almost instantaneous pressure rise, in which viscous effects and thermal conductivity lead to an increase in entropy. The shock wave compression therefore is an irreversible process.

Specific charge
>The total explosive mass in a blast divided with the total volume of rock fragmented.

Stemming
>Crushed rock, sand, or drill cuttings with which the top part of the drillhole is filled (see also stemming height).

Stemming height
    The hole length (closest to the collar) filled with crushed rock, sand, or drill cuttings to increase the rock fragmenting effect of the charge and decrease the risk for flyrock and air blast effects.

Stoping holes
    Holes in a blast (in tunneling or horizontal hole mining) which remove the burden either downwards or horizontally.

Subdrilling
    The length of drillhole that extends below the intended bottom of the bench.

Tensile strength
    Strength of a material in tension.

Toe
    The toe of the bench (in bench blasting) is the bottom part at the front of the bench.

Top charge
    Sometimes used for Column charge.

Toxic fumes
    Reaction product gases that may be toxic or contain toxic components.

Triaxial strength
    The strength of a material under a stress system in which all three principal stresses are different.

Tunnel excavation
    Usage in the US, see drifting.

Tunnel face
    The rock surface of a tunnel.

Unconfined compressive strength
    The strength of a material measured by compressing a specimen of the material without a confining casing.

V-cut
    See Plow cut.

Velocity Of Detonation
    Velocity of the detonation wavefront, VOD.

Vibration level
    Level of vibration velocity.

Weak plane
    Rock structure which introduces a weakness in the rock relative to the surrounding rock mass. Examples are cracks, joints and bedding planes.

Explosives Index

Page numbers for references within the text of the book are listed first (in the Times Roman font or, for the Foreword and Acknowledgments, in italic).

Table numbers are in boldface; e.g., **2.25** refers to Table 2.25 in Chapter 2.

Figure numbers are in italic, e.g., *4.15* refers to Figure 4.15 in Chapter 4.

Page numbers for the tables and figures are omitted since references are directly to the Table or Figure.

Table 3.2 lists properties of most single molecular explosive substances. Table 3.3 gives properties of the most common military explosives. Tables 4.14 through 4.18 list properties of a selection of commercial explosives and blasting agents.

2-ethylhexylacrylate, **3.3**.
304 (explosive), **4.12**.
Al + AN, 455
Al/AN, 485
AL/HTPB/AP, *3.6*
Aluminized ANFO, 118
Aluminized TNT slurry, 196, 358
Aluminized TNT-sensitized watergels, 431
Aluminum, **4.5**.
Aluminum oxide, **4.4**.
Amatol, **4.12**.
Ammonia dynamite, **9.2**.
Ammonium nitrate-based emulsion, 136, *4.25*.
Ammonium nitrate, AN, 59, 66, 73, 75, 76, 79, 86, 114, 128, 391, 454, 455, 456, 460, 461, 469, 473, 481, 484, 485, 514, **3.2**., **4.14**., **4.4**., **4.5**., **4.7**., **4.14**., *3.2*., *3.6*
Ammonium nitrate, prilled, see also AN prills, 55, 86
Ammonium nitrate-based emulsion, *4.23*., *4.24*.
Ammonium perchlorate, see AP
Ammonium picrate, **3.2**.
AN prills, 55, 81, 82, 86, 89, 431, *4.1*.
AN prills mixed with diesel oil, 457
AN, see Ammonium nitrate
AN/TNT/Al, 73

AN/water-in-oil emulsion, 85, 134
ANFO, 2, 55, 56, 63, 71, 72, 76, 79, 81, 82, 85, 86, 88, 89, 105, 118, 128, 129, 130, 131, 133, 134, 162, 178, 179, 196, 201, 212, 221, 225, 268, 281, 301, 358, 389, 395, 455, 456, 457, 460, 461, 462, 463, 473, 474, 475, 484, 486, **4.1**., **4.2**., **4.6**., **4.8**., **4.10**., **4.11**., **4.13**., **4.17**., **6.1**., **11.2**., **15.2**., *3.1*., *3.6*, *3.7*., *3.8*., *3.9*., *4.13*., *4.14*., *5.29*.
ANFO with and without *Al*, *4.14*.
ANFO/Al 90/10, **4.10**., **4.13**.
ANFO/Bulk emulsion, 474
ANFO/emulsion mixes, *3.7*.
ANFO/emulsion/styropore, **4.14**.
ANFO, heavy, **4.15**.
ANFO, prilled, **11.1**.
ANFO prills, *4.1*.
ANFO-K2Z, *5.28*.
ANFO/polystyrene, **9.2**.
AP, Ammonium perchlorate, 59, 67, 68, 73, 468, 469, **3.2**., **4.5**., *3.2*., *3.6*
Aquanal, 86
Aquanite, 86
Aquaram, 86
Arene, 72
BA, a highly aluminized slurry, *4.13*.
Balance substance and fuels, **4.5**.
Ballistite, 67

Barium nitrate, **4.5.**
Beeswax, 72
Black powder, 143, 144, 148, 430, 468, 469, *3.6, 5.1.*
Blastex 100, **4.15.**
BNT, **4.5.**
Bofors NSP-74, 280, 281
BP 474, **4.5.**
Bulk emulsion, 457, 463, 474, 486
Bulk emulsion/ANFO blend, **4.14.**
C-1, see Composition C-1
C-2, see Composition C-2
C-3, see Composition C-3
C-4, see Composition C-4
Calcium nitrate, **4.5.**
Calcium silicide, **4.5.**
Cartridged dynamite, 463
Cartridged emulsion, 463
Cast pentolite primer, *5.30.*
CBAN, *3.6*
Cellulose, **4.14.**
Chloroethylphosphate, **3.3.**
Class 1, 439
Class A Explosives, 76
Comp. A, see Composition A
Comp. B, see Composition B
Comp. C-1, see Composition C-1
Comp. C-2, see Composition C-2
Comp. C-3, see Composition C-3
Comp. C-4, see Composition C-4
Composition A, 72
Composition B, 72, 73, 284, **3.3.**, **4.14.**, **16.1.**, *3.6, 4.20.*
Composition C, 72
Composition C-1, 72
Composition C-2, 72
Composition C-3, 72, 415, **16.1.**
Composition C-4, 72, **3.3.**, *19.2.*
Cordite, 67
Crushed prills ANFO, 132
Crystalline ANFO, 82
Cupric azide, **3.2.**
Cyclotetramethylenetetranitramine, see HMX
Cyclotol, 73, **3.3.**, *3.6*
Cyclotrimethylenetrinitramine, see RDX
DEAN, *3.6*
DEN, *3.6*
Diaminodintophenol, *3.6*

Diaminotrinitrobenzene (DATB), **3.2.**, **4.4.**
Diazodinitrophenol (DDNP), **3.2.**
Diesel oil (see also Dodecane, FO), **4.5.**
Diethylene glycol dinitrate, **4.5.**
DINGU, 73
Dinitromethane, **3.2.**
Dinitrotoluene (DNT), **3.2.**
Dioctyl maleate, **3.3.**
DiTeU, **4.5.**
Ditrinitroethylurea (DiTeU), **3.2.**
Dodecan (Swedish) or Dodecane (English), FO (fuel oil), 114, 456, q461, **4.4.**, **4.5.**
DuPont's sheet explosive, 416
DxB-dynamite, 306
Dynamex, 383
Dynamex B, **12.1.**
Dynamex M, 284, **4.14.**, **6.1.**
Dynamit, **4.16.**
Dynamit, guhr dynamite, 514
Dynamite, 75, 358, 387, 474, 486, *3.6*
Dynamite I, **4.6.**
Dynamite II, **4.6.**
Dynamite, low velocity, high velocity, **4.1.**
Dynex 205, **4.16.**, **6.1.**
Dynex 300, **4.16.**
EGDN, Ethylene glycol dinitrate, Nitroglycol, 75, 387, 393, 475, **3.2.**, **4.4.**, **4.5.**, **4.8.**, **4.10.**, **4.14.**, *3.6*
EL506D (DuPont Sheet Explosive), **16.1.**
EL506L (DuPont Sheet Explosive), **16.1.**
Emulan, *3.13.*
Emulan 7000, **4.14.**
Emulet 20, **4.14.**, **6.1.**, **9.2.**
Emulet 50, **4.14.**, **9.2.**
Emulet 70, **4.14.**
Emulex, 486
Emuline, **9.2.**
Emulite, 314, **11.2.**, *3.13.*
Emulite 100 Gurit, **4.14.**, **9.2.**
Emulite 150, 284, **4.14.**, **6.1.**
Emulite 1200, **4.14.**
Emulite VCR, **11.1.**
Emulsion, 88, 456, 473, 474, 486, **3.1.**, **4.1.**, **9.2.**, **4.15.**, *4.13.*
Emulsion/ANFO/filler, 212
Emulsion/ANFO/polystyrene, **9.2.**

Emulsion explosive, **4.14.**
Emulsion matrix, 439
Emulsions, 212, *3.6*
Energan 2640, **4.17.**
Estane, **3.3.**
Ethylene glycol, **4.5.**
Ethylene glycol dinitrate, see EGDN
Ethylenediaminedinitrate (EDD), **3.2.**
Ethylenedinitramine (EDNA), **3.2.**
Ethylenedinitramine (EDNA), **3.2.**
Ethyleneglycoldinitrate (EGDN), see EGDN
Extra dynamite, **4.15.**
Extra gelatin, 274
Extra gelatin dynamite, **4.15.**
F-pipe charge, **9.2.**
FEFO (bis (2-fluro-2,2-dinitroethyl) formal), 73
FO + AN, 455
Forcite, **4.11.**
Fuel oil, 114
G. D., **4.12.**
GAP, glycidyl azide polymer, 73
G-Borenit, **4.12.**
Gelaprime F, **4.15.**
Gelatin dynamite, **4.15.**
Giant Gelatin, **4.11.**
Glycerine, **4.5.**
Glyceroltrinitrate, **3.2.**
Guanylnitrosoaminoguanyltetrazene, **3.2.**
Guargum (Jaguar 100), **4.5.**
Guhr dynamite, 514, **4.8.**
Gun-cotton, 67
Gurit, 268, 284, **9.2.**, *7.17.*, *9.12.*, *9.15.*
Gurit A, **4.14.**, **6.1.**
Gurit B, **4.14.**
HA, Red, **4.15.**
HAN, *3.6*
HAN/DEN/H20, *3.6*
Heavy ANFO, 134, *3.7.*, *4.22.*
Hercudet system, 168
Hexanitrohexaazaisowurzitan (HNIW), **3.2.**
Hexanitrobenzene, Hexyl (HNB) $C_6N_6O_{12}$, **3.2.**
Hexanitrostilbene (HNS), **3.2.**
Hexogen, see RDX
Hexyl, see Hexanitrobenzene
HMX, High Melting Explosive, see also Cyclotetramethylenetetranitramine,
 68, 72, 73, 74, 101, 182, 499, **3.2.**, **3.3.**, **4.4.**, **4.9.**, **16.1.**, *3.6*
HMX/TNT, 73
HNIW, 73
HNS, 74, **3.3.**
Hydrazine, **4.5.**, *3.6*
Hydrazine hydrate, **4.5.**
Hydrazine nitrate (HN), **3.2.**, **4.5.**
Hydrazoic acid, **3.2.**
Imatrex, **4.12.**
Irecoal E-1, **4.15.**
Irecogel B, **4.15.**
Iredyne 365, **4.15.**
Irefo 403, **4.15.**
Iregel 1.116, **4.15.**
Iregel 1135E, **4.15.**
Iregel 1135P, **4.15.**
Iregel RX Plus, **4.15.**
Iregel RX, **4.15.**, **6.1.**
Iremex 560, **4.15.**
Iremite 62, **4.15.**, **6.1.**
Iremite Presplit, **9.2.**
Iremite TX, **4.15.**
Ireseis, **4.15.**
IRESPLIT D, **9.2.**
Isanol 25/75, **9.2.**
Isopropyl nitrate (IPN), **3.2.**
Isopropyl nitrate, **4.5.**
Kel-F, **3.3.**
Kimit 80, **4.18.**, **6.1.**
Kimulux 42, *8.2.*
Kimulux 82, **4.18.**
Kimulux R, **4.18.**
KP-primer, **4.18.**
Larvikit, **9.2.**
Lead azide, 67, 144, 146, 147, 148, 149, 150, 173, **3.2.**, **4.7.**, *3.6*
Lead styphnate, 144, 147, **3.2.**, *3.6*
LFB, **4.2.**, **4.12.**
LFB Dynamite, 128, 129, 195, 196, **4.6.**, **6.1.**
LFIV, **4.12.**
LH2/LOX, *3.6*
Light slurry, **4.6.**
LOX/FO, *3.6*
Mannitolhexanitrate (MHN), **3.2.**
Mannitolhexanitrate, **4.5.**
Matrix, 431
Mercury fulminate, 144, 145, 147, **3.2.**, *3.6*

Methane, **4.4.**
Methylnitrate, **3.2.**
Microballoon sensitized explosives, 300
Mixed emulsion/ANFO, *3.13.*
MMAN, 75, **3.1. 3.2., 4.5.**
MNT, **4.5.**
Monomethylamine nitrate, see MMAN
Mononitrotoluene (MNT), **3.2.**, *3.6*
n-vynyl-2-pyrolidone, **3.3.**
Na 01, **4.12.**
Na 12, **4.12.**
Nabit, 284, **4.12., 6.1.,** *7.17.*
Nabit 2, **4.12.**
Nabit A, **4.14.**
Neat, liquid nitroglycerin, 143
NC, *3.6*
NC/NC, *3.6*
NC/NG/RDX, *3.6*
NC/NG/RDX/AL/AP, *3.6*
NG, see Nitroglycerin
NG Ammonia class, **4.15.**
NG/EGDN, 75, **4.5.**
NG/nitroglycol, **4.14.**
NG-powder, **9.2.**
Nitrate esters, 393
Nitrobenzene (NB), **4.5.**
Nitrocellulose (NC), 75, **3.2., 3.3., 4.5.**
Nitrocellulose gunpowder, 67
Nitrocellulose in nitroglycerin, 68
Nitroform, **3.2.**
Nitroglycerin (NG), 66, 67, 68, 89, 105, 143, 182, 196, 387, 393, 430, 431, 475, **3.2., 4.2., 4.4., 4.5., 4.6., 4.7., 4.8., 4.14., 6.1.,** *3.5., 3.6, 5.1.*
Nitroglycerin, neat, liquid, 143
Nitroglycol (EGDN), see EGDN
Nitroguanidine (NQ), **3.2.**
Nitrolit, **4.12.**
Nitromannit, **3.2.**
Nitromethane (NM), 66, 455, 456, 457, 458, 460, 473, 481, **3.2., 4.4., 4.5., 4.6., 16.1.,** *3.6, 4.13.*
Nitrostarch (Penta), **3.2.**
Nitrotriazolone, (NTO), **3.2.**
NM, see Nitromethane
NONEL, 145, 148, 149, 151, 161, 162, 163, 165, 168, 169, 181, 212, 282, 462, *5.2., 5.16., 5.30.*
NONEL blasting cap, 162
NONEL GT, **5.4.**
NONEL LP, 148, 163, **5.4.**
NONEL LP delay numbers, delay times, intervals, and scatter, **5.5.**
NONEL LP detonator delay times, table of, **5.6.**
NONEL MS, 148, 163, **5.4.**, *5.15.*
NONEL MS detonator delay times, table of, **5.6.**
NONEL tube, 149
NONEL UNIDET, *vii*, 148, 161, 165, *5.18.*
NONEL UNIDET detonator delay times, table of, **5.6.**
NPED, non-primary explosive detonators, 145
NTO, 73
Octanitrocubane, 74
Octogen (HMX), **4.5.**
Octogen, 72, **3.2.**
Octol 75/25, 73
Octol, 73, **3.3.**
Overdriven detonation, 179
Oxidizers, **4.5.**
Oxygen balanced explosives, **4.5.**
Paraffin, **4.5.**
PBX-9404, 73, **3.3., 16.1.**
PBX-9501, 73, **3.3.**
PBX-9502, 73, **3.3.**
PBXN-107, 73, **3.3.**
PBXW-115, 497
Penta, see also Nitrostarch, **3.2.**
Pentaerythritoltetranitrate, see PETN
Pentolite, 180, *3.6*
Permissible explosives, **4.15.**
PETN, Pentaerythritoltetranitrate, 59, 73, 74, 89, 144, 146, 148, 160, 162, 180, 182, 234, 259, 280, 320, 513, **3.2., 4.4., 4.5., 4.6., 4.7., 4.8., 4.9., 6.1., 16.1.,** *8.1.*
PETN/TNT, 73
Picric acid, 72, **3.2., 3.3.**
Picrylaminodinitropyridine, see PYX
Plastic dynamites, 196
Polyisobutylene, **3.3.**
Potassium chlorate, **4.5.**
Potassium nitrate, **4.5.**
Potassium perchlorate, **4.5.**
Power split, **9.2.**
Powergel 2131, **4.17., 6.1.**
Powergel 2540, **4.17.**

Powergel 2880, **4.17.**
Powergel 3151, **4.17.**
Powermex (water gel explosive), **3.1.**
Premix, **3.1.**
Prilled AN, 81, *5.28.*
Prilled ANFO, 82
Prillit A, **4.14.**
Primacord, 274, *9.8.*
Profile charge, **9.2.**
Pure emulsion blasting agents, 196
PYX, Picrylaminodinitropyridine, 74, **3.2.**
RDX (Hexogen), Research Department Explosive, Cyclotrimethylenetrinitramine, 72, 73, 74, 146, **3.2.**, **3.3.**, **4.4.**, **4.5.**, **4.6.**, **4.9.**, **4.14.**, **16.1.**
RDX/CBAN/AL/AP, *3.6*
RDX/HTPB/AL/AP, *3.6*
RDX/polyisobutylene/wax, 72
RDX/TNT, 72, 73, 144
RDX/TNT/wax, 72, 73
Red HA, **4.15.**
Red-E-split A, **9.2.**
Red-E-split B, **9.2.**
Red-E-split C, **9.2.**
Red-E-split D, **9.2.**
Reolit K, **11.1.**
Reolit, 314, **11.2.**
Securit, **4.12.**
Semi-gelatin dynamite, **4.15.**, **9.2.**
Silver azide, 147, **3.2.**, *3.6*
Slurries, see also slurry, *3.6*
Slurrit 110, **4.16.**
Slurrit 40, **4.16.**
Slurry (slurries), **3.1.**, **4.11.**, *3.6*
Slurry 1, **4.13.**
Slurry 2, **4.13.**
Slurry 3, **4.13.**
Slurry 4, **4.13.**
Smoothex, **9.2.**
SN, see sodium nitrate
Sodium chlorate, **4.5.**
Sodium nitrate (SN), 75, 86, **4.5.**
Span 80 (sorbitan mono-oleate), 85, 86
Starch (cellulose), **4.5.**
TACOT, **16.1.**

TATB, Triaminotrinitrobenzene, 59, 73, **3.2.**, **3.3.**, *3.6*
Teepol, **4.5.**
Tetranitromethane (TNM), **3.2.**, **4.4.**
Tetrazene, **3.2.**
Tetryl, 144, 146, **3.2.**, **4.5.**, **16.1.**
Thermalite connectors, 178
Thermite, 56, 128, *3.6*
Titan G booster, **4.15.**
TMETN, *3.6*
TMETN/HMX/AL/AP, *3.6*
TNM, **4.4.**
TNT, trinitrotoluene, 56, 59, 66, 72, 73, 75, 101, 104, 105, 180, 375, 376, 377, 383, 421, 431, 456, 464, 468, 473, **3.2.**, **4.2.**, **4.4.**, **4.5.**, **4.6.**, **4.7.**, **4.8.**, **4.9.**, **4.10.**, **4.14.**, **6.1.**, **16.1.**, *3.4.*, *3.6*, *14.1.*, *14.3.*
TNT slurry (slurries), 314, 319, 431
TNT-Al slurry, **4.6.**
TNT/RDX, 73
TNTAB, Triazidotrinitrobenzene, **3.2.**
Toluen, 316
Torpex, 73
Tovex, **3.1.**
Triaminotrinitrobenzene, see TATB
Triazidotrinitrobenzene, see TNTAB
Trimonite No. 1, **16.1.**
Trinitroazetidine (TNAZ), **3.2.**
Trinitrobenzene (TNB), **3.2.**
Trinitromethane, **3.2.**
Trinitrophenol, **3.2.**
Trinitrophenylmethylnitramine, **3.2.**
Trinitrotoluene, see TNT
Tritonal, 73
Tritonal (80/20), **16.1.**
Trotyl, **3.2.**
Unigel, **4.15.**
Urea, **4.5.**
Water/ethylene glycol solution desensitizer, 74
Waterbased ammonium nitrate emulsions, 71
Watergels, *3.6*
Watergel A, 201
Watergel/TNT slurry, 101
Woodmeal (cork), **4.5.**
Xylen, 316

Name Index

Page numbers for references within the text of the book are listed first (in the Times Roman font or, for the Foreword and Acknowledgments, in italic).

Table numbers are in boldface; e.g., **2.25** refers to Table 2.25 in Chapter 2.

Figure numbers are in italic, e.g., *4.15* refers to Figure 4.15 in Chapter 4.

Page numbers for the tables and figures are omitted since references are directly to the Table or Figure.

Abel, F. A. (Frederick), 67, 88, 102, 491
ACGIM, 393
AECI, 161, 178, **9.2.**
Aikov, K., 502
Akai, K., 491, *1.10.*
Al'tshuler, L. V., 107, 491, *4.10.*
Alma Ata, Kazakhstan, 317
Almgren, G. (Gunnar), *viii*, 304, 491, 494, 495, 498
Ambraseys, N. N., 491
American Conference of Governmental Industrial Hygienists (ACGIH), 491
Amherst, OH, USA, **4.1.**
Anderson, D. A., 174, 491, 504
Andersson, K., 393, 495
Andrews, A. B., 171, 491
Arbetarskyddsstyrelsen (Worker Safety Administration), Sweden, 393, 425, 491
Arnberg, P. W., 344, 491
Asay, B. W., 148, 499
Ashley, S., 421, 491
Atlas Copco AB, Sweden, *vii*, 288, 496, *2.25.*
Atlas Copco MCT, 283, *2.6.*
Atlas Copco, *2.1., 2.2., 2.8., 7.4., 7.6.*
Atlas Powder Company, *vii*, 491, *5.9.,* **9.2.**
Austin, **9.2.**
Axelson, O., 393, 495
Bachmann, W. E., 72, 491
BAI Inc., USA, 288*
Bakanova, A. A., 491

Baker, W. E., 384, 491, *14.6.*
Barker, D. B., 258, 491
Bauer, A., 491, **4.11.**
Bauer, P. A., 88, 491
Bdzil, J. B. (John), 89, 135, 148, 491, 492, 498, 499
Beard, J. A., 421, 492
Beattie, T. A., 127, 171, 492, 495
Belidor, B. F., 185, 492
Benedik, R., 491
Béranger, R., 492
Bergengren, E. (Erik), 67, 492
Bergmann, M., 503
Bergmann, O. R., 171, 418, 492, 493
Bergqvist, I. (Ingvar), *vii*, 501
Berning, W. W., 492
Berthelot, M., 88, 96, 492
Bertholf, L. D., 397, 497
Bhandari, S., 258, 492
Biasutti, G.S., 67, 492
Bickes, R. (Robert), 158
Birkhoff, G., 404, 492
Bjarnholt, G. (Gert), *vii*, 127, 130, 180, 276, 282, 492, 496, *4.20., 9.13., 9.15.*
Blindheim, O. T., 33, 492, 503
Bluhm, H. F., 86, 492
Bodås test mine, *7.7.*
Boddorff, R. D., 495
Bohuslän, **1.1.**
Boliden AB, *vii*
Boliden Aitik mine, *3.10.*
Boliden Mineral AB, 301
Bollinger, G. A., 337, 492

Borg, D. G., 171, 493
Borghamn, **1.1.**
Brännfors, S. (Sten), *viii*, 493
Brawner, C. O., 496
Bredseleforsen, **1.1.**
Brekke, T. (Tor), *viii*
Brinkley, S. R., 89, 493, **4.4.**
Brower, K. (Kay), *viii*, 454, 512
Brunnberg, B. (Bernt), *vii*
Buckley, P. (Pat), *viii*
Carmichael, R. S., 495
Cechanski, M. (Michael), *vii*
Center for Explosives Technology Research, New Mexico Institute of Mining and Technology, *v, vii*
Chapman, D. L., 88, 493
Chevkin, A. F., 256, 502
Chiappetta, R. F., 171, 493
Chile, 290
China, 72
Clark, D., 174, 493
Clark, G. B., 503
Clay, R. B., 86, 493
Coeur d'Alene mining district, USA, 178
Cole, R. H., 127, 493
Contestabile, E., 493
Cook, M. A., 89, 102, 121, 493
Cook, N. (Neville), *viii*
Cooper, J., 134, 499
Cooper, T., 20, 493, *1.16.*
Courant R., 493, *4.2.*
Cowan, G. R., 418, 493
Cowperthwaite, M., 103, 105, 493
Craig, B. G. (Bobby), *viii*
Craig, T. R., 493
CRC Press, *Handbook of Chemistry and Physics*, 493, **4.4.**
Crocker, C. S., 304, 493
Crouch, S., 496
Cundall, P. A., 20, 493, *1.16.*
CXA Ltd., *5.24.*
D'Andrea, D. V., 494
Dabora, E. K., 88, 491
Dally, J. W., 258, 368, 493, 502
Davis, W. C. (Bill, William C.), *vi, viii*, 89, 410, 411, 494
Detonics Laboratory, Nitro Nobel AB, Vinterviken, Stockholm, Sweden, *vii*, 182, 415
Dick, R.D., 496

Dixon, H., 494, 503
Döring, W., 88, 494
Douglas, J., 418, 493
Dowding, C. H., 174, 337, 494, 503
Dremin, A. N. (Anatolii Nikolajevich), 494
DuPont, (E. I. DuPont De Nemours & Co.), 161, 416, 418, 421, 493, 494, **5.2.**
Duvall, W. I., 500
DWL, **9.2.**
Dyno Explosives Group, Norway, 288, **4.16.**
DYNO Industrier, *vii*
DYNO international group of explosives companies, *v*, 429
DYNO Norway, **9.2.**
Eagle Laboratory, RCEM, New Mexico Tech, Socorro, NM, USA, *viii*
Edl, J. W., 492
Edwards, A. T., 174, 494
Egly, R. S., 85, 494
Ekeroth, R., 369, 494
Ekman, G., 496
Elitz, E., 388, 494
Engsbraaten, B. (Björn), *vii*
Ensign-Bickford, USA, 161, 165, *5.17.*
Eriksson, B., 377, 494, *14.3.*
Eriksson, S., 494
Ershov, A. P., 499
Europe, 72, 290
Explosives Fabricators, Inc., *17.5.*
Fabian Mine, Sweden, 162
Fairhurst, C., *vi*, 496
Falkdal, J. E., 301, 494
Federenko, P. I., 260, 494
Fedorott, B. T., 494
Feodoroff, P. I., 122, **4.7.**
Fickett, W., 89, 492, 494
Field, J. E., 235, 237, 274, 278, 494
Finger, M. (Milton), *viii*
Finnie, Iain, *viii*
Fischer, R. L., 494
Fogelson, D.E., 500
Forcit, **9.2.**
Fourney, W. L., 258, 491, 493, 496
Fraenkel, K. H., 197, 494
Friedrichs, K. O., 493, *4.2.*
From, T., 258, 494
Fuerstenau, D. (Douglas), *viii*

Fukui, H., 502
Fumanti, R. R., 503
Furth, W., 504
Gautojaure, **1.1.**
Gehrig, N. F., 85, 495
Gencorp Aerojet, *16.7.*, *16.8.*
Germany, 72, 393, **13.2.**
Gibbs, T. R., 495
Goodier, A., 495
Goodman, R. (Richard), *viii*
Gordon, W. E., 89, 493, **4.4.**
Gothenburg, Sweden, 202, 369
Gotland, **1.1.**
Graham, R. A., 420, 495
Granboforsen, **1.1.**
Grangesberg, **1.1.**
Granlund, L. (Lars), *vii*
Granström, S. A., 495, *14.1.*, *14.2.*
Grant, I., 174
Grant, J. R. (John), *viii*, 173, 492, 495
Great Britain, 72
Grebe, H. A. (Andrew), *viii*, 497
Greiner, N. R., 421, 422, 495
Gubareva, N. V., 503
Gurney, R. W. (Ronald), 397, 398, 495
Gustafsson, R. (Rune), *vii*, 495
Gyldén, N. (Nils), 401, 499
Hakunge (Nordkross AB Hakunge quarry), *9.16*
Hall, J. R., 502
Hansson, I., 165, 495
Harries, G., 127, 491, 495
Hartman, K. (Keith), 158
Hatheway, A. W., 495, **4.1**
Heinemeyer, F., 495
Henderson, Nevada, USA, 66
Hendrickson, A. D., 494
Hendron, A. J., Jr., 491
Hermes, R., 422
Hinzen, K.-G., 174, 495
Hjelmstad, K.E., 500
Höglund, P., 316, 501
Hogstedt, C. H., 393, 495
Holdo, J. (Jan), 496, *7.5.*
Holloway, D. C., 282, 491, 492, 496
Holmberg, R. (Roger), *i*, *v*, *vi*, 105, 127, 250, 258, 259, 276, 281, 304, 321, 330, 337, 354, 358, 383, 492, 493, 496, 497, 498, 499, 500, 501, 502, 504
Holston Defense Corporation, 72

Holtzman A. H., 418, 493
Hong Kong, 290
Hood, M. (Michael), *viii*
Hopkinson, B., 377, 497
Hopler, R. B., 497, *5.21.*
Hoskins, E. R., 497, *1.8.*
Hunter, D. (Dennis), *viii*
Hustrulid, W., 496
Ichikawa, K., 502
ICI Australia Ltd, Australia, **4.17.**
ICI Explosives, *vii*
ICI Explosives, Scotland, 288
ICI Nobel's Explosives, Ltd, 171
IME, see Institute of Makers of Explosives
India, 290, 508
Inspector of Explosives and Flammable Substances, Sweden, 450
Inspectorate of Explosives and Flammable Substances, Sweden, 432
Institute of Makers of Explosives (IME), 388, 497, *5.8.*
Institute of Mining and Metallurgy, London, 492
Institutt for Anleggsdrift, Geologisk Institutt, 502
Iqbal, Z., 497
IRECO Incorporated, USA, *vii*, **4.15.**, **9.2.**, *3.9.*, *3.11.*
Italy, 72
IVA, 497, *13.3.*
James, E., Jr., 421, 503
JANAF Thermochemical Tables, **4.4.**
Japan, 169
Jarlenfors, J., 259, 497
Jerberyd, L., 497
Jidestig, G. (Göran), *vii*
JKMRC, Australia, 288
Johansson, B. (Berthold), *viii*
Johansson, C. H. (Carl Hugo), *vii*, 89, 497, 501
Johnson, C. F., 500
Johnson, J. D., 495
Jones, D. A., 134, 497
Jones, G. E., 397, 497
Jones, H., 103
Jones, M. J., 497
Joshi, V. S. (Vasant), *viii*, 422, 497
Jouguet, E., 88, 497

Kallin, A. (Åke), *viii*, *2.25.*
Karnelo, E., 497
Kast, 122
Kazakhstan, 317
Keisu, M., 174, 500
Kennametal Corp., USA, 282
Kennedy, D. L., 134, 135, 497
Kennedy, J. E. (Jim), *viii*, 397, 497, **16.1.**
Kihlström, B. (Björn), *vii*, 132, 171, 184, 185, 190, 191, 213, 221, 237, 249, 251, 288, 290, 319, 331, 348, 361, 462, 498, 501, 515, *7.13.*, **13.1.**
Kimit AB, Sweden, **4.18.**
Kirby, I. J., 134, 497
Kirkwood, J. G., 134, 135, 504
Kirsch, G. A., 495, **4.1**
Kirunavaara, 301, *11.4.*
Klippen, Sweden, *2.8.*
Knasmillner, R., 98, 502
Kolsky, H., 97, 498
König, R., 172, 174, 498
Konya, C. J., 491, 493, 497, 500
Kormer, S. B., 491
Kovtun, I. N., 260, 494
Krauland, N. (Norbert), *viii*, 301, 498, *11.3.*
Kristineberg mine, 301
Labor Safety Board, Sweden, 393
Ladegaard-Pedersen, A. (Anders), *vii*, 235, 237, 252, 274, 278, 320, 323, 494, 498, 499, 501, **8.4.**
Laisvall, **1.1.**
Lande, G., 367, 371, 493, 498
Landvetter Airport, 202
Lang, G., *11.7.*
Lang, L. C., 302, 304, 305, 308, 498
Lang, R., 491
Langefors, U. (Ulf), *vii*, 128, 132, 171, 182, 184, 185, 190, 191, 213, 221, 249, 251, 252, 288, 290, 319, 331, 348, 361, 462, 498, 515, *7.13.*, **13.1.**
Larsson, B. (Bernt), *vii*, 249, 493, 498, 503
Larsson, G. (Gösta), 414, *17.1.*
Lavrentiev, V., 317, 498
Lawrence Livermore National Laboratory, CA, USA, 126, 505
Lawson, J. T., 175, 504
Le Chatelier, H., 88, 499
Lee, E. (Ed), *viii*, 103

Lee, J. (Jaimin), *i*, *v*, *vi*, 135, 136, 498, *4.23.*, *4.24.*
Leiper, G. A., 134, 497, 499
Libersky, L. (Larry), *viii*
Lindgren, L. (Lars), *2.3.*
Linehan, P., 382, 504
Lithner, G. (Gösta), *vii*
Lithonia, GA, **4.1.**
Livingston, C. W., 305, 307, 313, 331, 499
Ljungberg, S. (Sten), *vii*, *4.1.*
LKAB, *vii*, 496, *7.1.*
LKAB iron ore mine, Malmberget, Sweden, 209, 262, 389
LKAB Svappavaara open pit mine, 319
LKAB (underground) mine, Kirunavaara, Sweden, 301
Los Alamos National Laboratory, NM, USA, *viii*, 505
Los Angeles Subway, CA, USA, *2.7.*
Ludvig, B., 276, 500, *2.10.*
Lundberg, B. (Bengt), *viii*
Lundberg, L. (Lennart), 401, 499
Lundborg, N. (Nils), *vii*, 8, 9, 11, 13, 14, 288, 324, 329, 451, 499, 501, *1.7.*
Luossavaara, Sweden, 165, 314
Lyamkin, A. I., 421, 499, 503
Lüdeling, R., 495
MacDougall, D., 492
Mader, C. L. (Charles), *viii*, 89, 102, 105, 499
Mäki, K. (Kenneth), *vii*, 272, 311, 492, 496, 497, 499, 502
Mallard, E., 88, 499
Malmberget, Sweden, 209, 262, 389, **1.1.**
Manson, N., 88, 491
Marklund, I. (Ingemar), *viii*
McAfee, J. M., 148, 499
McCampbell, C. B., 158
McQueen, R. G., 502
Medearis, K., 174, 499
Meiser, F. J., 344, 502
Meyer, R., 122, 499, **4.8.**
Meyers, M. A., 420, 500
Miller, R. E., 304, 500
Mitchell, S. T., 304, 500
Mohr-Coulomb, 14, *1.7.*
Molin, V. (Verner), 414, 415, 416, 500, *17.2.*

Monahan, C. J., 304, 500
Mori, H., 491, *1.10.*
Mufilira Mine, Zaire, 85
Munroe, C. E. (Charles), 276, 415, 500
Murr, L. E., 420, 500
Naarttijärvi, T., 259, 276, 496, 497, 500
Nakano, M., 169, 500
National Swedish Road and Traffic Research Institute, 491
Natural History Museum, Gothenburg, Sweden, 369
Natural Science Museum, Albuquerque, New Mexico, USA, 416
Naturhistoriska Museet, Gothenburg, Sweden, 369
Naval Torpedo Station in Newport, RI., USA, 276
Neckar, A. E., 85, 494
New Mexico Tech (New Mexico Institute of Mining and Technology, Socorro, New Mexico, USA), *v, vii, viii,* 422
Nicholls, H. R., 360, 500
Niklasson, B. (Bengt), *vii,* 174, 304, 314, 498, 500, *11.13., 11.14., 11.15.*
Nilsson, L., 496
NISSAN, *5.22.*
Nitro Consult AB, 288
Nitro Nobel AB, Sweden, *v, vii,* 82, 148, 175, 176, 182, 288, 314, 389, 429, 431, 432, 439, **4.14., 9.2.,** *3.8., 3.12., 3.13., 4.1., 7.8.*
Nitroglycerin AB, 182
Nobel, A. (Alfred), 67, 88, 143, 145, 182, 415, 431, 514, *5.1.*
Noka Software Systems, Canada, 288
Nord, G. (Gunnar), *viii*
Nordkross AB Hakunge quarry, 282
Norell, B., 173, 500
Norén, C. H., 237, 248, 500, *8.12.*
Northwood, T. D., 174, 494
Norway, 349
Norway Institute of Technology (NTH), Trondheim, Norway, 33
Office of Safety and Health Administration (OSHA), 425
Office of Scientific Research and Development, 504
Office of Surface Mining (OSM), USA, 425
Olofsson, S. (Stig), *vii*
Olson, D. B. (Douglas), *viii,* 500, *3.2.*

Olson, J. J., 500, 503
Olsson, K., 500
Olsson, M. (Mats), 262, 501
Oppenheim, A.K. (Antoni), 88, 501
Öqvist, J., 276, 500
Osoko, M. N., 498
Ouchterlony, F. (Finn), *vii,* 276, 278, 492, 501
Oxley, J. (Jimmie), *viii*
Parmelee, R. A., 344, 504
Pershin, S. V., 501
Persson, A. (Algot), *vii,* 237, 320, 323, 492, 497, 498, 499, 501, *5.28.*
Persson, G. (Gunnar), *vii,* 316, 321, 330, 383, 389, 393, 496, 501, **15.3.**, *15.3., 15.4., 15.5.*
Persson, I. (Ingemar), *vii,* 418, 502, *4.10.*
Persson, P.-A. (Per-Anders), *i, v, vi,* 89, 135, 181, 182, 237, 252, 272, 497, 498, 499, 501, 502, 503, **8.4., 8.8.,** *4.10.*
Petrini, M. (Monica), 182
Petrov, E. A., 499
Petschek, A. (Albert), *viii*
Phillips, D. S., 495
Pickstown, S.D., USA, **4.1.**
Plaster of Paris, 416
Poncelet, E. F., 234, 502
Popolato, A., 495
Prader, D. (Duri), *viii,* 43
Precision Blasting Services, USA, 288
Prezios, L., 491
Pugh, E., 492
Reiher, H., 344, 502
Reinhardt, H. W., 368, 502
Research Mine of Luossavaara, Sweden, 165
Reynold Industries Inc., *5.12.*
Rice, M. H., 502, *4.10., 4.11.*
Richart, F. E., 337, 502
Riemann, B., 88, 502
Rifle, CO., USA, **4.1.**
Rinehart, J. S., 97, 502
Ritter, A. D., 491, 504
Rixo, **1.1.**
Roach, R. J., 498
Robbins Company, *vii, 2.7.*
Robbins, R. J. (Dick), *viii,* 502
Rogers, R., 504, *3.3., 3.4.*

Rosell, S. (Sven), *vii*
Rosenberg, E. (Evelyn), 414, 416, 504, *17.3.*
Rossmanith, H. P., 98, 501, 502
Roth, J., 330, 502
Rowlandson, P., 304, 502
Royal Swedish Academy of Engineering Science, 497
Russia, 72, 422
Rustan, A., 276, 493, 497, 500, 502, 504
Rutjebacken, **1.1.**
Röh P., 495
Sakamoto, M., 174, 502
Sakovich, G. V., 499
Sandia National Laboratories, Albuquerque, NM, USA, 158, 505
Sandström, F. (Fred), *viii*
Sandström, H., 496
Sandvik AB, 212
Sandvik Rock Tools AB, *vii*
Sandvik Rock Tools, *2.4., 2.5., 7.7.*
Sawaoka, A. B., 420, 495
SCB Technologies, Inc., 158
Schmidt, R. L., 502
Schörling, S., 500
Schwarz, A. (Alfred), 158
Seely, A. (Alice), 416
SEI, **9.2.**
Seinov, N. P., 256, 502
Seitz, F., 502
Selberg, H. L., 497
Selleck, D. J., 308, 491, 503
Selmer-Olsen, R., 33, 503
Sheenan, J. C., 72, 491
Sheffield, O. E., 122, 494, **4.7.**
Shuster, A., 88, 503
Sishen Mine, Sweden, *3.12.*
Siskind, D. E., 174, 348, 381, 382, 503
Sjöberg, C. (Conny), *vii*
Sjölin, T., 497
Skane, **1.1.**
Skanska, *vii*
Smith, L. C., 125, 421, 503, *4.9., 4.18.*
Soudan, MN, USA, **4.1.**
South Africa, 178
SRI International, 132
Stackora, V. J., 503
Staff, K. G., 491
Stagg, M. S., 346, 503
Staver, A. M., 422, 499, 503
Steckley, R. C., 503
Steiner, U., 495
Stevens, M. G., 503
Stewart, D. S., 89, 135, 492
Stockholm, Sweden, *12.16.*
Stora Vika, Sweden, 252
Stripa Mine, Sweden, 389, *15.1.*
Svanholm, B. O., 503
SveDeFo, see Swedish Detonic Research Foundation
Svärd, I., 174, 503
Swanholm, B. O., 503
Sweden, 129, 150, 266, 387, 393, 441, 450, 508
Swedish Board for Technical Development, *vii*
Swedish Detonic Research Foundation, SveDeFo, *v, vii*, 180, 259, 272, 288, 301, 314, 319, 335, 389, *8.2.*
Swedish Research Mine at Luossavaara, 314
Switzerland, 43
Symanski, H., 393
Taylor, G., 492
Taylor, J., 89, 102, 105, 503
Technical Museum, Stockholm, Sweden, 414
Texas City, TX, USA, 67
Thadhani, N. (Naresh), *viii*, 419, 420, 422, 497, 503
Thompson, D. E., 274, 503
Titov, V. M., 422, 499, 503
Tresca, 14, *1.7.*
Trondheim, Norway, 33
Trunin, R. F., 491
Tsaplin, D. N., 501
Turnbull, D., 502
Ueada, T., 169, 500
Ullern, Norway, *13.24.*
Umeå, Sweden, 252
United Nations Committee of Experts, 438, 441
University of California at Berkeley, *v, viii*
University of Luleå, Sweden, 288
Urizar, M. J., 421, 503
US Army Armament Research Development and Engineering Center (ARDEC), 132
US Bureau of Alcohol, Tobacco, and Firearms (ATF), 425

US Bureau of Mines (USBM), 348, 360, 361, 381, 388, 450, **15.1.**, *8.7.*, *13.16.*, *13.7.*
US Department of the Interior, 450
Valdemarsvik, **1.1.**
Valencia Co., NM, USA, **4.1.**
Västra Sund, Sweden, 389
Vestre, J., 127, 504
Video Camera Technology, 204
Vieille, P., 88, 492
Vinterviken, Stockholm, Sweden, *vii*, 182, 414, 415, **1.1.**, *17.1.*
Volchenko, N. G., 261, 504, *8.26.*
Volk, F., 495
von Mises, 14, **1.3.**
Von Neumann, J., 88, 504
Voreb'ev, I. T., 504
Wade, C. G., 86, 504
Walsh, J. M., 502
Walters, W. P., 404, 504
Weibull, W., 11, 13, 250, 504, *14.1.*

Westberg, J., 498
Wijk, G., 33, 37, 41, 504, *2.21.*, *2.22.*
Wilkins, M. L., 103, 500
Williams, P. (Pharis), *viii*
Wilson, W.H., 492, 496
Winzer, S. R., 151, 171, 491, 504
Wiss, J. F., 344, 382, 504
Wolkomir, R. (Richard), 416, 504
Wood, W. W., 134, 135, 504
Woods, R. D., 368, 502, 504
Worsey, P. N., 175, 504
Wright, G., 72, 504
Wu, F. E., 492
Yamamoto, M, 502
Zaire, 85, 290
Zeldovich, Y. B., 88, 504
Zienkiewcz, O. C., 491
Zimmerman, R., 388, 494
Zinn, J., 504, *3.3.*, *3.4.*
Zukas, J. A., 404, 504
Zwisler, 105, 493

Subject Index

Page numbers for references within the text of the book are listed first (in the Times Roman font or, for the Foreword and Acknowledgments, in italic).

Table numbers are in boldface; e.g., **2.25** refers to Table 2.25 in Chapter 2.

Figure numbers are in italic, e.g., *4.15* refers to Figure 4.15 in Chapter 4.

Page numbers for the tables and figures are omitted since references are directly to the Table or Figure.

Abrasion value (AV), 34, 35, *2.19.*
Abrasion value test, *2.17.*
Accelerated ageing, 345
Accidents with waterbased explosives, **3.1.**
Acetic anhydride, 72
Advance, 511
Air Blast, 511
Air pressure pulse, 382
Aluminum, 456, **1.4.**
Aluminum, finely powdered, 75
Aluminum (paint grade), 75
Aluminum shells, prohibited in coal mining underground, 151
Ammonia, 59, **15.3**
Ammonia in oxygen-deficient explosives, 391
Ammonium chloride, 78
Amphibolite, 282, 284
Angular deviation, 511
ANOL equipment for pneumatic loading of ANFO, *3.8.*
Ardeer Tank method, 388
Arrhenius equation, first order, 59
Atlas Copco Boomer H127, *7.4.*
Atlas Copco hydraulic rock drills, *7.6.*
Atlas Copco Jarva MK 27, *2.8.*
Auger feed loading of ANFO, *3.9.*
Bachmann process, 72
Bachmann process for RDX, 72
Balance substance and fuels, **4.5.**
Ballistic mortar, *4.15.*
BAM 50/60 Steel Tube Test, 439
Basalt, dense, **4.1.**

Becker-Kistiakowski-Wilson equation of state, 102, 105
Bench blasting, 511
Bench blasting using rubber mats, *12.16.*
Berea sandstone, **4.1.**
Bernoulli equation, 405
Bichel Gauge Test, 388
Bit penetration, 511
Bit wear index (BWI), 36, 43, *2.19., 2.20.*
BKW code, 389, 460, **4.2.**
Black powder fuse, *5.2.*
Black Slate, **1.1.**
Blast geometry, 511
Blast, 511
Blast wave, 511
Blue Ore, **4.11.**
Bohus granite, **2.1.**, *2.22., 9.10.*
Booster charge, 511
Boreability, 46
Borehole, 511
Borehole pressure, 281
Borehole pressure for a decoupled charge, 281
Boring, 511
Bottom charge, 512
Break-out angle, increase with burden, **8.1.**
Breaking index, 130, **4.13.**
Bridgewires, 158
Brisance, 512
Brisance value, 122
Brisance values for explosives, **4.7.**

Brower compression test (apparatus), 454, 512
Building damage, limit values for the vertical particle velocity, **13.1.**
Building factor for bridges, harbor piers, historical buildings, office buildings, houses, churches, and spans, **13.4.**
Bulk explosives, **4.15.**
Burden, 512
Bureau of Alcohol, Tobacco, and Firearms, 489
Button bits, 512
C-J pressure, 506
Cage molecules, 73
Calcium carbonate, 391
Calculated reaction product pressure, *8.4.*
Capacitive discharge, 152
Carbon dioxide, 104, 387, **15.3**
Carbon monoxide, 104, 387, 392, **15.3**
Cast blasting, 318
Channel effect, 512
Chapman-Jouguet (CJ) hydrodynamic theory of steady-state detonation, 88
Chapman-Jouguet detonation velocity, 101, 507, *4.25.*
Charge weight, 512
China, origin of black powder, 148
Code of Federal Regulations (CFR), 489
Coefficient of friction, **1.2.**
Collar flyrock, 320
Collar, 512
Collaring deviation, 512
Column charge, 513
Composite explosive composition, 110
Compressive strength (confined and unconfined), **1.2.**
Computer codes, BKW, 389
Computer codes, NITRODYNE, 389, **15.2.**
Computer codes, thermodynamic, 389
Computer codes, TIGER, 389
Computer codes: BLASTEC, "3 x 3" JKMRC, BLASTCALC, BLAST DESIGNER, CARE, DYNOVIEW, SABREX, SAROBLAST, SWEBENCH, TIGERWIN (Tiger for Windows), 288
Concrete, **4.1.**

Concussion, 513
Confined compressive strength, 513
Confined shear strength, 513
Connecting a round with redundancy, 161
Construction Material Factor, 508, **13.5.**
Construction Quality Factor, 508
Contour holes, 513
Copper Wire, **5.2.**
Covering, 513
Crack length, 507
Crater Volume, **2.1.**
Cratering, 513
Crawshaw-Jones Apparatus, 388
Cubic boron nitride cutting tools, 421
Cundall Block Program, 493
Current leakage, 155
Damage and fragmentation effects in Hard Scandinavian bedrock, **8.2.**
Decoupled charges, 513
Deflagration-to-Detonation Transition (DDT), 148
Density of rock, **4.1.**
Detaline, DuPont non-electric initiation system, 162
Detonating cord, 513
Detonation Shock Dynamics (DSD), 135
Detonation velocity, 506
Detonation wave velocity, **4.1.**
Detonography, 416
Diabase, **13.3.**
Diamond powder, 421
Diamond powders, shock consolidation of, 422
Diesel oil, 456
DIN 4150, see Standard, German
Disc cutter, 513
Displacement, **2.1.**
Distance Factor, 352, *13.8.*
DOT tests, 469
Double base propellants, 68
Dragline machines, 318
Drifting, 513
Drifting and tunneling, 146
Drill bit, 513
Drillhole, 513
Drilling, 514
Drilling deviation, 514

Drilling rate index (*DRI*), 33, 36, 43, *2.14.*, *2.15.*, *2.19.*, *2.23.*
DRM, see Measured drilling rate
Drop raising, 303
Drop-weight impact test, 454
DSD theory, see detonation shock dynamics
DuPont Electric Delay Blasting Caps, **5.2.**
Dynamite explosive, 514
EBW detonator, *5.12.*
Elastic Modulus, 508, **1.2.**
Electric delay detonator, *5.2.*
Electric detonator, 514
Electric detonator, instantaneous, 150
Electrical response characteristics detonators, *5.9.*
Electromagnetic firing method, *5.23.*
Electronic detonator, 514
Elongated bottom charges, *6.6.*
Emission of fumes from commercial explosives, **15.3.**
Emulsion explosive, 514
Energy Utilization Number, 307
Enthalpy, 93
Erosive burning, 65
Ethyl ether, 67
Excavator shield tunnel borer for soft rock, *2.7.*
Exploding bridge wire detonator, 56
Exploding bridgewire detonator (EBW), 147, 159
Explosive art, 415
Explosive casting, 318
Explosive consumption, summary, **7.1.**
Explosive engraving, 415
Explosive forming, 417
Explosive weight, 514
Explosively formed fragment, 406
Explosively synthesised polycrystalline diamond powder, 421
Explosives for smooth blasting and pre-splitting, listing by manufacturer, **9.2.**
Family housing, blasting directly below (in Ullern, Sweden), *13.24.*
Fan cut, 514
Far field stress waves (ground vibrations), *8.5.*
Fatal accidents, 423, **3.1.**

Fines, i.e., finely fragmented material, 249
First igniter cap for nitroglycerin, *5.1.*
First Law of Thermodynamics, 508
Fixation factor (f), 514
Flash x-ray shadowgraph, *16.7.*
Flegmatized gel, **4.14.**
Flintstone, **1.1.**
Flintstone, shear strength, 451
Flyrock, 318, 319, 323, 327, 335, 514, *12.4.*, *12.9.*
FM Transmitters, **5.1.**
Four Pillars (Major Elements) of the Science of Rock Blasting, 514
Four-section cut, 514
Fracture Force, *2.1.*
Fracture toughness values, **1.4.**
Fragment protective covering, 332
Fragmentation, 515
Free surface, 515
Fume class, **15.1.**
Fume classification of explosives, gas pendulum and gas mortar methods, *15.5.*
Fumes classification of explosives in the USA, **15.1.**
Gallium arsenide semiconductors, 421
Gas release pulse, 382
Geniss-granite, **1.1.**
Geometry for blasting with elongated bottom charges, *6.6.*
Glass, **1.1.**
Glass microballoons, 86, **4.14.**
Gneiss, 328, **13.3.**
Government regulations, 423
Grade strength, 515
Granite, 328, 452, **4.1.**, **4.11.**, **13.3.**
Granite gneiss, 282, 284
Granite I, **1.1.**, **1.4.**
Granite II, **1.1.**, **1.4.**
Granite III, **1.1.**
Granite, Westerley, **4.3.**
Granitized leptite, *7.7.*
Graphite, 421
Graphitic and diamond-like structures, 104
Gravity factor, 188
Grey Slate, **1.1.**
Greywacke (low phos.), **4.11.**
Ground faults, 154
Guidance level, 510

Gun-cotton, 67
Gurney energy, 398, 400, 401, 507, **16.1.**
Gurney Equations for metal plate acceleration, 397, 404, 405
Gurney solutions., *16.3.*
Gurney velocity, dimensionless, 401
HALF-PUSHER technique, 82, *3.14.*
Hard cherty magnetic iron formation, 308
Hard competent rock, 328
Hard rock tunnel boring machine, *2.8.*
Hard Scandinavian bedrock, **8.2.**
Heat of detonation, 506, 508
Heat of formation, 506, 508
Heats of formation of some explosives, **4.4.**
Hematite, **4.1.**
Henkin-McGill time-to-explosion test, 454
Hercudet non-electric initiation system, 167, 169, *5.19., 5.20.*
High pressure water jet notching system, 282
High-speed camera photograph of explosively formed projectile in flight, *16.8.*
High-temperature superconductors, 421
Hole deviation, 515
HOM equation of state of reacting explosives, *viii*
Homolite-100, 258
Hot wires (bridgewires), 158
Hot-wire actuators, **5.3.**
HS delay detonators, 150
HTPB, hydroxy-terminated polybutadiene, 68, *3.6*
Hugoniot curve(s), 95, 138, 419, 458, *4.4., 4.25.*
Hugoniot Elastic Limit, 452, **1.2.**
Hugoniot equation, 93
Hydrocode, 515
Hydrocode, definition of, 132
Hydrogen, **15.3**
Hypersonic airplanes, 421
Ideal detonation, 515
Igniter cords, 177
Impulse density, 509
Induced currents from AC transmission lines, 152

Influx angle and jet mass fraction, **16.2.**
Initial elastic modulus, 506
Inofficial recommendations, 423
Insensitive munitions, 515
Internal energy, 506
International System of Units (SI), 505
Iron Formation, **4.11.**
Iron ore, 328
Iron Pyrites, **1.1.**
Iron Wire, **5.2.**
Isothermal temperature change with altitude, 384
JCZ3, **4.2.**
Jet mass fraction, **16.2.**
JWL equation of state, 103
Kaolin, 256
Kerr-cell camera, 320
Kerr cell camera picture of a copper cylinder, *4.20.*
Kevlar canvas, 335
Kirby and Leiper reaction rate model, 134
Kirchhoff's Law, 152
Koenen Test, 439, 448
Lamp black, 456
Langefors weight strength, 130, 131
Langefors weight strength value s, 133
Langefors-Kihlström criteria, 349
Langefors-Kihlström weight strength, 515
Langefors-Kihlström weight strength concept, 195
Lead block Excavation values, **4.8.**
Lead dust, **15.3**
Lead Ore, **1.1.**
Leg wires, copper and iron, 151
Leptite, **1.1.**
Lifter holes, 515
Lifters, 515
Lightning discharges, thunderstorms, rapid electric field changes, 155
Limestone, **1.4., 13.3.**
Limestone I, **1.1.**
Limestone II, **1.1.**
Limestone, chalky, **4.1.**
Limestone, crystalline, 328
Limestone, Marly, **4.1.**
Limestone, Solenhofen, **4.3.**
Linear charge concentration, 192

Linear charge concentration or Linear loading density, 515
Linear shaped charge for notching, 9.15.
Liquid hydrogen, 456
Livingston theory, 313
Livingston's Breakage Process Equation, 307
Loading density, 192, 515
Lookout angle γ, 516
Lundborg's equation for the confined shear strength of brittle materials, 451
Lundborg's statistical theory of the strength of brittle materials, 451
Lundborg's theory, 14
Magnadet initiation system, CXA Ltd., 5.24.
Magnadet system, 171
Magnetite iron ore, 155
Magnetite, 1.1.
Magnetite, 4.11.
Manual of Tests for the Hazard Classification of Explosives, 438
Marble, 4.3.
Marble I, 1.4.
Marble II, 1.4.
Marlstone, 4.1.
Material behavior index, 309
Maximum cooperating charge, 354
Maximum throw, 516
Measured drilling rate (DRM), 34, 2.15.
Mechanical notching tool, 282, 9.14.
Methane, 469, 15.3
Metric system, 505
Micagneiss, 1.1.
Military explosives, 71, 101, 3.3.
Mine safety regulations, 151
Monocrystalline diamond, 421
Morain, 13.3.
Multiple-row round, 516
Multiple-row shot with detonating cord, 5.14.
National Aerospace Plane X-30, 421
Nearby free surfaces, 245
New rock blasting weight strength value, 133
Newton-Raphson iteration method, 198
NISSAN R.C.B. System for underwater blasting, 169, 5.22.
Nitric acid, 59

Nitric oxide NO_2 (Nitrogen dioxide), 387, 388, 390, 391, 392, 456, **15.4**, , *15.3*
Nitrocellulose, 67
NITRODYNE computer code, 281, 389, **4.2.**
Nitrogen, 387, 390, **15.3**
Nitrogen oxides NO_x, 387, 388, 390, 392, 393, 456, **15.4**
Nitrous oxide NO, 387, 388, 390, 391, 392, 456, **15.4**, *15.3*
Nitroglycerin explosives, **4.15.**
Nobel's invention of, and patents for, the detonator (1863-5), 143, 144, 145
Noise, 516
Noiseless Trunkline Delay system, *5.17.*
Nominal resistance of Dupont Blasting Caps, **5.2.**
NONEL, 145
NONEL detonator delay times, table, **5.6.**
NONEL LP, **5.4., 5.5., 5.6.**
NONEL MS, **5.4., 5.6.**
NONEL UNIDET, **5.6.**
NONEL UNIDET system, *5.18.*
Nonideal detonation, 516
Non-nitroglycerin explosives, **4.15.**
Non-primary explosive detonator (NPED), 145, 150, *5.6.*
Nordtest methods, 430
Noxious Gases, **15.1.**
OD detonators, 155
Ohm's Law, 152
One Man – One Machine, 516
One-row rounds, *8.16.*
Optimum depth ratio, 312
Orange Book, Volume 2, Tests and Criteria, 443
Oxygen balance, fumes from ANFO, **15.2.**
Oxygen balanced explosives, **4.5.**
Oxygen balanced explosives, **4.5.**
Oxygen deficiency, 104
P-wave, 516
P-wave velocity, **4.1.**
Packaged emulsions, **4.15.**
Packing density, 516
Paintrock, **4.11.**
Palladium, **4.3.**

Parallel hole cut, 516
Parallel plate explosive welding, 418
Parallel-guided feed beams, *7.2.*
Parallel-series blast, 154
Peak particle velocities (recommended in Germany, DIN 4150), **13.2.**
Peak particle velocities, *8.7.*
Peak vibration particle velocity, **8.2.**
Pegmatite, 328
Pegmatite-gneiss, **1.1.**
Permissible explosive, 516
Personal precautions, 423
Picric acid, 501
Plain or ordinary detonators, 177
Plaster of Paris, 416
Plate Dent Test, **4.9.**
Plate dent test, *4.18.*
Plate dent test, *4.19.*
Plexiglas, see PMMA
Plough cut, 516
Plow cut, 516
PMMA (plexiglas), 258, 259, 274, 320, 323, **1.4.**, *8.1.*, *8.23.*, *12.2.*, *12.3.*
Polybutadiene, hydroxy-terminated, HTBP, 68
Polycrystalline diamond, 420, 421
Polymethylmethacrylate, see PMMA
Practical burden, 516
Precambrian rock with a schistose and gneissic structure, 258
Pressed nitramines (RDX or HMX), 73
Primadet Noiseless Trunkline Delays (NTD), 165
Primary explosives, 469
Primary explosives, **3.2.**
Primer, 517
Project Time Factor, 508, **13.6.**
Propellant grain, 65
Pumpable blasting agent, 517
Pumping emulsion explosive, *3.13.*
Pyrotechnic delay detonator using a primary explosive, *5.4.*
Pyrotechnic delays, 162
Quartzite, **1.1.**, **4.11.**
Quartzite sandstone, **13.3.**
R-wave, 517
Radio frequency energy, 152
Rankine-Hugoniot shock relations, 91, 93, 95
Rayleigh line(s), 95, 458, *4.4.*
Rayleigh wave, 342, 367, 464, 517, *13.2.*

Reaction products of some explosives, **4.4.**
Reduction factors for detonators, **13.7.**
Redundancy, connecting a round with, 161
REFLEX Maxibor, 204
REFLEX Maxibor surveying equipment, *6.17.*
Relative Energies of deformation to fracture, **1.3.**
Risk analysis, *13.9.*
Robbins TBM 212 S-239, *2.7.*
Rock constant, 517
Rock mass damage, 517
Rock parameters for Bohus granite, **2.1.**
Rock pressure pulse, seismically induced, 382
Rock-breaking Capacity of Explosives, **4.12.**
Rocmec 2000 mechanized charging truck for drifting and tunneling, *7.8.*
Round robin series, 430
Round, 517
S-wave, 517
Safe distance, actual, 411
Safe distances from fragment generating explosive charges, **16.3.**
Safe operating procedures(SOPs), 423, 426
Safety Committee, 425
Safety fuse, 517
Sand-stemming, 181
Sandstone, **1.1.**, **1.4.**
Sandstone, Berea, **4.1.**
Sandvik AB test mine, 212
Scandinavian bedrock, 348
SCB actuators, **5.3.**
Schistosity, 356
Secondary explosives, 469
Secondary explosives, liquids, **3.2.**
Secondary explosives, solids, **3.2.**
Seismic cord, 161
Semiconductor bridge ignition, SCB, 158
Sequential blasting machine, 160, *5.13.*
Shake of a lamb's tail, 505
Shaped charge jet photograph, *16.7.*
Shear strength (unconfined and unconfined), **1.2.**

Shock adiabat, 95
Shock Hugoniot, 458
Shock travel time, **8.3.**
The Shock Tube Problem, 459
Shock wave, 517
Shock-induced reaction synthesis, 421
Shotcrete, 391
Sievers J-value, *2.14.*
Sievers J-value test, *2.13.*
Sievers test, 33
Sievers wear test, 33
Silicon Glass, **1.4.**
Silver plate, explosively formed, *17.1.*
SIN hydrocode, *viii*
Single molecular explosive(s), 66, 110
Single-base propellants, 68
Sintering, 421
Skarnbreccia, **1.1.**
Slate, **4.11.**
Sliding velocity, 506
Slurry explosives, **4.15.**
Snow-charging method, *3.15.*
Sodium nitrate, 78
Solid oxygen, 456
Solid silver, 414
SOP, Safe Operating Procedure
Specific charge, 191, 192, 517
Specific heat, 506
Specific impulse, 509
Specific impulses of selected explosives, **16.1.**
Specific volume, 506
Specularite, **4.11.**
Stamp Test, *2.21., 2.22.*
Stamp Test Strength, **2.1.**
Stamp Test Strength Index, 39
Standard, German, DIN 4150 (peak particle velocities as recommended in Germany), 347, **13.2.**
Standards, ISO (International) (ISO 3951-1981), 163
Standards, Swedish, 163, 349, 350
Steam H_2O, **15.3**
Steel, **1.4.**
Steel, shear strength, 451
Stemmed, fill with coarse sand or crushed rock, 190
Stemming, 518
Stemming height, 517
Stemming release pulse, 382
Stoping holes, 518

Strain Energy Factor, 312, **4.11.**
Stray currents, 154
Stress Distribution Number, 309
Stress waves, 97
Subdrilling, 518
Surlyn plastic tubing, 182
Swedish Brittleness Test, 33, *2.12.*
Swedish granite, 306, *7.6.*
Swedish limestone and granite quarries, 100
Symmetric acceleration of two metal plates by a slab of explosive, *16.1.*
Sympatetic detonation prediction, 432
Syncroballistic photo, *16.8.*
Temperature inversion, 384
Tensile strength, 518, **1.2.**
Tertiary explosives, 469, **3.2.**
Test blasting, *13.10.*
Testing, Air Gap Test, 432, *19.1.*
Testing, BAM 50/60 Steel Tube Test, 435, *19.4.*
Testing, BAM Fallhammer Test, 434
Testing, Cap Sensitivity Test, 434
Testing, Critical Diameter Test, 437
Testing, Deflagration to Detonation (DDT), 436, *19.5.*
Testing, External Fire Test for Hazard Division 1.5, 436
Testing, Koenen Test, 436
Testing, Minimum Booster Test, 433
Testing, Princess Incendiary Spark Test, 436
Testing, Projectile Impact Test, 434
Testing, Steel Tube Test (BAM 50/60), *19.4.*
Testing, Type 1(a) tests, 438
Testing, Type 1(b) tests, 439
Testing, Velocity of Detonation Test (VOD), 437
Testing, Woods Test, 436
Thermal conductivity, 506
Thermal decomposition rate coefficient, *3.2.*
Thermal expansion, 506
Threshold values for toxic fumes, **15.4.**
TIGER code, 389, *4.13.*
TIGER for Windows, TIGERWIN, 288
Titanium aluminide, 421
Titanium diboride ceramic plates, 421
Toe, 518

Top charge, 518
Toxic fumes, 456, 518
Tracer substances, 316
Transmission ratio for various foundation depths, **13.8.**
Trauzl lead block test, *4.17.*
Triaxial strength, 518
Tungsten carbide abrasion test, *2.16.*
Tungsten carbide inserts, 34, 284
Tungsten carbide tipped drillbits, 183
Tunnel excavation, 518
Tunnel excavation, calculation of advance rates, **2.2.**
Tunnel excavation, calculation of equipment costs and supplies, **2.4.**
Tunnel excavation, calculation of indirect costs, **2.5.**
Tunnel excavation, calculation of labor costs, **2.3.**
Tunnel excavation, drill hole volumes and costs, **2.7.**
Tunnel excavation, estimate summary, **2.6.**
Tunnel face, 518
Tunneling in Sweden, common explosives, **9.1.**
TV Transmitters, **5.1.**
Ultra-hard boron carbide, 421
UN classification, 435, 436
UN Compatibility Groups, Class 1, Explosives, **20.5.**
UN Dangerous Goods List, **20.1., 20.2.**
UN Division 1.5 substance, 436
UN HAZARD Division 1.5, 434, 436
UN Hazard Divisions, Class 1, Explosives, **20.3., 20.4.**
UN Recommendation on the Transport of Dangerous Goods — Tests and Criteria, 430, 489
UN Recommendations on the Transport of Danger Goods, 489
UN Recommended Tests and Criteria, 450
Unconfined compressive strength, 518
Uncorrected peak particle velocity, **13.3.**
Underdriven detonation, 179
Underwater detonation test, *4.21.*
Underwater detonation tests and calculated energies, **4.10.**
Underwater explosion tests, **4.13.**

Uniaxial compression, **1.3.**
Uniaxial strain, **1.3.**
Uniaxial tension, **1.3.**
Uranium, **4.3.**
US Code of Federal Regulations, 425
V-cut, 518
Vapor pressure, 506
Vaseline, 182
VCR, 302, 303, 304, 305, 307, 310, 311, 314, **11.1** *11.7. 11.13., 11.15.,*
VCR blasting, 314, *11.8., 11.13.*
VCR stope, 314, **11.2**
VCR stope, summary of, **11.2.**
Velocity of Detonation, see VOD
Vertical Crater Retreat blasting technique, see VCR
Vertical Retreat Mining (VRM), 304, *11.5*
VHF Transmitters, **5.1.**
Vibration level, 518
Vibration monitoring by unattended instruments, 373, *13.23.*
Virial type of equation of state, 105
VOD, Velocity of Detonation, 437, 518
VOD measurement, *19.6.*
Von Neumann point, 458
VRM, see vertical retreat mining
Waterproof materials, bitumen and wax, 177
Wave impedance in rock, **4.1.**
Weak plane, 518
Weibull distribution, 250
Weight strength, **4.6., 4.13.**
Weight strength concept, 196
Weight strength for some explosives, **6.1.**
Weight strength relative to LFB-dynamite and ANFO for some explosives, **4.6.**
Welding, parallel plate explosive, 418
Westerley granite, 460, 461
Wide Space Blasting Method, 291
Witness plate (in a Minimum Booster Test), *19.2.*
Wood and Kirkwood detonation theory, 135
Wooden plugs, 314
Woods metal, 436
Yellow Ore, **4.11.**
Young's modulus, 508